W9-BKX-159

The Sociology of Science

By the Same Author

Science, Technology, and Society in Seventeenth-Century England
Mass Persuasion [with Marjorie Fiske and Alberta Curtis]
Social Theory and Social Structure
The Focused Interview [with Marjorie Fiske and Patricia Kendall]
The Freedom to Read [with Richard McKeon and Walter Gellhorn]
On the Shoulders of Giants
On Theoretical Sociology

Continuities in Social Research [with Paul F. Lazarsfeld]
Reader in Bureaucracy [with Ailsa Gray, Barbara Hockey, and Hanan Selvin]
The Student-Physician [with George G. Reader and Patricia L. Kendall]
Sociology Today [with Leonard Broom and Leonard S. Cottrell, Jr.]
Contemporary Social Problems [with Robert A. Nisbet]

Robert K. Merton

Edited and
with an Introduction by
Norman W. Storer

The Sociology of Science

Theoretical and Empirical Investigations

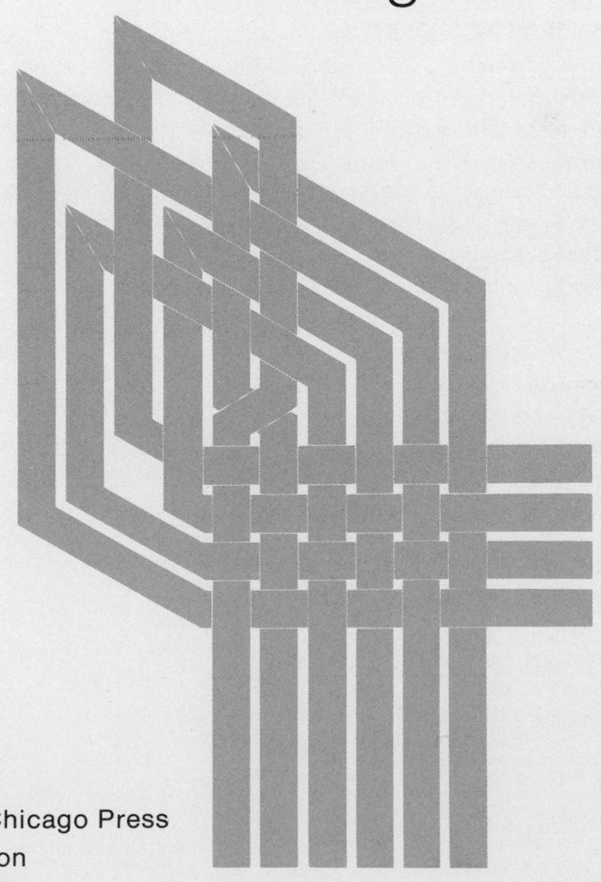

The University of Chicago Press
Chicago and London

ROBERT K. MERTON has taught for many years at Columbia University where he is the Giddings Professor of Sociology. A member of the National Academy of Sciences, he has received numerous awards and honorary degrees. His publications include the classic *Social Theory and Social Structure* and *On the Shoulders of Giants*.

NORMAN W. STORER is professor and chairman of the Department of Sociology and Anthropology, Baruch College, City University of New York.

[1973]

The University of Chicago Press, Chicago 60637
The University of Chicago Press, Ltd., London

Printed in the United States of America
International Standard Book Number: 0-226-52091-9
Library of Congress Catalog Card Number: 72-97623

To my teachers

Pitirim A. Sorokin
Talcott Parsons
George Sarton
L. J. Henderson
A. N. Whitehead

who together formed my interest in the sociological study of science

Contents

3 The Normative Structure of Science

4 The Reward System of Science

5 The Processes of Evaluation in Science

Author's Preface

After a long gestation, the sociology of science has finally emerged as a distinct sociological specialty. Having evolved a cognitive identity in the form of intellectual orientations, paradigms, problematics and tools of inquiry, it has begun to develop a professional identity as well, in the form of institutionalized arrangements for research and training, journals given over to the subject in part or whole, and invisible colleges of specialists engaged in mutually related inquiry and not infrequent controversy. In these as in its other aspects, the sociology of science exhibits a strongly self-exemplifying character: its own behavior as a discipline exemplifies current ideas and findings about the emergence of scientific specialties.

In the light of this development, there is now more point than before in taking up the suggestion of Michael Aronson of the University of Chicago Press to bring together some of my papers in the sociology of science which are presently scattered in various journals, symposia, and other books. Still, like Alfred Schutz facing a similar decision, I must recognize that few of us can bring to our own work the distance and hopefully exacting judgment of an informed editor. I am therefore indebted to Professor Norman W. Storer for agreeing to select and arrange the papers, to provide the general introduction and prefatory notes, and to eliminate repetition except when, in his opinion, it provides redundancy useful for highlighting continuities of theme and idea. Having contributed to the field for more than a decade, Professor Storer is thoroughly at home in it and able to put these perspectives on the sociological study of science into historical and intellectual context.

Reiteration would only dull the thanks I express in the individual papers to the many who have helped me get on with my work in this field. But there are other, current debts. I thank Richard Lewis for help in reading the proofs of this book, and Mary Miles and Hedda Garza for preparing the index. I owe special thanks to my colleagues Bernard Barber, Harriet Zuckerman, and Richard Lewis for allowing me to reprint our joint papers, and to Elinor Barber for allowing me to draw upon our published and unpublished collaborative work. I gladly acknowledge the help given me by a fellowship from the John Simon Guggenheim Memorial Foundation, by a term as Visiting Scholar of the Russell Sage Foundation and, more recently, by a grant from the National Science Foundation in support of the Columbia University Program in the Sociology of Science. I, for one, must testify to the growing worth of that program as I agreeably observe that my colleagues in it—Harriet Zuckerman, Stephen Cole, and Jonathan Cole—have come to teach me increasingly more than I have ever been capable of teaching them. I have also benefitted much from the thought and friendship of William J. Goode since those distant days when we first worked together in the sociology of the professions. And in this latest retrospect, I discover once again how much I have learned from Paul F. Lazarsfeld, in joint seminars, in other joint ventures and, most of all, from our continuing dialogue through the years.

R. K. M.

Introduction

By Norman W. Storer

If Robert K. Merton has not yet been publicly described as a founding father of the sociology of science, there is at least substantial agreement among those who know the field that its present strength and vitality are largely the result of his labors over the past forty years. His work has given the discipline its major paradigm. This judgment is perhaps most decisively affirmed when set forth not by the many whose work is guided by that paradigm but by those who find fault with some aspect of it. Barry Barnes, for instance, who with R. G. A. Dolby[1] has strongly argued the case against certain assumptions in the paradigm, sums things up by observing that

> A dominant influence in this development [of the sociology of science as a separate academic specialty] was the work of Robert Merton, both as writer and teacher. By 1945 Merton had laid down an approach which identified science as a social institution with a characteristic ethos, and subjected it to functional analysis. This was for a long period the only theoretical approach available to sociologists in the area, and it remains productive and influential today. Its central ideas have received detailed elaboration, modification and reinterpretation by, among others, Barber, Hagstrom, Storer and Merton himself, making it the only maturely developed framework for the sociological study of science.[2]

1. S. B. Barnes and R. G. A. Dolby, "The Scientific Ethos: A Deviant Viewpoint," *European Journal of Sociology*, 11 (1970): 3–25.

2. Barry Barnes, ed., *Sociology of Science: Selected Readings* (London and Baltimore: Penguin Books, 1972), pp. 9–10. And again, Barnes notes: "The only long standing tradition in the sociology of science derives from Robert K. Merton's insights into the nature of its institutional structure" (Ibid., p. 61). For similar observations, see the opening paragraph of the critical essay by Michael Mulkay, "Some Aspects of Cultural Growth in the Natural Sciences," *Social Research* 36 (1969): 22–52, and pages 244–46 of Kenneth J. Downey, "Sociology and the Modern Scientific Revolution" (*Sociological Quarterly* 8 [1971]: 239–54).

As a sociological specialty, the field has come alive only in the past fifteen years or so; the upward turn in the logistic curve describing its growth (which we know is typical of new, "hot" specialties in many fields of science) began in the mid-fifties. It would perhaps be a sign of premature senility, or at least of the flattening of the S-shaped curve, for any new field other than the sociology of science to begin so early to examine its own development. But this field has the peculiar character of being grist for its own mill. Yesterday's achievements—and failures—are data for today's research on the growth of scientific specialties, as is the case with no other specialized discipline. This unique property carries its own hazards. Too much thinking about one's own thinking can produce intellectual stasis; too much questioning of one's own questions can produce a kind of sociological anomie. Yet such difficulties can scarcely be allowed to dissuade us from trying to understand the character and development of this special field.

The papers collected here are intended to serve several purposes. Primarily, the volume brings together a number of articles that have been of central significance in the development of the sociology of science, together with others which are representative of certain stages in that process. At the same time, the collection may provide a sense of the intellectual continuity and coherence of the field; more clearly here than in some other fields of sociology, the seeds of future growth can be readily found in papers antedating this growth by ten years and more. In a more practical vein, enclosing these papers drawn from many different sources within a single cover will afford easy access to them for those wanting to make use of them in their own work. Finally, the collection pays tribute to the author; the substance and style of the papers themselves record, in a way mere panegyric could not, the enduring importance of his work.

The papers are not presented in strictly chronological order. The warp and woof of the entire corpus is drawn so tight—the intersections of different threads of thought are so frequent—that it has seemed better to separate and group the major elements in this mosaic for concentrated attention than to leave the task entirely to the reader. It is hoped that in this way the continued clarification of ideas and the ways they have been woven together to give added strength to this growing body of knowledge will be made more visible.

But the papers themselves, even with the extensive footnoting that has been characteristic of Merton's work since the beginning, cannot provide full perspective on the larger scene—the social and intellectual context within which they have been produced and to which they have contributed. It is the aim of this introduction to supply such perspective from the vantage point of 1973, aiming not at anything like a history of the sociology of science but rather at sketching the major landmarks and problems that have provided its broad outlines. Additional detail will be found in the prefatory notices to each of the five parts of the volume.

The sociology of science is sometimes defined as a part of the sociology of knowledge, and yet the multifaceted problem of the relations between knowledge and reality (not to speak of the reality of knowledge) is a more general one, at the heart of the larger part of sociology. Studies of religion and ideology, of the mass media and public opinion, and of norms and values, to say nothing of the methodological concerns of sociologists, all implicate the chicken-and-egg question of the interdependence of these two fundamental components of human life in groups. How do existential, everyday experiences mold the ways in which people conceptualize the world? How, in turn, do their conceptualizations influence their actions *in* the world, and how, further, do they react to discrepancies between what they "know" and what they experience?

It is perhaps because *Wissenssoziologie*, the sociology of knowledge, in a sense defined its concerns so narrowly in the beginning, focusing almost exclusively on trying to reason out the extent to which men's knowledge is shaped by their interests and experiences, that it had fallen into disarray by the 1930s. Indeed, as Merton's examination of the field in 1945 (included here as "Paradigm for the Sociology of Knowledge") demonstrates, this particular question contained within itself the petard by which it would eventually be hoist. To conclude that knowledge is *not* at all molded by men's experiences would undermine the raison d'être of the field, while to conclude that it is altogether so molded would seem tantamount to questioning, if not denying, the validity of all knowledge—including that conclusion. This restricted construction of the problem led to a maze of internal contradictions, a cul-de-sac from which escape had to be sought by beginning anew with different questions.

Such questions were, of course, vigorously pursued in different sectors of the sociological community. Weber's work on the importance of the Protestant *Weltanschauung* in producing capitalism in Europe had already had a long and effectively controversial history by the time Merton saw its relevance to his interest in the history of science. Durkheim's work on primitive religion and his orientation to problems in the sociology of knowledge was beginning to attract the notice even of some American sociologists. The task was to put the various problems back into some sort of orderly array.

In the early 1930s, however, Merton's interest was not primarily in the sociology of knowledge. During his graduate studies at Harvard, he undertook, at the suggestion of the economic historian E. F. Gay, an analytical book review of A. P. Usher's *History of Mechanical Invention*. Gay liked it and suggested that George Sarton, also at Harvard, publish it in *Isis*, the prime journal in the history of science which he had founded and still edited. Sarton did so, and he encouraged Merton's interest in the history of science by having him work in the renowned workshop in Widener Library. Noting his growing expertise in this field, Pitirim A. Sorokin recruited Merton to assist him in the studies of the development of science that would make up parts of his *Social and Cultural Dynamics*. This provided valuable experience in focusing on the development of quantitative measures of intellectual development and change, and perhaps paved the way to "prosopography"—"the study of the common background characteristics of a group of actors in history by means of a collective study of their lives"[3]—which Merton was to employ extensively in his later work.

Merton also studied with L. J. Henderson, the biochemist who had made a place for Sarton at Harvard and who was himself a gifted teacher of the history of science.[4] He attended the course of lectures in the philosophy of science given by Alfred North Whitehead and the unique course on comparative "animal sociology" in which specialists on a score of social species were brought together by William Morton Wheeler, the dean of entomologists whose omnivorous intellectual appetite included the history of science. And his early work on social aspects of science was monitored by the polymath E. B. Wilson, then associated with the new department of sociology. Merton was thus responding to the many opportunities at Harvard to develop various perspectives on science by going beyond the conventional boundaries of sociology, even though he continued in the department to be the student of Sorokin and, increasingly, of the young instructor, Talcott Parsons.

It was apparently this confluence of varied intellectual currents, rather than immediate developments in the sociology of knowledge, that led Merton to attempt a sociological analysis of the growth and development of science and that laid the foundation for his continuing interest in science as a distinctive social activity. Not that he was at this time unconcerned with the broader conceptual framework in which science could be located. Two papers[5] testify to this wider theoretical orientation. In 1935 he published in *Isis* a review of recent work in the sociology of knowledge by Max Scheler, Karl Mannheim, Alexander von Schelting, and Ernst Grünwald. In the next year he published "Civilization and Culture," a paper that located knowledge as a distinct focus of sociological interest in rela-

3. For an account of Merton's role in this development, see Lawrence Stone, "Prosopography," *Daedalus* 100 (1971): 46–79.

4. For an account of Henderson's role in sociology, see the introduction to L. J. Henderson, *On the Social System: Selected Writings*, ed. and with an introduction by Bernard Barber (Chicago: University of Chicago Press, 1970); for an account of Sarton's role in shaping the history of science, see Arnold Thackray and Robert K. Merton, "On Discipline-Building: The Paradoxes of George Sarton," *Isis* 63 (1972): 473–95.

5. The Bibliography lists all of Merton's writings cited here.

tion to concepts advanced by Alfred Weber and Robert MacIver. The concept of "culture" covered the realm of values and normative principles, while the concept of "civilization" included theoretical knowledge and practical technique, which tended—abstractly, not concretely—to be more accumulative than culture. In examining these concepts, Merton rejected the positivist interpretation of unilinear accumulation in science which was inherent in Alfred Weber's failure to deal adequately with the interdependence between culture and civilization. He claimed that this led Weber "virtually to revert to a theory of progress. What must be borne in mind is that accumulation is but an *abstractly* immanent characteristic of civilization. Hence *concrete* movements which always involve interaction with other spheres need not embody such a development."[6]

It was at this same time that Merton wrote his dissertation, *Science, Technology, and Society in Seventeenth-Century England* (begun in 1933 and completed two years later). Although the monograph did much to inaugurate the idea of systematic empirical investigation into the social matrix of science, it was not, of course, produced ab initio. In the United States, for instance, there had been W. F. Ogburn's major work, *Social Change*, and his paper with Dorothy S. Thomas, "Are Inventions Inevitable?" which developed basic conceptions about the social evolution of science and technology.[7] Ogburn's longtime research associate, S. Colum Gilfillan, published *The Sociology of Invention* in the mid-30s, setting forth almost forty "social principles of invention."[8] And in Europe a delegation from the Soviet Union, led by Bukharin, to the Second International Congress of the History of Science and Technology held in London (1931) had produced the volume of contributed papers, *Science at the Cross Roads*.[9] The most noticed contribution was the essay, "The Social and Economic Roots of Newton's 'Principia'," by Boris Hessen, the director of the Moscow Institute of Physics, which helped to reinforce and to focus interest in the social aspects of scientific knowledge. But, as noted by Robert S.

6. "Civilization and Culture," pp. 110–11.

7. William F. Ogburn, *Social Change* (New York: B. W. Huebsch, 1922; new ed., New York: Viking Press, 1950); W. F. Ogburn and Dorothy S. Thomas, "Are Inventions Inevitable? A Note on Social Evolution," *Political Science Quarterly* 37 (1922): 83–98. See also W. F. Ogburn, *On Culture and Social Change: Selected Papers*, ed. and with an introduction by Otis Dudley Duncan (Chicago: University of Chicago Press, 1964).

8. Chicago: Follett Publishing Company, 1935. The field of study of the social aspects of invention was so thinly populated at this time that Gilfillan was moved to include Merton, then the author of just a few papers in the field, among the eight "fellow students" of the subject to whom the book is dedicated. See also S. Colum Gilfillan, *Supplement to the Sociology of Invention* (San Francisco: San Francisco Press, 1971).

9. London: Kniga Ltd., 1931. Reprinted with a new foreword by Joseph Needham and a new introduction by P. G. Werskey (London: Frank Cass, 1971).

Cohen in his introduction to the recent separate reprinting of that essay,[10] its influence was mostly visible not in Stalin's Soviet Union, where Hessen soon disappeared from view, but in England, where it appeared in the far more discriminating historical work of scientists on the political left, such as Joseph Needham, J. D. Bernal, Lancelot Hogben, and J. B. S. Haldane, and in the rebuttals by such historians as Charles Singer, G. N. Clark, and Herbert Butterfield. In the United States, Hessen's essay and Clark's criticism of it were both taken into account in Merton's monograph.[11]

It was still too soon, in the mid-1930s, for that monograph to concentrate on the social structure of the emerging scientific community. As its title indicates, Merton's attention was directed to science *in society*, both its emergence as a social institution, fostered by the particular value-complex which was the hallmark of Puritanism, and its response to contemporary social interests (for example, practical problems of military technology, mining, and navigation).

But in the emphasis placed on the values that characterized the seventeenth-century practitioners of science, the foundations had been laid for later work that would trace the ethos of science and use it to define science as a subsystem of society and civilization. Contemporary events probably served to reinforce interest in the problem. The annihilative fate of "non-Aryan science" in Hitler's Germany during the 1930s directed attention to the various social conditions under which science can lose its autonomy; in the paper "Science and the Social Order" (presented in 1937), we find Merton's first allusions to the "norms of pure science" and signs of his developing interest in the structure and dynamics of the scientific community as distinguished from (and later related to) its substantive concerns. As Joseph Ben-David has noted,[12] the concept of the "scientific community" as a collectivity that evolves its own norms and policies was brought into sharp focus by Michael Polanyi from the early 1940s onward, was devel-

10. New York: Howard Fertig, Inc., 1971.

11. Joseph Needham, for example, reports that, having an interest in the history of science and having long been on the political left, it was only natural for him to attend the Congress and to be "very ready to give a sympathetic hearing to the Russian delegation." A few years later, in his *History of Embryology* (1934), he referred favorably to the Hessen essay as providing one model for historical research and, in his foreword to the recently reprinted volume of the Congress (see the reference in footnote 9), he notes that "with all its unsophisticated bluntness, [it] had a great influence during the subsequent forty years." The extent of that influence and the differences between the Hessen and Merton formulations are indicated in I. B. Cohen's review-essay on Merton's book in *Scientific American* 228 (1973): 117–20.

12. Joseph Ben-David, *The Scientist's Role in Society* (Englewood Cliffs, N.J.: Prentice-Hall, Inc., 1971), pp. 3–4; and "The Profession of Science and Its Powers," *Minerva* 10 (1972): 377.

oped by Edward Shils in the 1950s, and became a basic conception in the sociology of science in the 1960s.[13]

Recently, it should be noted, there has been renewed observation that the nature and direction of scientific growth cannot be adequately understood without dealing specifically with the contents of science—its concepts, data, theories, paradigms, and methods. The idea that the development of science can be analyzed at all effectively, apart from the concrete research of scientists, is said to have proven false.[14] The study of science, after all, begins with its product, scientific knowledge, rather than simply with those individuals who occupy the social position of "scientist." (This, incidentally, may account for the dearth of sociological studies focused on run-of-the-mill or relatively unproductive scientists: so long as science is defined by its research product, those who contribute little directly to that product are difficult to fit into the picture.)

Regarding the strategy of inquiry, however, it can be argued rather forcefully that it is of basic importance, especially in the beginning of sociological inquiry into the subject, to distinguish the behavior of scientists *as* scientists from the details of their "output"—if only to attend to the diverse aspects of doing science and to reduce the number of variables being considered at a given time. A comparable strategy is in fact employed by Thomas S. Kuhn in *The Structure of Scientific Revolutions*,[15] except that there the focus is on the formal organization of scientific knowledge and it is the social variables that need to be successively identified. Sociologically, it was necessary to identify the boundaries of the scientific community and to explore the bases of its place within society before the sociology of science could proceed to a range of other problems. (Indeed, the question of why science becomes established in any society, when most people can neither profit directly from the work of scientists nor comprehend and appreciate what they are doing, forms the central problem to which Joseph Ben-David addresses himself in his recent book on *The Scientist's Role in Society*.[16])

13. Polanyi's early paper of 1942, "Self-Government of Science," is included in his collection of essays, *The Logic of Liberty* (Chicago: University of Chicago Press, 1951), pp. 49–67; the general idea is developed in his many later books (see, for example, *Personal Knowledge* [Chicago: University of Chicago Press, 1958]). Edward Shils, "Scientific Community: Thoughts After Hamburg," *Bulletin of the Atomic Scientists* 10 (1954): 1151–55, reprinted in Edward Shils, *The Intellectuals and the Powers, Selected Papers*, vol. 1 (Chicago: University of Chicago Press, 1972), pp. 204–12. The developments in the 1960s are considered later in this introduction.

14. See, for instance, Barnes and Dolby, "The Scientific Ethos"; Mulkay, "Some Aspects of Cultural Growth"; and M. D. King, "Reason, Tradition, and the Progressiveness of Science," *History and Theory* 10 (1971): 3–32.

15. Second ed., enlarged (Chicago: University of Chicago Press, 1970).

16. Englewood Cliffs, N.J.: Prentice-Hall, Inc., 1971.

While there is no clear sign that Merton was then fully aware of the need for such a strategy, it does now seem that it was critically important to establish the relevance of distinctively sociological analyses to the study of science if the field were to develop at all. The social structure of an institution and the general orientations that characterize its participants, after all, *can* be separated from the specific concerns and activities that occupy their attention during particular periods of time. Thus, to take an analogous case, we assume that the central dynamics of public opinion are the same, whether its substantive focus is a war, an economic situation, a religious revival, or a fad.

Having explored the problematics of scientific knowledge in his monograph, with particular attention to the social as well as intellectual sources of foci of investigation in science, Merton evidently became persuaded that further sociological analysis required a more systematic conception of the social structure of science. It is significant that his early (1935) paper with Sorokin, "The Course of Arabian Intellectual Development, 700–1300 A.D.," is subtitled "A Study in Method." The fact was that without a sufficiently well-developed model of the social structure of science, there was no way to generate theoretically important questions that could use systematic data on scientific development to advantage. A research method is not much use if it cannot be coupled with theoretical questions (even though it may, through producing certain kinds of new data, encourage the subsequent development of theory).

So the decision was made, or perhaps evolved, to concentrate on the social structure of science rather than to continue with study of the social contexts that influence its substantive output of knowledge. The first phase of this work appeared in 1942 with the publication of "A Note on Science and Democracy" (reprinted here under the more appropriate title, "The Normative Structure of Science"). In this paper appeared the comprehensive statement of ideal norms to which scientists are oriented in their relations with each other: universalism, communism, organized skepticism, and disinterestedness.

Widely adopted as it has been, Merton's description of the "ethos of science" has not, it should be remarked, met with universal acceptance over the thirty years since its publication. Criticism, however, has been concentrated not so much on its having mistaken the components of this ethos, but on the question whether these norms in fact guide scientists' everyday behavior. No one has come forward with a radically different set of norms, but various critics have pointed out that scientists frequently violate one or more of the indicated norms. Thus, the treatment of the controversial Immanuel Velikovsky by members of the "scientific establishment" in the early 1950s is the one case repeatedly cited as an instance of widespread defection from the norms of universalism and disinterestedness.[17] There have been scattered attempts to measure the extent of commitment among scientists to the norms identified by Merton. The most recent of these, although its conclusions are limited by imperfect operationalization of some of these norms, finds substantial orientation to them in a sample of nearly a thousand American scientists, the extent of this varying somewhat by scientific discipline, scientific role, and organizational affiliation of scientists.[18]

It is, of course, the case that the behavior of scientists does not invariably adhere to the norms. But the implication sometimes drawn from this fact that the norms are therefore irrelevant stems from a misapprehension of the ways in which social norms operate. The theoretical problem is one of identifying the conditions under which behavior tends to conform to norms or to depart from them and to make for their change. Norms of this sort are associated primarily with a social role, so that even when they have been internalized by individuals, they come into play primarily in those situations in which the role is being performed and socially supported. When scientists are aware that their colleagues are oriented to these same norms—and know that these provide effective and legitimate rules for interaction in "routine" scientific situations—their behavior is the more likely to accord with them. These routine situations occur most frequently *within* an accepted universe of discourse, or paradigm; when there is general agreement on the ground-rules of the game (for example, basic concepts and problems, criteria of validity, etc.), acting in terms of the rules becomes personally rewarding and reinforces institutional bases for the development of knowledge. It is when such a universe of discourse is only slightly developed (as during the Kuhnian "pre-paradigm" stage in the development of a new discipline or during a "scientific revolution"), or when group loyalties outside the domain of science take over, that violations of the norms become more frequent, leading some to reject the norms entirely.

This analysis puts us somewhat ahead in the discussion of the development of the sociology of science, bringing in as it does explicit attention to the content of science (even though not at the level of specific data or theories). It was not until the late 1960s that the tactical advantages to be gained from drawing a sharp distinction between the social structure of science and its specific substantive output had been realized—the sociology of science had become established by then—and the time was ripe to pay

17. Alfred de Grazia, ed., *The Velikovsky Affair* (New York: University Books, 1966).

18. Marlan Blissett, *Politics in Science* (Boston: Little, Brown and Co., 1972), pp. 65–89, appendixes A, B, and C. An exploratory study by S. Stewart West ("The Ideology of Academic Scientists," *IRE Transactions on Engineering Management* EM-7 [1960]: 54–62) could be only suggestive at best, since it was based on responses from only fifty-seven scientists in one American university.

attention once more to the reciprocal relations between the social structure of science and scientific knowledge. The point will be taken up in more detail below.

Following the 1942 paper on the norms of science, there was a hiatus of about seven years in Merton's publications in the sociology of science, strictly conceived[19]—and there were few major contributions to the field from anyone else until 1952, when Bernard Barber's influential *Science and the Social Order* was published (with a foreword of considerable interest by Merton). Well before then, however, he had identified science as a special focus of interest within the sociology of knowledge. Certainly his analysis of Mannheim's work in "Karl Mannheim and the Sociology of Knowledge," appearing in 1941, exhibits intensive study of the topic, which was to lead to his more comprehensive discussion of the entire field in 1945 (reprinted here as "Paradigm for the Sociology of Knowledge"). Indeed, his interest in the social matrix of knowledge can be traced further back, to an article on "social time" published with Sorokin in 1937, which explores the question of how social processes influence the concepts and measurement of time.

The paper on Mannheim highlighted a number of unresolved difficulties in the study of how (and to what extent) existential conditions shape men's "knowledge" (which, for Mannheim, often seemed to include "every type of assertion and every mode of thought from folkloristic maxims to rigorous science"), but concluded with the courteous expectation that much enlightenment would be forthcoming from Mannheim's further explorations of the subject.

By 1945 Merton's dissatisfaction with the field was more clearly evident, and he undertook to chart some new directions through which further progress might be possible. It was at this point, too, that a separate chain of interests, also dating back to his years at Harvard, began to link with his interest in the sociology of knowledge. He had already worked with the general idea that available knowledge involves specifiable gaps in coping with social reality, developing it in various ways through his discussion in "The Unanticipated Consequences of Purposive Social Action" (1936), through the seminal paper "Social Structure and Anomie" (1938),

19. During this period, however, Merton did deal with various problems and conceptions relating to the sociology of science. For example: "The Role of the Intellectual in Public Bureaucracy" (1945), dealing with organizational constraints on policy-oriented knowledge; "The Machine, the Worker and the Engineer" (1947), treating the problem of the "rationalized abdication of social responsibilities" by technologists; "The Self-Fulfilling Prophecy" (1948), with its implications for social epistemology; "Election Polling Forecasts and Public Images of Social Science" (with Paul Hatt, 1949), examining a special case of failed claims to knowledge affecting the public standing of science; and "Patterns of Influence" (1949), with its concepts of local and cosmopolitan influentials.

and through the concept of latent dysfunctions. Although the topic could be viewed as a distinctive problem within the sociology of knowledge, it obviously had implications far beyond those evident in contemporary "mainstream" works in the field.

The next several years were a kind of harvest-time during which Merton brought to fruition a number of related interests—disparate though they may have appeared to others. Apart from his continuing codification of functional analysis during this period, which provided the theoretical background for his empirical interests, Merton was evidently centrally concerned with the various relationships that may exist between "knowledge" and "reality," and he worked on several subjects that served as what he describes as "strategic research sites" for the investigation of these relationships.

Studies in public opinion and personal influence, carried out with Paul F. Lazarsfeld in the Bureau of Applied Social Research of Columbia University, fitted in directly with the 1945 paradigm: the question of the social bases of knowledge was operationalized in empirical research on sources of people's decision-oriented knowledge. At the same time, exceptions to the postulate of "class-based consciousness" could be studied in the relationships between group membership and attitudes and could be partly accounted for in terms of a developing reference group theory.[20] The problem of how a member of the bourgeoisie could develop an active concern for the rights of the proletariat, for example, or how a member of the proletariat could maintain a "false consciousness" all his life, had been a major stumbling block in attempts to identify the social bases of knowledge; but once it was reconceptualized in terms of reference groups, it became amenable to systematic research.

March and Simon[21] have traced what they describe as the "Merton model" of bureaucracy focused on unanticipated organizational consequences as it was substantially developed in a series of outstanding empirical studies by graduates of the Columbia department (Selznick, Gouldner, Blau, Lipset, Trow, and Coleman). This model provided opportunity to wrestle with the problems that appear when "knowledge" (in this case, as exhibited in the formal structure and goals of a bureaucracy) is given and organizational "reality" (empirical patterns of interaction within the bureaucracy) becomes the dependent variable. Here social reality is adjusting (and reacting) to knowledge—in contrast to science, where knowledge must eventually be adjusted to fit reality.

20. See, for instance, Robert K. Merton and Alice S. Rossi, "Contributions to the Theory of Reference Group Behavior," and Robert K. Merton, "Continuities in the Theory of Reference Groups and Social Structure."

21. James G. March and Herbert A. Simon, *Organizations* (New York: Wiley, 1958), chapter 3.

The problem of adult socialization—including the processes through which the value-pervaded knowledge held by post-adolescents may be significantly altered by intensive exposure to a new set of "conditions"—was explored in research reported in *The Student-Physician* (1957), although Merton now maintains that a medical school was not a strategic research site for investigation of this general problem since the high degree of self-selection by students for this career reduced variability and limited the opportunity to uncover the bases and processes of change in their values and attitudes.

"The Self-Fulfilling Prophecy" (1948) represented yet another focus on the relationship between knowledge and reality, centering on the way in which emergent social reality is shaped by prior expectations and the conditions under which this occurs. Here the orientation deals with the role of ideas in the construction of social reality rather than with the social construction of knowledge.

But this is not the place to consider Merton's contributions to these several areas of research. It is enough to note that his foreword to Barber's *Science and the Social Order* is largely a rueful consideration of why the sociology of science had remained so conspicuously underdeveloped—even *un*developed—despite its promise as a field in which significant questions could be asked and answers to them found through empirical research. In this essay (reprinted here as "The Neglect of the Sociology of Science") he stops short of recommending this apparent paradox as itself a legitimate object of attention for the sociologist of science—but the implication is clearly there and may be taken as an early recognition of the peculiarly reflexive nature of the field[22] which was noted at the beginning of this introduction.

The foreword also notes a quickening of interest in the field and, perhaps based on the sociological understanding of science (at the intuitive if not the explicitly theoretical level), makes the forecast that under identifiable and probable conditions the field will soon attract the attention of more sociologists. It was nearly ten years before this forecast began to come true in any substantial form, but the insight that "when something is widely defined as a social problem in modern Western society, it becomes a proper object for study," and the sense that science was rapidly being defined in just this way, can be traced back to Merton's paper "Science and the Social Order" in 1937.

22. For discussion of Merton's formulations of "reflexive predictions" in his "Self-Fulfilling Prophecy," see May Brodbeck, ed., *Readings in the Philosophy of the Social Sciences* (New York: Macmillan, 1968), pp. 436–47; on self-exemplifying theories in the sociology of science, see "Insiders and Outsiders" in this volume, and Merton, *On the Shoulders of Giants* (1965); on "reflexive total relativism" in sociology, see his paper, "The Precarious Foundations of Detachment in Sociology," (1971).

It can be argued, though, that another element would be necessary before one could count on a blossoming of sociological interest in science. What was still to be achieved was, briefly, the development of a coherent theoretical orientation to science as a social phenomenon—what Merton described as an "analytical paradigm," capable of generating obvious, researchable questions and suggesting criteria by which answers could be evaluated. There are glimmerings, in the foreword, of what would be needed. Merton had already outlined the normative structure of science and had suggested how the four norms interact with each other in a functioning whole. But there was no clearly defined, distinctive source of "energy" in the system—no sense of why it should "move." The formulation was a state description rather than a process description. It was as if someone had described the physical construction of an electric motor but had not brought to it a clear concept of electricity; one could see what it *might* do, but could not understand why it *should* go round and round.

In his dissertation and in several subsequent papers, Merton had noted the often-recognized phenomenon of multiple discovery in science, and also the commemorative use of eponymy. He remarked on it again in the foreword to Barber's work, almost as though he *knew* there was additional significance to be drawn from it. This was to be another piece of the puzzle to be found and fitted into place, but it would be five years before his presidential address to the American Sociological Society would exploit its full meaning and cement this piece securely where we see it belonged. This was the conception that the institutionally reinforced drive for professional recognition, acquired almost exclusively in return for priority in scientific contributions and symbolized in the upper reaches of discovery by eponymy, constitutes the normatively prescribed reward for scientific achievement and thus the basis for a self-contained reward system of science.

Here was the energy that would drive the system, the distinctively institutionalized motivation that could account for scientists' orientations to the ethos of science and for their willingness to accept its often demanding strictures. (The norms of communism or communality and of disinterestedness seem obviously contrary to the workaday norms of an acquisitive, capitalistic Western civilization; and while universalism *is* an ideal of this civilization, even though it is more frequently breached than honored, to practice organized skepticism seriously often evokes public hostility by putting into question sacred verities of nature and society.)

The idea of an institutionally derived or reinforced individual need, linked directly to scientific accomplishment, also supplied a kind of urgency to the doing of science that had not been explained in earlier conceptions of science. Scientific knowledge is, in one sense, timeless. We are in the habit of writing "Aristotle *says*," and "Newton *points* out," and there is a widespread Platonic assumption that all the ideas of science coexist somewhere in a realm without clocks or calendars. Why, then, should a scientist ever be in a hurry to complete his work? Why should he feel obliged to rush into print to point out a colleague's error or to defend the validity of his own work when, under the optimistic assumptions of a positivistic outlook, sooner or later these perturbations in the path toward Truth would be straightened out anyway? The direct link between priority and recognition provided the outline of an answer.

We cannot, of course, be certain that the growth of the sociology of science had been hampered by the lack of a sufficiently apt "handle" by which it could be grasped and made to yield meaningful research questions. Perhaps it would have blossomed just as surely in the late 1950s even without this particular contribution, as a result of accretion of the factors Merton had noted in 1952. It is possible, too, that the position Merton had achieved within sociology by 1957 would have been sufficient to get the field going. The second edition of his *Social Theory and Social Structure* had just appeared and was widely taken to represent the best in modern American sociology; the prominence that "Priorities in Scientific Discovery" received because it was the presidential address (and thus achieved rapid publication in the *American Sociological Review*) must have added substantially to its impact, over and above that which it could rightfully claim on scholarly grounds alone.

Whatever the actual or most important reason for the rapid development of the sociology of science as a specialty after 1957 may have been, we can be sure that this fundamental formulation of the nature of the reward system of science was of critical importance in shaping the direction in which it evolved. The neat dovetailing of a functionally analyzed normative structure with an appropriately unique "scientific motivation" provided a paradigm (in the Mertonian sense) which opened the door to a wide range of meaningful research problems.

In the same year, the advent of Sputnik I and the ensuing "crisis" in American science and science education finally completed the definition of science as a "social problem." It had been progressively identified as a moral problem since at least Hiroshima, but apparently the problem had to be "practical" before the society could in effect treat it as a sphere of society and culture requiring intensive examination. As one offshoot, it became an area in which sociologists could legitimately specialize, with support slowly becoming available for its investigation. This is not to say that the sociological study of science had been entirely ignored before 1957, or that Merton's work set the tone for all studies of science thereafter. There had been a slowly growing stream of studies of scientists in organizations since the early 1950s, originating in the areas of management and industrial sociology, but a perceptible blending of this more practical, management-oriented tradition with the more theoretical Mertonian approach did not occur until after 1960.

The "organizational" study of scientists in the United States began in the early 1950s with the large-scale investigation of a government laboratory by Pelz and his associates and has continued ever since.[23] This and other works—published during the same period and later by Bennis; Gordon, Marquis, and Anderson; Kaplan; Kornhauser; Marcson; Shepard;[24] and others—focused largely on the problems of researchers' morale and productivity (and on the management of research in industrial research laboratories so as to enhance their work-satisfactions), particularly when their work was administered by nonscientists. The sociological questions posed came largely from the areas of bureaucracy and industrial sociology, and it was several years before the two separate research traditions were able to profit from each other.

An early conjunction of them came in the work of Barney G. Glaser, whose dissertation at Columbia in the early 1960s was based on a reanalysis of the original Pelz data and made effective use of the need for professional recognition in exploring some of the factors that shape the organizational careers of scientists.[25]

After this, three dissertations were written at Columbia under Merton's direction, appearing every other year, and their authors have continued to make substantial contributions to the field. Harriet Zuckerman's study of Nobel Prize winners became her dissertation in 1965,[26] and she has remained at Columbia as Merton's chief collaborator. Stephen Cole received his doctorate in 1967, and, from his present position at SUNY–Stony Brook, he has continued to participate in the work of the Columbia

23. For one overview, see Donald C. Pelz and Frank M. Andrews, *Scientists in Organizations* (New York: John Wiley, 1966). On similar research in Europe, see Hilary Rose and Steven Rose, *Science and Society* (London: Allen Lane, Penguin Press, 1969); and Stephen Cotgrove and Steven Box, *Science, Industry and Society: Studies in the Sociology of Science* (London: Allen and Unwin, 1970).

24. Representative publications include: Warren G. Bennis, "Values and Organizations in a University Social Research Group," *American Sociological Review* 21 (1956): 555–63; Gerald Gordon, Sue Marquis, and O. W. Anderson, "Freedom and Control in Four Types of Scientific Settings," *The American Behavioral Scientist* 6 (1962): 39–42; Norman Kaplan, "Professional Scientists in Industry: An Essay Review," *Social Problems* 13 (1965): 88–97; William Kornhauser, *Scientists in Industry* (Berkeley: University of California Press, 1953); Simon Marcson, *The Scientist in American Industry* (Princeton, N.J.: Industrial Relations Section, Princeton University, 1960); Herbert A. Shepard, "Nine Dilemmas in Industrial Research," *Administrative Science Quarterly* 1 (1956): 295–309.

25. The work is reported in Barney G. Glaser, *Organizational Scientists: Their Professional Careers* (Indianapolis: Bobbs-Merrill, 1964).

26. An outgrowth of this study will appear in her *Scientific Elites: Nobel Laureates in the United States* (Chicago: University of Chicago Press, in press).

group. Jonathan Cole, Stephen's younger brother, completed his degree in 1969 and has also remained at Columbia.[27] Diana Crane (Hervé) took her degree at Columbia, and although her dissertation was not directed by Merton, her research at Yale (where she worked with Derek J. de Solla Price and advised Jerry D. Gaston during the early stages of his study of the British physics community) and later at The Johns Hopkins University, blends well with the Mertonian approach.[28]

During this crucial decade and earlier, Merton's influence was by no means limited to the Columbia campus. Warren O. Hagstrom, who had worked at Berkeley with William Kornhauser on his 1962 study, *Scientists in Industry*, published *The Scientific Community*[29] in 1965. This is a detailed investigation of the internal social nature of science which, as he notes, drew very heavily upon the Mertonian paradigm. Working at Wisconsin since that time, Hagstrom has continued research in the sociology of science. The editor's own dissertation, directed by Norman Kaplan at Cornell (Kaplan had been a student of Merton in the early 1950s, but in the area of professions rather than the sociology of science as such),[30] was completed in 1960 and resulted later in an attempt to synthesize the Mertonian paradigm more completely which appeared as *The Social System of Science*[31] some years later.

A rather different focus of interest, developing out of public opinion studies in the late 1950s, was concerned with communication among scientists.[32] Herbert Menzel pioneered in this field while at the Bureau of Applied Social Research at Columbia; although he was in contact with Merton, the details of communication patterns in science did not at first seem to call for the broad perspective supplied by the paradigm, and it

27. Some of the Coles' work is synthesized in Jonathan Cole and Stephen Cole, *Social Stratification in Science* (Chicago: University of Chicago Press, 1973).

28. See Diana Crane, "Social Structure in a Group of Scientists: A Test of the 'Invisible College' Hypothesis," *American Sociological Review* 34 (1969): 345–51.

29. New York: Basic Books, Inc., 1965.

30. In his preface to *Science and Society* (Chicago: Rand McNally, 1965), Norman Kaplan nevertheless reports that "my own interest in the sociology of science first began to develop in a seminar I had with Professor Robert K. Merton at Columbia University nearly two decades ago." For his overview of the field through the early 1960s, see Norman Kaplan, "The Sociology of Science," in Robert E. L. Faris, ed., *Handbook of Modern Sociology* (Chicago: Rand McNally, 1964), pp. 852–81.

31. New York: Holt, Rinehart and Winston, 1966. On a special problem in the field, see also Norman W. Storer, "Relations Among Scientific Disciplines," in Saad Z. Nagi and Ronald G. Corwin, eds., *Social Contexts of Research* (New York: Wiley, 1972), pp. 229–68.

32. For overviews of research on scientific communication, see William J. Paisley, *The Flow of (Behavioral) Science Information: A Review of the Research Literature* (Stanford, Calif.: Stanford Institute for Communication Research, 1965), and "Information Needs and Uses," in Carlos Cuadra, ed., *Annual Review of Information Science and Technology*, vol. 3 (1968).

was not until these patterns were seen as fundamental to the development of new scientific specialties and not until the latent functions of planned communication came into focus that an explicit linkage was formed.[33] Crane's *Invisible Colleges*[34] brings the two interests together very effectively, and Nicholas C. Mullins' 1966 dissertation at Harvard, "Social Networks Among Biological Scientists," represents an independent approach to the problem.[35]

Another important source of information on the flow of communication in science was the American Psychological Association's project on communication among psychologists. This developed an extensive array of data during the 1960s, under the direction of William Garvey and Belver C. Griffith. Garvey then moved to Johns Hopkins, where the Center for Research in Scientific Communication undertook a notable series of studies of communication during and after the annual meetings of a number of scientific and technological associations.[36] These studies tended to be long on data and relatively short on theory, but they have proven to be quite compatible with questions derived from the paradigm. Griffith, now at the Drexel Institute, has also continued to study the communications and invisible colleges of scientists.

Three additional lines of development in the United States, largely independent of the Mertonian tradition and yet generally complementary, should be noted here. Joseph Ben-David, working partly at Chicago and partly at Hebrew University in Jerusalem, has produced since 1960 a series of important papers on the relation of different forms of academic organization to scientific developments, culminating in his recent important analysis of the growth of science in Western civilization since the time

33. Herbert Menzel, "The Flow of Information Among Scientists: Problems, Opportunities, and Research Questions," mimeographed (New York: Columbia University Bureau of Applied Social Research, 1958). On the functions of various patterns of communication, see Herbert Menzel, "Planned and Unplanned Scientific Communication," in International Conference on Scientific Information, *Proceedings* (Washington, D.C.: National Academy of Sciences, 1959), pp. 199–243. See also, Menzel, "Scientific Communication: Five Themes from Social Research," *American Psychologist* 21 (1966): 999–1004, and for other aspects of the communication process, the other papers in the same issue of this journal.

34. Diana Crane, *Invisible Colleges: Diffusion of Knowledge in Scientific Communities* (Chicago: University of Chicago Press, 1972).

35. For example, see Nicholas C. Mullins, "The Distribution of Social and Cultural Properties in Informal Communication Networks among Biological Scientists," *American Sociological Review* 33 (October 1968): 786–97.

36. See, for example, the Johns Hopkins University Center for Research in Scientific Communication, "Scientific Exchange-Behavior at the 1966 Annual Meeting of the American Sociological Association," report no. 4, Baltimore, September 1967, pp. 209–50; and "The Dissemination of Scientific Information, Informal Interaction, and the Impact of Information Associated with the 48th Annual Meeting of the American Geophysical Union," report no. 5, Baltimore, October 1967, pp. 251–92.

of the Greeks.[37] His emphasis has been continually on the social-structural factors that have influenced the growth or decline of science or of scientific specialties, and in this sense he has been closer to those studying communication and invisible colleges than to those working directly from the normative structure of science.

Trained as a historian of science, Thomas S. Kuhn has drawn ever closer to the sociology of science since the appearance of his extremely influential book, *The Structure of Scientific Revolutions*, in 1962. It is from this book that the term paradigm has gained its present currency, even though the concept had, of course, been employed in a less general sense much earlier (e.g., in Merton's 1945 review of the sociology of knowledge and in his analysis, four years later, of the functions of paradigms as distinct from theories).[38] In contrast to Ben-David, Kuhn's emphasis is on the substantive content of science—with how changes in the focus and organization of scientific knowledge come about. But more than most historian-philosophers of science, he has dealt with the social structure of "the scientific community" as basic to the operation of paradigms and, more generally, to the development of science. When the central elements of a body of knowledge are fairly stable and widely accepted (as, for instance, was the case with Newtonian physics between 1700 and 1900), a paradigm exists, and research on the questions that flow naturally from this basic definition of a discipline's work is called "normal science." A scientific revolution, then, occurs when the coherence of this paradigm breaks down under an accumulation of new theories, new questions, and new data which throw its validity into doubt, and a new paradigm develops in its place.

It is explicitly with Kuhn's work in mind that some of Merton's recent critics such as Mulkay and King have suggested that he has erred in not taking more direct account of the substantive content of science in his own formulations. These criticisms, however, seldom refer to any of Merton's work since 1957 and (with the exception of the difficulties over the "reality" of the norms of science, to which Kuhn's ideas provide a useful resolution) may be characterized as belatedly premature. To say nothing of the earlier work on the foci of attention in science, the fact is that Merton and his colleagues at Columbia have been working on problems at the interface of the social structure and cognitive structure of science—that is, the grounds on which it is organized and the forms and extent of

37. Ben-David, *The Scientist's Role in Society*. References to Ben-David's earlier work, including papers with associates such as Randall Collins and A. Zloczower, can be found in his book.

38. As noted by Robert W. Friedrichs in his discussion of paradigms and exemplars: "Dialectical Sociology: An Exemplar for the 1970s," *Social Forces* 50 (1972): 447–55. For Merton's account of the uses of paradigms, see *Social Theory and Social Structure*, 1968 edition, pp. 69–72.

consensus and dissensus that attend various aspects of claims to knowledge under different circumstances.

A third line of development has been vigorously explored by Derek J. de Solla Price, the historian of science at Yale, who has become more and more sociologically oriented over the past two decades. Since 1951 he has been increasingly concerned with the quantification of broad parameters of world science, as evidenced by topics dealt with in his *Science Since Babylon* and *Little Science, Big Science*: growth rates since 1600 (in numbers of scientists and scientific discoveries, new journals and societies, and gross publications per year); patterns of national investment in research and development (which he finds to equal roughly .07 percent of all countries' gross national product, regardless of size or level of economic development); and various aspects of communication.[39] Price also introduced the seventeenth-century term *invisible colleges* in its present conceptual sense—informal clusters of scientists collaborating at newly developing research frontiers—and has been highly effective in demonstrating the importance of the *Science Citation Index* in tracing the give-and-take between national science communities and in assessing the intellectual influence of specific scientific writings over time.

Price's work has thus served largely to establish some of the demographic and other "material" parameters of the scientific community over time and has not had occasion to tackle problems to which the Mertonian paradigm would be directly relevant. Two indirect linkages exist, however —through his work on invisible colleges, a topic with which Diana Crane has been effectively concerned, and through his encouragement of the use of the *Science Citation Index* as a research tool; both Price and Merton are members of the SCI Advisory Board, and the Columbia group, notably Stephen and Jonathan Cole, has made intensive systematic use of science citations.

Work in the sociology of science outside the United States has lately acquired impressive momentum, and no effort is made here to catalogue even its principal contributions to the field. A short, obviously incomplete inventory must suffice. There have been founded in Britain a science studies unit at the University of Edinburgh, under the direction of David O. Edge, and a program in the history and social studies of science at the University of Sussex, under the guidance of Roy M. MacLeod and, in science policy studies, Christopher Freeman. In 1971 MacLeod and Edge joined forces as joint editors of *Science Studies*, a journal specializing in "research in the social and historical dimensions of science and technol-

39. Derek J. de Solla Price, *Science Since Babylon* (New Haven: Yale University Press, 1961); *Little Science, Big Science* (New York: Columbia University Press, 1963); and among his many papers, "Nations Can Publish or Perish," *International Science and Technology* 70 (October 1967): 84–89.

ogy." Others are at work in the field at Leeds (J. R. Ravetz and R. G. A. Dolby), at Cambridge (N. J. Mulkay), at London (Hilary Rose), at Cardiff (Paul Halmos), and at Manchester (Richard D. Whitley).

In Sweden, Stevan Dedijer has assembled a research group at the University of Lund. Considerable interest in the sociology of science exists in the Soviet Union as A. Zvorikin and S. R. Mikulinskii, among others, testify, with the work of Gennady Dobrov at Kiev being along the lines developed by Derek Price in this country. Increasing interest in the field has been evidenced by social scientists in Poland, where years ago, in the mid-1930s, Maria Ossowka and Stanislaw Ossowski introduced the "science of science"; in Czechoslovakia where the Academy of Sciences has coordinated disciplinary work on the social and human implications of science and technology; in France, in the work of such men as Jean-Jacques Salomon, Serge Moscovici, and Bernard Lécuyer (whose work in the sociology of science, chiefly under the guidance of Paul Lazarsfeld, began in a joint seminar given by Lazarsfeld and Merton); and in Germany, Israel, Holland, Japan, and an array of other countries. A number of important investigations have been carried out by UNESCO and the Union Internationale d'Histoire et de Philosophie des Sciences in Paris, and a Research Committee on the Sociology of Science has been established by the International Sociological Association. Any comprehensive survey of the present state of the sociology of science that examined the contributions of these groups would surely conclude that it is becoming less and less an American specialty.

Since the emergence of the Mertonian paradigm in the early 1960s, most research in the field appears to fit Kuhn's definition of "normal science." Not only Merton's own work but that of many others in the field have focused primarily on problems which, once elucidated, turn out to be directly relevant to questions explicit or implicit in the paradigm. In short, the sociology of science has matured to the point where much research involves "puzzle-solving." As Kuhn emphasizes,[40] to describe research as "puzzle-solving" does not imply that it falls short of being imaginative, satisfying, or important. Filling out the areas which a paradigm can only identify—what Merton has described as "specified ignorance"—is as necessary to the development of scientific knowledge as is the scientific revolution; without the yin of normal science, there would be no basis for the yang of scientific revolution—and the latter is comparatively rare.

As the five divisions of this volume indicate, several fundamental questions generated by the paradigm have led to substantial research. For example, the effort to work out a comprehensive concept of the reward system in science—in part by intensively investigating the meanings in-

40. Kuhn, *Structure of Scientific Revolutions*, chapter 4.

volved in the quest for priority—helped to focus attention on how professional recognition is achieved in science and to indicate how the reward system is linked to the normative structure. Further related to this line of investigation are the social organization and processes of evaluation that are seen as central to science as an intellectual enterprise. This leads to research on such empirical problems as the ways in which the quality of scientific contributions is assessed and the general adequacy or inadequacy of this process in facilitating the equitable allocation of rewards for these contributions. Finally, as we have seen, problems of this sort have led to closer scrutiny of the criteria by which scientific excellence is determined and to an explicit consideration of the intellectual variables involved: the degree of consensus that exists in a given discipline and several aspects of the organization of its body of knowledge.

There is no completely satisfactory way to close this introduction, for the field described here which Merton and others have thus far advanced is still in a stage of rapid expansion. Presented here are many of the foundation stones of the sociology of science as it is presently constituted, together with numerous examples of work on questions built upon this foundation. One of the facts that is lost when one categorizes all science as "normal" or "revolutionary," however, is that the shape of the paradigms—the facets prominent at a given time—changes as "normal" questions are solved and new ones take their place. Such changes hardly constitute a revolution, any more than the invasion of a new type of tree changes the essential character of a forest, but those of us in the midst of the forest may look forward to different foci of attention and to different research methods and data as this particular forest continues to flourish.

For all of this, the enduring value of the papers collected here cannot be mistaken. It will be several decades at least before the Whiteheadian maxim, "A science which hesitates to forget its founders is lost," has any relevance at all to the sociology of science—and the editor mentions this only as a bet-hedging counter to his own conviction that these papers will not lose their basic value so long as the perspectives of sociology are applied to any science, including itself.

Norman W. Storer

The Sociology of Knowledge

Part 1

Prefatory Note

The five papers comprising part one delineate Merton's continuing interest in the sociology of knowledge, and the dates of their appearance attest to the fruitful reciprocity between work in that field and in the sociology of science. Since "knowledge" is more inclusive than "*scientific* knowledge," the latter must constitute the more specialized focus of attention; one conclusion that can be drawn from these (and other) papers is that Merton has been at pains since the first of his work to keep the distinction and connection between the two explicit and analytically useful.

The papers are arranged here, without regard for chronology, to trace a line of thought that extends from a concern with pure theory to a vivid awareness of the concrete moral dilemmas faced by individual men and women of knowledge. After all, it is not often the case that scholars and scientists see, from the outset, all of the phases in a developing line of reasoning and then focus on each one in logical sequence. Rather, the stages are "filled in" as occasion and opportunity allow, and it is the task of hindsight to discern the underlying order that knits them together.

The section begins with Merton's examination of the condition of the sociology of knowledge circa 1945, in which he argues that its fixation on the one problem of the "existential basis of mental productions" leads to an impasse. The essay still stands as a landmark in the field, and it was perhaps fitting that it should come just after World War II, at a time when new perspectives were emerging in much of social science. A program for relating philosophical conceptions of the sources of knowledge in society to the empirical investigation of specified problems is sketched out in the explicit "paradigm" laid out early in the essay. Merton himself, in fact, had an identifiable part in the process he describes in the last paragraph (perhaps as much an effort to induce a self-fulfilling prophecy as to picture contemporary reality): ". . . the sociology of knowledge is fast outgrowing a prior tendency to confuse provisional hypothesis with unimpeachable dogma; the plenitude of speculative insights which marked its early stages are now being subjected to increasingly rigorous test." Among the first fruits of such an orientation were Merton's introduction of the concepts of "local and cosmopolitan influentials,"[1] subsequently adapted by Gouldner[2] in the academic realm; the Merton and Kitt[3] "contributions to reference group theory" based on findings reported in Stouffer *et al.*, *The American Soldier* (1949) and Merton's "continuities in reference group theory."[4] As was noted by Herbert H. Hyman,[5] the social scientist who had introduced the concept of "reference groups" years before, these writings systematized problems and ideas that deal with the role of nonmembership reference groups in shaping values and intellectual perspectives.

The next paper, on Znaniecki's book, *The Social Role of the Man of Knowledge*, is one of those widely appreciated commentaries in which Merton goes well beyond the role of passive responder to elucidate the book's central points and to place them provocatively within a broader sociological framework. Here the focus is entirely theoretical and atemporal, so that the categorization of the various roles played by scholars and scientists and the winnowing out of researchable hypotheses give the paper enduring value. It also holds interest for the historian of ideas in its outline

1. Robert K. Merton, "Patterns of Influence: A Study of Interpersonal Influence and of Communications Behavior in a Local Community," in Paul F. Lazarsfeld and Frank Stanton, eds., *Communications Research 1948–49* (New York: Harper, 1949), and in Merton, *Social Theory and Social Structure* (1949; rev. ed., N.Y.: The Free Press, 1968).

2. Alvin W. Gouldner, "Cosmopolitans and Locals: Toward an Analysis of Latent Social Roles," *Administrative Science Quarterly* 2 (December 1957 and March 1958): 281–306, 444–80.

3. Robert K. Merton and Alice Kitt (Rossi), "Contributions to the Theory of Reference Group Behavior," in Robert K. Merton and Paul F. Lazarsfeld, *Continuities in Social Research* (N.Y.: The Free Press, 1950), and in *Social Theory and Social Structure*.

4. Robert K. Merton, "Continuities in the Theory of Reference Groups and Social Structure," in *Social Theory and Social Structure*.

5. Herbert H. Hyman, "Reference Groups," in David L. Sills, ed., *International Encyclopedia of the Social Sciences* (New York: Macmillan, 1968), 13: 353–61.

of an analytical approach that would later be embodied in the "paradigm for the sociology of knowledge." To reiterate a point made above, the path of theoretical development that can be traced out well after a series of interrelated papers has appeared need not reflect their chronology; it is as though the theoretical progression exists in two dimensions and the scholar in a third, for he is free to move back and forth along this line of development as opportunities present themselves, with little regard for the logic inherent in the theory itself.

In the following paper, "Social Conflict over Styles of Sociological Work," Merton still finds insufficient "sustained and methodical investigation" in the sociology of knowledge, some sixteen years after he had hopefully noted its acceleration. The paper was originally presented at the Fourth World Congress of Sociology in 1959 at Stresa, Italy, which was devoted to the theme of "the Sociology of Sociology." (This is a special topic within the sociology of science that occasionally threatens, quite naturally, to run away with sociologists' interests. However, Merton takes the sociological community as partly representative of scholarly communities in general rather than simply as an arena in which to present practical advice for practicing sociologists.) The focus is on the ways in which relationships among sociologists may be as important in shaping their shared body of knowledge as are new data and new theoretical perspectives. For the most part, Merton manages to keep his attention focused on those aspects of intradisciplinary conflicts which are parallelled by similar phenomena in other disciplines at a comparable stage of development, rather than succumbing to the familiar and distracting theme of the peculiar vulnerability of sociologists to involvement in the ideological concerns of the larger society.

The next paper, "Technical and Moral Dimensions of Policy Research" (1949), together with the related pages drawn from *Mass Persuasion* (1946), moves on from the relationships between scientists to another problem in the sociology of knowledge. This analyzes the conflict between the demands of technique and morality to which social scientists, especially those engaged in policy-oriented research, are often subject. It examines the conditions leading social scientists to become "routinized in the role of [a] bureaucratic technician" who "does not question policies, state problems and formulate alternatives." Merton expressly rejects "the standpoint of the positivist," observing that "the investigator may naïvely suppose that he is engaged in the value-free activity of research, whereas in fact he may simply have so defined his research problems that the results will be of use to one group in the society, and not to others. His very choice and definition of a problem reflect his tacit values."

These writings take on an added interest in view of recent vagaries in the history of sociological ideas. Clearly, they nullify some current assertions[6] that Merton is essentially a positivist, concerned only with the technical aspects of knowledge. Moreover, as Lord Simey has noted,[7] Merton went beyond theoretical and empirical analysis in a subsequent paper with Daniel Lerner—"Social Scientists and Research Policy" (in *The Policy Sciences*, 1951)—to argue that social scientists have a normative obligation to assert their scholarly values against the short-range and self-interested objectives often found in research requests coming from policy makers. This suggests that the current discovery of such problems and perspectives in "the New Sociology" may rather be a reaffirmation under social conditions now more propitious to such views than before; in fact, the "cryptomnesia" (see the last paper in section 4 of this volume) suffered by successive generations of social scientists on this subject is itself a specific problem for the sociology of knowledge.

The last paper in this section links up with the first, published a quarter-century before. It examines in broad perspective the problem of the existential bases of knowledge in the social sciences as this was set out in the paradigm for the sociology of knowledge and in "continuities in the theory of reference groups." Immediately occasioned by the rise in collective consciousness of ethnic, racial, and other identities, the central question of whether "only blacks can understand blacks" or "only women can understand women" is generalized into a crucial issue in the sociology of knowledge. The issue is this: whether monopolistic or privileged access to knowledge, or exclusion from it, derives from one's group membership or social position. Merton systematically dissects the general problem into its component issues and implications, staying carefully in touch with the social structural concomitants of each one. He reminds us that such claims, growing out of a burgeoning ethnocentrism, are by no means new: whenever different sectors of a society, or subcultures within it, find themselves in full-scale ideological conflict, such claims are bound to arise. A coordinate prediction, made obvious by the very appearance of this paper, would be that the sociology of knowledge itself is likely to experience spurts of growth shortly after such problems come to center-stage within the intellectual community.

N. W. S.

6. See, for instance, M. D. King, "Reason, Tradition and the Progressiveness of Science," *History and Theory: Studies in the Philosophy of History* 10 (1971): 3–32.

7. T. S. Simey, *Social Science and Social Purpose* (London: Constable and Co., 1968), pp. 59–62, 178–80.

1 Paradigm for the Sociology of Knowledge

1945

The last generation has witnessed the emergence of a special field of sociological inquiry: the sociology of knowledge (*Wissenssoziologie*). The term "knowledge" must be interpreted very broadly indeed, since studies in this area have dealt with virtually the entire gamut of cultural products (ideas, ideologies, juristic and ethical beliefs, philosophy, science, technology). But whatever the conception of knowledge, the orientation of this discipline remains largely the same: it is primarily concerned with the relations between knowledge and other existential factors in the society or culture. General and even vague as this formulation of the central purpose may be, a more specific statement will not serve to include the diverse approaches which have been developed.

Manifestly, then, the sociology of knowledge is concerned with problems that have had a long history. So much is this the case, that the discipline has found its first historian, Ernst Gruenwald.[1] But our primary concern is not with the many antecedents of current theories. There are indeed few present-day observations which have not found previous expression in suggestive aperçus. King Henry IV was being reminded that "Thy wish was father, Harry, to that thought" only a few years before Bacon was writing that "The human understanding is no dry light but receives an

Originally published as "Sociology of Knowledge," in Georges Gurvitch and Wilbert E. Moore, eds., *Twentieth-Century Sociology* (New York: Philosophical Library, 1945), pp. 366–405. Reprinted with permission.

1. Nothing will be said of this history in this paper. Ernst Gruenwald provides a sketch of the early developments, at least from the so-called era of Enlightenment in *Das Problem der Soziologie des Wissens* (Vienna-Leipzig: Wilhelm Braumueller, 1934). For a survey, see H. Otto Dahlke, "The Sociology of Knowledge," in H. E. Barnes, Howard Becker, and Frances B. Becker, eds., *Contemporary Social Theory* (New York: Appleton-Century, 1940), pp. 64–89.

infusion from the will and affections; whence proceed sciences which may be called 'sciences as one would.' " And Nietzsche had set down a host of aphorisms on the ways in which needs determined the perspectives through which we interpret the world so that even sense perceptions are permeated with value-preferences. The antecedents of *Wissenssoziologie* only go to support Whitehead's observation that "to come very near to a true theory, and to grasp its precise application, are two very different things, as the history of science teaches us. Everything of importance has been said before by somebody who did not discover it."

The Social Context

Quite apart from its historical and intellectual origins, there is the further question of the basis of contemporary interest in the sociology of knowledge. As is well known, the sociology of knowledge, as a distinct discipline, has been especially cultivated in Germany and France. Only within the last decades have American sociologists come to devote increasing attention to problems in this area. The growth of publications and, as a decisive test of its academic respectability, the increasing number of doctoral dissertations in the field partly testify to this rise of interest.

An immediate and obviously inadequate explanation of this development would point to the recent transfer of European sociological thought by sociologists who have lately come to this country. To be sure, these scholars were among the culture-bearers of *Wissenssoziologie*. But this merely provided availability of these conceptions and no more accounts for their actual acceptance than would mere availability in any other instance of culture diffusion. American thought proved receptive to the sociology of knowledge largely because it dealt with problems, concepts, and theories that are increasingly pertinent to our contemporary social situation, because our society has come to have certain characteristics of those European societies in which the discipline was initially developed.

The sociology of knowledge takes on pertinence under a definite complex of social and cultural conditions.[2] With increasing social conflict, differences in the values, attitudes, and modes of thought of groups develop to the point where the orientation which these groups previously had in common is overshadowed by incompatible differences. Not only do there develop distinct universes of discourse, but the existence of any one universe challenges the validity and legitimacy of the others. The coexistence of these conflicting perspectives and interpretations within the same society leads to an active and reciprocal *distrust* between groups.

2. See Karl Mannheim, *Ideology and Utopia,* pp. 5–12; Pitirim A. Sorokin, *Social and Cultural Dynamics,* 4 vols. (New York: American Book Co., 1937), 2: 412–13.

Within a context of distrust, one no longer inquires into the content of beliefs and assertions to determine whether they are valid or not, one no longer confronts the assertions with relevant evidence, but introduces an entirely new question: how does it happen that these views are maintained? Thought becomes functionalized; it is interpreted in terms of its psychological or economic or social or racial sources and functions. In general, this type of functionalizing occurs when statements are doubted, when they appear so palpably implausible or absurd or biased that one need no longer examine the evidence for or against the statement but only the grounds for its being asserted at all.[3] Such alien statements are "explained by" or "imputed to" special interests, unwitting motives, distorted perspectives, social position, and so on. In folk thought, this involves reciprocal attacks on the integrity of opponents; in more systematic thought, it leads to reciprocal ideological analyses. On both levels, it feeds upon and nourishes collective insecurities.

Within this social context, an array of interpretations of man and culture which share certain common presuppositions finds widespread currency. Not only ideological analysis and *Wissenssoziologie*, but also psychoanalysis, Marxism, semanticism, propaganda analysis, Paretanism, and, to some extent, functional analysis have, despite their other differences, a similar outlook on the role of ideas. On the one hand, there is the realm of verbalization and ideas (ideologies, rationalizations, emotive expressions, distortions, folklore, derivations), all of which are viewed as expressive or derivative or deceptive (of self and others), all of which are functionally related to some substratum. On the other hand are the previously conceived substrata (relations of production, social position, basic impulses, psychological conflict, interests and sentiments, interpersonal relations, and residues). And throughout runs the basic theme of the unwitting determination of ideas by the substrata; the emphasis on the distinction between the real and the illusory, between reality and appearance in the sphere of human thought, belief, and conduct. And whatever the intention of the analysts, their analyses tend to have an acrid quality: they tend to indict, secularize, ironicize, satirize, alienate, devalue the intrinsic content of the avowed belief or point of view. Consider only the overtones of terms chosen in

3. Freud had observed this tendency to seek out the "origins" rather than to test the validity of statements which seem palpably absurd to us. Thus, suppose someone maintains that the center of the earth is made of jam. "The result of our intellectual objection will be a *diversion of our interests; instead of their being directed on to the investigation itself*, as to whether the interior of the earth is really made of jam or not, *we shall wonder what kind of man it must be who can get such an idea into his head. . . .*" Sigmund Freud, *New Introductory Lectures* (New York: W. W. Norton, 1933), p. 49 (italics added). On the social level, a radical difference of outlook of various social groups leads not only to ad hominem attacks, but also to "functionalized explanations."

these contexts to refer to beliefs, ideas, and thought: vital lies, myths, illusions, derivations, folklore, rationalizations, ideologies, verbal façade, pseudo-reasons, and so on.

What these schemes of analysis have in common is the practice of discounting the *face value* of statements, beliefs, and idea-systems by reexamining them within a new context which supplies the "real meaning." Statements ordinarily viewed in terms of their manifest content are debunked, whatever the intention of the analyst, by relating this content to attributes of the speaker or of the society in which he lives. The professional iconoclast, the trained debunker, the ideological analyst and their respective systems of thought thrive in a society where large groups of people have already become alienated from common values; where separate universes of discourse are linked with reciprocal distrust. Ideological analysis systematizes the lack of faith in reigning symbols which has become widespread; hence its pertinence and popularity. The ideological analyst does not so much create a following as he speaks for a following to whom his analyses "make sense," that is, conform to their previously unanalyzed experience.[4]

In a society where reciprocal distrust finds such folk-expression as "what's in it for him?"; where "buncombe" and "bunk" have been idiom for nearly a century and "debunk" for a generation; where advertising and propaganda have generated active resistance to the acceptance of statements at face-value; where pseudo-*Gemeinschaft* behavior as a device for improving one's economic and political position is documented in a best seller on how to win friends who may be influenced; where social relationships are increasingly instrumentalized so that the individual comes to view others as seeking primarily to control, manipulate, and exploit him; where growing cynicism involves a progressive detachment from significant group relationships and a considerable degree of self-estrangement; where uncertainty about one's own motives is voiced in the indecisive phrase, "I may be rationalizing, but . . ."; where defenses against traumatic disillusionment may consist in remaining permanently disillusioned by reducing expectations about the integrity of others through discounting their motives and abilities in advance;—in such a society, systematic ideological analysis and a derived sociology of knowledge take on a

4. The concept of *pertinence* was assumed by the Marxist harbingers of *Wissenssoziologie*. "The theoretical conclusions of the Communists are in no way based on ideas or principles that have been invented, or discovered, by this or that would-be universal reformer. *They merely express, in general terms, the actual relations* springing from an existing class struggle, from a historical movement going on under our very eyes" (Karl Marx and Friedrich Engels, *The Communist Manifesto,* in Karl Marx, *Selected Works,* 2 vols. [Moscow: Co-operative Publishing Society, 1935], 1: 219 [italics added]).

socially grounded pertinence and cogency. And American academicians, presented with schemes of analysis which appear to order the chaos of cultural conflict, contending values, and points of view, have promptly seized upon and assimilated these analytical schemes.

The "Copernican revolution" in this area of inquiry consisted in the hypothesis that not only error or illusion or unauthenticated belief but also the discovery of truth was socially (historically) conditioned. As long as attention was focused on the social determinants of ideology, illusion, myth, and moral norms, the sociology of knowledge could not emerge. It was abundantly clear that in accounting for error or uncertified opinion, some extratheoretic factors were involved, that some special explanation was needed, since the reality of the object could not account for error. In the case of confirmed or certified knowledge, however, it was long assumed that it could be adequately accounted for in terms of a direct object-interpreter relation. The sociology of knowledge came into being with the signal hypothesis that even truths were to be held socially accountable, were to be related to the historical society in which they emerged.

To outline even the main currents of the sociology of knowledge in brief compass is to present none adequately and to do violence to all. The diversity of formulations—of a Marx or Scheler or Durkheim; the varying problems—from the social determination of categorical systems to that of class-bound political ideologies; the enormous differences in scope—from the all-encompassing categorizing of intellectual history to the social location of the thought of Negro scholars in the last decades; the various limits assigned to the discipline—from a comprehensive sociological epistemology to the empirical relations of particular social structures and ideas; the proliferation of concepts—ideas, belief-systems, positive knowledge, thought, systems of truth, superstructure, and so on; the diverse methods of validation—from plausible but undocumented imputations to meticulous historical and statistical analyses—in the light of all this, an effort to deal with both analytical apparatus and empirical studies in a few pages must sacrifice detail to scope.

To introduce a basis of comparability among the welter of studies which have appeared in this field, we must adopt some scheme of analysis. The following paradigm is intended as a step in this direction. It is, undoubtedly, a partial and, it is to be hoped, a temporary classification which will disappear as it gives way to an improved and more exacting analytical model. But it does provide a basis for taking an inventory of extant findings in the field; for indicating contradictory, contrary, and consistent results; setting forth the conceptual apparatus now in use; determining the nature of problems which have occupied workers in this field; assessing the character of the evidence which they have brought to bear upon these problems; ferreting out the characteristic lacunae and weaknesses in current types of interpretation. Full-fledged theory in the sociology of knowledge lends itself to classification in terms of the following paradigm.

Paradigm for the Sociology of Knowledge

1. Where *is the existential basis of mental productions located?*

a. *social bases:* social position, class, generation, occupational role, mode of production, group structures (university, bureaucracy, academies, sects, political parties), "historical situation," interests, society, ethnic affiliation, social mobility, power structure, social processes (competition, conflict, and so on).

b. *cultural bases:* values, ethos, climate of opinion, *Volksgeist, Zeitgeist,* type of culture, culture mentality, *Weltanschauungen,* and so on.

2. What *mental productions are being sociologically analyzed?*

a. *spheres of:* moral beliefs, ideologies, ideas, the categories of thought, philosophy, religious beliefs, social norms, positive science, technology, and so on.

b. *which aspects are analyzed:* their selection (foci of attention), level of abstraction, presuppositions (what is taken as data and what as problematical), conceptual content, models of verification, objectives of intellectual activity, and so on.

3. How *are mental productions related to the existential basis?*

a. *causal or functional relations:* determination, cause, correspondence, necessary condition, conditioning, functional interdependence, interaction, dependence, and so on.

b. *symbolic or organismic or meaningful relations:* consistency, harmony, coherence, unity, congruence, compatibility (and antonyms); expression, realization, symbolic expression, *Strukturzusammenhang*, structural identities, inner connection, stylistic analogies, logicomeaningful integration, identity of meaning, and so on.

c. *ambiguous terms to designate relations:* correspondence, reflection, bound up with, in close connection with, and so on.

4. Why *related? Manifest and latent functions imputed to these existentially conditioned mental productions.*

a. to maintain power, promote stability, orientation, exploitation, obscure actual social relationships, provide motivation, canalize behavior, divert criticism, deflect hostility, provide reassurance, control nature, coordinate social relationships, and so on.

5. When *do the imputed relations of the existential base and knowledge obtain?*

a. historicist theories (confined to particular societies or cultures).
b. general analytical theories.

There are, of course, additional categories for classifying and analyzing studies in the sociology of knowledge, which are not fully explored here. Thus, the perennial problem of the implications of existential influences upon knowledge for the epistemological status of that knowledge has been hotly debated from the very outset. Solutions to this problem, which assume that a sociology of knowledge is necessarily a sociological theory of knowledge, range from the claim that the "genesis of thought has no necessary relation to its validity" to the extreme relativist position that truth is "merely" a function of a social or cultural basis, that it rests solely upon social consensus and, consequently, that any culturally accepted theory of truth has a claim to validity equal to that of any other.

But the foregoing paradigm serves to organize the distinctive approaches and conclusions in this field sufficiently for our purposes.

The chief approaches to be considered here are those of Marx, Scheler, Mannheim, Durkheim, and Sorokin. Current work in this area is largely oriented toward one or another of these theories, either through a modified application of their conceptions or through counterdevelopments. Other sources of studies in this field indigenous to American thought, such as pragmatism, will be advisedly omitted, since they have not yet been formulated with specific reference to the sociology of knowledge nor have they been embodied in research to any notable extent.

The Existential Basis

A central point of agreement in all approaches to the sociology of knowledge is the thesis that thought has an existential basis insofar as it is not immanently determined and insofar as one or another of its aspects can be derived from extra-cognitive factors. But this is merely a formal consensus, which gives way to a wide variety of theories concerning the nature of the existential basis.

In this respect, as in others, Marxism is the storm center of *Wissenssoziologie*. Without entering into the exegetic problem of closely identifying Marxism—we have only to recall Marx's *"je ne suis pas Marxiste"*—we can trace out its formulations primarily in the writings of Marx and Engels. Whatever other changes may have occurred in the development of their theory during the half-century of their work, they consistently held fast to the thesis that "relations of production" constitute the "real foundation" for the superstructure of ideas. "The mode of production in material life determines the general character of the social, political and intellectual processes of life. It is not the consciousness of men that determines their existence, but on the contrary, their social existence determines their consciousness."[5] In seeking to functionalize ideas, that is, to relate the ideas of individuals to their sociological bases, Marx locates them within the class structure. He assumes, not so much that other influences are not at all operative, but that class is a primary determinant and, as such, the single most fruitful point of departure for analysis. This he makes explicit in his first preface to *Capital*: ". . . here individuals are dealt with *only in so far* as they are the personifications of economic categories, embodiments of particular class-relations and class-interests."[6] In abstracting from other variables and in regarding men in their economic and class roles, Marx hypothesizes that these roles are primary determinants and thus leaves as an open question *the extent to which they adequately account for thought and behavior in any given case.* In point of fact, one line of development of Marxism, from the early *German Ideology* to the latter writings of Engels, consists in a progressive definition (and delimitation) of the extent to which the relations of production do in fact condition knowledge and forms of thought.

However, both Marx and Engels, repeatedly and with increasing insistence, emphasized that the ideologies of a social stratum need not stem only from persons who are *objectively* located in that stratum. As early as the *Communist Manifesto*, Marx and Engels had indicated that as the ruling class approaches dissolution, "a small section . . . joins the revolutionary class. . . . Just as, therefore, at an earlier period, a section of the nobility went over to the bourgeoisie, so now a portion of the bourgeoisie goes over to the proletariat, and in particular, a portion of *the bourgeois ideologists*, who have *raised themselves* to the level of comprehending theoretically the historical movement as a whole."[7]

Ideologies are socially located by analyzing their perspectives and presuppositions and determining how problems are construed: from the standpoint of one or another class. Thought is not mechanistically located by merely establishing the class position of the thinker. It is attributed to that class for which it is "appropriate," to the class whose social situation with its class conflicts, aspirations, fears, restraints, and objective possibilities within the given sociohistorical context is being expressed. Marx's most explicit formulation holds:

> One must not form the narrow-minded idea that the petty bourgeoisie wants on principle to enforce an egoistic class interest. It believes, rather, that the

5. Karl Marx, *A Contribution to the Critique of Political Economy* (Chicago: C. H. Kerr, 1904), pp. 11–12.

6. Karl Marx, *Capital,* 1: 15 (italics added); cf. Marx and Engels, *The German Ideology* (New York: International Publishers, 1939), p. 76; cf. Max Weber, *Gesammelte Aufsaetze zur Wissenschaftslehre,* p. 205.

7. Marx and Engels, *The Communist Manifesto*, in Marx, *Selected Works*, 1:216 (italics added).

special conditions of its emancipation are the *general* conditions through which alone modern society can be saved and the class struggle avoided. Just as little must one imagine that the democratic representatives are all shopkeepers or are full of enthusiasm for them. *So far as their education and their individual position are concerned,* they may be as widely separated from them as heaven from earth. *What makes them representatives of the petty bourgeosie is the fact that in their minds* [*im Kopfe*] *they do not exceed the limits which the latter do not exceed in their life activities,* that they are consequently driven to the same problems and solutions in theory to which material interest and social position drive the latter in practice. *This is ueberhaupt, the relationship of the political and literary representatives of a class to the class which they represent.*[8]

But if we cannot derive ideas from the objective class position of their exponents, this leaves a wide margin of indeterminacy. It then becomes a further problem to discover why some identify themselves with the characteristic outlook of the class stratum in which they objectively find themselves whereas others adopt the presuppositions of a class stratum other than "their own." An empirical description of the fact is no adequate substitute for its theoretical explanation.

In dealing with existential bases, Max Scheler characteristically places his own hypothesis in opposition to other prevalent theories.[9] He draws a distinction between cultural sociology and what he calls the sociology of real factors (*Realsoziologie*). Cultural data are "ideal," in the realm of ideas and values: "real factors" are oriented toward effecting changes in the reality of nature or society. The former are defined by ideal goals or intentions; the latter derive from an "impulse structure" (*Triebstruktur*, for example, sex, hunger, power). It is a basic error, he holds, of all naturalistic theories to maintain that real factors—whether race, geopolitics, political power structure, or the relations of economic production—unequivocally determine the realm of meaningful ideas. He also rejects all ideological, spiritualistic, and personalistic conceptions which err in viewing the history of existential conditions as a unilinear unfolding of the history of mind. He ascribes complete autonomy and a determinate sequence to these real factors, though he inconsistently holds that value-laden ideas serve to guide and direct their development. Ideas as such initially have no social effectiveness. The "purer" the idea, the greater its

8. Karl Marx, *Der achtzehnte Brumaire des Louis Bonaparte* (Hamburg, 1885), p. 36 (italics inserted).

9. This account is based upon Max Scheler's most elaborate discussion, "Probleme einer Soziologie des Wissens," in his *Die Wissensformen und die Gesellschaft* (Leipzig: Der Neue-Geist Verlag, 1926), pp. 1–229. This essay is an extended and improved version of an essay in his *Versuche zu einer Soziologie des Wissens* (Munich: Duncker und Humblot, 1924), pp. 5–146. For further discussions of Scheler, see P. A. Schillp, "The Formal Problems of Scheler's Sociology of Knowledge," *The Philosophical Review* 36 (March 1927): 101–20; Howard Becker and H. Otto Dahlke, "Max Scheler's Sociology of Knowledge," *Philosophy and Phenomenological Research* 2 (March 1942): 310–22.

impotence, so far as dynamic effect on society is concerned. Ideas do not become actualized, embodied in cultural developments, unless they are bound up in some fashion with interests, impulses, emotions, or collective tendencies and their incorporation in institutional structures.[10] Only then—and in this limited respect, naturalistic theories (for example, Marxism) are justified—do they exercise some definite influence. Should ideas not be grounded in the immanent development of real factors, they are doomed to become sterile Utopias.

Naturalistic theories are further in error, Scheler holds, in tacitly assuming the *independent variable* to be one and the same throughout history. There is no constant independent variable but there is, in the course of history, a definite sequence in which the primary factors prevail, a sequence which can be summed up in a "law of three phases." In the initial phase, blood-ties and associated kinship institutions constitute the independent variable; later, political power and, finally, economic factors. There is, then, no constancy in the effective primacy of existential factors but rather an ordered variability. Thus, Scheler sought to relativize the very notion of historical determinants.[11] He claims not only to have confirmed his law of the three phases inductively but to have derived it from a theory of human impulses.

Scheler's conception of *Realfaktoren*—race and kinship, the structure of power, factors of production, qualitative and quantitative aspects of population, geographical and geopolitical factors—hardly constitutes a usefully defined category. It is of small value to subsume such diverse elements under one rubric, and, indeed, his own empirical studies and those of his disciples do not profit from this array of factors. But in suggesting a variation of significant existential factors, though not in the ordered sequence which he failed to establish, he moves in the direction which subsequent research has followed.

Thus, Mannheim derives from Marx primarily by extending his conception of existential bases. Given the *fact* of multiple group affiliation, the problem becomes one of determining *which* of these affiliations are decisive in fixing perspectives, models of thought, definitions of the given, and so on. Unlike "a dogmatic Marxism," he does not assume that class position is alone ultimately determinant. He finds, for example, that an organically integrated group conceives of history as a continuous movement toward the realization of its goals, whereas socially uprooted and loosely integrated groups espouse a historical intuition which stresses the fortuitous and imponderable. It is only through exploring the variety of group formations

10. Scheler, *Die Wissensformen*, pp. 7, 32.

11. Ibid., pp. 25–45. It should be noted that Marx had long since rejected out of hand a similar conception of shifts in independent variables which was made the basis for an attack on his *Critique of Political Economy*; see *Capital*, 1: 94n.

—generations, status groups, sects, occupational groups—and their characteristic modes of thought that there can be found an existential basis corresponding to the great variety of perspectives and knowledge which actually obtain.[12]

Though representing a different tradition, this is substantially the position taken by Durkheim. In an early study with Mauss of primitive forms of classification, he maintained that the genesis of the categories of thought is to be found in the group structure and relations and that the categories vary with changes in the social organization.[13] In seeking to account for the social origins of the categories, Durkheim postulates that individuals are more directly and inclusively oriented toward the groups in which they live than they are toward nature. The primarily significant experiences are mediated through social relationships, which leave their impress on the character of thought and knowledge.[14] Thus, in his study of primitive forms of thought, he deals with the periodic recurrence of social activities (ceremonies, feasts, rites), the clan structure, and the spatial configurations of group meetings as among the existential bases of thought. And, applying Durkheim's formulations to ancient Chinese thought, Granet attributes their typical conceptions of time and space to such bases as the feudal organization and the rhythmic alternation of concentrated and dispersed group life.[15]

In sharp distinction from the foregoing conceptions of existential bases is Sorokin's idealistic and emanationist theory, which seeks to derive every aspect of knowledge, not from an existential social basis, but from varying "culture mentalities." These mentalities are constructed of "major premises": thus, the ideational mentality conceives of reality as "non-material, ever-lasting Being"; its needs as primarily spiritual and their full satisfaction through "self imposed minimization or elimination of most physical needs."[16] Contrariwise, the sensate mentality limits reality to what can be perceived through the senses, it is primarily concerned with physical needs which it seeks to satisfy to a maximum, not through self-modification, but

12. Karl Mannheim, *Ideology and Utopia*, pp. 247–48. In view of the recent extensive discussions of Mannheim's work, it will not be treated at length in this essay.

13. Emile Durkheim and Marcel Mauss, "De quelques formes primitives de classification," *L'Année Sociologique* 6 (1901–2): 1–72: "... even ideas as abstract as those of time and space are, at each moment of their history, in close relation with the corresponding social organization." As Marcel Granet has indicated, this paper contains some pages on Chinese thought which have been held by specialists to mark a new era in the field of sinological studies.

14. Emile Durkheim, *The Elementary Forms of the Religious Life*, pp. 443–44; see also Hans Kelsen, *Society and Nature* (Chicago: University of Chicago Press, 1943), p. 30.

15. Marcel Granet, *La pensée chinoise* (Paris: La Renaissance du Livre, 1934), for example, pp. 84–104.

16. Sorokin, *Social and Cultural Dynamics*, 1: 72–73.

through change of the external world. The chief intermediate type of mentality is the idealistic, which represents a virtual balance of the foregoing types. It is these mentalities, that is, the major premises of each culture, from which systems of truth and knowledge are derived. And here we come to the self-contained emanationism of an idealistic position: it appears plainly tautological to say, as Sorokin does, that "in a sensate society and culture the Sensate system of truth based upon the testimony of the organs of senses has to be dominant."[17] For sensate mentality has already been *defined* as one conceiving of "reality as only that which is presented to the sense organs."[18]

Moreover, an emanationist phrasing such as this bypasses some of the basic questions raised by other approaches to the analysis of existential conditions. Thus, Sorokin considers the failure of the sensate "system of truth" (empiricism) to monopolize a sensate culture as evidence that the culture is not "fully integrated." But this surrenders inquiry into the bases of those very differences of thought with which our contemporary world is concerned. This is true of other categories and principles of knowledge for which he seeks to apply a sociological accounting. For example, in our present sensate culture, he finds that "materialism" is less prevalent than "idealism," and that "temporalism" and "eternalism" are almost equally current; so, too, with "realism" and "nominalism," "singularism" and "universalism," and so on. Since there are these diversities within a culture, the overall characterization of the culture as sensate provides no basis for indicating which groups subscribe to one mode of thought, and which to another. Sorokin does not systematically explore varying existential bases *within* a society or culture; he looks to the "dominant" tendencies and imputes these to the culture as a whole.[19] Our contemporary society, quite apart from the *differences* of intellectual outlook of divers classes and groups, is viewed as an integral exemplification of sensate culture. On its own premises, Sorokin's approach is primarily suited for an overall characterization of cultures, not for analyzing connections between varied existential conditions and thought within a society.

Types of Knowledge

Even a cursory survey is enough to show that the term "knowledge" has been so broadly conceived as to refer to every type of idea and every mode

17. Ibid., 2: 5.
18. Ibid., 1: 73.
19. One "exception" to this practice is found in his contrast between the prevalent tendency of the "clergy and religious landed aristocracy to become the leading and organizing classes in the Ideational, and the capitalistic bourgeoisie, intelligentsia, professionals, and secular officials in the Sensate culture" (ibid., 3: 250). And see his account of the diffusion of culture among social classes (ibid., 4: 221 ff).

of thought ranging from folk belief to positive science. Knowledge has often come to be assimilated to the term "culture" so that not only the exact sciences but ethical convictions, epistemological postulates, material predications, synthetic judgments, political beliefs, the categories of thought, eschatological doxies, moral norms, ontological assumptions, and observations of empirical fact are more or less indiscriminately held to be "existentially conditioned."[20] The question is, of course, whether these diverse kinds of "knowledge" stand in the same relationship to their sociological basis, or whether it is necessary to discriminate between spheres of knowledge precisely because this relationship differs for the various types. For the most part, there has been a systematic ambiguity concerning this problem.

Only in his later writings did Engels come to recognize that the concept of ideological superstructure included a variety of "ideological forms" which differ *significantly*, that is, are not equally and similarly conditioned by the material basis. Marx's failure to take up this problem systematically[21] accounts for much of the initial vagueness about *what* is comprised by the superstructure and how these several "ideological" spheres are related to the modes of production. It was largely the task of Engels to attempt this clarification. In differentiating the blanket term "ideology," Engels granted a degree of autonomy to law.

> As soon as the new division of labor which creates professional lawyers becomes necessary, another new and independent sphere is opened up which, for all its general dependence on production and trade, still has its own capacity for reacting upon these spheres as well. In a modern state, law must not only correspond to the general economic position and be its expression, but must also be an expression which is *consistent in itself*, and which does not, owing to inner contradictions, look glaringly inconsistent. And in order to achieve this, the faithful reflection of economic conditions is more and more infringed upon. All the more so the more rarely it happens that a code of law is the blunt, unmitigated, unadulterated expression of the domination of a class—this in itself would already offend the "conception of justice."[22]

If this is true of law, with its close connection with economic pressures, it is all the more true of other spheres of the "ideological superstructure."

20. Cf. R. K. Merton, "Karl Mannheim and the Sociology of Knowledge," *Journal of Liberal Religion* 2 (1941): 133–35; Kurt H. Wolff, "The Sociology of Knowledge: Emphasis on an Empirical Attitude," *Philosophy of Science* 10 (1943): 104–23; Talcott Parsons, "The Role of Ideas in Social Action," in *Essays in Sociological Theory*, chapter 6.

21. This is presumably the ground for Scheler's remark: "A specific thesis of the economic conception of history is the subsumption of the laws of development of *all* knowledge under the laws of development of ideologies." *Die Wissensformen*, p. 21.

22. Engels, letter to Conrad Schmidt, 27 October 1890, in Marx, *Selected Works*, 1: 385.

Philosophy, religion, science are particularly constrained by the preexisting stock of knowledge and belief, and are only indirectly and ultimately influenced by economic factors.[23] In these fields, it is not possible to "derive" the content and development of belief and knowledge merely from an analysis of the historical situation:

> Political, juridical, philosophical, religious, literary, artistic, etc., development is based on economic development. But all these react upon one another and also upon the economic base. It is not that the economic position is the *cause and alone active*, while everything else only has a passive effect. There is, rather, interaction on the basis of the economic necessity, which *ultimately* always asserts itself.[24]

But to say that the economic basis "ultimately" asserts itself is to say that the ideological spheres exhibit some degree of independent development, as indeed Engels goes on to observe: "The further the particular sphere which we are investigating is removed from the economic sphere and approaches that of pure abstract ideology, the more shall we find it exhibiting accidents [that is, deviations from the "expected"] in its development, the more will its curve run in zig-zag."[25]

Finally, there is an even more restricted conception of the sociological status of natural science. In one well-known passage, Marx expressly distinguishes natural science from ideological spheres.

> With the change of the economic foundation the entire immense superstructure is more or less rapidly transformed. In considering such transformations the distinction should always be made between the material transformation of the economic conditions of production *which can be determined with the precision of natural science*, and the legal, political, religious, aesthetic or philosophic—in short, ideological forms in which men become conscious of this conflict and fight it out.[26]

Thus, natural science and political economy, which can match its precision, are granted a status quite distinct from that of ideology. The conceptual content of natural science is not imputed to an economic base: merely its "aims" and "material." "Where would natural science be without industry and commerce? Even this "pure" natural science is provided with an aim, as with its material, *only* through trade and industry, through

23. Ibid., 1: 386.

24. Engels, letter to Heinz Starkenburg, 25 January 1894, ibid., 1: 392.

25. Ibid., 1: 393; cf. Engels, *Feuerbach* (Chicago: C. H. Kerr, 1903), pp. 117 ff. "It is well known that certain periods of highest development of art stand in *no direct connection* with the general development of society, nor with the material basis and the skeleton structure of its organization" (Marx, introduction to *Critique of Political Economy*, pp. 309–10 [italics added]).

26. Marx, *Critique of Political Economy*, p. 12 (italics added).

the sensuous activity of men."[27] Along the same lines, Engels asserts that the appearance of Marx's materialistic conception of history was itself determined by "necessity," as is indicated by similar views appearing among English and French historians at the time and by Morgan's independent discovery of the same conception.[28]

He goes even further to maintain that socialist theory is itself a proletarian "reflection" of modern class conflict, so that here, at least, the very content of "scientific thought" is held to be socially determined,[29] without vitiating its validity.

There was an incipient tendency in Marxism, then, to consider natural science as standing in a relation to the economic base different from that of other spheres of knowledge and belief. In science, the focus of attention may be socially determined but not, presumably, its conceptual apparatus. In this respect, the social sciences were sometimes held to differ significantly from the natural sciences. Social science tended to be assimilated to the sphere of ideology, a tendency developed by later Marxists into the questionable thesis of a class-bound social science which is inevitably tendentious[30] and into the claim that only "proletarian science" has valid insight into certain aspects of social reality.[31]

Mannheim follows in the Marxist tradition to the extent of exempting the "exact sciences" and "formal knowledge" from existential determina-

27. Marx and Engels, *The German Ideology,* p. 36 (italics added). See also Engels, *Socialism: Utopian and Scientific* (Chicago: C. H. Kerr, 1910), pp. 24–25, where the needs of a rising middle class are held to account for the revival of science. The assertion that "only" trade and industry provide the aims is typical of the extreme, and untested, statements of relationships which prevail especially in the early Marxist writings. Such terms as "determination" cannot be taken at their face value; they are characteristically used very loosely. The actual *extent* of such relationships between intellectual activity and the material foundations were not investigated by either Marx or Engels.

28. Engels, in Marx, *Selected Works,* 1: 393. The occurrence of parallel independent discoveries and inventions as "proof" of the social determination of knowledge was a repeated theme throughout the nineteenth century. As early as 1828, Macaulay in his essay on Dryden had noted concerning Newton's and Leibniz's invention of the calculus: "Mathematical science, indeed, had reached such a point, that if neither of them had existed, the principle must inevitably have occurred to some person within a few years." He cites other cases in point. Victorian manufacturers shared the same view with Marx and Engels. In our own day, this thesis, based on independent duplicate inventions, has been especially emphasized by Dorothy Thomas, Ogburn, and Vierkandt.

29. Engels, *Socialism: Utopian and Scientific,* p. 97.

30. V. I. Lenin, "The Three Sources and Three Component Parts of Marxism," in Marx, *Selected Works,* 1: 54.

31. Nikolai Bukharin, *Historical Materialism* (New York: International Publishers, 1925), pp. xi–xii; B. Hessen in *Science at the Cross-Roads* (London: Kniga, 1932), p. 154; A. I. Timeniev in *Marxism and Modern Thought* (New York: Harcourt, Brace, 1935), p. 310; "Only Marxism, only the ideology of the advanced revolutionary class is scientific."

tion but not "historical, political and social science thinking as well as the thought of everyday life."[32] Social position determines the "perspective," that is, "the manner in which one views an object, what one perceives in it, and how one construes it in his thinking." The situational determination of thought does not render it invalid; it does, however, particularize the scope of the inquiry and the limits of its validity.[33]

If Marx did not sharply differentiate the superstructure, Scheler goes to the other extreme. He distinguishes a variety of forms of knowledge. To begin with, there are the "relatively natural *Weltanschauungen*": that which is accepted as given, as neither requiring nor being capable of justification. These are, so to speak, the cultural axioms of groups; what Joseph Glanvill, some three hundred years ago, called a "climate of opinion." A primary task of the sociology of knowledge is to discover the laws of transformation of these *Weltanschauungen*. And since these outlooks are by no means necessarily valid, it follows that the sociology of knowledge is not concerned merely with tracing the existential bases of truth but also of "social illusion, superstition and socially conditioned errors and forms of deception."[34]

The *Weltanschauungen* constitute organic growths and develop only in long time-spans. They are scarcely affected by theories. Without adequate evidence, Scheler claims that they can be changed in any fundamental sense only through race-mixture or conceivably through the "mixture" of language and culture. Building upon these very slowly changing *Weltanschauungen* are the more "artificial" forms of knowledge which may be ordered in seven classes, according to degree of artificiality: (1) myth and legend; (2) knowledge implicit in the natural folk-language; (3) religious knowledge (ranging from the vague emotional intuition to the fixed dogma of a church); (4) the basic types of mystical knowledge; (5) philosophical-metaphysical knowledge; (6) positive knowledge of mathematics, the natural and cultural sciences; (7) technological knowledge.[35] The more artificial these types of knowledge, the more rapidly they change. It is evident, says Scheler, that religions change far more slowly than the various metaphysics, and the latter persist for much longer periods than the results of positive science, which change from hour to hour.

This hypothesis of rates of change bears some points of similarity to Alfred Weber's thesis that civilizational change outruns cultural change and to the Ogburn hypothesis that "material" factors change more rapidly

32. Mannheim, *Ideology and Utopia,* pp. 150, 243; Mannheim, "Die Bedeutung der Konkurrenz im Gebiete des Geistigen," in *Verhandlungen des 6. deutschen Soziologentages* (Tuebingen: 1929), p. 41.

33. Mannheim, *Ideology and Utopia,* pp. 256, 264.

34. Scheler, *Die Wissensformen,* pp. 59–61.

35. Ibid., p. 62.

than the "nonmaterial." Scheler's hypothesis shares the limitations of these others as well as several additional shortcomings. He nowhere indicates with any clarity what his principle of classification of types of knowledge —so-called artificiality—actually denotes. Why, for example, is "mystical knowledge" conceived as more "artificial" than religious dogmas? He does not at all consider what is entailed by saying that one type of knowledge changes more rapidly than another. Consider his curious equating of new scientific "results" with metaphysical systems; how does one compare the degree of change implied in neo-Kantian philosophy with, say, change in biological theory during the corresponding period? Scheler boldly asserts a sevenfold variation in rates of change and, of course, does not empirically confirm this elaborate claim. In view of the difficulties encountered in testing much simpler hypotheses, it is not at all clear what is gained by setting forth an elaborate hypothesis of this type.

Yet only certain aspects of this knowledge are held to be sociologically determined. On the basis of certain postulates, which need not be considered here, Scheler goes on to assert:

> The sociological character of all knowledge, of all forms of thought, intuition and cognition is unquestionable. Although the *content* and even less the objective validity of all knowledge is not determined by the *controlling perspectives of social interests*, nevertheless this is the case with the *selection* of the objects of knowledge. Moreover, the "forms" of the mental processes by means of which knowledge is acquired are always and necessarily codetermined sociologically, i.e. by the social structure.[36]

Since explanation consists in tracing the relatively new to the familiar and known and since society is "better known" than anything else,[37] it is to be expected that the modes of thought and intuition and the classification of knowable things generally are codetermined (*mitbedingt*) by the division and classification of groups which comprise the society.

Scheler flatly repudiates all forms of sociologism. He seeks to escape a radical relativism by resorting to a metaphysical dualism. He posits a realm of "timeless essences" which in varying degrees enter into the *content of judgments*; a realm utterly distinct from that of historical and social reality which determines the *act* of judgments. As Mandelbaum has aptly summarized this view:

> The realm of essences is to Scheler a realm of possibilities out of which we, bound to time and our interest, first select one set and then another for consideration. Where we as historians turn the spotlight of our attention depends upon our own sociologically determined valuations; what we see there is deter-

36. Ibid., p. 55 (italics added).
37. See the same assumption of Durkheim, cited in fn. 14 of this essay.

mined by the set of absolute and timeless values which are implicit in the past with which we are dealing.[38]

This is indeed counterrelativism by fiat. Merely asserting the distinction between essence and existences avoids the incubus of relativism by exorcising it. The concept of eternal essences may be congenial to the metaphysician; it is wholly foreign to empirical inquiry. It is noteworthy that these conceptions play no significant part in Scheler's empirical efforts to establish relations between knowledge and society.

Scheler indicates that different types of knowledge are bound up with particular forms of groups. The content of Plato's theory of ideas required the form and organization of the platonic academy; so, too, the organization of Protestant churches and sects was determined by the content of their beliefs which could exist only in this and in no other type of social organization, as Troeltsch has shown. And, similarly, *Gemeinschaft* types of society have a traditionally defined fund of knowledge which is handed down as conclusive; they are not concerned with discovering or extending knowledge. The very effort to test the traditional knowledge, in so far as it implies doubt, is ruled out as virtually blasphemous. In such a group, the prevailing logic and mode of thought is that of an "*ars demonstrandi*" not of an "*ars inveniendi.*" Its methods are prevailingly ontological and dogmatic, not epistemologic and critical; its mode of thought is that of conceptual realism, not nominalistic as in the *Gesellschaft* type of organization; its system of categories, organismic and not mechanistic.[39]

Durkheim extends sociological inquiry into the social genesis of the categories of thought, basing his hypothesis on three types of presumptive evidence. (1) The fact of cultural variation in the categories and the rules of logic "prove that they depend upon factors that are historical and consequently social."[40] (2) Since concepts are imbedded in the very language the individual acquires (and this holds as well for the special terminology of the scientist) and since some of these conceptual terms refer to things which we, as individuals, have never experienced, it is clear that they are a product of the society.[41] And (3), the acceptance or rejection of concepts is not determined *merely* by their objective validity but also by their consistency with other prevailing beliefs.[42]

38. Maurice Mandelbaum, *The Problem of Historical Knowledge* (New York: Liveright, 1938), p. 150; Sorokin posits a similar sphere of "timeless ideas," e.g., in his *Sociocultural Causality, Space, Time* (Durham: Duke University Press, 1943), pp. 215, passim.

39. Scheler, *Die Wissensformen*, pp. 22–23; compare a similar characterization of "sacred schools" of thought by Florian Znaniecki, *The Social Role of the Man of Knowledge* (New York: Columbia University Press, 1940), chap. 3.

40. Durkheim, *Elementary Forms*, pp. 12, 18, 439.

41. Ibid., pp. 433–35.

42. Ibid., p. 438.

Yet Durkheim does not subscribe to a type of relativism in which there are merely competing criteria of validity. The social origin of the categories does not render them wholly arbitrary so far as their applicability to nature is concerned. They are, in varying degrees, adequate to their object. But since social structures vary (and with them, the categorical apparatus) there are inescapable "subjective" elements in the particular logical constructions current in a society. These subjective elements "must be progressively rooted out, if we are to approach reality more closely." And this occurs under determinate social conditions. With the extension of intercultural contacts, with the spread of intercommunication between persons drawn from different societies, with the enlargement of the society, the local frame of reference becomes disrupted. "Things can no longer be contained in the social moulds according to which they were primitively classified; they must be organized according to principles which are their own. So logical organization differentiates itself from the social organization and becomes autonomous. Genuinely human thought is not a primitive fact; it is the product of history."[43] Particularly those conceptions which are subjected to scientifically methodical criticism come to have a greater objective adequacy. Objectivity is itself viewed as a social emergent.

Throughout, Durkheim's dubious epistemology is intertwined with his substantive account of the social roots of concrete designations of temporal, spatial, and other units. We need not indulge in the traditional exaltation of the categories as a thing set apart and foreknown to note that Durkheim was dealing not with them but with conventional divisions of time and space. He observed, in passing, that differences in these respects should not lead us to "neglect the similarities, which are no less essential." If he pioneered in relating variations in systems of concepts to variations in social organization, he did not succeed in establishing the social origin of the categories.

Like Durkheim, Granet attaches great significance to language as constraining and fixing prevalent concepts and modes of thought. He has shown how the Chinese language is not equipped to note concepts, analyze ideas, or to present doctrines discursively. It has remained intractable to formal precision. The Chinese word does not fix a notion with a definite degree of abstraction and generality, but evokes an indefinite complex of particular images. Thus, there is no word which simply signifies "old man." Rather, a considerable number of words "paint different aspects of old age": *k'i*, those who need a richer diet; *k'ao*, those who have difficulty in breathing, and so on. These concrete evocations entail a multitude of other similarly concrete images of every detail of the mode of life of the aged: those who should be exempt from military service; those for

43. Ibid., pp. 444–45, 437.

whom funerary material should be held in readiness; those who have a right to carry a staff through the town, and so on. These are but a few of the images evoked by *k'i* which, in general, corresponds to the quasi-singular notion of old persons, some sixty to seventy years of age. Words and sentences thus have an entirely concrete, emblematic character.[44]

Just as the language is concrete and evocative, so the most general ideas of ancient Chinese thought were unalterably concrete, none of them comparable to our abstract ideas. Neither time nor space was abstractly conceived. Time proceeds by cycles and is round; space is square. The earth which is square is divided into squares; the walls of towns, fields, and camps should form a square. Camps, buildings, and towns must be oriented and the selection of the proper orientation is in the hands of a ritual leader. Techniques of the division and management of space—surveying, town development, architecture, political geography—and the geometrical speculations which they presuppose are all linked with a set of social regulations. Particularly as these pertain to periodic assemblies, they reaffirm and reinforce in every detail the symbols which represent space. They account for its square form, its heterogeneous and hierarchic character, a conception of space which could only have arisen in a feudal society.[45]

Though Granet may have established the social grounds of concrete designations of time and space, it is not at all clear that he deals with data comparable to Western conceptions. He considers traditionalized or ritualized or magical conceptions and implicitly compares these with our matter-of-fact, technical, or scientific notions. But in a wide range of actual *practices*, the Chinese did not *act* on the assumption that "time is round" and "space, square." When comparable spheres of activity and thought are considered it is questionable that this radical cleavage of "categorical systems" occurs, in the sense that there are no common denominators of thought and conception. Granet has demonstrated qualitative differences of concepts in *certain contexts*, but not within such comparable contexts as, say, that of technical practice. His work testifies to different foci of intellectual interests in the two spheres and within the ritualistic sphere, basic differences of outlook, but not unbridgeable gaps in other spheres. The fallacy which is most prominent in Levy-Bruhl's concept of the "pre-logicality" of the primitive mind thus appears in the work of Granet as well. As Malinowski and Rivers have shown, when comparable spheres of thought and activity are considered, no such irreconcilable differences are found.[46]

44. Granet, *La pensée chinoise,* pp. 37–38, 82, and the whole of chap. 1.
45. Ibid., pp. 87–95.
46. Cf. B. Malinowski in *Magic, Science & Religion* (Glencoe: The Free Press, 1948), p. 9: "Every primitive community is in possession of a considerable store of

Sorokin shares in this same tendency to ascribe entirely disparate criteria of truth to his different culture types. He has cast into a distinctive idiom the fact of shifts of attention on the part of intellectual élites in different historical societies. In certain societies, religious conceptions and particular types of metaphysics are at the focus of attention, whereas in other societies, empirical science becomes the center of interest. But the several "systems of truth" coexist in each of these societies within certain spheres; the Catholic church has not abandoned its "ideational" criteria even in this sensate age.

Insofar as Sorokin adopts the position of radically different and disparate criteria of truth, he must locate his own work within this context. It may be said, though an extensive discussion would be needed to document it, that he never resolves this problem. His various efforts to cope with a radically relativistic impasse differ considerably. Thus, at the very outset, he states that his constructions must be tested in the same way "as any scientific law. First of all the principle must by nature be logical; second, it must successfully meet the test of the 'relevant facts,' that is, it must fit and represent the facts."[47] In Sorokin's own terminology, he has thereby adopted a scientific position characteristic of a "sensate system of truth." When he confronts his own epistemologic position directly, however, he adopts an "integralist" conception of truth which seeks to assimilate empirical and logical criteria as well as a "supersensory, superrational, metalogical act of 'intuition' or 'mystical experience.' "[48] He thus posits an integration of these diverse systems. In order to justify the "truth of faith"—the only item which would remove him from the ordinary criteria used in current scientific work—he indicates that "intuition" plays an important role as a *source* of scientific discovery. But does this meet the issue? The question is not one of the psychological *sources* of valid conclusions, but of the *criteria* and *methods of validation*. Which criteria would Sorokin adopt when "supersensory" intuitions are not consistent with empirical observation? In such cases, presumably, so far as we can judge from his work rather than from his comments about his work, he accepts the facts and rejects the intuition. All this suggests that Sorokin is discussing under the generic label of "truth" quite distinct and not comparable types of judgments: just as the chemist's analysis of an oil painting is neither consistent nor inconsistent with its aesthetic evaluation, so Sorokin's systems of truth refer to quite different kinds of judgments. And, indeed, he is finally led to say as much, when he remarks that "each of

---

knowledge, based on experience and fashioned by reason." See also Emile Benoit-Smullyan, "Granet's *La pensée chinoise*," *American Sociological Review* 1 (1936): 487–92.

47. Sorokin, *Social and Cultural Dynamics*, 1: 36; cf. 2: 11–12n.

48. Ibid., vol. 4, chap. 16; idem, *Sociocultural Causality*, chap. 5.

the systems of truth, within its legitimate field of competency, gives us genuine cognition of the respective aspects of reality."[49] But whatever his private opinion of intuition he cannot draw it into his sociology as a *criterion* (rather than a source) of valid conclusions.

Relations of Knowledge to the Existential Basis

Though this problem is obviously the nucleus of every theory in the sociology of knowledge, it has often been treated by implication rather than directly. Yet each type of imputed relation between knowledge and society presupposes an entire theory of sociological method and social causation. The prevailing theories in this field have dealt with one or both of two major types of relation: causal or functional, and the symbolic or organismic or meaningful.[50]

Marx and Engels, of course, dealt solely with some kind of causal relation between the economic basis and ideas, variously terming this relation as "determination, correspondence, reflection, outgrowth, dependence," and so on. In addition, there is an "interest" or "need" relation; when strata have (imputed) needs at a particular stage of historical development, there is held to be a definite pressure for appropriate ideas and knowledge to develop. The inadequacies of these divers formulations have risen up to plague those who derive from the Marxist tradition in the present day.[51]

Since Marx held that thought is not a mere "reflection" of objective class position, as we have seen, this raises anew the problem of its imputation to a determinate basis. The prevailing Marxist hypotheses for coping with this problem involve a theory of history which is the ground for determining whether the ideology is "situationally adequate" for a given stratum in the society: this requires a hypothetical construction of what men *would think and perceive* if they were able to comprehend the historical situation adequately.[52] But such insight into the situation need not *actually* be widely current within particular social strata. This, then, leads to the further problem of "false consciousness," of how ideologies which are neither in conformity with the interests of a class nor situationally adequate come to prevail.

49. *Sociocultural Causality,* 230–31n.

50. The distinctions between these have long been considered in European sociological thought. The most elaborate discussion in this country is that by Sorokin in *Social and Cultural Dynamics*; see, for example, vol. 1, chaps. 1–2.

51. Cf. the comments of Hans Speier, "The Social Determination of Ideas," *Social Research* 5 (1938): 182–205; C. Wright Mills, "Language, Logic and Culture," *American Sociological Review* 4 (1939): 670–80.

52. Cf. the formulation by Mannheim, *Ideology and Utopia,* pp. 175 ff.; Georg Lukács, *Geschichte und Klassenbewusstsein* (Berlin, 1923), pp. 61 ff.; Arthur Child, "The Problem of Imputation in the Sociology of Knowledge," *Ethics* 51 (1941): 200–214.

A partial empirical account of false consciousness, implied in the *Manifesto*, rests on the view that the bourgeoisie control the content of culture and thus diffuse doctrines and standards alien to the interests of the proletariat.[53] Or, in more general terms, "the ruling ideas of each age have ever been the ideas of its ruling class." But this is only a partial account; at most it deals with the false consciousness of the subordinated class. It might, for example, partly explain the fact noted by Marx that even where the peasant proprietor "does belong to the proletariat by his position he does not believe that he does." It would not, however, be pertinent in seeking to account for the false consciousness of the ruling class itself.

Another, though not clearly formulated, theme which bears upon the problem of false consciousness runs throughout Marxist theory. This is the conception of ideology as being an *unwitting, unconscious* expression of "real motives," these being in turn construed in terms of the objective interests of social classes. Thus, there is repeated stress on the unwitting nature of ideologies: "Ideology is a process accomplished by the so-called thinker consciously indeed but with a false consciousness. The real motives impelling him remain unknown to him, otherwise it would not be an ideological process at all. Hence he imagines false or apparent motives."[54]

The ambiguity of the term "correspondence" to refer to the connection between the material basis and the idea can only be overlooked by the polemical enthusiast. Ideologies are construed as "distortions of the social situation";[55] as merely "expressive" of the material conditions;[56] and, whether "distorted" or not, as motivational support for carrying through real changes in the society.[57] It is at this last point, when "illusory" beliefs are conceded to provide motivation for action, that Marxism ascribes a measure of independence to ideologies in the historical process. They are no longer merely epiphenomenal. They enjoy a measure of autonomy. From this develops the notion of interacting factors in which the superstructure, though interdependent with the material basis, is also assumed

53. Marx and Engels, *The German Ideology*, p. 39: "In so far as they rule as a class and determine the extent and compass of an epoch, it is self-evident that they do this in their whole range, hence among other things rule also as thinkers, as producers of ideas, and regulate the production and distribution of the ideas of their age."

54. Engels' letter to Mehring, 14 July 1893, in Marx, *Selected Works*, 1: 388–89; cf. Marx, *Der achtzehnte Brumaire*, p. 33; idem, *Critique of Political Economy*, p. 12.

55. Marx, *Der achtzehnte Brumaire*, p. 39, where the democratic Montagnards indulge in self-deception.

56. Engels, *Socialism: Utopian and Scientific*, pp. 26–27. Cf. Engels, *Feuerbach*, pp. 122–23: "The failure to exterminate the Protestant heresy *corresponded* to the invincibility of the rising bourgeoisie. . . . Here Calvinism proved itself to be the true religious disguise of the interests of the bourgeoisie of that time" (italics added).

57. Marx grants motivational significance to the "illusions" of the burgeoning bourgeoisie, in *Der achtzehnte Brumaire*, p. 8.

to have some degree of independence. Engels explicitly recognized that earlier formulations were inadequate in at least two respects: first, that both he and Marx had previously overemphasized the economic factor and understated the role of interaction;[58] the second, that they had "neglected" the formal side—the way in which these ideas develop.[59]

The Marx-Engels views on the connectives of ideas and economic substructure hold, then, that the economic structure constitutes the framework which limits the range of ideas that will prove socially effective; ideas which do not have pertinence for one or another of the conflicting classes may arise, but will be of little consequence. Economic conditions are necessary, but not sufficient, for the emergence and spread of ideas which express either the interests or outlook, or both, of distinct social strata. There is no strict determination of ideas by economic conditions, but a definite predisposition. Knowing the economic conditions, we can predict the kinds of ideas which can exercise a controlling influence in a direction that can be effective. "Men make their own history, but they do not make it just as they please; they do not make it under circumstances chosen by themselves, but under circumstances directly found, given and transmitted from the past." And in the making of history, ideas and ideologies play a definite role: consider only the view of religion as "the opiate of the masses"; consider further the importance attached by Marx and Engels to making those in the proletariat "aware" of their "own interests." Since there is no fatality in the development of the total social structure, but only a development of economic conditions which make certain lines of change *possible* and probable, idea-systems may play a decisive role in the selection of one alternative which "corresponds" to the real balance of power rather than another alternative which runs counter to the existing power situation and is therefore destined to be unstable, precarious, and temporary. There is an ultimate compulsive which derives from economic development, but this compulsive does not operate with such detailed finality that no variation of ideas can occur at all.

The Marxist theory of history assumes that, *sooner or later*, idea-systems which are inconsistent with the actually prevailing and incipient power structure will be rejected in favor of those which more nearly express the actual alignment of power. It is this view that Engels expresses in his metaphor of the "zig-zag course" of abstract ideology: ideologies may temporarily deviate from what is compatible with the current social relations of production, but they are ultimately brought back in line. For this reason, the Marxist analysis of ideology is always bound to be concerned with the "total" historical situation, in order to account both for the tem-

58. Engels, letter to Joseph Bloch, 21 September 1890, in Marx, *Selected Works*, 1: 383.

59. Engels, letter to Mehring, 14 July 1893, ibid., 1: 390.

porary deviations and the later accommodation of ideas to the economic compulsives. But for this same reason, Marxist analyses are apt to have an excessive degree of "flexibility," almost to the point where *any* development can be explained away as a temporary aberration or deviation; where "anachronisms" and "lags" become labels for the explaining away of existing beliefs which do not correspond to theoretical expectations; where the concept of "accident" provides a ready means of saving the theory from facts which seem to challenge its validity.[60] Once a theory includes concepts such as "lags," "thrusts," "anachronisms," "accidents," "partial independence," and "ultimate dependence," it becomes so labile and so indistinct that it can be reconciled with virtually any configuration of data. Here, as in several other theories in the sociology of knowledge, a decisive question must be raised in order to determine whether we have a genuine theory: how can the theory be invalidated? In any given historical situation, which data will contradict and invalidate the theory? Unless this can be answered directly, unless the theory involves statements which can be controverted by definite types of evidence, it remains merely a pseudo-theory which will be compatible with any array of data.

Though Mannheim has gone far toward developing actual research procedures in the substantive sociology of knowledge, he has not appreciably clarified the connectives of thought and society. As he indicates, once a thought structure has been analyzed, there arises the problem of imputing it to definite groups. This requires not only an empirical investigation of the groups or strata which prevalently think in these terms but also an interpretation of why these groups, and not others, manifest this type of thought. This latter question implies a social psychology which Mannheim has not systematically developed.

The most serious shortcoming of Durkheim's analysis lies precisely in his uncritical acceptance of a naïve theory of correspondence in which the categories of thought are held to "reflect" certain features of the group organization. Thus "there are societies in Australia and North America where space is conceived in the form of an immense circle, *because* the camp has a circular form . . . the social organization has been the model for the spatial organization and a reproduction of it."[61] In similar fashion, the general notion of time is derived from the specific units of time differentiated in social activities (ceremonies, feasts, rites).[62] The category of class and the modes of classification, which involve the notion of a hierarchy, are derived from social grouping and stratification. Those social categories are then "projected into our conception of the new world."[63] In

60. Cf. Weber, *Gesammelte Aufsaetze zur Wissenschaftslehre*, pp. 166–70.
61. Durkheim, *Elementary Forms*, pp. 11–12.
62. Ibid., pp. 10–11.
63. Ibid., p. 148.

summary, then, categories "express" the different aspects of the social order.[64] Durkheim's sociology of knowledge suffers from his avoidance of a social psychology.

The central relation between ideas and existential factors for Scheler is interaction. Ideas interact with existential factors which serve as selective agencies, releasing or checking the extent to which potential ideas find actual expression. Existential factors do not "create" or "determine" the content of ideas; they merely account for the *difference* between potentiality and actuality; they hinder, retard, or quicken the actualization of potential ideas. In a figure reminiscent of Clerk Maxwell's hypothetical daemon, Scheler states: "in a definite fashion and order, existential factors open and close the sluice-gates to the flood of ideas." This formulation, which ascribes to existential factors the function of selection from a self-contained realm of ideas is, according to Scheler, a basic point of agreement between such otherwise divergent theorists as Dilthey, Troeltsch, Max Weber, and himself.[65]

Scheler operates as well with the concept of "structural identities" which refers to common presuppositions of knowledge or belief on the one hand, and of social, economic, or political structure on the other.[66] Thus, the rise of mechanistic thought in the sixteenth century, which came to dominate prior organismic thought is inseparable from the new individualism, the incipient dominance of the power-driven machine over the hand-tool, the incipient dissolution of *Gemeinschaft* into *Gesellschaft*, production for a commodity market, rise of the principle of competition in the ethos of western society, and so on. The notion of scientific research as an endless process through which a store of knowledge can be accumulated for practical application as the occasion demands and the total divorce of this science from theology and philosophy was not possible without the rise of a new principle of infinite acquisition characteristic of modern capitalism.[67]

In discussing such structural identities, Scheler does not ascribe primacy either to the socioeconomic sphere or to the sphere of knowledge. Rather, and this Scheler regards as one of the most significant propositions in the field, both are determined by the impulse structure of the élite which is closely bound up with the prevailing ethos. Thus, modern technology is not merely the application of a pure science based on observation, logic, and mathematics. It is far more the product of an orientation toward the control of nature which defined the purposes as well as the conceptual structure of scientific thought. This orientation is largely implicit and is not to be confused with the personal motives of scientists.

64. Ibid., p. 440.
65. Scheler, *Die Wissensformen*, p. 32.
66. Ibid, p. 56.
67. Ibid., p. 25; cf. pp. 482–84.

With the concept of structural identity, Scheler verges on the concept of cultural integration or *Sinnzusammenhang*. It corresponds to Sorokin's conception of a "meaningful cultural system" involving "the identity of the fundamental principles and values that permeate all its parts," which is distinguished from a "causal system" involving interdependence of parts.[68] Having constructed his types of culture, Sorokin's survey of criteria of truth, ontology, metaphysics, scientific and technologic output, and so on, finds a marked tendency toward the meaningful integration of these with the prevailing culture.

Sorokin has boldly confronted the problem of how to determine the *extent* to which such integration occurs, recognizing, despite his vitriolic comments on the statisticians of our sensate age, that to deal with the extent or degree of integration necessarily implies some statistical measure. Accordingly, he developed numerical indexes of the various writings and authors in each period, classified these in their appropriate category, and thus assessed the comparative frequency (and influence) of the various systems of thought. Whatever the technical evaluation of the validity and reliability of these cultural statistics, he has directly acknowledged the problem overlooked by many investigators of integrated culture or *Sinnzusammenhaengen*, namely, the approximate degree or extent of such integration. Moreover, he plainly bases his empirical conclusions very largely upon these statistics.[69] And these conclusions again testify that his approach leads to a statement of the problem of connections between existential bases and knowledge, rather than to its solution. Thus, to take a case in point, "empiricism" is defined as the typical sensate system of truth. The last five centuries, and more particularly the last century represent "sensate culture par excellence!"[70] Yet, even in this flood-tide of sensate culture, Sorokin's statistical indices show only some 53 percent of influential writings in the field of "empiricism." And in the earlier centuries of this sensate culture—from the late sixteenth to the mid-eighteenth—the indices of empiricism are consistently lower than those for rationalism (which is associated, presumably, with an idealistic rather than a sensate culture).[71] The object of these observations is not to raise the question whether

68. Sorokin, *Social and Cultural Dynamics*, vol. 4, chap. 1; vol. 1, chap. 1.

69. Despite the basic place of these statistics in his empirical findings, Sorokin adopts a curiously ambivalent attitude toward them, an attitude similar to the attitude toward experiment imputed to Newton: a device to make his prior conclusions "intelligible and to convince the vulgar." Note Sorokin's approval of Park's remark that his statistics are merely a concession to the prevailing sensate mentality and that "if they want 'em, let 'em have 'em." See Sorokin, *Sociocultural Causality*, p. 95n. Sorokin's ambivalence arises from his effort to integrate quite disparate "systems of truth."

70. Sorokin, *Social and Cultural Dynamics*, 2: 51.

71. Ibid., 2: 30.

Sorokin's conclusions coincide with his statistical data: it is not to ask why the sixteenth and seventeenth centuries are said to have a dominant "sensate system of truth" in view of these data. Rather, it is to indicate that even on Sorokin's own premises, overall characterizations of historical cultures constitute merely a first step, which must be followed by analyses of deviations from the central tendencies of the culture. Once the notion of *extent* of integration is introduced, the existence of types of knowledge which are not integrated with the dominant tendencies cannot be viewed merely as "congeries" or as "contingent." Their *social* bases must be ascertained in a fashion for which an emanationist theory does not provide.

A basic concept which serves to differentiate generalizations about the thought and knowledge of an entire society or culture is that of the "audience" or "public" or what Znaniecki calls "the social circle." Men of knowledge do not orient themselves exclusively toward their data nor toward the total society, but to special segments of that society with their special demands, criteria of validity, of significant knowledge, of pertinent problems, and so on. It is through anticipation of these demands and expectations of particular audiences, which can be effectively located in the social structure, that men of knowledge organize their own work, define their data, seize upon problems. Hence, the more differentiated the society, the greater the range of such effective audiences, the greater the variation in the foci of scientific attention, of conceptual formulations, and of procedures for certifying claims to knowledge. By linking each of these typologically defined audiences to their distinctive social position, it becomes possible to provide a *wissenssoziologische* account of variations and conflicts of thought within the society, a problem that is necessarily bypassed in an emanationist theory. Thus, the scientists in seventeenth-century England and France who were organized in newly established scientific societies addressed themselves to audiences very different from those of the savants who remained exclusively in the traditional universities. The direction of their efforts, toward a "plain, sober, empirical" exploration of specific technical and scientific problems differed considerably from the speculative, unexperimental work of those in the universities. Searching out such variations in effective audiences, exploring their distinctive criteria of significant and valid knowledge,[72] relating these to their position within the society, and examining the sociopsychological processes through which these operate to constrain certain modes of thought constitutes a procedure which promises to take research in the

72. The Rickert-Weber concept of "Wertbeziehung" (relevance to value) is but a first step in this direction; there remains the further task of differentiating the various sets of values and relating these to distinctive groups or strata within the society.

sociology of knowledge from the plane of general imputation to that of testable empirical inquiry.[73]

The foregoing account deals with the main substance of prevailing theories in this field. Limitations of space permit only the briefest consideration of one other aspect of these theories singled out in our paradigm: functions imputed to various types of mental productions.[74]

Functions of Existentially Conditioned Knowledge

In addition to providing causal explanations of knowledge, theories ascribe social functions to knowledge, functions which presumably serve to account for its persistence or change. These functional analyses cannot be examined in any detail here, though a detailed study of them would undoubtedly prove rewarding.

The most distinctive feature of the Marxist imputation of function is its ascription, not to the society as a whole, but to distinct strata within the society. This holds not only for ideological thinking but also for natural science. In capitalist society, science and derivative technology are held to become a further instrument of control by the dominant class.[75] Along these same lines, in ferreting out the economic determinants of scientific development, Marxists have often thought it sufficient to show that the scientific results enabled the solution of some economic or technological need. But the application of science to a need does not necessarily testify that the need has been significantly involved in leading to the result. Hyperbolic functions were discovered two centuries before they had any practical significance and the study of conic sections had a broken history of two millennia before being applied in science and technology. Can we infer, then, that the "needs" which were ultimately satisfied through such applications served to direct the attention of mathematicians to these

73. This is perhaps the most distinctive variation in the sociology of knowledge now developing in American sociological circles, and may almost be viewed as an American acculturation of European approaches. This development characteristically derives from the social psychology of G. H. Mead. Its pertinence in this connection is being indicated by C. Wright Mills, Gerard de Gré, and others. See Znaniecki's conception of the "social circle," in *Social Role*. See also the beginnings of empirical findings along these lines in the more general field of public communications: Paul F. Lazarsfeld and R. K. Merton, "Studies in Radio and Film Propaganda," *Transactions*, New York Academy of Sciences, 2d ser., 6 (1943): 58–79.

74. An appraisal of historicist and ahistorical approaches is necessarily omitted. It may be remarked that this controversy definitely admits of a middle ground.

75. For example, Marx quotes from the nineteenth century apologist of capitalism, Ure, who, speaking of the invention of the self-acting mule, says: "A creation destined to restore order among the industrious classes. . . . This invention confirms the great doctrine already propounded, that when capital enlists science into her service, the refractory hand of labor will always be taught docility" (*Capital*, 1: 477).

fields, that there was, so to speak, a retroactive influence of some two to twenty centuries? Detailed inquiry into the relations between the emergence of needs, recognition of these needs by scientists or by those who direct their selection of problems, and the consequences of such recognition are required before the role of needs in determining the thematics of scientific research can be established.[76]

In addition to his claim that the categories are social emergents, Durkheim also indicates their social functions. The functional analysis, however, is intended to account not for the particular categorical system in a society but for the existence of a system common to the society. For purposes of intercommunication and for coordinating men's activities, a common set of categories is indispensable. What the apriorist mistakes for the constraint of an inevitable, native form of understanding is actually "the very authority of society, transferring itself to a certain manner of thought which is the indispensable condition of all common action."[77] There must be a certain minimum of "logical conformity" if joint social activities are to be maintained at all; a common set of categories is a functional necessity. This view is further developed by Sorokin who indicates the several functions served by different systems of social space and time.[78]

Further Problems and Recent Studies

From the foregoing discussion, it becomes evident that a wide diversity of problems in this field require further investigation.[79]

Scheler had indicated that the social organization of intellectual activity is significantly related to the character of the knowledge which develops under its auspices. One of the earliest studies of the problem in this country was Veblen's caustic, impressionistic, and often perceptive account of the pressures shaping American university life.[80] In more systematic fashion, Wilson has dealt with the methods and criteria of recruitment, the assignment of status, and the mechanisms of control of the

76. Compare B. Hessen, *Science at the Cross-Roads*; R. K. Merton, *Science, Technology and Society in Seventeenth-Century England* (Bruges: Osiris History of Science Monographs, 1938), chaps. 7–10; J. D. Bernal, *The Social Function of Science* (New York: The Macmillan Co., 1939); J. G. Crowther, *The Social Relations of Science* (New York: The Macmillan Co., 1941); Bernard Barber, *Science and the Social Order* (Glencoe, Illinois: The Free Press, 1952); Gerard De Gré, *Science as a Social Institution* (New York: Doubleday & Company, 1955).

77. Durkheim, *Elementary Forms*, pp. 17, 10–11, 443.

78. Sorokin, *Sociocultural Causality*, passim.

79. For further summaries, see Louis Wirth's preface to Mannheim, *Ideology and Utopia*, xxviii–xxxi; J. B. Gittler, "Possibilities of a Sociology of Science," *Social Forces* 18 (1940): 350–59.

80. Thorstein Veblen, *The Higher Learning in America* (New York: Huebsch, 1918).

academic man, thus providing a substantial basis for comparative studies.[81] Setting forth a typology of the roles of men of knowledge, Znaniecki developed a series of hypotheses concerning the relations between these roles and the types of knowledge cultivated; between types of knowledge and the bases of appraisal of the scientist by members of the society; between role-definitions and attitudes toward practical and theoretical knowledge; and so on.[82] Much remains to be investigated concerning the bases of class identifications by intellectuals, their alienation from dominant or subordinate strata in the population, their avoidance of or indulgence in researches which have immediate value-implications challenging current institutional arrangements inimical to the attainment of culturally approved goals,[83] the pressures toward technicism and away from dangerous thoughts, the bureaucratization of intellectuals as a process whereby problems of policy are converted into problems of administration, the areas of social life in which expert and positive knowledge are deemed appropriate and those in which the wisdom of the plain man is alone considered necessary—in short, the shifting role of the intellectual and the relation of these changes to the structure, content, and influence of his work require growing attention, as changes in the social organization increasingly subject the intellectual to conflicting demands.[84]

Increasingly, it has been assumed that the social structure does not influence science merely by focusing the attention of scientists upon certain problems for research. In addition to the studies to which we have already referred, others have dealt with the ways in which the cultural and social context enters into the conceptual phrasing of scientific problems. Darwin's theory of selection was modeled after the prevailing notion of a competitive economic order, a notion which in turn has been assigned an ideological function through its assumption of a natural identity of interests.[85] Rus-

81. Logan Wilson, *The Academic Man;* cf. E. Y. Hartshorne, *The German Universities and National Socialism* (Harvard University Press, 1937).

82. Florian Znaniecki, *Social Role.*

83. Gunnar Myrdal in his treatise, *An American Dilemma,* repeatedly indicates the "concealed valuations" of American social scientists studying the American Negro and the effect of these valuations on the formulation of "scientific problems" in this area of research; see especially 2: 1027–64.

84. Mannheim refers to an unpublished monograph on the intellectual; general bibliographies are to be found in his books and in Roberto Michels's article on "Intellectuals," *Encyclopedia of the Social Sciences.* Recent papers include C. Wright Mills, "The Social Role of the Intellectual," *Politics,* vol. 1 (April 1944); R. K. Merton, "Role of the Intellectual in Public Policy," presented at the annual meeting of the American Sociological Society, 4 December 1943; Arthur Koestler, "The Intelligentsia," *Horizon* 9 (1944):162–75.

85. Keynes observed that "The Principle of the Survival of the Fittest could be regarded as one vast generalization of the Ricardian economics" (quoted by Talcott Parsons in *The Structure of Social Action,* p. 113); cf. Alexander Sandow, "Social Factors in the Origin of Darwinism," *Quarterly Review of Biology* 13 (1938): 316–26.

sell's half-serious observation on the national characteristics of research in animal learning points to a further type of inquiry into the relations between national culture and conceptual formulations.[86] So, too, Fromm has attempted to show that Freud's "conscious liberalism" tacitly involved a rejection of impulses tabooed by bourgeois society and that Freud himself was in his patricentric character, a typical representative of a society which demands obedience and subjection.[87]

In much the same fashion, it has been indicated that the conception of multiple causation is especially congenial to the academician, who has relative security, is loyal to the status quo from which he derives dignity and sustenance, who leans toward conciliation and sees something valuable in all viewpoints, thus tending toward a taxonomy which enables him to avoid taking sides by stressing the multiplicity of factors and the complexity of problems.[88] Emphases on nature or nurture as prime determinants of human nature have been linked with opposing political orientations. Those who emphasize heredity are political conservatives whereas the environmentalists are prevalently democrats or radicals seeking social change.[89] But even environmentalists among contemporary American writers on social pathology adopt conceptions of "social adjustment" which implicitly assume the standards of small communities as norms and characteristically fail to assess the possibility of certain groups achieving their

86. Bertrand Russell, *Philosophy* (New York: W. W. Norton and Co., 1927), pp. 29–30. Russell remarks that the animals used in psychological research "have all displayed the national characteristics of the observer. Animals studied by Americans rush about frantically, with an incredible display of hustle and pep, and at last achieve the desired result by chance. Animals observed by Germans sit still and think, and at last evolve the solution out of their inner consciousness." Witticism need not be mistaken for irrelevance; the possibility of national differences in the choice and formulation of scientific problems has been repeatedly noted, though not studied systematically. Cf. Richard Mueller-Freienfels, *Psychologie der Wissenschaft* (Leipzig: J. A. Barth, 1936), chap. 8, which deals with national, as well as class, differences in the choice of problems, "styles of thought," and so on, without fully acquiescing in the echt-deutsch requirements of a Krieck. This type of interpretation, however, can be carried to a polemical and ungrounded extreme, as in Max Scheler's debunking 'analysis' of English cant. He concludes that, in science, as in all other spheres, the English are incorrigible 'cantians.' Hume's conception of the ego, substance, and continuity as biologically useful self-deceptions was merely purposive cant; so, too, was the characteristic English conception of working hypotheses (Maxwell, Kelvin) as aiding the progress of science but not as truth—a conception which is nothing but a shrewd maneuver to provide momentary control and ordering of the data. All pragmatism implies this opportunistic cant, says Scheler in *Genius des Krieges* (Leipzig: Verlag der Weissenbuecher, 1915).

87. Erich Fromm, "Die gesellschaftliche Bedingtheit der psychoanalytischen Therapie," *Zeitschrift fuer Sozialforschung* 4 (1935):365–97.

88. Lewis S. Feuer, "The Economic Factor in History," *Science and Society* 4 (1940):174–75.

89. N. Pastore, "The Nature-Nurture Controversy: A Sociological Approach," *School and Society* 57 (1943):373–77.

objectives under the prevailing institutional conditions.[90] The imputations of perspectives such as these require more systematic study before they can be accepted, but they indicate recent tendencies to seek out the perspectives of scholars and to relate these to the framework of experience and interests constituted by their respective social positions. The questionable character of imputations which are not based on adequate *comparative* material is illustrated by a recent account of the writings of Negro scholars. The selection of analytical rather than morphological categories, of environmental rather than biological determinants of behavior, of exceptional rather than typical data is ascribed to the caste-induced resentment of Negro writers, without any effort being made to compare the frequency of similar tendencies among white writers.[91]

Vestiges of any tendency to regard the development of science and technology as *wholly* self-contained and advancing irrespective of the social structure are being dissipated by the actual course of historical events. An increasingly visible control and, often, restraint of scientific research and invention has been repeatedly documented, notably in a series of studies by Stern[92] who has also traced the bases of resistance to change in medicine.[93] The basic change in the social organization of Germany has provided a virtual experimental test of the close dependence of the direction and extent of scientific work upon the prevailing power structure and the associated cultural outlook.[94] And the limitations of any unqualified assumption that science or technology represents the basis to which the social structure must adjust become evident in the light of studies showing how science and technology have been put in the service of social or economic demands.[95]

90. C. Wright Mills, "The Professional Ideology of Social Pathologists," *American Journal of Sociology* 49 (1943):165–90.

91. William T. Fontaine, "'Social Determination' in the Writings of Negro Scholars," *American Journal of Sociology* 49 (1944):302–15.

92. Bernard J. Stern, "Resistances to the Adoption of Technological Innovations," in National Resources Committee, *Technological Trends and National Policy* (Washington, D.C.: U. S. Government Printing Office, 1937), pp. 39–66; "Restraints upon the Utilization of Inventions," *The Annals* 200 (1938):1–19, and further references therein; Walton Hamilton, *Patents and Free Enterprise,* TNEC Monograph no. 31 (1941).

93. Bernhard J. Stern, *Social Factors in Medical Progress* (New York: Columbia University Press, 1927); idem, *Society and Medical Progress* (Princeton: Princeton University Press, 1941); cf. Richard H. Shryock, *The Development of Modern Medicine* (Philadelphia: University of Pennsylvania Press, 1936); Henry E. Sigerist, *Man and Medicine* (New York: W. W. Norton and Co., 1932).

94. Hartshorne, *German Universities and National Socialism.*

95. Only most conspicuously in time of war; see Sorokin's observation that centers of military power tend to be the centers of scientific and technologic development (*Dynamics,* vol. 4, pp. 249–51); cf. I. B. Cohen and Bernard Barber, *Science and War* (ms.); R. K. Merton, "Science and Military Technique," *Scientific Monthly* 41(1935):542–45; Bernal, *Social Function of Science*; Julian Huxley, *Science and Social Needs* (New York: Harper and Bros., 1935).

To develop any further the formidable list of problems which require and are receiving empirical investigation would outrun the limits of this chapter. There is only this to be said: the sociology of knowledge is fast outgrowing a prior tendency to confuse provisional hypothesis with unimpeachable dogma; the plenitude of speculative insights which marked its early stages are now being subjected to increasingly rigorous test. Though Toynbee and Sorokin may be correct in speaking of an alternation of periods of fact-finding and generalization in the history of science, it seems that the sociology of knowledge has wedded these two tendencies in what promises to be a fruitful union. Above all, it focuses on problems which are at the very center of contemporary intellectual interest.[96]

96. For extensive bibliographies, see Bernard Barber, *Science and the Social Order;* Mannheim, *Ideology and Utopia;* Barnes, Becker, and Becker, eds., *Contemporary Social Theory.*

2 Znaniecki's The Social Role of the Man of Knowledge

1941

Florian Znaniecki is in many respects the most distinguished exponent of sociology as a special rather than an encyclopedic social science. In a remarkable series of books, he has for some twenty years consistently demonstrated the *special* contributions of sociology to the analysis of human interaction and culture. The books evidence a notable theoretical integration that derives not from dogmatic convictions but from the exploration of new ranges of data guided by a conceptual framework which has proved conspicuously useful. It is peculiarly fitting that Znaniecki's latest volume in this series, the Julius Beer Foundation Lectures at Columbia University, should deal with the sociology of the scientist, for until September 1939 Poland was the home of *Nauka Polska* and of *Organon*, journals devoted exclusively to the "science of science," that is, the psychology, sociology, history, and philosophy of science.

Znaniecki sets himself two main types of problems in this study of specialists in knowledge. (Throughout his book, the terms scientist, savant, and man of knowledge are used synonymously and broadly to designate such specialists.) The first of these problems is taxonomic: what is the composition and structure of the various types of scientists' social roles; what are their interrelations; their lines of development? Secondly, how, if at all, are the systems of knowledge and methods of savants influenced by the normative patterns which define their behavior in a social order? The very formulation of these questions is clear evidence that Znaniecki has not confused problems in the sociology of knowledge with a sociological theory of knowledge, that is, with a special epistemology. This is a study

Reprinted with permission from *American Sociological Review* 6 (February 1941): 111–15.

in substantive *Wissenssoziologie*, not an essay on the foundations of valid knowledge.

Znaniecki conceives a social role as a dynamic social system involving four interacting components: (1) the *social circle:* a set of persons who interact with the actor and estimate his performance (that is, the effective audience); (2) the actor's *self:* the physical and psychological characteristics attributed to him by virtue of his position; (3) the actor's *social status:* the permissions and immunities assigned to him as inherent in his position; (4) the actor's *social functions:* his contributions to his social circle. This paradigm defines the minimal elements which must be examined in the systematic comparison of social roles.

A scant outline of Znaniecki's typology of scientists' roles will not, of course, set forth the analytical uses to which this typology is put. It will, however, indicate the classificatory framework within which his analyses are expressed. Znaniecki's reconstructions of the presumable lines of development of one role into another are not included in this outline.

Types of Social Roles of Men of Knowledge

A. Technological Advisers

1. *Technological expert:* the diagnostician who defines the relevant data in the situation, their essential components and interrelations, and the theoretic foundations for planned collective tasks; he performs the "staff" or advisory function.
2. *Technological leader:* the executive-director who devises the plan and selects the instrumentalities for its execution on the basis of a complex of practically oriented, heterogeneous knowledge.

B. Sages[1] provide intellectual justification of collective tendencies of their party, sect, stratum.

	Apologists for existing tendencies	Idealists with norms not contained in the existing order or in the opposition party
1. *Conservative:*	(a) "Standpatter"	(b) Meliorist
2. *Novationist:*	(a) Oppositionist	(b) Revolutionary

C. Scholars (that is, Schoolmen)

1. *Sacred scholar:* perpetuates sacred truths through exact and faithful reproduction of their symbolic expressions; he is charged with the

1. Attention should be called to the instructive comparison between these roles and Mannheim's concepts of ideologists and utopianists. The fourfold table and resultant types, supplied by the reviewer, are clearly implicit in Znaniecki's text (pp. 72–77).

maintenance of a self-contained, stable, unchallengeable, sacred system of unchanging truths.

2. *Secular scholar:* with the following subtypes:
   a. *discoverer of truth:* initiates a "school of thought" with a claim to "absolute truth" validated by the certainty of rational evidence.
   b. *systematizer:* tests and organizes the total existing knowledge in certain fields into a coherent system by means of deduction from the self-evident first principles established by the discoverer.
   c. *contributor:* furnishes new findings which are implicitly or explicitly expected to furnish new proof that experience accords with the master's system; revises "unsatisfactory" inductive evidence until it is so integrated or is "justifiably" rejected.
   d. *fighter for truth:* ensures the logical victory of one school over another by convincing scholars in a polemical situation that his school has a truth-claim validated by rationalistic evidence. (Differs from tendentious partisan sage by confining polemics to a closed arena accessible only to those who hold truth as dominant value.)
   e. *disseminator of knowledge*
      (1). *popularizer:* cultivates amateur interests among adults, thus aiding popular support of learning, especially in democratized society.
      (2). *educating teacher:* imparts theoretic knowledge to youth as part of their non-occupational education.

D. Creators of Knowledge (Explorers)

1. *Discoverer of facts (fact-finder):* discovers hitherto unknown and unanticipated empirical data, largely as a basis for modifications in existing systems of knowledge.
2. *Discoverer of problems (inductive theorist):* discovers new and unforeseen theoretic problems which are to be solved by new theoretical constructions.

It should be noted at once that this is a classification of social *roles* and not of persons, and that individual men of knowledge may incorporate several of these analytically distinguishable roles. A further development of Znaniecki's analysis would lead to a statement of the circumstances under which shifts from one role to another occur.

Znaniecki skillfully traces a variety of relations between the components of these classified roles; relations between role-definitions and types of knowledge cultivated; types of knowledge and bases of positive estimation of the scientist by members of the society; normative role-definitions and attitudes toward practical and theoretical knowledge; and so forth. These relations are examined genetically and functionally. A brief review cannot even list these relations, but one or two instances will serve to illustrate the systematic findings.

A convincing demonstration of the value of Znaniecki's approach is found in his suggestive though brief resume of the various attitudes toward "new unanticipated facts" of those who perform different intellectual roles. It should be noted that *these divers attitudes can be "understood" (or "derived") from the particular role-systems in which the men of knowledge participate*; it is, in other words, an analysis of the ways in which various social structures exert pressures for the adoption of certain attitudes toward new empirical data. The *specialized interest* in the finding of new facts is construed as a revolt against established systems of thought which have persisted largely because they have not been confronted with fresh stubborn facts. Later, to be sure, even this "rebellious" activity becomes institutionalized, but it arises initially in opposition to established and vested intellectual systems. The technological leader regards genuinely new facts with suspicion, for they may destroy belief in the rationality of his established plans, or show the inefficiency of his plans, or disclose undesirable consequences of his program. New facts within the compass of his activity threaten his status. The technological expert, under the control of the leader, is circumscribed in new fact-finding lest he discover facts which are unwelcome to the powers that be. (See, for example, the suppression of new but "unwanted" inventions.) The sage, with his predetermined conclusions, has no use for the impartial observer of new facts which might embarrass his tendentious views. Scholars have positive or negative attitudes toward genuinely new facts, depending upon the extent to which the schools' system is established: in the initial stages new facts are at least acceptable, but once the system is fully formulated the intellectual commitment of the school precludes a favorable attitude toward novel findings. Thus, "a discoverer of facts, freely roaming in search of the unexpected, has no place in a milieu of scientists with well-regulated traditional roles." Znaniecki provides a pioneering analysis of the kind of intellectual neophobia which Pareto largely treated as given rather than problematical.

In similar fashion, Znaniecki shows how rivalry between schools of sacred thought leads to secularization. The most general theorem holds that conflict, as a type of social interaction, leads to the partial secularization of sacred knowledge in at least three ways. First, the usual appeal to sacred authority cannot function in the conflict situation, inasmuch as the rival schools either accept different sacred traditions or interpret the same tradition diversely. "Rational analysis" is adopted as an impartial arbiter. Secondly, members of the outgroup (nonbelievers) must be persuaded that their own faiths are suspect and that another faith has more to commend it. This again involves rational or pseudorational argument, since there is no other common unchallenged authority. Finally, the battle of the sacred schools awakens skepticism on the part of intellectual onlookers, and such skepticism must be curbed lest it subvert the authority of the sacred school among the "public." One such safeguard is again rational persuasion. A body of empirical data to which this analysis is peculiarly appropriate, though Znaniecki does not explicitly deal with it, is the situation of the contending Protestant sects during the sixteenth and seventeenth centuries. These, in the process of validating their claims to sacred authority for their conflicting views, gradually adopted an elaborate set of rationalistic and empirical bases for legitimacy.[2] The forces conducing to the secularization of sacred knowledge in this historical period are readily conceptualized in Znaniecki's terms. When, however, it becomes manifest that the multiplicity of schools, dogmas, and power structures precludes dominance by any one school, a *modus vivendi* is found in a doctrine of mutual toleration.

In summary, then, this little book presents a conceptual framework for organizing varied materials in one sphere of the sociology of knowledge. It contributes a rich store of hypotheses which often derive from Znaniecki's earlier work, and so have a measure of empirical confirmation at the outset. It should be said, however, as Znaniecki would doubtlessly be the first to acknowledge, that this book is simply a prolegomenon to the sociology of men of knowledge; an introduction, moreover, liable to several criticisms. It includes no systematic documentation, although it may be inferred from the text that a considerable body of empirical data was the basis for much of the work. It would have been especially desirable to include systematic evidence in the generalized account of the ways in which the various roles presumably developed from earlier structures. At present, Znaniecki's account is simply a plausible reconstruction, with all the liabilities to which such developmental schemes are subject. His leading hypothesis that these roles develop by successive differentiation is amenable to empirical test; until it is so tested it can be considered only conjectural. The value of the work would have been considerably enhanced, also, if the role-paradigm (social circle, self, status, function) had been more fully exploited in the analysis of each of the roles actually discussed. As it is, most attention is devoted to the functions of each role and not enough to the structural relations between the other components. Perhaps this is only tantamount to saying that Znaniecki's conceptions are so fertile that he has found it possible to gather only the ripest of the first fruits. Such

2. Cf. Richard Baxter, *Christian Directory* (London, 1825), 1:171, in a passage written in 1665: "They that believe, and know not why, or know no sufficient reason to warrant their faith, do take a fancy, or opinion, or a dream for faith." Or, Henry More, *Brief Discourse of the True Grounds of the Certainty of Faith in Point of Religion* (London, 1688), p. 578: "To take away all the certainty of sense rightly circumstantiated, is to take away all the certainty of belief in the main points of our religion."

forthcoming empirical studies as Logan Wilson's *Academic Man* will doubtless profit by the conceptual framework which Znaniecki has built for handling such subjects. His classification is of course provisional and lends itself to necessary modifications. In short, this is a prospectus which no future student of the subject dare neglect; it is a promise of things to come and a promise which is in part its own fulfillment.

3 Social Conflict Over Styles of Sociological Work

1961

After enjoying more than two generations of scholarly interest, the sociology of knowledge remains largely a subject for meditation rather than a field of sustained and methodical investigation. This has resulted in the curious condition that more monographs and papers are devoted to discussions of what the sociology of knowledge is and what it ought to be than to detailed inquiries into specific problems.

What is true of the sociology of knowledge at large is conspicuously true of the part concerned with the analysis of the course and character taken by sociology itself. This, at least, is the composite verdict of the jury of twelve who have reviewed for us the social contexts of sociology in countries all over the world. Almost without exception, the authors of these papers report (or intimate) that, for their own country, they could find only fragmentary evidence on which to draw for their account. They emphasize the tentative and hazardous nature of interpretations based on such slight foundations. It follows that my own paper, drawing upon the basic papers on national sociologies, must be even more tentative and conjectural.

In effect, these authors tell us that they have been forced to resort to loose generalities rather than being in a position to report firmly grounded generalizations. Generalities are vague and indeterminate statements that bring together particulars which are not really comparable; generalizations report definite though general regularities distilled from the methodical comparison of comparable data. We all know the kind of generalities found in the sociology of knowledge: that societies with sharp social

Originally published in Fourth World Congress of Sociology, *Transactions* (Louvain, Belgium: International Sociological Association, 1961), 3:21–46; reprinted with permission.

cleavages, as allegedly in France, are more apt to cultivate sociology intensively than societies with a long history of a more nearly uniform value-system, as allegedly in England; that a rising social class is constrained to see the social reality more authentically than a class long in power but now on the way out; that an upper class will focus on the static aspects of society and a lower one on its dynamic, changing aspects; that an upper class will be alert to the functions of existing social arrangements and a lower class to their dysfunctions; or, to take one last familiar generality, that socially conservative groups hold to multiple-factor doctrines of historical causation and socially radical groups to monistic doctrines. These and comparable statements may be true or not, but as the authors of the national reports remind us, we cannot say, for these are not typically the result of systematic investigations. They are, at best, impressions derived from a few particulars selected to make the point.

It will be granted that we sociologists cannot afford the dubious luxury of a double standard of scholarship; one requiring the systematic collection of comparable data when dealing with complex problems, say, of social stratification, and another accepting the use of piecemeal illustrations when dealing with the no less complex problems of the sociology of knowledge. It might well be, therefore, that the chief outcome of this first session of the congress will be to arrange for a comparative investigation of sociology in its social contexts similar to the investigation of social stratification that the Association has already launched. The problems formulated in the national papers and the substantial gaps in needed data uncovered by them would be a useful prelude to such an undertaking.

The growth of a field of intellectual inquiry can be examined under three aspects: as the historical filiation of ideas considered in their own right; as affected by the structure of the society in which it is being developed; and as affected by the social processes relating the men of knowledge themselves. Other sessions of the congress will deal with the first when the substance and methods of contemporary sociology are examined. In his overview, Professor Aron considers the second by examining the impact on sociology of the changing social structure external to it: industrialization, the organization of universities, the role of distinctive cultural traditions, and the like. He goes on to summarize the central tendencies of certain national sociologies, principally those of the United States and the Soviet Union, and assesses their strengths and weaknesses. Rather than go over much the same ground to arrive at much the same observations, I shall limit myself to the third of these aspects. I shall say little about the social structure external to sociologists and focus instead on some social processes internal to the development of sociology and in particular on the role in that development played by social conflict between sociologists.

There is reason to believe that patterns of social interaction among sociologists, as among other men of science and learning, affect the changing contours of the discipline just as the cultural accumulation of knowledge manifestly does. Juxtaposing the national papers gives us an occasion to note the many substantial similarities if not identities in the development of sociology in each country that underlie the sometimes more conspicuous if not necessarily more thoroughgoing differences. These similarities are noteworthy if only because of the great variability and sometimes profound differences of social structure, cultural tradition, and contemporary values among the twelve nations whose sociology has been reviewed. These societies differ among themselves in the size of the underlying population, in the character of their systems of social stratification, in the number, organization, and distribution of their institutions of higher learning, in their economic organization and the state of their technology, in their current and past political structure, in their religious and national traditions, in the social composition of their intellectuals, and so on through other relevant bases of comparison. In view of these diversities of social structure, it is striking that there are any similarities in the course sociology has taken in these societies. All this suggests that a focus on the social processes internal to sociology as a partly autonomous domain can help us to understand a little better the similarities of sociological work in differing societies. It may at the least help us identify some of the problems that could be profitably taken up in those monographs on the sociological history of sociology that have yet to be written.[1]

Phases of Sociological Development

From the national reports, we can distinguish three broad phases in the development of sociology: first, the differentiation of sociology from antecedent disciplines with its attendant claim to intellectual legitimacy; second, the quest to establish its institutional legitimacy or academic autonomy; and third, when this effort has been moderately successful, a movement toward the reconsolidation of sociology with selected other social sciences. These well-known phases are of interest here insofar as they derive from processes of social interaction between sociologists and between them and scholars in related fields, processes that have left their distinctive mark on the kinds of work being done by sociologists.

1. One last introductory word: we have been put on notice that since the papers on national sociologies could not be circulated in advance, we should keep our general remarks to a minimum. I shall therefore omit much of the concrete material on which my paper is based.

Differentiation from other disciplines

The beginnings of sociology are of course found in the antecedent disciplines from which it split off. The differentiation differs in detail but has much the same general character in country after country. In England, we are told, sociology derived chiefly from political economy, social administration, and philosophy. In Germany, it shared some of these antecedents as well as an important one in comparative law. In France, its roots were in philosophy and, for a time, in the psychologies that were emerging. Its varied ancestry in the United States included a concern with practical reform, economics, and, in some degree, anthropology. Or, to turn to some countries which have been described by their reporters as "sociologically underdeveloped," in Yugoslavia, sociology became gradually differentiated from ethnology, the history of law, and anthropogeography; in Spain, it was long an appendage of philosophy, especially the philosophy of history. The Latin American countries saw sociology differentiated from jurisprudence, traditionally bound up as it was with an interest in the social contexts of law and the formation of law that came with the creation, in these states, of governments of their own.

The process of differentiation had direct consequences for the early emphasis in sociology. Since the founding fathers were self-taught in sociology—the discipline was, after all, only what they declared it to be—they each found it incumbent to develop a classification of the sciences in order to locate the distinctive place of sociology in the intellectual scheme of things. Virtually every sociologist of any consequence throughout the nineteenth century and partly into the twentieth proposed his own answers to the socially induced question of the scope and nature of sociology and saw it as his task to evolve his own system of sociology.

Whether sociology is said to have truly begun with Vico (to say nothing of a more ancient lineage) or with St. Simon, Comte, Stein, or Marx is of no great moment here, though it may be symptomatic of current allegiances in sociology. What is in point is that the nineteenth century—to limit our reference—was the century of sociological systems not necessarily because the pioneering sociologists happened to be system-minded men but because it was their role, at that time, to seek intellectual legitimacy for this "new science of a very ancient subject." In the situation confronting them, when the very claim to legitimacy of a new discipline had to be presented, there was little place for a basic interest in detailed and delimited investigations of specific sociological problems. It was the framework of sociological thought itself that had to be built and almost everyone of the pioneers tried to fashion one for himself.

The banal flippancy tempts us to conclude that there were as many sociological systems as there were sociologists in this early period. But of course this was not so. The very multiplicity of systems, each with its claim to being the genuine sociology, led naturally enough to the formation of schools, each with its masters, disciples, and epigoni. Sociology not only became differentiated from other disciplines but became internally differentiated. This was not in terms of specialization but in the form of rival claims to intellectual legitimacy, claims typically held to be mutually exclusive and at odds. This is one of the roots of the kinds of social conflict among sociologists today that we shall examine in a little detail.

Institutional legitimacy of sociology

If it was the founding fathers who initiated and defended the claim of sociology to intellectual legitimacy—as having a justifiable place in the culture—it was their successors, the founders of modern sociology, who pressed the claim to institutional legitimacy, by addressing themselves to those institutionalized status-judges of the intellect: the universities. Here again, the pattern in different nations differs only in detail. Whether ultimate control of the universities was lodged in the state or the church, it was their faculties that became the decisive audience for a Weber, Durkheim, or Simmel. Sociology was variously regarded by the faculties as an illegitimate upstart, lacking warrant for a recognized place in the collegial family, or sometimes as an institutional competitor. And this social situation repeatedly led to a limited number of responses by sociologists of the time.

They directed themselves, time and again (as some still do), to the questions that, satisfactorily answered, would presumably make the case for sociology as an autonomous academic discipline. They continued to deal with the question: is a science of society possible? And having satisfied themselves (and, it is hoped, others in the university) that it is, they turned above all to the further question, whose relevance was reinforced by the social condition of being on trial: what is sociology? that is to say, what is its distinctive scope, its distinctive problems, its distinctive functions; in short, its distinctive place in the academic world.

I do not try to enumerate the many answers to these questions, which we can all readily call to mind. What I do want to suggest is that the long-lasting focus on these questions seemed peculiarly pertinent, not only because of an immediate intellectual interest in them but because these were generations of sociologists seeking but not yet finding full academic legitimation. This sort of public search for an identity becomes widespread in a group rather than being idiosyncratic to a few of its members whenever a status or a way of life has yet to win acceptance or is under attack.

The socially induced search for an institutional identity led sociologists to identify a jurisdiction unshared by other disciplines. Simmel's notion of a geometry of social interaction and his enduring attention to the so-called molecular components of social relations is only one of the best-known efforts to center on elements of social life that were not systematically treated by other disciplines. It would be too facile to "derive" his interest in the distinctive sociology of everyday life from his experience of having been excluded, until four years before his death, from a professorship in a field that was still suspect. But this kind of individual experience may have reinforced an interest that had other sources. The early sociologists in the United States were responding to a comparable social situation in much the same way, locating such subjects of life in society as "corrections and charities" that had not yet been "preempted" for study.

A related consequence of the quest for academic legitimacy was the motivated separation of sociology from the other disciplines: the effort to achieve autonomy through self-isolation. We have only to remember, for example, Durkheim's taboo on the use of systematic psychology which, partly misunderstood, for so long left its stamp on the work stemming from this influential tradition in sociology.

The struggle for academic status may have reinforced the utilitarian emphasis found in sociology, whether in its positivistic or Marxist beginnings. However much the dominant schools disagreed in other respects, they all saw sociology as capable of being put to use for concerted objectives. The differences lay not in the repudiation or acceptance of utility as an important criterion of sociological knowledge but in the conception of what was useful.

As sociology achieved only limited recognition by the universities, it acquired peripheral status through the organizational device of research institutes. These have been of various kinds: as adjuncts to universities; as independent of universities but state-supported or aided; and, in a few cases, as private enterprises. Socially, they tended to develop where the university system was felt to provide insufficient recognition. Just as in the seventeenth century, when no one arrived at the seemingly obvious thought of basing research laboratories for the physical sciences in the university, so we have witnessed a comparable difficulty, now overcome in many quarters, in arriving at the idea that the universities should house research organizations in the social sciences. They are now to be found in just about every country represented here. With their prevalently apprentice system of research training and, as the national papers report, with their greater readiness to try out new orientations in sociology, these institutes might well turn out to be a major force in the advancement of sociology. If so, they would represent an intellectual advance substantially responsive to the social situation of institutional exclusion or underrecognition.

Reconsolidation with other disciplines

As the institutional legitimacy of sociology becomes substantially acknowledged—which does not mean, of course, that it is entirely free from attack—the pressure for separatism from other disciplines declines. No longer challenged seriously as having a right to exist, sociology links up again with some of its siblings. But since new conceptions and new problems have meanwhile emerged, this does not necessarily mean reconsolidation with the same disciplines from which sociology drew its origins in a particular country.

Patterns of collaboration between the social sciences differ somewhat from country to country and it would be a further task for the monographs on the sociology of sociology to try to account for these variations. Some of these patterns are found repeatedly. In France, we are told, the long-lasting connection between sociology and ethnology, which the Durkheim group had welded together, has now become more tenuous, with sociologists being increasingly associated with psychologists, political scientists, and geographers. In the United States, as another example, the major collaboration is with psychology—social psychology being the area of convergence—and with anthropology. Another cluster links sociology with political science and, to some extent, with economics. There are visible stirrings to renew the linkage, long attenuated in the United States, of sociology with history. The events long precede their widespread recognition. At the very time that American graduate students of sociology are learning to repeat the grievance that historical contexts have been lost to view by systematic sociology, the national organization of sociologists is devoting annual sessions to historical sociology and newer generations of sociologists, such as Bellah, Smelser, and Diamond are removing the occasion for the grievance through their work and their program.

Each of the various patterns of interdisciplinary collaboration has its intellectual rationale. They are not merely the outcome of social forces. However, these rationales are apt to be more convincing, I suggest, to sociologists who find that their discipline is no longer on trial. It has become sufficiently legitimized that they no longer need maintain a defensive posture of isolation. Under these social circumstances, interdisciplinary work becomes a self-evident value and may even be exaggerated into a cultish requirement.

Summary

In concluding this sketch of three phases in the development of sociology, I should like to counter possible misunderstandings.

It is not being said that sociology in every society moves successively through these phases, with each promptly supplanted by the next. Concretely, these phases overlap and coexist. Nevertheless, it is possible to detect in the national reports a distinct tendency for each phase to be dominant for a time and to become so partly as a result of the social processes of opposition and collaboration that have been briefly examined.

It is not being said, also, that the social processes internal to sociology and related disciplines fully determine the course sociology has taken. But it is being said that together with *culturally* induced change in the contours of sociology, resulting from the interplay of ideas and cumulative knowledge there is also *socially* induced change, such that particular preoccupations, orientations, and ideas that come to "make sense" to sociologists in one phase elicit little interest among them in another. The concrete development of sociology is of course not the product only of social processes immanent to the field. It is the resultant of social and intellectual forces internal to the discipline with both of these being influenced by the environing social structure, as the reports on national sociologies and the companion piece by Professor Aron have noted. The emphasis on social processes internal to sociology is needed primarily because the sociology of knowledge has for so long centered on the relations between social structures, external to intellectual life, and the course taken by one or another branch of knowledge.

Continuing with this same restriction of focus on social processes internal to the discipline, I turn now to some of the principal occasions for conflict between various styles of sociological work. In doing so, I am again mindful of the need for monographs on the sociological history of sociology emphasized in the papers presented to this session. If the linkages between sociology and social structure are to be seriously investigated, then it is necessary to decide which aspects of sociology might be so related. These would presumably include, as Professor Aron has indicated, the questions it asks, the concepts it employs, the objects it studies, and the types of explanations it adopts. One way of identifying the alternative orientations, commitments, and functions ascribed to sociology is by examining, however briefly, the principal conflicts and polemics that have raged among sociologists. For these presumably exhibit the alternative paths that sociology might have taken in a particular society, but did not, as well as the paths it has taken. In reviewing some of these conflicts, I do not propose to consider the merits of one or another position. These are matters that will be examined in the other sessions of the congress that deal with the various specialties and with the uses of sociology. I intend to consider them only as they exhibit alternative lines of development in sociology that are influenced by the larger social structure and by social processes internal to sociology itself.

Some Uniformities in the Conflict of Sociological Styles

A few general observations may provide a guide through the jungle of sociological controversy.

First, the reports on national sociologies naturally center on the dominant kinds of sociological work found in each country; on the modes rather than on the less frequent variants. But to judge from the reports, these sociologies differ not only in their central tendencies but also in the *extent of variation* around these tendencies. Each country provides for different degrees of heterodoxy in sociological thought, and these differences are probably socially patterned. In the Soviet Union, for example, there appears to be a marked concentration in the styles of sociological work with little variability: a heavy commitment to Marxist-Leninist theory with divergence from it only in minor details; a great concentration on the problem of the forces making for sequences of historical development of total societies; and a consequent emphasis, with little dispersion, upon historical evidence as the major source material. It would be instructive to compare the extent of dispersion around the dominant trends of sociological work in the United States, which are periodically subjected to violent attacks from within, as in the formidable book by Sorokin, *Fads and Foibles in Modern Sociology*, and in the recent little book by C. Wright Mills which, without the same comprehensive and detailed citation of seeming cases in point, follows much the same lines of arguments as those advanced by Sorokin. As we compare the national sociologies, we should consider how the social organization of intellectual life affects the extent to which the central tendencies of each country's sociology are concentrated.

Much of the controversy among sociologists involves social conflict and not only intellectual criticism. Often, it is less a matter of contradictions between sociological ideas than of competing definitions of the role considered appropriate for the sociologist. Intellectual conflict of course occurs; an unremitting Marxist sociology and an unremitting Weberian or Parsonian sociology do make contradictory assumptions. But in considering the cleavages among a nation's sociologists, or among those of different nations, we should note whether the occasion for dispute is this kind of substantive or methodological contradiction or rather the claim that this or that sociological problem, this or that set of ideas, is not receiving the attention it allegedly deserves. I suggest that very often these polemics have more to do with the allocation of intellectual resources among different kinds of sociological work than with a closely formulated opposition of sociological ideas.

These controversies follow the classically identified course of social conflict.[2] Attack is followed by counterattack, with progressive alienation of each party to the conflict. Since the conflict is public, it becomes a status battle more nearly than a search for truth. (How many sociologists have publicly admitted to error as a result of these polemics?) The consequent polarization leads each group of sociologists to respond largely to stereotyped versions of what is being done by the other. As Professor Germani says, Latin American sociologists stereotype the North Americans as mere nose-counters or mere fact-finders or merely descriptive sociographers. Or others become stereotyped as inveterately speculative, entirely unconcerned with compelling evidence, or as committed to doctrines that are so formulated that they cannot be subjected to disproof.

Not that these stereotypes have no basis in reality at all, but only that, in the course of social conflict, they become self-confirming stereotypes as sociologists shut themselves off from the experience that might modify them. The sociologists of each camp develop selective perceptions of what is actually going on in the other. They see in the other's work primarily what the hostile stereotype has alerted them to see, and then promptly mistake the part for the whole. In this process, each group of sociologists become less and less motivated to study the work of the other, since there is manifestly little point in doing so. They scan the out-group's writings just enough to find ammunition for new fusillades.

The process of reciprocal alienation and stereotyping is probably reinforced by the great increase in the bulk of sociological publication. Like many other scholars, sociologists can no longer keep up with all that is being published in their field. They must become more and more selective in their reading. And this selectivity readily leads those who are hostile to a particular line of sociological work to give up studying the very publications that might possibly have led them to abandon their stereotype.

All this tends to move towards the emergence of an all-or-none doctrine. Sociological orientations that are not substantively contradictory are regarded as if they were. Sociological inquiry, it is said, must be statistical *or* historical in character; only the great issues of the time must be the objects of study *or* these refractory issues of freedom or compulsion must be avoided because they are not amenable to scientific investigation; and so on.

The process of social conflict would more often be halted in mid-course and instead turn into intellectual criticism if there were nonreciprocation of affect, if a stop were put to the reciprocity of contempt that typically marks these polemics. But we do not ordinarily find here the social setting

2. For an unsolemn but serious extension of these observations on social conflict as distinct from cognitive controversy in science, see R. K. Merton, *On The Shoulders of Giants* (New York: The Free Press, 1965; New York: Harcourt Brace Jovanovich, 1967), pp. 25–29.—Ed.

that seems required for the nonreciprocation of affect to operate with regularity. This requires a differentiation of status between the parties, at least with respect to the occasion giving rise to the expression of hostility. When this status differentiation is present, as with the lawyer and his client or the psychiatrist and his patient, the nonreciprocity of expressed feeling is governed by a technical norm attached to the more authoritative status in the relationship. But in scientific controversies, which typically take place among a company of equals for the occasion (however much the status of the parties might otherwise differ) and, moreover, which take place in public, subject to the observation of peers, this structural basis for nonreciprocation of affect is usually absent. Instead, rhetoric is met with rhetoric, contempt with contempt, and the intellectual issues become subordinated to the battle for status.

In these polarized controversies, also, there is usually little room for the third, uncommitted party who might convert social conflict into intellectual criticism. True, some sociologists in every country will not adopt the all-or-none position that is expected in social conflict. They will not be drawn into what are essentially disputes over the definition of the role of the sociologist and over the allocation of intellectual resources though put forward as conflicts of sociological ideas. But typically, these would-be noncombatants are caught in the crossfire between the hostile camps. Depending on the partisan vocabulary of abuse that happens to prevail, they become tagged either as "mere eclectics," with the epithet, by convention, making it unnecessary to examine the question of what it asserts or how far it holds true; or, they are renegades, who have abandoned the sociological truth; or, perhaps worst of all, they are mere middle-of-the-roaders or fence-sitters who, through timidity or expediency, will not see that they are fleeing from the fundamental conflict between unalloyed sociological good and unalloyed sociological evil.

We all know the proverb that "conflict is the gadfly of truth." Now, proverbs, that abiding source of social science for the millions, often express a part-truth just as they often obscure that truth by not referring to the conditions under which it holds. This seems to be such a case. As we have noted, in social conflict cognitive issues become warped and distorted as they are pressed into the service of "scoring off the other fellow." Nevertheless, when the conflict is regulated by the community of peers, it has its uses for the advancement of the discipline. With some regularity, it seems to come into marked effect whenever a particular line of investigation—say, of small groups—or a particular set of ideas—say, functional analysis—or a particular mode of inquiry—say, historical sociology or social surveys—has engrossed the attention and energies of a large and growing number of sociologists. Were it not for such conflict, the reign of orthodoxies in sociology would be even more marked than it sometimes is. Self-assertive claims that allegedly neglected problems, methods, and theoretical orientation merit more concerted attention than they are receiving may serve to diversify the work that gets done. With more room for heterodoxy, there is more prospect of intellectually productive ventures, until these develop into new orthodoxies.

Even with their frequent intellectual distortions (and possibly, sometimes because of them), polemics may help redress accumulative imbalances in scientific inquiry. No one knows, I suppose, what an optimum distribution of resources in a field of inquiry would be, not least of all, because of the ultimate disagreement over the criteria of the optimum. But progressive concentrations of effort seem to evoke counterreactions, so that less popular but intellectually and socially relevant problems, ideas, and modes of inquiry do not fade out altogether. In social science as in other fields of human effort, a line of development that has caught on—perhaps because it has proved effective for dealing with certain problems—attracts a growing proportion of newcomers to the field who perpetuate and increase that concentration. With fewer recruits of high caliber, those engaged in the currently unpopular fields will have a diminished capacity to advance their work and with diminished accomplishments, they become even less attractive. The noisy claims to underrecognition of particular kinds of inquiry, even when accompanied by extravagantly rhetorical attacks on the work that is being prevalently done, may keep needed intellectual variants from drying up and may curb a growing concentration on a narrowly limited range of problems. At least, this possibility deserves study by the sociologist of knowledge.

These few observations on social conflict, as distinct from intellectual criticism, are commonplace enough, to begin with. It would be a pity if they were banalized as asserting that peace between sociologists should be sought at any price. When there is genuine opposition of ideas—when one set of ideas plainly contradicts another—then agreement for the sake of peaceful quiet would mean abandoning the sociological enterprise. I am suggesting only that when we consider the current disagreements among sociologists, we find that many of them are not so much cognitive oppositions as contrasting evaluations of the worth of one and another kind of sociological work. They are bids for support by the social system of sociologists. For the sociologist of knowledge, these conflicts afford clues to the alternatives from which the sociologists of each country are making their deliberate or unwitting selection.

Types of Polemics in Sociology

These general remarks are intended as a guide to the several dozen foci of conflict between sociologists. Let me comfort you by saying that I shall not consider all of them here, nor is it necessary. Instead, I shall review two or three of them in a little detail and then merely identify some of the rest for possible discussion.

The trivial and the important in sociology

Perhaps the most pervasive polemic, the one which, as I have implied, underlies most of the rest, stems from the charge by some sociologists that others are busily engaged in the study of trivia, while all about them the truly significant problems of human society go unexamined. After all, so this argument goes, while war and exploitation, poverty, injustice, and insecurity plague the life of men in society or threaten their very existence, many sociologists are fiddling with subjects so remote from these catastrophic troubles as to be irresponsibly trivial.

This charge typically assumes that it is the topic, the particular objects under study, that fixes the importance or triviality of the investigation. This is an old error that refuses to stay downed, as a glance at the history of thought will remind us. To some of his contemporaries, Galileo and his successors were obviously engaged in a trivial pastime, as they watched balls rolling down inclined planes rather than attending to such really important topics as means of improving ship construction that would enlarge commerce and naval might. At about the same time, the Dutch microscopist, Swammerdam, was the butt of ridicule by those far-seeing critics who knew that sustained attention to his "tiny animals," the microorganisms, was an unimaginative focus on patently trivial minutiae. These critics often had authoritative social support. Charles II, for example, could join in the grand joke about the absurdity of trying to "weigh the ayre," as he learned of the fundamental work on atmospheric pressure which to his mind was nothing more than childish diversion and idle amusement when compared with the Big Topics to which natural philosophers should attend. The history of science provides a long if not endless list of instances of the easy confusion between the seemingly self-evident triviality of the object under scrutiny and the cognitive significance of the investigation.

Nevertheless, the same confusion periodically turns up anew in sociology. Consider the contributions of a Durkheim for a moment: his choice of the division of labor in society, of its sources and consequences, would no doubt pass muster as a significant subject, but what of the subject of suicide? Pathetic as suicide may be for the immediate survivors, it can seldom be included among the major troubles of a society. Yet we know that Durkheim's analysis of suicide proved more consequential for sociology than his analysis of social differentiation; that it advanced our understanding of the major problem of how social structures generate behavior that is at odds with the prescriptions of the culture, a problem that confronts every kind of social organization.

You can add at will, from the history of sociology and other sciences, instances which show that there is no *necessary* relation between the socially ascribed importance of the object under examination and the scope of its implications for an understanding of how society or nature works. The social and the scientific significance of a subject matter can be poles apart.

The reason for this is, of course, that ideally that empirical object is selected for study which enables one to investigate a scientific problem to particularly good advantage. Often, these intellectually strategic objects hold little intrinsic interest, either for the investigator or anyone else.

Again, there is nothing peculiar to sociology here. Nor is one borrowing the prestige of the better-established sciences by noting that all this is taken for granted there. It is not an intrinsic interest in the fruit fly or the bacteriophage that leads the geneticist to devote so much attention to them. It is only that they have been found to provide strategic materials for working out selected problems of genetic transmission. Comparing an advanced field with a retarded one, we find much the same thing in sociology. Sociologists centering on such subjects as the immigrant, the stranger, small groups, voting decisions, or the social organization of industrial firms need not do so because of an intrinsic interest in them. They may be chosen, instead, because they strategically exhibit such problems as those of marginal men, reference group behavior, the social process of conformity, patterned sources of nonconformity, the social determination of aggregated individual decisions, and the like.

When the charge of triviality is based on a common-sense appraisal of the outer appearance of subject matter alone, it fails to recognize that a major part of the intellectual task is to find the materials that are strategic for getting to the heart of a problem. If we want to move toward a better understanding of the roots and kinds of social conformity and the socially induced sources of nonconformity, we must consider the types of concrete situations in which these can be investigated to best advantage. It does not mean a commitment to a particular object. It means answering questions such as these: which aspects of conformity as a social process can be observed most effectively in small, admittedly contrived, and adventitious groups temporarily brought together in the laboratory but open to detailed observations? which aspects of conformity can be better investigated in established bureaucracies? and which require the comparative study of organizations in different societies? So with sociological problems of every kind: the forms of authority; the conditions under which power is converted into authority and authority into power; limits on the range of variability among social institutions within particular societies; processes of self-defeating and self-fulfilling cultural mandates; and so on.

If we ask, in turn, how we assess the significance of the sociological problem (rather than that of the object under scrutiny), then, it seems to me, sociologists have found no better answer than that advanced by Max Weber and others in the notion of *Wertbeziehung*. It is the relevance of the problem to men's values, the puzzles about the workings of social structure and its change that engage men's interests and loyalties. And the fact is that this rough-and-ready criterion is so loose that there is ample room for differing evaluations of the worth, as distinct from the validity and truth, of a sociological investigation even among those who ostensibly have the same general scheme of values. The case for the significance of problems of reference-group behavior, for example, stems from the cumulative recognition, intimated but not followed up by sociologists from at least the time of Marx, that the behavior, attitudes, and loyalties of men are not uniformly determined by their current social positions and affiliations. Puzzling inconsistencies in behavior are becoming less puzzling by systematically following up the simple idea that people's patterned selection of groups other than their own provide frames of normative reference which intervene between the influence of their current social position and their behavior.

In short, the attack on the alleged triviality of much sociological work, found apparently in all the national sociologies, is something less than the self-evident case it is made out to be. It often derives from a misconception of the connection between the selection of an object for study, the object having little intrinsic significance for people in the society, and the strategic value of that object for helping to clarify a significant sociological problem. In saying this, I assume that I will not be misunderstood. I am not saying that there is no genuinely trivial work in contemporary sociology any more than it can be said that there was no trivial work in the physical science of the seventeenth century. Quite otherwise: it may be that our sociological journals during their first fifty years have as large a complement of authentic trivia as the *Transactions* of the Royal Society contained during their first fifty years (to pursue the matter no further). But these are trivia in the strict rather than the rhetorical sense: they are publications which are both intellectually and socially inconsequential. But much of the attack on alleged trivia in today's sociology is directed against entire classes of investigation solely because the objects they examine do not enjoy widespread social interest.

This most pervasive of polemics sets problems for those prospective monographs on the sociological history of sociology. As I have repeatedly said, we are here not concerned with the substantive merit of the charges and rejoinders involved in any particular polemic of this kind. These can be and possibly will be discussed in the later sessions of this congress. But for the sociological analysis of the history of sociology, there remains the task of finding out the social sources and consequences of assigning triviality or importance to particular lines of inquiry. It seems improbable that the angels of light are all on one side and the angels of darkness, all on the other. If the division is not simply between the wise and the foolish, there must be other bases, some of them presumably social, for the various distributions of evaluation. The discussions that are to follow in this session might usefully be devoted to interpretations that might account for the opposed positions taken up in the assignment of merit to particular kinds of sociological work.

The alleged cleavage between substantive sociology and methodology

Another deep-seated and long-lasting conflict, requiring the same kind of interpretation, has developed between those sociologists who are primarily or exclusively concerned with inquiry into substantive problems of society and those who are primarily or exclusively concerned with solving the methodological problems entailed by such inquiry. Unlike the kind of intellectual criticism often developed within each of these camps, designed to clarify cognitive issues, this debate has the earmarks of social conflict, designed to best the opponent.

The main lines of attack on methodology and the replies to these are familiar enough to need only short summary.

Concern with methodology, it is said, succeeds only in diverting the attention of sociologists from the major substantive problems of society. It does so by turning from the study of society to the study of how to study society.

To this, it is replied, in the words of one philosopher: "you cannot know too much of methods which you always employ." Responsible inquiry requires intellectual self-awareness. Whether they know it or not, the investigators speak methodological prose and some specialists must work out its grammar. To try to discover the rates of social mobility and some of their consequences, for example, first requires solving the methodological problems of devising suitable classifications of classes, appropriate measures of rates, and the like, as some sociologists have learned, to their discomfiture.

Again, it is charged, that a concern with the logic of method quickly deteriorates into "mere technicism." These would-be precisionists strain at a gnat and swallow a camel: they are exacting in details and careless about their basic assumptions. For an interest in substantive questions they substitute an interest in seeming precision for its own sake. They try to use a razor blade to hack their way through forests. These technical virtuosos are committed to the use of meticulous means to frivolous ends.

The rebuttal holds that it is the methodologically naïve, those knowing little or nothing of the foundations of procedure, who are most apt to misuse precise measures on materials for which they are not suited. Further, that it is the assumptions underlying the quick and ready use of verbal constructs by investigators of substantive problems which need, and receive, critical scrutiny and clarification by the methodologist.

It is argued that the methodologist turns research technician in spite of himself, and becomes an aimless itinerant, moving in whatever direction his research techniques summon him. He studies changing patterns of voting because these are readily accessible to his techniques rather than the workings of political institutions and organizations for which he has not evolved satisfying techniques of investigation.

The rejoinder holds that the selection of substantive problems is not the task of specialists in methodology. Once the problem is selected, however, the question ensues of how to design an inquiry so that it can contribute to a solution of the problem. The effort to answer such questions of design is part of the business of methodology.

During at least the last half-century, ideological significance has also been ascribed to methodological work. The methodologist is said to choose a politically "safe" focus of work rather than attend to substantive inquiries that might catch him up in criticism of the social institutions about him.

This allegation is treated by methodologists as not only untrue, but irrelevant. Practically all disciplines, even the strictly formal ones of logic and mathematics, have at one time or another been assigned political or ideological import. As we have been told here, even certain procedures of sociological research, such as "large-scale fieldwork" and the use of attitude scales, have been regarded as politically suspect in some nations. The irrelevance of the charge lies on its surface where the indefensible effort is made to merge intellectual and political criteria of scientific work.

The complaint is heard that the methodologist supposes knowledge to consist only of that which can be measured or at least counted. He is addicted to numbers. As a result, he retreats from historical inquiry and from all other forms of sociological inquiry where even crude measures have not been devised or where, in principle, they cannot be.

To the methodologist, this is a distorted image, fashioned by the uninformed who run as they read. He regards himself as no more committed to working out the logic of tests and measurements than the logic of historical and institutional analysis. This he points out, has been understood by sociologists of consequence, at least from the time of Max Weber who, as Professor Adorno reminds us, "devoted a large part of his work to methodology, in the form of philosophical reflections on the nature and procedures of sociology," and who considered the methodology of historical inquiry, in particular, an important part of the sociological enterprise.

Since the opponents in this controversy show no trace of being either vanquished or converted, this raises anew the question of the grounds, other than intellectual, for maintaining their respective positions. Like the other persistent conflicts I shall now summarize far more briefly, this one sets a problem for the sociologist of knowledge.

The lone scholar and the research team

Until the last generation or so, the sociologist, like most other academic men, worked as an individual scholar (or, as the idiom has it, as a "lone scholar"). Since then, as the national reports inform us, institutes for sociological research have multiplied all over the world. This change in the social organization of sociological work has precipitated another conflict, with its own set of polarized issues.

The new forms of research are characterized, invidiously rather than descriptively or analytically, as the bureaucratization of the sociological mind. The research organization is said to stultify independent thought, to deny autonomy to members of the research staff, to suffer a displacement of motive such that researches are conducted in order to keep the research team or organization in operation rather than have the organization provide the facilities for significant research; and so on through the familiar calendar of indictments.

In return, it is pointed out that the individual scholar has not been as much alone as the description may imply. He was (and often is) at the apex of a group of research assistants and graduate students who follow his lead. Moreover, he has had to limit his problems for serious research to those for which the evidence lay close to hand, principally in libraries. He cannot deal with the many problems that require the systematic collection of large-scale data which are not provided for him by the bureaucracies that assemble census data and other materials of social bookkeeping. The research institute is said to extend and to deepen kinds of investigation that the individual scholar is foreclosed from tackling. Finally, it is suggested that close inspection of how these institutes actually work will find that many of them consist of individual scholars with associates and assistants, each group engaged in pursuing its own research bents.

This continuing debate affords another basis for inquiry, this time into the ways in which the social organization of sociological research in fact affects the character of the research. This would require the kind of systematic comparison of the work being done by individual scholars and by research teams, a methodical comparison which, so far as I know, has yet to be made. Not that the results of this inquiry will necessarily do away with the conflict but only that it will contribute to that as yet largely unwritten sociological history of sociology whose outlines all of us here aim to sketch out.

Cognitive agreement and value disagreement

A particularly instructive type of case is provided by seeming intellectual conflict that divides sociologists of differing ideological persuasion. Upon inspection, this often (not, of course, always) turns out to involve cognitive agreements that are obscured by a basic opposition of values and interests.

To illustrate this type of conflict, we can draw upon a few observations by Marx and by so-called bourgeois sociologists. You will recall Marx's observation that in a capitalist society, social mobility "consolidates the rule of capital itself, enabling it to recruit ever new forces for itself out of the lower layers of society." This general proposition has won independent assent from all manner of non-Marxist sociologists, not least of all from one such as Pareto. The lines of disputation are not therefore drawn about the supposed fact of these systematic consequences of social mobility. The conflict appears only in the evaluation of these consequences. For, as Marx went on to say, the "more a ruling class is able to assimilate the most prominent men of the dominated class the more stable and dangerous is its rule." A Pareto could agree with the stabilizing function of such mobility while rejecting the judgment of it as "dangerous." What empirical investigations by "bourgeois sociologists" can do, and are doing, is to find out how far the cognitively identical assumption of a Marx and a Pareto holds true. To what extent do these mobile men identify themselves with their newfound class? Who among them retain loyalty to the old? When does it result in a consolidation of power and when, under conditions of retained values, does it modify the bases of cleavage between classes?

You can readily add other instances of agreement in sociological ideas being mistaken for disagreement, owing to an overriding conflict of values or interests between sociologists. When the functionalists examine religion as a social mechanism for reinforcing common sentiments that make for social integration, they do not differ significantly in their analytic framework from the Marxists who, if the metaphor of the opium of the masses is converted into a neutral statement of alleged consequences, assert the same sort of thing, except that they evaluate these consequences differently. Religion is then seen as a device for social exploitation.

Again, it has often been noted that Marx, in his theory, underrated the social significance of his own moral ideas. The emphasis on communist doctrine and ideology is perhaps the best pragmatic testimony that, whatever Marxist theory may say in general of the role of ideas in history, Marxists in practice ascribe great importance to ideas as movers, if not as prime movers, in history. If this were not so, the communist emphasis on a proper ideological commitment would be merely expressive rather than instrumental behavior.

Or, to take one last instance, Marx repeatedly noted that the patterns of production—for example, in large-scale industry and among small-holding peasants—have each a distinctive social ecology. The spatial distribution of men on the job was held to affect the frequency and kind of social interaction between them and this, in turn, to affect their political outlook and the prospects of their collective organization. In these days, a large body of investigation by non-Marxists, both in industrial and in rural sociology, is centered on this same variable of the social ecology of the job, together with its systemic consequences. But again, this continuity of problem and of informing idea tends to be obscured by conflicts in political orientation. Detailed monographic study is needed to determine the extent to which lines of sociological development fail to converge and instead remain parallel because of ideological rather than theoretical conflict.

Formal (abstract) and concrete sociology

Time and again, in the papers on national sociology, reference is made to the dangers of a "merely" formal sociology. This signals another familiar cleavage, that between concrete and abstract sociology. The first centers on interpreting particular historical constellations and developments. Sometimes these are society-wide in character; sometimes they are more limited social formations. The problem may be to explain the rise and transformation of Christianity or of capitalism, of particular class structures, family-systems or social institutions of science. The second, the formal orientation, is directed toward formulating general propositions and models of interpretation that cut across a variety of historically concrete events. Here the focus is on such abstract matters as role-theory, social processes of legitimation, the effect of the size of a group on its characteristic patterns of social interaction, and so on.

To some, formal sociology is an invidious epithet. It is ascribed to "defenders of the established order" who expressly neglect social change and deny that there are discoverable uniformities of social change. For these critics, formal sociology is like a sieve that strains out all the awkward facts that fail to suit its theory. To others, concrete sociology is seen as having some utility but at the price of abdicating the search for those social regularities that presumably occur in cultures of most different kind.

It would serve little purpose to note the obvious at this point, for it is precisely the obvious that gets lost in this conflict between commitments to primarily concrete and primarily abstract sociologies. Little will be gained in repeating, therefore, that concrete sociological investigations of course make at least implicit use of abstract models—that, for example, in order even to depict social change, let alone account for it, one must identify the formally defined elements and patterns of social structure that are changing—and conversely, that these models often grow out of and are modified and judged by their applicability to selected aspects of concrete social events. With respect to this conflict, the sociology of knowledge confronts such problems as that of finding out whether, as is commonly said, formal sociology is linked with politically conservative orientations and concrete sociology with politically radical orientations. Furthermore, how this social cleavage affects the prospects of methodical interplay between the two types of sociology.

A short miscellany of sociological conflicts

There is time only to list and not at all to discuss a few more of the current conflicts in sociology.

The microscopic and the macroscopic. More than ever before, conflict is focused on the social units singled out for investigation. This is often described by the catchwords of "microscopic" and "macroscopic" sociology. The industrial firm is said to be studied in isolation from the larger economic and social system or, even more, particular groups within the single plant are observed apart from their relations with the rest of the organization and the community. A microscopic focus is said to lead to "sociology without society." A counteremphasis centers on the laws of evolution of "the total society." Here, the prevailing critique asserts that the hypotheses are put so loosely that no set of observations can be taken to negate them. They are invulnerable to disproof and so, rather a matter of faith than of knowledge.

Experiment and natural history in sociology. A parallel cleavage has developed between commitment to experimental sociology, typically though not invariably dealing with contrived or "artificial" small groups, and commitment to study of the natural history of groups or social systems. Perhaps the instructive analogue here is to be found in the well-known fact that Darwin and Wallace found certain problems forced upon their attention when they reflected on what they saw in nature "on the large, on the outdoor scale" but that they failed to see other related problems that came into focus for the laboratory naturalists. Polarization into mutually exclusive alternatives served little purpose there and it remains to be seen whether it will prove any more effective in the advancement of sociology.

Reference-groups of sociologists. Conflict is found also in the sometimes implicit selection of reference-groups and audiences by sociologists. Some direct themselves primarily to the literati or to the "educated general public"; others, to the so-called men of affairs who manage economic or political organizations; while most are oriented primarily to their fellow academicians and professionals. The recurrent noise about jargon, cults of unintelligibility, the overly abundant use of statistics or of mathematical models is largely generated by the sociologists who have the general public as their major reference-group. The work of these outer-oriented sociologists, in turn, is described by their academic critics as sociological journalism, useful more for arousing public interest in sociology than for advancing sociological knowledge. They are said to persuade by rhetoric rather than to instruct by responsible analysis—and so on. It would be instructive to study the actual social roles and functions of these diversely oriented sociologists, rather than to remain content with offhand descriptions such as these, even though again we cannot expect that the results of such sudy would modify current alignments.

Sociology vs. social psychology. One last debate requires mention, at least. It is charged that many sociologists, especially in the United States, are converting sociology into social psychology, with the result that the study of social institutions is fading into obscurity. The trend toward social psychology is said to be bound up with an excessive emphasis on the subjective element in social action, with a focus on men's attitudes and sentiments at the expense of considering the institutional conditions for the emergence and the effective or ineffective expression of these attitudes. To this, the polarized response holds that social institutions comprise an idle construct until they are linked up empirically with the actual attitudes and values and the actual behavior of men, whether this is conceived as purposive or as also unwitting, as decisions or as responses. These sociologists consider the division between the two disciplines an unfortunate artifact of academic organization. And again, apart from the merits of one or the other position, we have much to learn about the social bases for their being maintained by some and rejected by others.

A Concluding Observation

In the final remark on these and the many other lines of cleavage among sociologists, I should like to apply a formulation about the structure of social conflict in relation to the intensity of conflict that was clearly stated by Georg Simmel and Edward Ross. This is the hypothesis, in the words of Ross, that

> a society . . . which is riven by a dozen . . . (conflicts) along lines running in every direction, may actually be in less danger of being torn with violence or falling to pieces than one split along just one line. For each new cleavage contributes to narrow the cross clefts, so that one might say that society *is sewn togeher* by its inner conflicts.

It is a hypothesis borne out by its own history, for since it was set forth by Simmel and by Ross, it has been taken up or independently originated by some scores of sociologists, many of whom take diametrically opposed positions on some of the issues we have reviewed. (I mention only a few of these: Wiese and Becker, Hiller, Myrdal, Parsons, Berelson, Lazarsfeld and McPhee, Robin Williams, Coser, Dahrendorf, Coleman, Lipset and Zelditch, and among the great number of recent students of "status-discrepancy," Lenski, Adams, Stogdill, and Hemphill.)

Applied to our own society of sociologists, the Simmel-Ross hypothesis has this to say. If the sociologists of one nation take much the same position on each of these many issues while the sociologists of another nation consistently hold to the opposed position on them all, then the lines of cleavage will have become so consolidated along a single axis that any conversation between the sociologists of these different nations will be pointless. But if, as I believe is the case, there is not this uniformity of outlook among the sociologists of each nation; if individual sociologists have different combinations of position on these and kindred issues, then effective intellectual criticism can supplant social conflict.

That is why the extent of heterodoxies among the sociologists of each nation has an important bearing on the future development of world sociology. The heterodoxies in one nation provide intellectual linkages with orthodoxies in other nations. On the worldwide scale of sociology, this bridges lines of cleavage and makes for the advance of sociological science rather than of sociological ideologies.

4 Technical and Moral Dimensions of Policy Research

1949

1. Rationale of the Inquiry

Although the application of social science to practical problems of policy and action is still in its early stages, a large body of experience has been accumulated. Social science *has* been applied, in diverse spheres and with diverse results. The experience is there, but it has not been systematically reviewed and codified. Consequently, no one knows the present status of applied social science or, more importantly, its potentialities.

Social scientists have been so busy examining the behavior of others that they have largely neglected the study of their own situation, problems, and behavior. Foundations, government, and commercial enterprises have been so concerned with research directed toward pressing problems that they have failed to take systematic inventory of the achievements and potentialities represented by this large body of researches. The hobo and the saleslady have been singled out for close study, but not the social science expert. Sociological monographs document the problems and performance of the professional thief and the professional beggar but not the problems and performance of the professional social scientist. Yet it would seem that clarity might well begin at home.

Originally published as "The Role of Applied Social Science in the Formation of Policy," in *Philosophy of Science* 16, no. 3 (July 1949): 161–81; reprinted with permission.

This paper is based upon a document prepared under the auspices of the Columbia University Council for Research in the Social Sciences and presented to a conference of the Social Science Research Council. I am indebted for useful suggestions to the following who attended that conference: Donald Young, Charles Dollard, E. P. Herring, Lyman Bryson, Leland DeVinney, Carl Hovland, R. V. Bowers, Paul F. Lazarsfeld, Lincoln Gordon, Alexander Leighton, Don Price, Glen Heathers, Douglas MacGregor.

Quite apart from the direct intellectual merits of the problem, the most varied groups have a stake in an analysis of the present and potential role of applied social science in American society. Most prominently, social scientists themselves stand to gain by such inquiry. Perhaps owing to the absence of any systematic appraisal of their role, social scientists are sometimes beset with exaggerated doubts and harassed by exaggerated claims concerning their contributions to solutions for the problems of our day. The actual workaday relations between basic and applied social science must for them be largely matters of opinion, sometimes well founded, at other times not, simply because these relations have not been made the object of systematic investigation.

Foundations and other philanthropic agencies engaged in endowing social science research have their stake in the inquiry as well. For until the *actual,* not the *supposed* or *ideal,* relations between basic and applied research are clarified, policies governing a program of endowed research must be based on rule-of-thumb experience. Yet it would seem the most elementary rule of intelligent administration to examine, from time to time, the consequences of diverse decisions. Are there types of applied research in social science which fructify basic theory? Do other types of applied research deflect scientific talent from fundamental inquiries into theory and methodology? Under which conditions does there occur a fruitful reciprocity between applied and basic research? A preliminary inventory may not succeed in providing circumstantial answers to these questions, but it can scarcely fail to lighten the fog of ignorance which, one must admit, now settles about the role of applied social science.

This inventory promises much the same returns for the maker of policy in government, business, and industry. To a large and growing but precisely unknown extent, applied social science does find a place in the world of practical decision. Much experience therefore exists, but this experience has not been codified. What are the obstacles to the effective utilization of applied social science? For which types of practical problems is the introduction of applied social science presently pointless and for which is it prerequisite to the formation of intelligent policy? Are there circumstances in which men of affairs have a direct stake in endowing basic research rather than calling for immediate applications of preexisting knowledge? After all, the decision to utilize or to forego applied social science is itself a matter of policy and it would seem useful to have this policy based on available, though presently uncoordinated, information.

It is long since time, finally, for the intelligent layman who does not directly utilize applied social science himself to learn something of this current in contemporary life. His preconceptions of social science may range from unshakable skepticism to equally ill-founded fetishism. In either case, how is he to arrive at an appropriate opinion? He is subjected to varied propagandas. One day he is told by seemingly unimpeachable authority that social science is merely gobbledegook. The next, he hears from other authorities that science alone, including its social divisions, can build the road to salvation. His choices are thereby limited. He may remain in a state of suspended judgment, which, in the present instance, may be only a euphemism for a state of confusion. Or he may cast his vote for one or another conflicting authority and emerge with clear and erroneous images of the present-day role of applied social science.

No preliminary inquiry can satisfy the diverse interests of social scientists, foundations, policy makers in government and business, and the men-in-the-street. But it can manifestly gather up presently scattered materials, coordinate these and seek to lay a basis for an instructive appraisal of applied social science. Through a review of cases in point, it should provide some new perspectives on (1) the achievements of applied social science; (2) the conditions limiting and making for these achievements; (3) the scientific (that is, theoretic and methodological) by-products of research in applied social science.

2. Scope of the Inquiry

We shall center our attention upon the disciplines commonly regarded as comprising the field of "human relations," namely, anthropology, psychology, social psychology, and sociology. For as is well known, a given practical problem ordinarily requires the *collaborative* researches of several social sciences. And it is precisely these four disciplines among the social sciences which have most consistently entered into collaborative research. This has in part resulted in a system of interlocking theory and implications for policy. The marked tendency toward coordination of these disciplines should therefore facilitate an immediate focusing of the inquiry.

A second delimitation should be noted. Although all applied social science research involves *advice* (recommendations for policy), not all advice on social policy is based on *research.* An analogy with medicine may help clarify the distinction. In medicine, advice may be based on anything from sheer empiricism to systematic applied research, thus:

1. *empiricism:* the experienced herb doctor finds empirically that cinchona bark (for some unknown reason) is a specific antidote for malaria and advises his patients accordingly;

2. *standardized therapies derived from previous cumulative medical research:* the physician treats well-identified cases of malaria with standardized quinine treatment;

3. *advice based on specific researches oriented toward new problems:* investigating a high rate of malaria in a specified area to identify the local

factors which must be brought under control, and advising alternative modes of treatment.

In matters of social policy, advice may likewise be based on anything from empiricism to systematic applied research, thus:

1. *empiricism:* Henry Ford finds in 1914 that he can pay the then-extraordinary minimum wage of five dollars a day for unskilled labor, with consequent rise in output and profit;

2. *standardized practices derived from previously cumulated research:* "scientific" wage policies based on Taylorism, and so on;

3. *advice based on specific researches:* detailed analyses of an industrial plant to determine the "most effective" wage policy.

The passing analogy with medicine suggests a further point. In contrast to medical advice, advice on social policy appears most often based on rule-of-thumb experience and only infrequently on generalized knowledge or on specific researches directed toward the problems in hand. Moreover, it may not always be possible to identify the lines between "sheer empiricism" and "cumulative scientific knowledge" in the realm of social policy. But this presents no large difficulty. Since the proposed inquiry will be centered on *research* in applied social science, the role of the social science expert who proffers advice out of his general fund of knowledge will receive only secondary attention.

3. Orientation of the Inquiry

Ultimately, cases of policy-oriented and action-oriented research must be collected and analyzed to determine the problems involved in the formulation of the research and its application. Among these materials are documentary sources (correspondence, memoranda, drafts, minutes of conferences, reports and so forth) and interviews with the researchers and policy makers (interviews aimed to clarify and elaborate the impressions gained from the documentary data).

Commonly underlying the occasional published accounts of the role of applied social science has been the presupposition that the applied research itself has been entirely adequate to the occasion, and that the "essential problem" has been one of persuading the policy maker to utilize these adequate results. In other words, the intellectual adequacy of the research does not typically come into question, but merely the organizational and interpersonal problems of "selling" the research. This emphasis is understandable enough. These accounts have usually been written by social scientists, and it is to be expected that they would be more sensitive to the inadequacies of the policy maker and his organization than to the possible inadequacies of the research. This orientation cannot be adopted here.

Instead, we shall distinguish the two distinct, though interrelated, types of problems attending the utilization of policy-oriented social research:

1. *interpersonal and organizational problems:* stemming from the relations between the research worker and the "clientele" (operating agency, administrator, and so on);

2. *scientific problems:* involving the difficulty of developing scientific researches adequate to the practical demands of the situation.

Although these types of problems are closely interrelated, and although we shall want to consider these interrelations, they should not be telescoped into one set of problems with the result that their distinctive aspects are lost to view. Interrelation should not be mistaken for identity. The sources of the organizational and scientific problems are different; the available means for coping with them are different; and their consequences for the development of applied social science are different.

Throughout our discussion, therefore, we shall attempt consistently to distinguish between the broadly "organizational" and the broadly "scientific" problems involved in the utilization of social science research.

4. The Culture Context

The repute of applied social science, as of any other intellectual resource, is in part a product of its accomplishments. This is an interlocking system in which social status and utilization interact endlessly. Not only does utilization affect esteem but esteem also affects utilization. The higher the social standing of a discipline, the more likely it will recruit able talents, the greater the measure of its financial support, and the greater its actual accomplishments. And closing the circle, the greater its utilization, the higher, ordinarily, its social standing. Even a cursory examination of the history of medicine or of physics will suggest the same pattern of interplay between cultural evaluations of the discipline, intellectual development of the discipline, and utilization of the findings.

The cultural context of evaluation, therefore, has a basic place in any analysis of the utilization of applied social science. And here we find ourselves limited by an impressive gap in available data. What are the prevailing evaluations of social science? How do they differ among various groups and strata in the population? And how have they been changing in the course of time? Manifestly, we do not know. No systematic inquiries into the cultural evaluations of social science have been made.

In the absence of the facts, we must speculate on the prevailing public images of applied social science and on the determinants of these images. All this is premised on the view that these prevailing images in part determine the extent to which policy-oriented research in social science is sought, by whom it is sought, and the purposes for which it is sought.

Of the numerous dimensions which may be found in public images of social science, only a few can be itemized and fewer still, briefly discussed. Experience suggests at least the following dimensions of these images:

Objectivity: ranging from the view that social science is merely private opinion masquerading as science to the faith in its rigorously objective quality;

Adequacy: ranging from belief in its unmitigated futility to belief in social science as the means of social salvation;

Political relevance: ranging from belief in its inherently "subversive" nature to belief that democracy can function adequately only if social science data are at hand;

"Costs": ranging from the naïve view that scientific results can be obtained with little expense (of time and funds) to the view that usable results are so costly as to be "uneconomic."

Other possible aspects of prevailing images will readily come to mind, but these may suffice to set the problem. Of these, I should like to consider the first two in brief outline.

The dimension of objectivity

We do not know the frequency of these images ranging from the view that social researches can be (and have been) "used to prove almost anything" to the view that they are wholly objective, uncontaminated by the researcher's predilections.

The fact that clients often, perhaps typically, publicize the findings of applied social science only when these are in accord with their own interests probably helps spread this belief in the unobjective nature of this research. Thus, the *New York Times* has seized upon the curious coincidence between the interests of clients and social science findings to conclude, in effect, that the winds of social science bloweth where they listeth. When an applied economist files a research report for the CIO which differs basically in its findings from a comparable report filed by experts of the NAM, the *Times* not only stresses the discrepancies, but notes that, oddly enough, the disparate findings coincide with the rival economic positions of the sponsors. Competing interest-groups attack and counterattack with their own social science researches. This is not merely a problem of "who shall decide when doctors disagree?" Since they are ostensibly based on research, the disagreements may activate a disbelief in the objectivity of applied social research *in general.* The specific instance may be generalized with consequent deterioration of the status of the disciplines involved.

The difficulty of distinguishing between "genuine" and "spurious" social science research further supports this skepticism of objectivity. The layman (often including the administrator and potential client for research) cannot always discriminate between the "research" which has all the outward trappings of rigorous investigation (sampling, design, controls, and so on) but which is defective in basic respects, and the genuinely disciplined investigation. The outward appearance is mistaken for the reality: "all social researches look alike" to many laymen.

Since careless, undisciplined, irresponsible "research" may promise larger returns at less expense, there may be a tendency for "bad research to drive out good research." And when these spurious investigations are tested in the crucible of experience, the resulting disappointment may lead to a repudiation of social science in general.

The dimension of adequacy

There are apparently some enthusiasts who would seek in social science knowledge the vade mecum to a scientifically planned and altogether desirable world. There are others who view applied social science as only an elaboration of the obvious, and who therefore consider it entirely dispensable as a basis for policy and action. Still others hold that social research is adequate when it deals with picayune problems and inadequate when it deals with "significant" problems. Here again, more information on the diverse images of adequacy and the comparative frequency of those images would be of value in helping to shape the future of applied social research.

Obviously, existing social science knowledge may be sufficient to deal with certain types of practical problems and wholly inadequate to deal with others. Thus, specific types of market researches may quite typically satisfy the needs of clients, whereas researches on, say, propaganda may prove typically unsatisfactory. The demands now made of applied social scientists may far outrun the *present* capacity and equipment of social science knowledge. As long as there is no roughly established inventory of our present knowledge such that laymen and scientists alike may have some approximate idea of applied researches which are and which are not promising for policy decisions, this will continue to provide a flow of disappointment and a consequent devaluation of the adequacy of applied social research *in general*. It is unwise to permit exaggerated public images of the immediately attainable achievements of applied social science to go unchecked.[1]

1. This general observation now takes on added force since the subsequent public reactions to the erroneous election forecasts on 2 November 1948 by the major polling organizations. It remains to be seen if the reaction against empiristic polling forecasts is generalized to the discredit of social science. For further implications of this polling episode, see footnote 9. [This subject was investigated by R. K. Merton and P. K. Hatt, "Election Polling Forecasts and Public Images of Social Science," *Public Opinion Quarterly* 13 (1949): 185–222.—Ed.]

Reacting against *under*estimates of the potentialities of applied social science, social scientists themselves may inadvertently supply exaggerated conceptions of what is now possible. Such propaganda for applied social science may boomerang and produce the excessive expectations which lead to subsequent disillusionment and popular reaction against the use of social science to any degree.

The preceding examples only touch upon the probably rich array of public images of applied social science. At this point, we can be confident only of two things. First, that we do not have adequate information on the range and comparative frequencies of these images and second, that this information would be useful. The NORC poll on the social status of occupations has some suggestive findings on the comparative status of some types of social scientists (economist, sociologist, psychologist, and so forth). Apart from limited material of this sort, literally no systematic data exist on prevailing conceptions of social science in general and of applied social science in particular. Interestingly enough, the very social scientists engaged in studying standardized images of ethnic and racial groups, labor unions, business, and so on, have not yet begun studies of prevailing images of themselves.

There is plainly a need for an "applied social research on applied social research" to ferret out the public images of social science, particularly among makers of policy in government, labor, and business.

The proposed inquiry may suggest appropriate lines of action. The role of the expert always includes an important fiduciary component. Laymen must be placed in a position where they can count on the responsible exercise of specialized competence by experts. Yet in contrast to the medical and legal professions, applied social scientists have not explored these problems of their own professional group. If the proliferation of irresponsible agencies of social research, for example, is found to be a major source of unfavorable and unrealistic images of applied social science generally, this may lead to recommendations for the regulation of these agencies.

5. The Organizational Context

The problems of utilizing applied social science research in policy-formation probably differ according to both the social position of the research agency and the client (or sponsor).[2] Each type of research agency may have diverse types of clients and each type of client may utilize diverse types of agencies.

2. The importance of this was long since recognized by Walter Lippmann, in his perceptive chapters on the potential role of the social science expert. See *Public Opinion* (New York: Macmillan, 1922), chaps. 25, 26.

To obtain a systematic picture of the various structures of social relations between researcher and clientele, we have only to cross-classify the two variables of research agency and of clients (as in the following specimen classification).

TABLE 1
Synopsis of Social Structures of Researcher and Client

Type of Research Agency	Types of Clientele				
	(1) Government Agency	(2) Foundations	(3) Business Corporations	(4) Welfare Agencies	(5) Etc.
1. Research agencies independent of operating agency					
A. "Academic" (endowed in part or whole)	A–1		A–3		
B. "Commercial" (dependent on research income)					
2. Research agencies incorporated in operating agency					
C. Permanent research staff	C–1				
D. Ad hoc research staff (for limited period)					

Each of the type-relations generated by this cross-classification can be readily identified. Thus, row A, column 1, "academic research agencies" with a governmental agency as client would include, for example, the contracted research conducted by the University of Michigan for the Treasury Department; row C, column 1 would include the Division of Agricultural Surveys in the Department of Agriculture; A–3, the Hawthorne studies conducted for the Western Electric Company by the Graduate School of Business Administration at Harvard; and so on.

Starting with some such classification, it should be possible to determine, through comparative analysis, the distinctive problems, procedures, and effects upon research of these several structures of social relations between researchers and clients. Do these structures characteristically differ, for example, with respect to the role of the researcher in defining the problem, in the types of research problem at the focus of attention, in the type of interaction between researcher and client, in the relevance of the research for policy and action, in the degree to which the research findings are utilized for policy purposes, in the methodological and theoretical by-products of the research, and so on.

This would provide a beginning toward systematically analyzing the part played in the formulation and utilization of applied social research by the organizational context. Thus, it may be found that when the organization of a client is itself the object of study (for example, a corporation, governmental bureau or division, and so on), the research findings are more likely to be taken as a basis of policy when the research is done by "independent" outside agencies than by a research staff which is itself part of the organization. Or, it may be found that applied social research for welfare agencies tends to have fewer methodological by-products than research for, say, business corporations.

It is in any case necessary to explore the assumption that the problems of making social research applicable will vary according to the organizational contexts. And to test this assumption, it is necessary to have some working classification of these contexts.

6. The Situational Context

There appears to be no literature which collates the types of situations leading to the decision to conduct a research in applied social science. Which occasions call applied research into being? And how do these different types of situations affect the nature of the research and its utilization?

The conventional picture of how this comes to pass is clear enough: a "problem" arises and the research worker, as a professional solver of problems, is asked to discover a solution. But who originally perceives the problem? Is it invariably the practical man of affairs, or, at times, the social scientist himself? And which types of "problems" are subjected to applied research and which are characteristically dealt with, without recourse to research? What are the functions of the research as conceived by the sponsor? And how does all this relate to the utilization and development of applied social science?

No systematic inventory of situational contexts is attempted here, but at least several can be identified. We can first consider the situations in which the need for applied research is initially perceived by policy makers or by social scientists.

Functions of research originated by policy makers

1. *Individuals or organizations confront the problem of "influencing" or "persuading" others to a given course of action.* They seek "objective data" to aid in persuasion. For example, an advertising agency has a research conducted in the hope of convincing a client of the greater effectiveness of their advertising program over alternatives proposed by rival agencies; a pressure group sponsors an applied research to obtain data in support of proposed legislation; a corporation vice-president solicits a research in defense of his policies as against those advocated by another vice-president;

a group of public-spirited citizens advocate a research on racial segregation to demonstrate the dangers of segregation to the general public, and so forth.

Since the chief function of these researches is persuasion, they are perhaps more subject to the tendency to have the research findings exploited for propagandistic aims. In these instances, the research findings are not likely to be subjected to the test of experience. They serve primarily to lend support to predetermined courses of action.

2. *Individuals or organizations confront problems requiring action by them, and find that they do not have sufficient information for "intelligent" action.* An industrial plant is repeatedly strike-bound; it tries a variety of expedients which are unsuccessful, and then turns to a research to suggest new alternatives.

Under which conditions is social science research sought? How does the pattern of action-oriented research differ from the pattern of persuasion-oriented research?

3. *Individuals or organizations wish to delay action to the point where the pressure for action from others is eliminated.* In such contexts, the applied research is intended not to lead to action, but to preclude it. The function of the research is to allay criticism of inaction. Public officials not infrequently authorize a "thorough study" of a problem on which they do not wish to take action.

In different situations, then, the policy maker may utilize applied research for quite different functions. We have mentioned three broad functions—persuasion, action, inaction. It is, of course, important to learn how each of these affects the nature of the research.

Functions of research originated by social scientists

1. *Social scientists may seek to sensitize policy makers to new types of achievable goals.* Some applied research has its origins in the work of the academic social scientist. He may detect what he considers a "practical problem" which has not yet been so identified by the maker of policy. In these instances, it is the first task of the research worker *to create* a practical problem for the policy maker.

For what is a "practical problem"? It represents a gap between aspiration and achievement, and holds out a challenge for closing this gap. If a policy maker has certain aspirations which are moderately well met, he, of course, perceives no "practical problem." But the social scientist may at times detect the possibility of at once heightening or extending these aspirations and of realizing the new goals. This requires him to serve as a gadfly, stinging contented policy makers into a state of discontent by widening their horizons, by introducing new criteria of the achievable, and by orienting applied research toward ways of reaching these new goals. Thus, the manager of a housing community may feel that it is running smoothly and well. He experiences no acute "problem." Rents are paid promptly, tenant turnover is low, few complaints reach him. An inquiring social scientist may find that there is little organized community life in the housing development and that the level of residents' satisfaction is less than it could be if specific provision were made for community organization. In effect, the researches of the social scientist are here aimed at introducing new and more demanding criteria of a "satisfactory state of affairs," of extending the goals of the housing manager.

A major function of research emanating from social science circles, then, may be to establish new goals and bench marks of the attainable.

2. *Social scientists may seek to sensitize policy makers to more effective means of reaching established goals.* In much the same fashion, administrators may assume that their organization is operating at a satisfactory level of effectiveness. The social scientist may discover more effective instrumentalities for approximating present goals. The task here is one of modifying criteria of effectiveness of ways and means. Thus, output in an industrial plant may be judged satisfactory by the policy maker. Further inquiry may show that this is at the expense of a rigorous regimen which puts considerable strain upon the work force. Alternative methods may produce the same high level of output without exacting this price of workers. It is altogether likely, as these casual instances suggest, that the modification of criteria of effectiveness of ways and means will characteristically involve a modification of goals, as well. The pattern is the same in both types of instance: sensitizing policy makers to a wider range of realizable potentialities.

Practical problems are many-faceted. They can be examined from the perspectives of several different disciplines. Increasingly, policy makers have been weaned from the naïve view that a practical problem is invariably in the orbit of one specialized body of science. High labor turnover, for example, is no longer automatically assumed to be in the province of "applied economics." Psychology and sociology may find partial determinants of rates of labor turnover in the human relations and social organization of the plant, or in the inadequacies of the local community from which the work force is drawn. On what grounds, then, does the policy maker select certain disciplines rather than others as most appropriate for studying the problems at hand?

This question introduces several considerations which can only be mentioned here. It points to the fact that for many, if not most, practical problems demanding applied research, collaboration among several disciplines is required. It suggests the role of the specialized research worker himself in acquainting the policy maker with the need for such collaboration. It points to the major organizational and scientific problems of providing for collaboration between the several applied social scientists. (The experience of the Tennessee Valley Authority should be especially instructive in this connection.) And, anticipating a later section of this discussion, it suggests that a major function of applied research is to provide occasions and pressures for interdisciplinary investigations and for the development of a theoretic system of "basic social science," rather than discrete bodies of uncoordinated specialized theory.

7. Defining the Practical Problems and the Research Problems

Experience suggests that the policy maker seldom formulates his practical problem in terms sufficiently precise to permit the researcher to design an appropriate investigation. Characteristically, the problem is so stated as to result in the possibility of the researcher being seriously misled as to the "basic" aspects of the problem which gives rise to a contemplated research. This initial clarification of the *practical* problem, therefore, is the first crucial step in applied social science.

Some types of unwitting misstatement of the practical problem by the client can be itemized here. Further inquiry will undoubtedly disclose others.

Overspecification of the problem

The policy maker often assumes that he has precisely identified his particular problem and comes to the researcher with a specific request for research. But this may be premature specification. The researcher has the task of ascertaining the *central* pragmatic problem rather than passively accepting its initial specifications by the policy maker. Thus, a Jewish "defense agency" requests a research to determine which of alternative types of mass propaganda will probably be most effective in curbing anti-Semitism. This does not represent the *prime* objective which is "reduction of anti-Semitism." The policy maker has prematurely included in his statement of the problem a specification of *means* as well as the end-in-view. The expert redefines the practical problem. On the basis of previous researches, he indicates that deep-seated prejudices are not markedly vulnerable to propaganda campaigns. The problem becomes reformulated: it is no longer an inquiry into efficiencies of alternative propaganda, but the comparative efficacy of a given propaganda campaign and of inter-religious voluntary organizations.

Overgeneralization of the problem

Or, the maker of policy may assume that he has sufficiently stated his problem when he indicates his general objective. He may seek fuller participation of the rank-and-file in a labor union or reduction of race tensions or increase in college attendance. But each of these general objectives may be approached through very different types of procedures, requiring different types of research.

When the policy maker overspecifies his practical problem, the expert must clarify by searching out the prime objective, thus often redefining the problem. When the policy maker overgeneralizes his practical problem, the expert must clarify by searching out the various alternative instrumentalities, and determine the consequences of each of these.

8. The Framework of Values in Definition of Problems

Value framework of the policy maker

We assume that the policy maker always has a set of values, tacit or explicit, and that this places limits upon the scope and nature of the applied research directed toward his problem. These "value constants" circumscribe the alternative lines of action to be investigated. It is the task of the researcher to search out these values in order to know in advance the limits set upon the investigation by the policy maker's values. (This is not only an ethical task but also a technical task. If it is true that the policy maker always assumes certain features of his problem-situation as *given*, as *constant*, as items which he would not under any circumstances consider modifying, this at once limits the range and type of research which will be done with his support, thus affecting the social scientist's decision to undertake the research.) Thus, policy-oriented research is requested on ways and means of improving morale of Negro workers in an industrial plant. The constant assumed by the policy maker: continued segregation of jobs, sanitary facilities, and so on. Policy-oriented research is requested on means for increasing sales of a product. The constant assumed by the policy maker: no change in the product itself.

These value-constants are probably of limited types. Two major types are noted here:

Objective factors of the situation shall remain unchanged, while attitudes toward the situation are modified. (For example, not changing objective fact of segregation but modifying Negro workers' morale; not changing product, but increasing sales; research may show that the proportion of Negroes and whites in an interracial housing project must be administratively stabilized if it is not to become an all-Negro project, but the policy maker rejects this research conclusion since it implies a "quota system" which offends his values; and so forth.)

Objective factors in the situation shall be changed, but no arrangements are to be made to modify attitudes. (For example, eliminating racial segregation in a housing community but not providing for ways and means for local acceptance of this change.)

Value framework of the research worker

The research worker also has his values, tacit or explicit, which affect his definition of the problem, the lines of investigation which seem to him most fruitful, the alternative policies to be explored, and so on. These values can be detected by determining the researcher's self-image of his role:

As a technician, he will accept alternative proposals for policy as a basis for research, providing only that these alternatives be technically amenable to research. Since it is feasible to test symbolic ("psychological") measures for improving the morale of Negro workers without eliminating segregation, the technician finds this definition of the problem adequate and *confines* himself accordingly. The researcher is asked to determine how a given radio program can reach a larger audience; since this is a feasible problem, he searches out strategic listening periods, and so forth and is content to accept the policy maker's constant of increasing audience without exploring effects upon audience size of changing the program content.

As a "socially oriented" scientist, he will explore only those policy alternatives which do not violate his own values. He not only includes in his study symbolic means of improving worker morale (for example, symbolic awards for performance, recreation groups, and so forth) but also "realistic" changes in situation (modified wage-policies, and so forth).

Study of the actual role played by the values of policy maker and researcher in the formulation of the research should help carry this question from the exclusively ethical context to that of the impact of values upon the relevance, scope, and utility of the research itself.

Moral and technical dimensions of research

The interaction between these aspects of social research is examined in the following passage drawn from the study of a radio War Bond Drive in World War II.[3]

Our primary concern with the social psychology of mass persuasion should not obscure its moral dimension. The technician or practitioner in mass opinion and his academic counterpart, the student of social psychology, cannot escape the moral issues which permeate propaganda as a means of social control. The character of these moral issues differ some-

3. This section on moral and technical dimensions of research is drawn from Robert K. Merton, Marjorie Fiske, and Alberta Curtis, *Mass Persuasion* (New York: Harper & Row, 1946), pp. 185–89.

what for the practitioner and the investigator, but in both cases the issues themselves are inescapable.

The practitioner in propaganda is at once confronted by a dilemma: he must either forego the use of certain techniques of persuasion which will help him obtain the immediate end-in-view or violate prevailing moral codes. He must choose between being a less than fully effective technician and a scrupulous human being or an effective technician and a less than scrupulous human being. The pressure of the immediate objective tends to push him toward the first of these alternatives.[4] For when effective mass persuasion is sought, and when "effectiveness" is measured solely by the number of people who can be brought to the desired action or the desired frame of mind, then the choice of techniques of persuasion will be governed by a narrowly technical and amoral criterion. And this criterion exacts a price of the prevailing morality, for it expresses a manipulative attitude toward man and society. It inevitably pushes toward the use of whatsoever techniques "work."

The sense of power that accrues to manipulators of mass opinion, it would appear, does not always compensate for the correlative sense of guilt. The conflict may lead them to a flight into cynicism. Or it may lead to uneasy efforts to exonerate themselves from moral responsibility for the use of manipulative tecnniques by helplessly declaring, to themselves and to all who will listen, that "unfortunately, that's the way the world is. People are moved by emotions, by fear and hope and anxiety, and not by information or knowledge." It may be pointed out that complex situations must be simplified for mass publics and, in the course of simplification, much that is relevant must be omitted. Or, to take the concrete case we have been examining, it may be argued that the definition of war bonds as a device for curbing inflation is too cold and too remote and too difficult a conception to be effective in mass persuasion. It is preferable to focus on the sacred and sentimental aspects of war bonds, for this "copy slant" brings "results."

Like most half-truths, the notion that leaders of mass opinion must traffic in sentiment has a specious cogency. Values *are* rooted in sentiment and values *are* ineluctably linked with action. But the whole truth extends beyond this observation. Appeals to sentiment within the context of relevant information and knowledge are basically different from appeals to sentiment which blur and obscure this knowledge. Mass persuasion is not manipulative when it provides access to the pertinent facts; it is manipulative when the appeal to sentiment is used to the exclusion of pertinent information.

4. R. K. Merton, "Social Structure and Anomie," *American Sociological Review* 3 (1938): 672–82.

The technician, then, must decide whether or not to use certain techniques which though possibly "effective" violate his own sentiments and moral codes. He must decide whether or not he should devise techniques for exploiting mass anxieties, for using sentimental appeals in place of information, for masking private purpose in the guise of common purpose.[5] He faces the moral problem of choosing not only among social ends but also among propaganda means.

Although less conspicuous and less commonly admitted, a comparable problem confronts the social scientist investigating mass opinion. He may adopt the standpoint of the positivist, proclaim the ethical neutrality of science, insist upon his exclusive concern with the advancement of knowledge, explain that science deals only with the discovery of uniformities and not with ends and assert that in his role as a detached and dispassionate scientist, he has no traffic with values. He may, in short, affirm an occupational philosophy which appears to absolve him of any responsibility for the use to which his discoveries in methods of mass persuasion may be put. With its specious and delusory distinction between "ends" and "means" and its insistence that the intrusion of social values into the work of scientists makes for special pleading, this philosophy fails to note that the investigator's social values do influence his choice and definition of problems. The investigator may naïvely suppose that he is engaged in the value-free activity of research, whereas in fact he may simply have so defined his research problems that the results will be of use to one group in the society, and not to others. His very choice and definition of a problem reflects his tacit values.

To illustrate: the "value-free" investigator of propaganda proceeds to the well-established mode of scientific formulations, and states his findings: "*If* these techniques of persuasion are used, *then* there will be (with a stated degree of probability) a given proportion of people persuaded to take the desired action." Here, then, is a formulation in the honored and successful tradition of science—apparently free of values. The investigator

5. During the war, imagination triumphed over conscience among advertisers who "ingeniously" related their products to the war effort. Radio commercials were not immune from this technique. A commercial dentist, for example, suggests that a victory smile helps boost morale and that we can have that smile by purchasing our dentures from him. So, too, a clothing manufacturer reminds listeners that morale is a precious asset in time of war and that smart clothes, more particularly Selfridge Lane Clothes, give a man confidence and courage. Even ice cream becomes essential to the war effort. "Expecting your boys back from an army camp? Give them JL Ice Cream. They get good food in the army and it's your job to give them the same at home." And a manufacturer of cosmetics becomes solicitous about the imbalance in the sex ratio resulting from the war. "Fewer men around because of the war? Competition keen? Keep your skin smooth. Keep attractive for the boys in the service when they come marching home." See R. K. Merton, Office of Radio Research, *Broadcasting the War* (Washington, D.C.: Bureau of Intelligence, Office of War Information, 1943), p. 37.

takes no moral stand. He merely reports his findings, and these, if they are valid, can be used by any interested group, liberal or reactionary, democratic or fascistic, idealistic or power-hungry. But this comfortable solution of a moral problem by the abdication of moral responsibility happens to be no solution at all, for it overlooks the crux of the problem: the initial formulation of the scientific investigation has been conditioned by the implied values of the scientist.

Thus, had the investigator been oriented toward such democratic values as respect for the dignity of the individual, he would have framed his scientific problem differently. He would not only have asked which techniques of persuasion produce the *immediate result* of moving a given proportion of people to action, but also, what are the *further, more remote* but not necessarily less significant, *effects* of these techniques upon the individual personality and the society? He would be, in short, sensitized to certain questions stemming from his democratic values which would otherwise be readily overlooked. For example, he would ask, Does the unelaborated appeal to sentiment which displaces the information pertinent to assessing this sentiment blunt the critical capacities of listeners? What are the effects upon the personality of being subjected to virtual terrorization by advertisements which threaten the individual with social ostracism unless he uses the advertised defense against halitosis or B.O.? Or, more relevantly, what are the effects, in addition to increasing the sale of bonds, of terrorizing the parents of boys in military service by the threat that only through their purchase of war bonds can they ensure the safety of their sons and their ultimate return home? Do certain types of war bond drives by celebrities do more to pyramid their reputations as patriots than to further the sale of bonds which would otherwise not have been purchased? No single advertising or propaganda campaign may significantly affect the psychological stability of those subjected to it. But a society subjected ceaselessly to a flow of "effective" half-truths and the exploitation of mass anxieties may all the sooner lose that mutuality of confidence and reciprocal trust so essential to a stable social structure. A "morally neutral" investigation of propaganda will be less likely than an inquiry stemming from democratic values to address itself to such questions.

The issue has been drawn in its most general terms by John Dewey: "Certainly nothing can justify or condemn means except ends, results. But we have to include consequences impartially. . . . It is wilful folly to fasten upon some single end or consequence which is liked, and permit the view of that to blot from perception all other undesired and undesirable consequences."[6] If this study has one major implication for the under-

6. John Dewey, *Human Nature and Conduct* (New York: Henry Holt & Co., 1922), pp. 228–29. Cf. R. K. Merton, "The Unanticipated Consequences of Purposive Social Action," *American Sociological Review* 1 (1936): 894–904.

standing of mass persuasion, it consists in this recognition of the intimate interrelation of technique and morality.

9. The Economic Framework of the Research

Whether "applied" or "pure," empirical research in social science is costly in time and money. But the economics of empirical research may affect the patterns of applied research and of basic research in quite different fashion. To be sure, the applied and the basic research alike may have a fixed budget and a definite deadline. But this is not to say that the degree to which and the ways in which these affect the research are alike in the two instances.

The tempo of applied research and of policy decisions

The tempo of policy decisions and action is often much more rapid than the tempo of applied research. Since action cannot always wait upon the completion of a research, varying degrees of urgency in decision affect research in various ways.

When there is great pressure for almost immediate decision, the *research* expert comes to be converted into the expert *adviser*. The policy maker draws upon the cumulative knowledge of the expert and foregoes an actual research. At this extreme, urgency is lethal for research, though not necessarily for other social utilities.

When there is need for decision at a definite, but more distant, occasion, a research may be designed to supply appropriate information. But since the "key" problems cannot be adequately investigated within this limited period, the research is necessarily confined to "practicable" though secondary problems. Furthermore, it becomes evident that data other than those needed for the immediate problem-in-hand may be expeditiously collected at the same time. But since this would prolong the period of fieldwork, these collateral materials are not included. The potential theoretic usefulness of the research is thus further circumscribed. As the research proceeds, fresh implications, not closely related to the present practical problem, are sensed by the research worker. These provocative clues, barely crystallized and wholly unformulated, are lost to view as the researcher bends to his immediate task of meeting the unalterable deadline. How often does the researcher return to the materials, after the deadline has been met (or not met), in an effort to recapture the fresh perceptions experienced during the research?

Occasionally, policy-oriented research escapes any marked time pressure. Certain types of data are periodically needed in order to take appropriate action. Or the research staff itself may at times reduce the pressure by anticipating future needs for decision, or by planning continuous researches in given areas so that the time required for new researches on a specific problem in that area is somewhat reduced.

The costs of applied research

Just as the pressure for immediate decision tends to eliminate research in favor of the considered judgments of expert advisers, so does the pressure to reduce costs. The comparative expensiveness of certain investigations leads to the substitution of advice for research. It would be useful to determine the grounds for opinions on the amount of money which can be justifiably expended for research on a given problem. How often is a given appropriation made first and the research tailored to fit this budget, and how often does the researcher plan the seemingly most appropriate research, and then have the estimated budget accepted? How does this differ as between researches in applied and basic social science? Since there exists no social bookkeeping for determining the "economic value" of *basic* research, the criteria for allocating funds to basic social science cannot be narrowly "economic" in character. But what of applied researches? Are the economic returns of specific applied researches typically estimated by sponsors or clients? And do these economic calculations determine their appropriations for research? Is there a tendency for applied researches to be diverted to peripheral problems when it is clear that research on the central problems-in-hand would be "too costly?" And since costs are inevitably increased by following up purely scientific leads developed in the research—leads which can have no value for the immediate practical problem—does this practice not limit the "nonpractical" by-products of applied research?

10. Types of Research Problems in Applied Social Science

Just as we suggested that patterns of applied social research may differ according to the organizational context (see section 5), so we know that they will differ according to the type of problem in hand. The most fruitful bases for classifying these problems are by no means clear. From several possible classifications, one developed by the Columbia University Bureau of Applied Social Research is here presented for discussion.

Research problems classified according to practical purpose

1. *Diagnostic:* Determining whether action is required. Magnitude and extent of problem; changes and trends since last appraisal of situation (for example, changes in level of race tensions); differentials in affected groups, areas, institutions.

2. *Prognostic:* Forecasting trends to plan for future needs. Predicting behavior of individuals and groups from stated intentions (postwar plans

of demobilized soldiers; people's disposition of liquid assets); predicting needs by trend analysis and other hypothetical means (unemployment, wage, price trends from business-cycle analysis; predicting housing needs by analysis of birth and marriage rates and trends in size of family).

3. *Differential Prognosis:* Determining choice between alternative policies, (for example, public reaction to rent control or rationing).

4. *Evaluative:* Appraising effectiveness of action program (assessing effectiveness of information and propaganda campaigns; of Emergency Maternal and Infant Care program in reducing infant and maternal mortality).

5. *General Background Data:* Of general utility or serving diverse purposes (for example, censuses of population, housing, business, manufacturing).

6. *"Educative" Research:* Informing publics upon pertinent data and particularly countering misconceptions.

"Strategic Fact-Finding": this involves the systematic assembling of descriptive data pertinent for popular conceptions and controversial beliefs. Thus, facts pertinent to stereotypes: "labor-leaders-are-foreign" stereotype confronted with facts on place of birth of labor leaders; "United States-remains-the-land-of-increasing-personal-opportunity" conception confronted with periodic data on social mobility; "you-don't-need-a-college-education-to-get-ahead" conception confronted with data on correlations between education and income, occupation, and so on.

To assess the current and potential role of applied social science, it is necessary to note the scope and scale of the practical problems with which it has dealt. These might range from broad, generic problems (generalized means of reducing crime, race hostilities) to highly circumscribed problems in a specific setting (the comparative effectiveness of two propaganda campaigns). It may develop that the extremes represent the least promising sectors of applied social science research. With the excessively large problem only failure can presently be reported and with the excessively limited problem, the results are often trivial. It would be important to identify the *strategic, intermediate range of problems*, namely, those which have generalized theoretical and practical significance, but which are not too large in scope to be subjected to disciplined research.[7]

11. Scientific Gaps between Research and Policy

Several of the circumstances which seemingly make for applied researches *not* affecting policy have been considered. The values of the policy maker,

7. Further observations on this point are presented by S. A. Stouffer, "The Strategy of the Social Sciences" (Address before the Harvard Graduate Forum, 20 April 1948) and by R. K. Merton, "Discussion of 'The Position of Sociological Theory,'" *American Sociological Review* 13 (April 1948): 164–68.

questions of time-and-cost, inadequacies in the formulation of the problem, and so forth, conduce to discrepancies between research-based recommendations and actual policies. As suggested previously, these gaps are of two interrelated types—the "scientific" and the "organizational and interpersonal." Since each type raises distinct problems for the research worker, it is advisable to consider them separately.

The research is not adequately focused on the practical problem (cf. section 7)

When the research worker inadvertently accepts the "overly specified" or "overgeneralized" statement of his problem by the policy maker, the resulting research will ultimately be found partly irrelevant to the actual problems of decision by the client. Alternative lines of action which have *not* been explored by the research may come to the later attention of the policy maker and he will conclude that the choice between the explored alternatives is spurious.

Concrete forecasts are contingent upon uncontrolled conditions

Many, if not most, applied researches involve forecasts. These concrete forecasts in applied science differ significantly from abstract predictions in basic science.

Basic research typically deals with "abstract predictions," that is, with predictions in which a large number of "other factors" are, conveniently enough, assumed to remain constant. The prediction will of course include a statement of the conditions under which the predicted consequences will probably occur. Ceteris paribus is an indispensable concept in basic research.

In applied research, ceteris paribus is often an embarrassing obstacle—for what if the "other factors" do not remain constant? As a matter of well-known fact, the research worker in applied research is not permitted the luxury of the *supposition* that other pertinent factors will remain equal. If action is to be based on his findings, he must indicate whether relevant "other factors" *will* remain constant. And since they typically will not, he has the further large task of assessing the changes in these factors and their effect upon contemplated action.

In short, applied research requires the greatly complicated study of the interaction of many interrelated factors comprising the *concrete situation*. The research cannot be confined entirely to the interplay of a severely limited number of variables under severely limited conditions.

This requirement of applied research has several consequences:

a. Every applied research must include some speculative inquiry into the role of diverse factors which can only be roughly assessed, not meticulously studied.

b. The validity of the concrete forecast depends upon the degree of (noncompensated) error in any phase of the total inquiry. The weakest links in the chain of applied research may typically consist of the *estimates* of contingent conditions under which the investigated variables will *in fact* operate.

c. To this degree, the recommendations for policy do not flow directly and exclusively from the *research*. Recommendations are the product of the research *and* the estimates of contingent conditions, these estimates not being of the same order of probability or precision as the more abstract interrelations examined in the research itself.

d. Such contingencies make for indeterminacy of the recommendations derived from the research and thus create a gap between research and policy.

Diverse utility of samples for different types of practical problems

Though this is not peculiar to *applied* social research, it should be noted that adequate samples are not readily obtained in the study of certain types of problems. Public opinion and market researches typically sample aggregates of individuals and the findings can be readily extrapolated to the universes which have been sampled. But much greater difficulties are encountered in other spheres—for example, in studies of social organization. The units here are *not* individuals, but *organized aggregates of interrelated individuals*. And since the study of *one* such unit is ordinarily a major research enterprise, this leaves open the question to which the *single* unit under examination is representative of the universe of organized units. Thus, recommendations for policy based upon the detailed study of one biracial housing community may not be adopted in other such communities because the policy maker feels that the communities are significantly different.

Further clarification is needed of the problems in which sampling problems can be met through available procedures and those in which the research, though involving hundreds or thousands of individuals, is essentially a case study of one social unit. We have further to determine when a case study is and is not regarded as an adequate basis for shaping policy.

12. Interpersonal and Organizational Gaps between Research and Policy

Other sections of the memorandum have touched upon some possible sources of gaps between research and policy which are interpersonal and organizational rather than strictly intellectual in character. There are others, some of which are here tentatively identified.

The framework of values precludes examination of some practicable courses of action (cf. section 8)

It appears that practicable policy alternatives are not explored because they run counter to the values of the policy maker or the research worker. (Thus, determining the most stable proportion of Negroes and whites in a biracial community may be rejected since it implies an objectionable "quota system.") In some cases, it is precisely the policy thus eliminated which most fully meets the requirements of the practical situation. Since these are ruled out, the resultant alternatives deriving from the research may be of dubious utility, and the research eventuates in inaction.

The economic framework may lead to the premature conclusion of a research (cf. section 9)

It is evident that limitations of time and funds at times condemn an applied research to practical futility. In most investigations, there emerge alternative lines of inquiry which are not followed through simply because of budgetary fiat. In such cases, it often happens that the research findings may not be entirely adequate to arrive at the most appropriate recommendations for action. The gap between research and action could only be closed or narrowed by following up the emerging implications.

Attitudes of the policy maker toward risk-bearing

Policy makers differ in their attitudes toward the taking of risks. No matter how circumstantial and meticulous the research, there is an element of risk in following the recomendations which seem to flow from the research. The policy maker may be more willing to take the risks involved in decisions based on his past experience than risks found in research-based recommendations. The applied scientist may be more often willing to support certain policies than the policy maker, since it is the latter who takes ultimate responsibility for the decision.

In some instances, a given research, however competent, may seem too slender a basis for running the large risk. Thus, a bank or insurance company may hesitate to invest in an interracial housing development despite researches which suggest that the resulting problems can probably be "managed." The economic investment is large; deep-seated public attitudes are involved; once made, the decision cannot be easily modified. In such instances, it would not be expected that research, however sound intellectually, will appreciably modify prevailing policies. Correlatively, when risks are more limited—for example, the decision to introduce a new personnel selection policy or a new advertising campaign—a far-from-conclusive research may affect a decision.

Lack of continuing communication between policy maker and research staff

Once mentioned, this need not be elaborated. The problem is generally recognized and it is likely that data bearing on this problem are abundantly available.

Status of researcher vis-à-vis operating agency

It is possible that the quality of the research does not completely determine its use: the status of the research worker may play a large part. Systematic inquiry into this possibility is indicated.

The foregoing account is far from exhaustive. It does, however, suggest leads for determining how and why applied research does or does not provide a direct mandate for policy and does or does not eventuate in policy-formation. A key set of problems centers in the determinants of this leap from research to practice.

13. Theory and Applied Social Science

Everyone who has read a textbook on scientific method knows the ideally constructed relations between scientific theory and applied research. Basic theory embraces key concepts (variables and constants), postulates, theorems, and laws. Applied science consists simply in ascertaining (a) the variables relevant to the problem in hand, (b) the values of the variables, and (c) in accordance with previous knowledge, setting forth the uniform relationships between these variables.

It will be instructive to discover how often this ideal pattern actually occurs in the application of social science. We anticipate finding that it is the exceptional rather than the typical pattern. In one sense, a major objective of our proposed inquiry is to account for the discrepancies and coincidences between the "ideal pattern" and the "actual pattern" of relations between basic and applied social science.

In the present section, we confine ourselves to some remarks on the role of preliminary conceptualization in applied research. There is no danger that this will be mistaken for a comprehensive discussion.

Conceptualization at work: the "overlooked variable"

Perhaps the most striking role of conceptualization in applied social research is its transformation of practical problems by introducing concepts which refer to variables *overlooked* in the common-sense view of the policy maker. At times, the concept leads to a statement of the problem diametrically opposed to that of the policy maker.

Types of frequently overlooked variables will be ascertained through further inquiry, but a few can be set forth now.

Concept of the definition of the situation. Not all policy makers have the practice of viewing policies from the perspective of others affected by the policy. As a result, they periodically find their decisions leading to a train of unanticipated, and often undesired, consequences.

Case: The policy maker in the field of colonial administration may seek to "educate" the "native" by building and staffing schools for him. He, the administrator, sees this as beneficent activity. Education is a positive value, and he is making education available to the native. He is subsequently shocked by a nativistic reaction; the "ungrateful" natives rebel against this policy. The expert introduces the concept of definition of the situation and of culture differences. He indicates that western education, defined as an asset by the administrator in terms of his cultural values, is defined by the natives as a device for cutting their children off from their traditional tribal values. The key concept brought to bear upon the problem by the expert is the different definition of the "same" situation by members of different cultural groups.

Case: illustrating *a variable below the level of awareness of the administrator.* An industrial manager hopes to achieve better employee morale and higher output by introducing a general rise in wages. He is disturbed when the expected results do not occur. The expert approaches the problem with a clarifying conception: wage *differentials* are of central concern to workers. Previous low morale had been a product of differentials conceived as unfair by some groups of workers; the general rise in wages did not change the differentials.

The concept of a social system. Naïve common sense seldom thinks in terms of total systems of interrelated variables. Behavior is construed as a series of isolated events. Yet many of the untoward consequences of policy decisions stem from the interaction between variables in a system.

As Wesley Mitchell has remarked in this general connection: "When some change in existing arrangements is proposed, our minds [that is, of the economist] fasten immediately upon the effects this change will have upon other factors directly or indirectly, immediately or after a time: we think also about how these consequences will react upon the initial change. . . . *Obvious as the concept of the interdependence of all economic activities seems to us, it is not part of the working equipment of many lawyers, business men, or engineers,* if the able and patriotic dollar-a-year men I have collaborated with are a fair sample" [italics added].

The theoretic by-products of applied research.[8]

In passing, we note two major relations between applied research and theory.

Applied research tests the assumptions underlying theory. As noted earlier (section 11), basic research includes certain assumptions (ceteris paribus) in its abstract formulation of a problem. Since applied research is conceived as a basis for action, and since action must always occur in a *concrete* situation and not under abstractly envisaged conditions, the applied researcher is continuously engaged, *nolens volens,* in testing the assumptions contained in basic theory. This is perhaps a key function of applied research.

Immediate pragmatic success postpones theoretic analysis. Not infrequently, applied research leads to an empirical finding which may be at once successfully applied, although the finding itself is not "understood" (that is, located) in theoretical terms. Thus, it may be found that provision for several rest periods in an industrial plant reduces labor turnover, raises employee morale, and so forth. The plant manager who finds that this program "works" may see no occasion for further research. If the research worker is not theoretically sensitized, he, too, may be content with this "successful" application of an empirical finding. The fact remains that he has not yet identified the critical variable in this result: was it that rest periods reduced fatigue? Or was it, possibly, that the degree of managerial concern with employees' problems (as symbolized by the rest pauses) was the decisive variable? Or, again, was it the part played by employee representatives in arriving at the decision regarding rest periods—in short, the manner in which this policy was introduced—that proved basic? Unless the crucial theoretical variable in the concrete practice of rest periods can be identified, there is no basis for assuming that the same results will be obtained on other occasions. It will be of interest to learn if such practical successes tend to vitiate the continuance of inquiry until the theoretically significant findings have been extracted from the empirical results.[9] It is at

8. Since I have provisionally discussed this in a paper presented to the American Sociological Society in 1946, I shall not comment on it further at this time. "The Bearing of Empirical Research upon the Development of Social Theory," *American Sociological Review* 13 (October 1948): 505–15. More study of actual cases is required to outline the conditions under which applied research in social science leads to theoretical by-products.

9. Since this was written, the 1948 election forecasts have provided the most dramatic recent instance of the danger of operating with wholly empirical and theoretically ungrounded uniformities. In previous national elections it had been found that virtually no net shift in vote-intentions occurred during the last weeks of the political campaign. The extrapolation of this empirical pattern to the 1948

least possible that specific practical successes may invite theoretical failures.[10]

14. Methodology and Applied Social Science

Just as with theory, so with the logic of procedure. The skeletonized version of relations between methodology and applied research found in textbooks is logically impeccable, but not always descriptive of what actually occurs. It will be necessary to review cases in point to determine the respects in which the ideal and actual patterns coincide or differ.

Without attempting any systematic discussion, we raise several questions which require study.

To what extent does the applied scientist's thorough familiarity with certain types of procedures and relative lack of familiarity with others, predetermine the design of the applied research? Do such predispositions toward procedures sometimes deflect attention from more appropriate though less well-known procedures?

Do applied researches more often call for quantitative treatment than do "pure" researches? Is the policy-maker's concern with "how much" and "when" a prod to quantification? What are some of the scientific consequences of this pressure for quantification?

For which types of practical problems has nonquantified case study proved most appropriate?

One has the impression that the practical demands laid upon the applied research worker result in a continuing pressure for improvement of methods. The development of sampling procedure in social science, for example, appears to have been markedly advanced by applied researches in public opinion, market studies, and so on.

---

campaign by the polling organizations led to now familiar and unfortunate consequences.

10. Since this was first written, James Bryant Conant has set forth apposite remarks (in his presidential address before the A.A.A.S., 28 December 1947): "To my mind we need to analyze the present situation, not by attempting to classify the various sciences and their subdivisions into pure and applied science, but by examining closely each separate undertaking. I have suggested in a paper on 'Science and the Practical Arts' that we need to inquire as to the *degree of empiricism* now present in any branch of science. The cases I quoted as examples were classical optics and chemotherapy. In the former *the conceptual scheme employed has wide validity, the degree of empiricism is very low. In the latter the concepts are few and of limited application,* progress toward a new drug is still very much of a 'cut and dry' affair, *the degree of empiricism is high.* . . . I should like to suggest that unless progress is made in reducing the degree of empiricism in any area, the rate of advance of the practical arts connected with that area will be relatively slow and highly capricious" (italics supplied). There remains for us the question of the factors which make for the retention of a high degree of empiricism in much of applied social science, and it is in connection with this question that we make the suggestions found in the text at this point.

We should like to learn whether the applied researcher is subjected to a greater variety of rigorous criticism by diverse "interested parties," leading him perhaps to search out increasingly effective tools of analysis.

An inventory of methodologic by-products of applied social science would be similarly instructive.

In any event, the reciprocal relations between theory and methodology on the one hand, and applied social science on the other should constitute a major focus of inquiry.

5 The Perspectives of Insiders and Outsiders

1972

The sociology of knowledge has long been regarded as a complex and esoteric subject, remote from the urgent problems of contemporary social life. To some of us, it seems quite the other way.[1] Especially in times of great social change, precipitated by acute social conflict and attended by much cultural disorganization, the perspectives provided by the various

A first version of this paper was read on 6 November 1969 to the seminar celebrating the fiftieth anniversary of the department of sociology at the University of Bombay, India. A second version was read at the Centennial Symposium of Loyola University (of Chicago) on 5 January 1970 and at the annual meetings of the Southwestern Sociological Association in Dallas, Texas, on 25 March 1971. The third, and present, version was presented at the annual meetings of the American Sociological Association in Denver, Colorado, 1 September 1971, and was published as "Insiders and Outsiders: A Chapter in the Sociology of Knowledge" in *American Journal of Sociology* 77 (July 1972): 9–47; reprinted with permission. Any errors I have retained after the critical examinations of the paper by Walter Wallace and Harriet Zuckerman are of course entirely my own. Aid from the National Science Foundation is gratefully acknowledged, as is indispensable help of quite another kind provided by Hollon W. Farr, M.D.

1. As witness the spate of recent writings in and on the sociology of knowledge, including far too many to be cited here. Some essential discussions and bibliography are provided by Peter L. Berger and Thomas Luckmann, *The Social Construction of Reality* (Garden City, N.Y.: Doubleday, 1966); Werner Stark, *The Sociology of Knowledge* (London: Routledge & Kegan Paul, 1958); Kurt H. Wolff, "Ernst Grünwald and the Sociology of Knowledge: A Collective Venture in Interpretation," *Journal of the History of the Behavioral Sciences* 1 (1965): 152–64; and James E. Curtis and John W. Petras, *The Sociology of Knowledge* (New York and Washington: Praeger, 1970). The application of the sociology of knowledge to the special case of sociology itself has also burgeoned since 1959 when the Fourth World Congress of Sociology held by the International Sociological Association focused on the social contexts of sociology. See, for prime examples, Alvin W. Gouldner, *The Coming Crisis of Western Sociology* (New York: Basic Books, 1970); Robert W. Friedrichs, *A Sociology of Sociology* (New York: Free Press, 1970); and Edward A. Tiryakian, ed., *The Phenomenon of Sociology* (New York: Appleton-Century-Crofts, 1971).

sociologies of knowledge bear directly upon problems agitating the society. It is then that differences in the values, commitments, and cognitive orientations of conflicting groups become deepened into basic cleavages, both social and cultural. As the society becomes polarized, so do the contending claims to truth. At the extreme, an active and reciprocal distrust between groups finds expression in intellectual perspectives that are no longer located within the same universe of discourse. The more deep-seated the mutual distrust, the more does the argument of the other appear so palpably implausible, even absurd, that one no longer inquires into substance or logical structure to assess its truth claims. Instead, one confronts the other's argument with an entirely different question: how does it happen to be advanced at all? Thought thus becomes altogether functionalized, interpreted only in terms of their presumed social or economic or psychological sources and functions. In the political arena, where the rules of the game often condone and sometimes support the practice, this involves reciprocated attacks on the integrity of the opponent; in the academic forum, where the norms are somewhat more restraining, it leads to reciprocated ideological analyses (which easily decline into innuendo). In both, the process feeds upon and nourishes collective insecurities.[2]

Social Change and Social Thought

This conception of the social sources of the intensified interest in the sociology of knowledge and some of the theoretical difficulties which they foster plainly has the character, understandably typical in the sociology of scientific knowledge, of a self-exemplifying idea. It posits reciprocal connections between thought and society, in particular the social conditions that make for or disrupt a common universe of intellectual discourse within which the most severe disagreements can take place. Michael Polanyi[3] has noted, more perceptively than anyone else I know,[4] how the growth of

2. This passage on the conditions making for intensified interest in the sociology of knowledge and for derivative problems of theoretical analysis in the field has not been written for this occasion. It is largely drawn from my paper which was printed in Georges Gurvitch and Wilbert E. Moore, eds., *Twentieth Century Sociology* (New York: Philosophical Library, 1945 [now out of print]) (and reprinted as chapter 1 in this volume.—Ed.) Since the cognitive orientation of group members and nonmembers has long been a problem of enduring interest to me, I shall have occasion to refer to my writings throughout this paper.

3. See his *Personal Knowledge* (London: Routledge & Kegan Paul, 1958); *The Study of Man* (London: Routledge & Kegan Paul, 1959); *Science, Faith and Society* (Chicago: University of Chicago Press, 1964); and *The Tacit Dimension* (London, Routledge & Kegan Paul 1967).

4. Polanyi's detailed development of this theme over the years represents a basic contribution to the sociology of science by providing a model of the various overlapping cognitive and social structures of intellectual disciplines. John Ziman (*Public Knowledge* [Cambridge: At the University Press, 1968]) has useful observations

knowledge depends upon complex sets of social relations based on a largely institutionalized reciprocity of trust among scholars and scientists. In one of his many passages on this theme, he observes that

> in an ideal free society each person would have perfect access to the truth: to the truth in science, in art, religion, and justice, both in public and private life. But this is not practicable; each person can know directly very little of truth and must trust others for the rest. Indeed, to assure this process of mutual reliance is one of the main functions of society. It follows that such freedom of the mind as can be possessed by men is due to the services of social institutions, which set narrow limits to man's freedom and tend to threaten it even within those limits. The relation is analogous to that between mind and body: to the way in which the performance of mental acts is restricted by limitations and distortions due to the medium which makes these performances possible.[5]

But as cleavages deepen between groups, social strata, or collectivities of whatever kind, the social network of mutual reliance is at best strained and at worst broken. In place of the vigorous but cognitively disciplined mutual checking and rechecking that operates to a significant extent, though never of course totally, within the social institutions of science and scholarship, there develops a strain toward separatism, in the domain of the intellect as in the domain of society. Partly grounded mutual suspicion increasingly substitutes for partly grounded mutual trust. There emerge claims to group-based truth: Insider truths that counter Outsider untruths and Outsider truths that counter Insider untruths.

In our time, vastly evident social change is being initiated and funneled through a variety of social movements. These are formally alike in their objectives of achieving an intensified collective consciousness, a deepened solidarity, and a new or renewed primary or total allegiance of their members to certain social identities, statuses, groups, or collectivities. Inspecting the familiar list of these movements centered on class, race, ethnicity, age, sex, religion, and sexual disposition, we note two other instructive similarities among them. First, the movements are principally formed on the basis of ascribed rather than acquired statuses and identities, with eligibility for inclusion being in terms of who you are rather than what you are (in the sense of status being contingent on role performance). And second, the movements largely involve the public affirmation of pride in statuses and solidarity with collectivities that have long been

along these lines, and Donald T. Campbell ("Ethnocentrism of Disciplines and the Fish-Scale Model of Omniscience," in *Interdisciplinary Relationships in the Social Sciences,* ed. Muzafer Sherif and Carolyn W. Sherif [Chicago: Aldine Press, 1969]) has contributed some typically Campbellian (that is, imaginative and evocative) thinking on the subject, in developing his "fish-scale model" of overlapping disciplines.

5. Polanyi, *Study of Man,* p. 68.

socially downgraded, stigmatized, or otherwise victimized in the social system. As with group affiliations generally, these newly reinforced social identities find expression in affiliative symbols of distinctive speech, bodily appearance, dress, public behavior, and, not least, assumptions and foci of thought.

The Insider Doctrine

Within this context of social change, we come upon the contemporary relevance of a long-standing problem in the sociology of knowledge: the problem of patterned differentials among social groups and strata in access to knowledge. In its strong form, the claim is put forward as a matter of epistemological principle that particular groups in each moment of history have *monopolistic access* to certain kinds of knowledge. In the weaker, more empirical form, the claim holds that some groups have *privileged access,* with other groups also being able to acquire that knowledge for themselves but at greater risk and cost.

Claims of this general sort have been periodically introduced. For one imposing and consequential example, Marx, a progenitor of the sociology of knowledge as of much else in social thought, advanced the claim that after capitalistic society had reached its ultimate phase of development, the strategic location of one social class would enable it to achieve an understanding of the society that was exempt from false consciousness.[6] For another, altogether unimposing but also consequential example involving ascribed rather than achieved status, the Nazi *Gauleiter* of science and learning, Ernest Krieck,[7] expressed an entire ideology in contrasting the access to authentic scientific knowledge by men of unimpeachable Aryan ancestry with the corrupt versions of knowledge accessible to non-Aryans. Krieck could refer without hesitation to "Protestant and Catholic science, German and Jewish science." And, in a special application of the Insider doctrine, the Nazi regime could introduce the new racial category of "white Jews" to refer to those Aryans who had defiled their race by actual or symbolic contact with non-Aryans. Thus, the Nobel Prize physicist, Werner Heisenberg, became the most eminent member of this new race by persisting in his declaration that Einstein's theory of relativity constituted "an obvious basis for further research." While another Nobel

6. Observations on the advantaged position of the proletariat for the perception of historical and social truth are threaded throughout Marx's writings. For some of the crucial passages, see his *Poverty of Philosophy* (Moscow: Foreign Languages Press, n.d.), pp. 125–26, for example. On Marx's thinking along these lines, Georg Lukács, in spite of his own disclaimers in the new introduction to his classic work, *History and Class Consciousness* (1923; reprint ed., Cambridge, Mass.: M.I.T. Press, 1971), remains of fundamental importance; see especially pages 47–81 and 181–209.

7. See his *Nationalpolitische Erziehung* (Leipzig: Armanen Verlag, 1935).

laureate in physics, Johannes Stark, could castigate not only Heisenberg but his other great scientific contemporaries—Planck, von Laue, and Schrödinger—for accepting what Stark described as "the Jewish physics of Einstein."[8]

For our purposes, we need not review the array of elitist doctrines which have maintained that certain groups have, on biological or social grounds, monopolistic or privileged access to new knowledge. Differing in detail, the doctrines are alike in distinguishing between Insider access to knowledge and Outsider exclusion from it.

Social Bases of Insider Doctrine

The ecumenical problem of the interaction between a rapidly changing social structure and the development of Insider and Outsider doctrines is examined here in a doubly parochial fashion. Not only are my observations largely limited to the United States in our time but they are further limited to the implications of doctrines advocated mainly by spokesmen for certain black social movements, since these movements have often come to serve as prototypical for the others (women, youth, homosexuals, and other ethnic collectivities.).

Although Insider doctrines have been intermittently set forth by white elitists through the centuries, white male Insiderism in American sociology during the past generations has largely been of the tacit or de facto rather than doctrinal or principled variety. It has simply taken the form of patterned expectations about the appropriate selection of specialities and of problems for investigation. The handful of Negro sociologists were in large part expected, as a result of social selection and self-selection, to study problems of Negro life and relations between the races just as the handful of women sociologists were expected to study problems of women, principally as these related to marriage and the family.

In contrast to this de facto form of Insiderism, an explicitly doctrinal form has in recent years been put forward most clearly and emphatically by some black intellectuals. In its strong version, the argument holds that, as a matter of social epistemology, *only* black historians can truly understand black history, *only* black ethnologists can understand black culture, *only* black sociologists can understand the social life of blacks, and so on. In the weaker form of the doctrine, some practical concessions are made. With regard to programs of black studies, for example, it is proposed that some white professors of the relevant subjects might be brought in since there are not yet enough black scholars to staff all the proliferating pro-

8. See Merton, *Social Theory and Social Structure,* pp. 538–41; also see chapter 12 of this volume.

grams of study. But as Nathan Hare, the founding publisher of the *Black Scholar,* stated several years ago, this is only on temporary and conditional sufferance: "Any white professors involved in the program would have to be black in spirit in order to last. The same is true for 'Negro' professors."[9] Apart from this kind of limited concession, the Insider doctrine maintains that there is a body of black history, black psychology, black ethnology, and black sociology which can be significantly advanced only by black scholars and social scientists.

In its fundamental character, this represents a major claim in the sociology of knowledge that implies the balkanization of social science, with separate baronies kept exclusively in the hands of Insiders bearing their credentials in the shape of one or another inherited status. Generalizing the specific claim, it would appear to follow that if only black scholars can understand blacks, then only white scholars can understand whites. Generalizing further from race to nation, it would then appear, for example, that only French scholars can understand French society and, of course, that only Americans, not their external critics, can truly understand American society. Once the basic principle is adopted, the list of Insider claims to a monopoly of knowledge becomes indefinitely expansible to all manner of social formations based on ascribed (and, by extension, on some achieved) statuses. It would thus seem to follow that only women can understand women—and men, men. On the same principle, youth alone is capable of understanding youth just as, presumably, only the middle-aged are able to understand their age peers.[10] So, too, as we shift to the hybrid cases of ascribed and acquired statuses in varying mix, on the Insider principle, proletarians alone can understand proletarians, and presumably capitalists, capitalists; only Catholics, Catholics; Jews, Jews; and, to halt the inventory of socially atomized claims to knowledge with a limiting case that on its face would seem to have some merit, it would then plainly follow that only sociologists can possibly understand their fellow sociologists.[11]

9. Nathan Hare as quoted by John H. Bunzel in "Black Studies at San Francisco State," *Public Interest* 13 (Fall 1968): 32.

10. Actually, the case of age status is structurally different from that of other ascribed statuses. For although, even in this time of advanced biotechnology, a few men become transformed into women and vice versa, this remains a comparatively rare instance of the ordinarily ascribed status of sex becoming an achieved status. But in contrast to sex and other ascribed statuses, each successive age status has been experienced by suitably long-lived social scientists (within the limits of their own inexorably advancing age cohorts). On the basis of a dynamic Insider doctrine, then, it might even be argued that older social scientists are better able than very young ones to understand the various other age strata. As context, see the concept of the reenactment of complementary roles in the life cycle of scientists in chapter 22 of this volume.

11. As we shall see, this is a limiting type of case that merges into quite another type, since as a fully acquired status, rather than an ascribed one, that of the soci-

In all these applications, the doctrine of extreme Insiderism represents a new credentialism.[12] This is the credentialism of ascribed status, in which understanding becomes accessible only to the fortunate few or fortunate many who are to the manner born. It thus contrasts with the credentialism of achieved status that characterizes systems of meritocracy.[13]

Extreme Insiderism moves toward a doctrine of *social* solipsism that is isomorphic to *individual* solipsism.[14] In the first, the group or collectivity has a monopoly of knowledge about itself, just as in the second, the individual person has absolute privacy of knowledge about himself (as we recognize in that proverbial and exemplary aching tooth which he and he alone can authentically experience). And like individual solipsists who address themselves to the other minds whose existence their doctrine denies, the group solipsists (as we shall see) have an abiding commitment, not a merely empirical proclivity, to deny in continued practice what they affirm in basic principle.

The Insider doctrine can be put in the vernacular with no great loss in meaning: you have to be one in order to understand one. In less idiomatic language, the doctrine holds that one has monopolistic or privileged access to knowledge, or is excluded from it, by virtue of group membership or social position. For some, the notion appears in the form of a question-begging pun: Insider as Insighter, one endowed with special insight into matters necessarily obscure to others, thus possessed of penetrating discernment. Once adopted, the pun provides a specious solution, but the serious Insider doctrine has its own rationale.

We can quickly pass over the trivial version of that rationale: the argu-

---

ologist (or physician or physicist) presumably presupposes functionally relevant expertise.

12. I am indebted to Harriet Zuckerman for these observations on the new credentialism of ascribed status. The classic source on meritocracy remains Michael Young's *Rise of Meritocracy, 1870–2033* (London: Thames & Hudson, 1958); on the dysfunctions of educational credentialism, see S. M. Miller and Pamela A. Roby, *The Future of Inequality* (New York: Basic Books, 1970), chapter 6.

13. But, as we shall see, when the extreme Insider position is transformed from a doctrine of assumptions-treated-as-established-truth into a set of questions about the distinctive roles of Insiders and Outsiders in intellectual inquiry, there develops a convergence though not coincidence between the assumptions underlying credentials based on ascribed status and credentials based on achieved status. In the one, early socialization in the culture or subculture is taken to provide readier access to certain kinds of understanding; in the other, the component in adult socialization represented by disciplined training in one or another field of learning is taken to provide a higher probability of access to certain other kinds of understanding.

14. As Joseph Agassi ("Privileged Access," *Inquiry* 12 [Winter 1969]: 420–26) reminds us, the term "methodological solipsism" was introduced by Rudolf Carnap to designate the theory of knowledge known as sensationalism: "the doctrine that all knowledge—of the world and of one's own self—derives from sensation." The belief that all one *really* knows is one's subjective experience is sometimes described as the "egocentric predicament."

ment that the Outsider may be incompetent, given to quick and superficial forays into the group or culture under study and even unschooled in its language. That this kind of incompetence can be found is beyond doubt but it holds no principled interest for us. Foolish men (and women) or badly trained men (and women) are to be found everywhere, and anthropologists and sociologists and psychologists and historians engaged in study of groups other than their own surely have their fair share of them.[15] But such cases of special ineptitude do not bear on the Insider *principle*. It is not merely that Outsiders have their share of incompetents. The Insider principle does not refer to stupidly designed and stupidly executed inquiries that happen to be made by stupid Outsiders; it advances a far more fundamental position. According to that position, the Outsider, no matter how careful and talented, is excluded in principle from gaining access to the social and cultural truth.

In short, the doctrine holds that the Outsider has a structurally imposed incapacity to comprehend alien groups, statuses, cultures, and societies. Unlike the Insider, the Outsider has neither been socialized in the group nor has engaged in the run of experience that makes up its life, and therefore cannot have the direct, intuitive sensibility that alone makes empathic understanding possible. Only through continued socialization in the life of a group can one become fully aware of its symbolisms and socially shared realities: only so can one understand the fine-grained meanings of behavior, feelings, and values; only so can one decipher the unwritten grammar of conduct and the nuances of cultural idiom. Or, to take a specific expression of this thesis by Ralph W. Conant (1968): "Whites are not and never will be as sensitive to the black community precisely because they are not part of that community." Correlatively, Abd-l Hakimu Ibn Alkalimat (Gerald McWorter) draws a sharp contrast between the concepts of "a black social science" and "a white social science."[16]

A somewhat less stringent version of the doctrine maintains only that Insider and Outsider scholars have significantly different foci of interest. The argument goes somewhat as follows. The Insiders, sharing the deepest concerns of the group or at the least being thoroughly aware of them, will so direct their inquiries as to have them be relevant to those concerns. So, too, the Outsiders will inquire into problems relevant to the distinctive values and interests which they share with members of *their* group. But

15. As I have noted in the first edition of this paper, the social scientists of India, for one example, have long suffered the slings and arrows of outrageously unprepared and altogether exogenous social scientists engaging in swift, superficial inquiries into matters Indian; see "Insiders and Outsiders," in *Essays on Modernization of Underdeveloped Societies*, ed. A. R. Desai (Bombay: Thacker, 1971).

16. See "The Ideology of Black Social Science," *Black Scholar* 1 (December 1969): 35.

these are bound to differ from those of the group under study if only because the Outsiders occupy different places in the social structure.

This is a hypothesis which has the not unattractive quality of being readily amenable to empirical investigation. It should be possible to compare the spectrum of research problems about, say, the black population in the country that have been investigated by black sociologists and by white ones, or say, the spectrum of problems about women that have been investigated by female sociologists and by male ones, in order to find out whether the foci of attention in fact differ and if so, to what degree and in which respects. The only inquiry of this kind I happen to know of was published more than a quarter-century ago. William Fontaine[17] found that Negro scholars tended to adopt analytical rather than morphological categories in their study of behavior, that they emphasized environmental rather than biological determinants of that behavior, and tended to make use of strikingly dramatic rather than representative data. All this was ascribed to a caste-induced resentment among Negro scholars. But since this lone study failed to examine the frequency of subjects, types of interpretation, and uses of data among a comparable sample of white scholars at the time, the findings are somewhat less than compelling. All the same, the questions it addressed remain. For there is theoretical reason to suppose that the foci of research adopted by Insiders and Outsiders and perhaps their categories of analysis as well will tend to differ. At least, Max Weber's notion of *Wertbeziehung* suggests that differing social locations, with their distinctive interests and values, will affect the selection of problems for investigation.[18]

Unlike the stringent version of the doctrine which maintains that Insiders and Outsiders must arrive at different (and presumably incompatible) findings and interpretations even when they do examine the same problems, this weaker version argues only that they will not deal with the same questions and so will simply talk past one another. With the two versions combined, the extended version of the Insider doctrine can also be put in the vernacular: one must not only be one in order to understand one; one must be one in order to understand what is most worth understanding.

Clearly, the social epistemological doctrine of the Insider links up with what Sumner long ago defined as ethnocentrism: "the technical name for [the] view of things in which one's own group is the center of everything, and all others are scaled and rated with reference to it." Sumner then goes on to include as a component of ethnocentrism, rather than as

17. " 'Social Determination' in the Writings of Negro Scholars," *American Journal of Sociology* 49 (Winter 1944): 302–15.

18. See Weber's *Gesammelte Aufsätze zur Wissenschaftslehre* (1922; reprint ed., Tubingen: J. C. B. Mohr [P. Siebeck], 1951), pp. 146–214.

a frequent correlate of it (thus robbing his idea of some of its potential analytical power), the belief that one's group is superior to all cognate groups: "each group nourishes its own pride and vanity, boasts itself superior, exalts its own divinities, and looks with contempt on outsiders."[19] For although the practice of seeing one's own group as the center of things is empirically correlated with a belief in its superiority, centrality and superiority need to be kept analytically distinct in order to deal with patterns of alienation from one's membership group and contempt for it.[20]

Supplementing the abundance of historical and ethnological evidence of the empirical tendency for belief in one's group or collectivity as superior to all cognate groups or collectivities—whether nation, class, race, region, or organization—is a recent batch of studies of what Theodore Caplow, in *Principles of Organization*, has called the aggrandizement effect: the distortion upward of the prestige of an organization by its members. Caplow examined thirty-three kinds of organizations—ranging from dance studios to Protestant and Catholic churches, from skid row missions to big banks, and from advertising agencies to university departments—and found that members overestimated the prestige of their organization some "eight times as often as they underestimated it" (when compared with judgments by Outsiders). More in point for us, while members tended to disagree with Outsiders about the standing of their own organization, they tended to agree with them about the prestige of the other organizations in the same set. These findings can be taken as something of a sociological

19. William Graham Sumner, *Folkways* (Boston: Ginn & Co., 1907), p. 13.

20. By introducing their useful term "xenocentrism" to refer to both basic *and* favorable orientations to groups other than one's own, Donald P. Kent and Robert G. Burnight ("Group Centrism in Complex Societies," *American Journal of Sociology* 57 [November 1951]: 256–59) have retained Sumner's unuseful practice of prematurely combining centrality and evaluation in the one concept rather than keeping them analytically distinct. The analytical distinction can be captured terminologically by treating "xenocentrism" as the generic term, with the analytically distinct components of favorable orientation to nonmembership groups (as with the orientation of many white middle-class Americans toward blacks) being registered in the term "xenophilia" and the unfavorable orientation by Pareto's term "xenophobia." The growing theoretical interest in nonmembership reference groups (a concept implying a type of Outsider)—see Herbert H. Hyman, "Reference Groups," in the *International Encyclopedia of the Social Sciences* (New York: Macmillan and Free Press, 1968), vol. 13; and Robert K. Merton and Alice Kitt Rossi, "Contributions to the Theory of Reference Group Behavior," in *Continuities in Social Research,* ed. R. K. Merton and P. F. Lazarsfeld (New York: Free Press, 1950 [now out of print and reprinted in Merton, *Social Theory and Social Structure*])—and the intensified spread of both ethnocentrism and xenocentrism in our times have given the term xenocentrism greater relevance than ever, and yet, for obscure reasons, it has remained largely sequestered in the pages of the *American Journal of Sociology,* where it first appeared twenty years ago. Caplow and Horton are the only ones I know to have made good use of the term, but their unaccustomed behavior only emphasizes its more general neglect; see Theodore Caplow, *Principles of Organization* (New York: Harcourt Brace Jovanovich, 1964), p. 216, and Paul B. Horton, *Sociology and the Health Sciences* (New York: McGraw-Hill, 1965).

allegory. In these matters at least, the judgments of "Insiders" are best trusted when they assess groups other than their own; that is, when members of groups judge as Outsiders rather than as Insiders.

Findings of this sort of course do not testify that ethnocentrism and its frequent spiritual correlate, xenophobia, fear and hatred of the alien, are incorrigible. They do, however, remind us of the widespread tendency to glorify the ingroup, sometimes to that degree in which it qualifies as chauvinism: the extreme, blind, and often bellicose extolling of one's group, status, or collectivity. We need not abandon the useful concept "chauvinism" merely because it has lately become adopted as a vogue word, blunted in meaning through indiscriminate use as a rhetorical weapon in intergroup conflict. Nor need we continue to confine the scope of the concept, as it was confined in its origins and later by Lasswell[21] in his short, incisive discussion of it, to the special case of the state or nation. The concept can be usefully, not tendentiously, extended to designate the extreme glorification of *any* social formation.

Chauvinism finds fullest ideological expression when groups are subject to the stress of acute conflict. Under the stress of war, for example, scientists have been known to violate the values and norms of universalism in which they were socialized, allowing their status as nationals to dominate over their status as scientists. Thus, at the outset of World War I, almost a hundred German scholars and scientists—including many of the first rank, such as Brentano, Ehrlich, Haber, Eduard Meyer, Ostwald, Planck, and Schmoller—could bring themselves to issue a manifesto that impugned the contributions of the enemy to science, charging them with nationalistic bias, logrolling, intellectual dishonesty, and, when you came right down to it, the absence of truly creative capacity. The English and French scientists were not far behind in advertising their own brand of chauvinism.[22]

Ethnocentrism, then, is not a historical constant. It becomes intensified under specifiable conditions of acute social conflict. When a nation, race, ethnic group, or other powerful collectivity has long extolled its own admirable qualities and, expressly or by implication, deprecated the qualities of others, it invites and provides the potential for counter-ethnocentrism. And when a once largely powerless collectivity acquires a socially vali-

21. "Chauvinism," in the *Encyclopedia of the Social Sciences* (New York: Macmillan, 1937), 3:361.

22. Current claims of Insiderism still have a distance to go, in the academic if not the political forum, to match the chauvinistic claims of those days. For collections of such documents, see Gabriel Pettit and Maurice Leudet, *Les allemands et la science* (Paris 1916); Pierre Duhem, *La science allemande* (Paris: Hermann, 1915); Hermann Kellermann, *Der Krieg der Geister* (Weimar, 1915); and Karl Kherkhof, *Der Krieg gegen die Deutsche Wissenschaft* (Halle, 1933). Also see chapter 13 of this volume.

dated sense of growing power, its members experience an intensified need for self-affirmation. Under such circumstances, collective self-glorification, found in some measure among all groups, becomes a predictable and intensified counter-response to long-standing belittlement from without.[23]

So it is that, in the United States, the centuries-long institutionalized premise that "white (and for some, presumably only white) is true and good and beautiful" induces, under conditions of revolutionary change, the counterpremise that "black (and for some, presumably only black) is true and good and beautiful." And just as the social system has for centuries operated on the tacit or explicit premise that in cases of conflict between whites and blacks, the whites are presumptively right, so there now develops the counterpremise, finding easy confirmation in the long history of injustice visited upon American Negroes, that in such cases of conflict today, the blacks are presumptively right.

What is being proposed here is that the epistemological and ontological claims of the Insider to monopolistic or privileged access to social truth develop under particular social and historical conditions. Social groups or strata on the way up develop a revolutionary élan. The new thrust to a larger share of power and control over their social and political environment finds various expressions, among them claims to a unique access to knowledge about their history, culture, and social life.

On this interpretation, we can understand why this Insider doctrine does not argue for a Black Physics, Black Chemistry, Black Biology, or Black Technology. For the new will to control their fate deals with the social environment, not the environment of nature. There is, moreover, nothing in the segregated life experience of Negroes that is said to sensitize them to the subject matters and problematics of the physical and life sciences. An Insider doctrine would have to forge genetic assumptions about racial modes of thought in order to claim, as in the case of the Nazi version it did claim, monopolistic or privileged access to knowledge in those fields of science. But the black Insider doctrine adopts an essentially social-environmental rationale, not a biologically genetic one.

The social process underlying the emergence of Insider doctrine is reasonably clear. Polarization in the social structure becomes reflected in the polarization of claims in the cognitive and ideological domain, as groups or collectivities seek to capture what Heidegger called the "public interpretation of reality."[24] With varying degrees of intent, groups in

23. This is not a prediction after the fact. E. Franklin Frazier repeatedly made the general point, and I have examined this pattern in connection with the self-fulfilling prophecy; see Frazier, *The Negro in the United States* (New York: Macmillan, 1949), and *Black Bourgeoisie* (New York: Free Press, 1957); and Merton, *Social Theory and Social Structure*, p. 485.

24. Martin Heidegger, *Sein und Zeit* (Halle: Max Niemeyer, 1927), as cited and discussed by Karl Mannheim in *Essays on the Sociology of Knowledge* (New York: Oxford University Press, 1952), pp. 196ff.

conflict want to make their interpretation the prevailing one of how things were and are and will be. The critical test occurs when the interpretation moves beyond the boundaries of the ingroup to be accepted by Outsiders. At the extreme, it then gives rise, through identifiable processes of reference-group behavior, to the familiar case of the converted Outsider validating himself, in his own eyes and in those of others, by becoming even more zealous than the Insiders in adhering to the doctrine of the group with which he wants to identify himself, if only symbolically.[25] He then becomes more royalist than the king, more papist than the pope, or, as their creator had Buck Mulligan say of Leopold Bloom, "Greeker than the Greeks." Some white social scientists, for example, vicariously and personally guilt-ridden over centuries of white racism, are prepared to outdo the claims of the group they would symbolically join. They are ready even to surrender their hard-won expert knowledge if the Insider doctrine seems to require it. This type of response was perhaps epitomized in a televised educational program in which the white curator of African ethnology at a major museum engaged in discussion with a black who, as it happens, had had no prolonged ethnological training. All the same, at a crucial juncture in the public conversation, the distinguished ethnologist could be heard to say: "I realize, of course, that I cannot begin to understand the black experience, in Africa or America, as you can. Won't you tell our audience about it?" Here, in the spontaneity of an unrehearsed public discussion, the Insider doctrine has indeed become the public interpretation of reality.

The black Insider doctrine links up with the historically developing social structure in still another way. The dominant social institutions in this country have long treated the racial identity of individuals as actually if not doctrinally relevant to all manner of situations in every sphere of life. For generations, neither blacks nor whites, though with notably differing consequences, were permitted to forget their race. This treatment of a social status (or identity) as relevant when intrinsically it is functionally irrelevant constitutes the very core of social discrimination. As the once firmly rooted systems of discriminatory institutions and prejudicial ideology began to lose their hold, this meant that increasingly many judged the worth of ideas on their merits, not in terms of their racial pedigree.

What the Insider doctrine of the most militant blacks proposes on the level of social structure is to adopt the salience of racial identity in every sort of role and situation, a practice long imposed upon the American Negro, and to make that identity a total commitment issuing from within the group rather than one imposed upon it from without. By thus affirming the universal saliency of race and by redefining race as an abiding source of pride rather than stigma, the Insider doctrine in effect models itself after doctrine long maintained by white racists.

25. Merton, *Social Theory and Social Structure*, pp. 405–6.

Neither this aspect of the Insider doctrine nor the statement on its implications is at all new. Almost a century ago, Frederick Douglass hinged his observations along these lines on the distinction between collective and individual self-images based on ascribed and achieved status:

> One of the few errors to which we are clinging most persistently and, as I think, most mischievously has come into great prominence of late. It is the cultivation and stimulation among us of a sentiment which we are pleased to call race pride. I find it in all our books, papers, and speeches. For my part I see no superiority or inferiority in race or color. Neither the one nor the other is a proper source of pride or complacency. Our race and color are not of our own choosing. We have no volition in the case one way or another. The only excuse for pride in individuals or races is in the fact of their own achievements. . . . I see no benefit to be derived from this everlasting exhortation of speakers and writers among us to the cultivation of race pride. On the contrary, I see in it a positive evil. It is building on a false foundation. Besides, what is the thing we are fighting against, and what are we fighting for in this country? What is the mountain devil, the lion in the way of our progress? What is it, but American race pride; an assumption of superiority upon the ground of race and color? Do we not know that every argument we make, and every pretension we set up in favor of race pride is giving the enemy a stick to break over our heads?[26]

In rejecting racial chauvinism, Douglass addressed the normative rather than the cognitive aspect of Insiderism. The call to total commitment requiring one collective loyalty to be unquestionably paramount is most often heard when the collectivity is engaged in severe conflict with others. Just as conditions of war between nations have long produced hyperpatriotism among national ethnocentrics, so current intergroup conflicts have produced a strain toward hyperloyalty among racial or sex or age or religious ethnocentrics. Total commitment easily slides from the solidarity doctrine of "our group, right or wrong" to the morally and cognitively preemptive doctrine of "our group, always right, never wrong."

Turning from the normative aspect, with its ideology exhorting prime loyalty to this or that group, to the cognitive, specifically epistemological aspect, we note that the Insider doctrine presupposes a particular imagery of social structure.

Social Structure of Insiders and Outsiders

From the discussion thus far, it should be evident that I adopt a structural conception of Insiders and Outsiders. In this conception, Insiders are the

26. From "The Nation's Problem," a speech delivered in 1889 before the Bethel Literary and Historical Society in Washington, D.C., and now published in *Negro Social and Political Thought*, ed. Howard Brotz (New York: Basic Books, 1966).

members of specified groups and collectivities or occupants of specified social statuses; Outsiders are the nonmembers.[27] This structural concept comes closer to Sumner's usage in his *Folkways* than to various meanings assigned the Outsider by Nietzsche, Kierkegaard, Sartre, Camus, or, for that matter, by Colin Wilson, just as, to come nearer home, it differs from the usages adopted by Riesman, Denny, and Glazer, Price or Howard S. Becker.[28] That is to say, Insiders and Outsiders are here defined as categories in social structure, not as inside dopesters or the specially initiated possessors of esoteric information on the one hand and as social-psychological types marked by alienation, rootlessness, or rule breaking, on the other.

In structural terms, we are all, of course, both Insiders and Outsiders, members of some groups and, sometimes derivatively, not of others; occupants of certain statuses which thereby exclude us from occupying other cognate statuses. Obvious as this basic fact of social structure is, its implications for Insider and Outsider epistemological doctrines are apparently not nearly as obvious. Else, these doctrines would not presuppose, as they typically do, that human beings in socially differentiated societies can be sufficiently located in terms of a single social status, category, or group affiliation—black or white, men or women, under thirty or older—or of several such categories, taken seriatim rather than conjointly. This neglects the crucial fact of social structure that individuals have not a single status but a status set: a complement of variously interrelated statuses which interact to affect both their behavior and perspectives.

The structural fact of status sets, in contrast to statuses taken one at a time, introduces severe theoretical problems for total Insider (and Outsider) doctrines of social epistemology. The array of status sets in a population means that aggregates of individuals share some statuses and not others; or, to put this in context, that they typically confront one another simultaneously as Insiders and Outsiders. Thus, if only whites can understand whites and blacks, blacks, and only men can understand men, and women, women, this gives rise to the paradox which severely limits both premises: for it then turns out, by implication, that some Insiders are excluded from understanding other Insiders with white women being condemned not to

27. This is not the place to go into the theoretical problems of identifying the boundaries of groups, the criteria of group membership, and the consequent varieties of members and nonmembers. For an introduction to the complexities of these concepts, see Merton, *Social Theory and Social Structure,* pp. 338–54, 405–7.

28. Colin Wilson, *The Outsider* (Boston: Houghton Mifflin, 1956); David Riesman, Reuel Denny, and Nathan Glazer, *The Lonely Crowd* (New Haven: Yale University Press, 1950); Don K. Price, *The Scientific Estate* (Cambridge: Harvard University Press, 1965), pp. 83–84; and Howard S. Becker, *Outsiders: Studies in the Sociology of Deviance* (New York: Free Press, 1963).

understand white men, and black men, not to understand black women,[29] and so through the various combinations of status subsets.

Structural analysis in terms of shared and mutually exclusive status sets will surely not be mistaken either as advocating divisions within the ranks of collectivities defined by a single prime criterion or as predicting that such collectivities cannot unite on many issues, despite their internal divisions. Such analysis only indicates the bases of social divisions that stand in the way of enduring unity of any of the collectivities and so must be coped with, divisions that are not easily overcome as new issues activate statuses with diverse and often conflicting interests. Thus, the obstacles to a union of women in England and North Ireland resulting from national, political, and religious differences between them are no less formidable than the obstacles, noted by Marx, confronting the union of English and Irish proletarians. So, too, women's liberation movements seeking unity in the United States find themselves periodically contending with the divisions between blacks and whites within their ranks, just as black liberation movements seeking unity find themselves periodically contending with the divisions between men and liberated women within their ranks.[30]

The problem of achieving unity in large social movements based on any one status when its members are differentiated by crosscutting status sets is epitomized in these words about women's liberation by a black woman where identification with race is dominant: "Of course there have been women who have been able to think better than they've been trained and have produced the canon of literature fondly referred to as 'feminist literature': Anais Nin, Simone de Beauvoir, Doris Lessing, Betty Friedan, etc. And the question for us arises: how relevant are the truths, the experiences, the findings of white women to Black women? Are women after all simply women? I don't know that our priorities are the same, that our concerns and methods are the same, or even similar enough so that we can afford to depend on this new field of experts (white, female). It is rather obvious that we do not. It is obvious that we are turning to each other."[31]

29. The conflicts periodically reported by black women—for example, the debate between Mary Mebane [Liza] and Margaret Sloan (in defense of Gloria Steinem)—between identification with black liberation and the women's liberation movement, reflect this sociological fact of cross-cutting status sets. The problem of coping with these structurally induced conflicts is epitomized in Margaret Sloan's "realization that I was going to help the brothers realize that as black women we cannot allow black men to do [to] us what white men have been doing to their women all these years" ("What We Should Be Doing, Sister," *New York Times,* 8 December 1971, Op-Ed).

30. See Shirley Chisholm, "Racism and Anti-Feminism," *Black Scholar* 1 (Jan.–Feb. 1970): 40–45; and Linda La Rue, "The Black Movement and Women's Liberation," *Black Scholar* 1 (May 1970): 36–42.

31. Toni Cade, ed., *The Black Woman: An Anthology* (New York: New American Library, 1970), p. 9.

Correlatively, the following passage epitomizes the way in which internal differentiation works against unity of the black liberation movement where dominant identification with sex status is reinforced by further educational differentiation:

Seems to me the Brother does us all a great disservice by telling her to fight the man with the womb. Better to fight with the gun and the mind. . . . The all too breezy no-pill/have-kids/mess-up-the-man's-plan notion these comic-book-loving Sisters find so exciting is very seductive because it's a clear-cut and easy thing for her to do for the cause since it nourishes her sense of martyrdom. If the thing is numbers merely, what the hell. But if we are talking about revolution, creating an army for today and tomorrow, I think the Brothers who've been screaming these past years had better go do their homework.[32]

The internal differentiation of collectivities based on a single status thus provides structural bases for diverse and often conflicting intellectual and moral perspectives within such collectivities. Differences of religion or age or class or occupation work to divide what similarities of race or sex or nationality work to unite. That is why social movements of every variety that strive for unity—whether they are establishmentarian movements whipped up by chauvinistic nationals in time of war or antiestablishmentarian movements designed to undo institutionalized injustice—press for total commitments in which all other loyalties are to be subordinated, on demand, to the dominant one.

This symptomatic exercise in status-set analysis may be enough to indicate that the idiomatic expression of total Insider doctrine—one must be one in order to understand one—is deceptively simple and sociologically fallacious (just as we shall see is the case with the total Outsider doctrine). For, from the sociological perspective of the status set, "one" is not a man *or* a black *or* an adolescent *or* a Protestant, *or* self-defined and socially defined as middle class, and so on. Sociologically, "one" is, of course, all of these and, depending on the size of the status set, much more. Furthermore, as Simmel[33] taught us long ago, the individuality of human beings can be sociologically derived from social differentiation and not only psychologically derived from intrapsychic processes. Thus, the greater the number and variety of group affiliations and statuses distributed among individuals in a society, the smaller, on the average, the number of individuals having precisely the same social configuration.

Following out the implications of this structural observation, we note that, on its own assumptions, the total Insider doctrine should hold only

32. Ibid., pp. 167–68.

33. Georg Simmel, *Soziologie* (Leipzig: Duncker und Humblot, 1908), pp. 403–54; see also Lewis A. Coser, *Georg Simmel* (Englewood Cliffs, N.J.: Prentice-Hall, 1965), pp. 18–20.

for highly fragmented small aggregates sharing the same status sets. Even a truncated status set involving only three affiliations—WASPs, for example—would greatly reduce the number of people who, under the Insider principle, would be able to understand their fellows (WASPs). The numbers rapidly decline as we attend to more of the shared status sets by including such social categories as sex, age, class, occupation, and so on, toward the limiting case in which the unique occupant of a highly complex status set is alone qualified to achieve an understanding of self. The tendency toward such extreme social atomization is of course damped by differences in the significance of statuses which vary in degrees of dominance, saliency, and centrality.[34] As a result, the fragmentation of the capacity for understanding that is implied in the total Insider doctrine will not empirically reach this extreme. The structural analysis in terms of status sets, rather than in the fictional terms of individuals being identified in terms of single statuses, serves only to push the logic of Insiderism to its ultimate methodological solipsism.

The fact of structural and institutional differentiation has other kinds of implications for the effort to translate the Insider claim to solidarity into an Insider epistemology. Since we all occupy various statuses and have group affiliations of varying significance to us, since, in short, we individually link up with the differentiated society through our status sets, this runs counter to the abiding and exclusive primacy of any one group affiliation. Differing situations activate different statuses which then and there dominate over the rival claims of other statuses.

This aspect of the dynamics of status sets can also be examined from the standpoint of the differing margins of functional autonomy possessed by various social institutions and other social subsystems. Each significant affiliation exacts loyalty to values, standards, and norms governing the given institutional domain, whether religion, science, or economy. Sociological thinkers such as Marx and Sorokin, so wide apart in many of their other assumptions, agree in assigning a margin of autonomy to the sphere of knowledge even as they posit their respective social, economic, or cultural determinants of it. The alter ego of Marx, for example, declares the partial autonomy of spheres of thought in a well-known passage that bears repetition here:

34. This is not the place to summarize an analysis of the dynamics of status sets that takes up variation in key statuses (dominant, central, salient) and the conditions under which various statuses tend to be activated, along lines developed in unpublished lectures by Merton (1955–71). For pertinent uses of these conceptions in the dynamics of status sets, particularly with regard to functionally irrelevant statuses, see Cynthia Epstein, *Woman's Place: Options and Limits in Professional Careers* (Berkeley: University of California Press, 1970), esp. chap. 3.

According to the materialist conception of history the determining element in history is *ultimately* the production and reproduction in real life. More than this neither Marx nor I have ever asserted. If therefore somebody twists this into the statement that the economic element is the *only* determining one, he transforms it into a meaningless, abstract and absurd phrase. The economic situation is the basis, but the various elements of the superstructure—political forms of the class struggle and its consequences, constitutions established by the victorious class after a successful battle, etc.—forms of law—and then even the reflexes of all these actual struggles in the brains of the combatants: political, legal, philosophical theories, religious ideas and their further development into systems of dogma—also exercise their influence upon the course of the historical struggles and in many cases preponderate in determining their *form*. There is an interaction of all these elements in which . . . the economic movement finally asserts itself as necessary. Otherwise the application of the theory to any period of history one chooses would be easier than the solution of a simple equation of the first degree.[35]

We can see structural differentiation and institutional autonomy at work in current responses of scholars to the extreme Insider doctrine. They reject the monopolistic doctrine of the Insider that calls for total ideological loyalty in which efforts to achieve scholarly detachment and objectivity become redefined as renegadism just as ideological reinforcement of collective self-esteem becomes redefined as the higher objectivity. It is here, to continue with our case in point, that Negro scholars who retain their double loyalty—to the race and to the values and norms of scholarship—part company with the all-encompassing loyalty demanded by the Insider doctrine. Martin Kilson, for example, repudiates certain aspects of the doctrine and expresses his commitment to both the institutionalized values of scholarship and to the black community in these words:

I am opposed to proposals to make Afro-American studies into a platform for a particular ideological group, and to restrict these studies to Negro *students and teachers*. For, and we must be frank about this, what this amounts to is racism in reverse—black racism. I am certainly convinced that it is important for the Negro to know of his past—of his ancestors, of their strengths and weaknesses—and they should respect this knowledge, when it warrants respect, and they should question it and criticize it, when it deserves

35. Friedrich Engels to Joseph Bloch, 21 September 1890, in Karl Marx, *Selected Works*, ed. V. Adoratsky, 2 vols. ([1890]; Moscow: Cooperative Publishing Society 1936), 1: 381, see also p. 392. For a detailed discussion of the partial autonomy of subsystems in the conceptions of Marx and Sorokin, see chapters 1 and 6 of this volume. On the general notion of functional autonomy as advanced by Gordon W. Allport in psychology, see the discussion and references in Merton, *Social Theory and Social Structure*, pp. 15–16; on functional autonomy in sociology, see Alvin W. Gouldner, "Reciprocity and Autonomy in Functional Theory," in *Symposium on Social Theory*, ed. L. Z. Gross (Evanston, Ill.: Row, Peterson, 1958), and "Organizational Analysis," in *Sociology Today*, ed. Robert K. Merton, Leonard Broom, and L. S. Cottrell, Jr. (New York: Basic Books, 1959).

criticism. But it is of no advantage to a mature and critical understanding or appreciation of one's heritage if you approach that heritage with the assumption that it is intrinsically good and noble, and intrinsically superior to the heritage of other peoples. That is, after all, what white racists have done; and none of my militant friends in the black studies movement have convinced me that racist thought is any less vulgar and degenerate because it is used by black men. . . . What I am suggesting here is that the serious study of the heritage of any people will produce a curious mixture of things to be proud of, things to criticize and even despise and things to be perpetually ambivalent toward. And this is as it should be: only an ideologically oriented Afro-American studies program, seeking to propagate a packaged view of the black heritage, would fail to evoke in a student the curious yet fascinating mixture of pride, criticism and ambivalence which I think *is, or ought to be the product of serious intellectual and academic activity.*[36]

Along with the faults of neglecting the implications of structural differentiation, status sets, and institutional autonomy, the Insider (and comparable Outsider) doctrine has the further fault of assuming, in its claims of monopolistic or highly privileged status-based access to knowledge, that social position wholly determines cognitive perspectives. In doing so, it affords yet another example of the ease with which truths can decline into error merely by being extended well beyond the limits within which they have been found to hold. (There *can* be too much of a good thing.)

A long-standing conception shared by various "schools" of sociological thought holds that differences in the social location of individuals and groups tend to involve differences in their interests and value orientations (as well as the sharing of some interests and values with others). Certain traditions in the sociology of knowledge have gone on to assume that these structurally patterned differences should involve, on the *average,* patterned differences in perceptions and perspectives. And these, so the convergent traditions hold—their convergence being often obscured by diversity in vocabulary rather than in basic concept—should make for discernible differences, on the average, in the definitions of problems for inquiry and in the types of hypotheses taken as points of departure. So far, so good. The evidence is far from in, since it has also been a tradition in the sociology of scientific knowledge during the greater part of this century to prefer speculative theory to empirical inquiry. But the idea, which can be taken as a general orientation guiding such inquiry, is greatly transformed in Insider doctrine.

For one thing, that doctrine assumes total coincidence between social position and individual perspectives. It thus exaggerates into error the conception of structural analysis which maintains that there is a *tendency for, not a full determination of,* socially patterned differences in the per-

36. Martin Kilson, "Black Studies Movement: A Plea for Perspective," *Crisis* 76 (October 1969): 329–30, italics added.

spectives, preferences, and behavior of people variously located in the social structure. The theoretical emphasis on tendency, as distinct from total uniformity, is basic, not casual or niggling. It provides for a range of variability in perspective and behavior among members of the same groups or occupants of the same status (differences which, as we have seen, are ascribable to social as well as psychological differentiation). At the same time, this structural conception also provides for patterned differences, *on the whole,* between the perspectives of members of different groups or occupants of different statuses. Structural analysis thus avoids what Dennis Wrong has aptly described as "the oversocialized conception of man in modern sociology."[37]

Important as such allowance for individual variability is for general structural theory, it has particular significance for a sociological perspective on the life of the mind and the advancement of science and learning. For it is precisely the individual differences among scientists and scholars that are often central to the development of the discipline. They often involve the differences between good scholarship and bad; between imaginative contributions to science and pedestrian ones; between the consequential ideas and the stillborn ones. In arguing for the monopolistic access to knowledge, Insider doctrine can make no provision for individual variability that extends beyond the boundaries of the ingroup which alone can develop sound and fruitful ideas.

Insofar as Insider doctrine treats ascribed rather than achieved statuses as central in forming perspectives, it adopts a static orientation. For with

37. Dennis Wrong, "The Oversocialized Conception of Man in Modern Sociology," *American Sociological Review* 26 (April 1961): 183–93. Wrong's paper is an important formulation of the theoretical fault involved in identifying structural position with individual behavior. But, in some cases, he is preaching to the long-since converted. It is a tenet in some forms of structural analysis that differences in social location *make for* patterned differences in perspectives and behavior *between* groups while still allowing for a range of variability *within* groups and thus, in structurally proximate groups, for considerably overlapping ranges of behavior and perspective. On the general orientation of structural analysis in sociology, see Filippo Barbano, "Social Structures and Social Functions: The Emancipation of Structural Analysis in Sociology," *Inquiry* 11 (1968): 40–84; for some specific terminological clues to the fundamental distinction between social position and actual behavior or perspective as this is incorporated in structural analysis, see Merton, *Social Theory and Social Structure,* pp. 175–278 passim, for the key theoretical expressions that *"structures exert pressures"* and structures "tend" to generate perspectives and behaviors. For specific examples: "people in the various occupations *tend* to take different parts in the society, to have different shares in the exercise of power, both acknowledged and unacknowledged, and to *see* the world differently" (ibid., p. 180); "Our primary aim is to discover how some *social structures exert a definite pressure upon certain persons in the society to engage in nonconforming rather than conforming conduct.* If we can locate groups peculiarly subject to such pressures, we should expect to find fairly high *rates* of deviant behavior in those groups" (ibid., p. 186). And for immediate rather than general theoretical bearing on the specific problems here under review, see footnote 18 in chapter 14 of this volume.

the glaring exception of age status itself, ascribed statuses are generally retained throughout the life span. Yet sociologically considered, there is nothing fixed about the boundaries separating Insiders from Outsiders. As situations involving different values arise, different statuses are activated and the lines of separation shift. Thus, for a large number of white Americans, Joe Louis was a member of an outgroup. But when Louis defeated the Nazified Max Schmeling, many of the same white Americans promptly redefined him as a member of the (national) ingroup. National self-esteem took precedence over racial separatism. That this sort of drama in which changing situations activate differing statuses in the status set is played out in the domain of the intellect as well is the point of Einstein's ironic observation in an address at the Sorbonne: "If my theory of relativity is proven successful, Germany will claim me as a German and France will declare that I am a citizen of the world. Should my theory prove untrue, France will say that I am a German and Germany will declare that I am a Jew."[38]

Like earlier conceptions in the sociology of knowledge, recent Insider doctrines maintain that, in the end, it is a special category of Insider—a category that generally manages to include the proponent of the doctrine—that has sole or privileged access to knowledge. Mannheim,[39] for example, found a structural warranty for the validity of social thought in the "classless position" of the "socially unattached intellectuals" (*sozialfreischwebende Intelligenz*). In his view, these intellectuals can comprehend the conflicting tendencies of the time since, among other things, they are "recruited from constantly varying social strata and life-situations." (This is more than a little reminiscent of the argument in the *Communist Manifesto* which emphasizes that "the proletariat is recruited from all classes of the population.")[40] Without stretching this argument to the breaking point, it can be said that Mannheim in effect claims that there is a category of socially free-floating intellectuals who are both Insiders and Outsiders.

38. On the general point of shifting boundaries, see Merton, *Social Theory and Social Structure,* pp. 338–42, 479–80. Einstein was evidently quite taken with the situational determination of shifts in group boundaries. In a statement written for the London *Times* at a time (28 November 1919) when the animosities of World War I were still largely intact, he introduced slight variations on the theme: "The description of me and my circumstances in the *Times* shows an amusing flare of imagination on the part of the writer. By an application of the theory of relativity to the taste of the reader, today in Germany I am called a German of science and in England I am represented as a Swiss Jew. If I come to be regarded as a 'bête noire' the description will be reversed, and I shall become a Swiss Jew for the German and a German for the English" (quoted in Philipp Frank, *Einstein: His Life and Times* [New York: Alfred A. Knopf, 1963], p. 144).

39. Karl Mannheim, *Ideology and Utopia* (New York: Harcourt Brace Jovanovich, 1936), pp. 10, 139, 232.

40. For further discussion of the idea of social structural warranties of validity, see Merton, *Social Theory and Social Structure,* pp. 560–62.

Benefiting from their collectively diverse social origins and transcending group allegiances, they can observe the social universe with special insight and a synthesizing eye.

Insiders as "Outsiders"

In an adaptation of this same kind of idea, what some Insiders profess as Insiders they apparently reject as Outsiders. For example, when advocates of black Insider doctrine engage in analysis of "white society," trying to assay its power structure or to detect its vulnerabilities, they seem to deny in practice what they affirm in principle. At any rate, their behavior testifies to the tacit assumption that it is possible for self-described "Outsiders" to diagnose and to understand what they describe as an alien social structure and culture.

The contradiction may be specious. For this implies the conception that there is a special category of people in the system of social stratification who have distinctive, if not exclusive, perceptions and understanding in their capacities as *both* Insiders and Outsiders. We need not review again the argument for special access to knowledge that derives from being an Insider. What is of interest here is the idea that special perspectives and insights are available to that category of Outsiders who have been systematically frustrated by the social system: the disinherited, deprived, disenfranchised, dominated, and exploited Outsiders. Their run of experience in trying to cope with these problems serves to sensitize them—and in a more disciplined way, particularly the trained social scientists among them—to the workings of the culture and social structure that are more readily taken for granted by Insider social scientists drawn from social strata who have either benefited from the going social system or have not greatly suffered from it.

This reminder that Outsiders are not all of a kind and the derived hypothesis in the sociology of knowledge about socially patterned differences in perceptiveness is plausible and deserving of far more systematic investigation than it has received. That the white-dominated society has long imposed social barriers which excluded Negroes from anything remotely like full participation in that society is now evident to intransigent and innocently neglectful whites alike. But many of them have evidently not noticed that the high walls of segregation do not at all separate whites and blacks symmetrically from intimate observation of the social life of the other. As socially invisible men and women, blacks at work in white enclaves have for centuries moved through or around the walls of segregation to discover by the way what was on the other side. This was tantamount to their having access to a one-way screen. In contrast, the highly visible whites characteristically did not want to find out the actualities of life in the black community and typically could not, even in those rare cases where they would. The structure of racial segregation meant that the whites who prided themselves on "understanding" Negroes knew little more than their stylized role behaviors in relation to whites and next to nothing of their private lives. As Arthur Lewis has noted, something of the same sort still obtains with the "integration" of many blacks into the larger society during the day coupled with segregation at night as blacks and whites return to their respective ghettos. In these ways, segregation can make for asymmetrical sensitivities across the divide.

Although there is a sociological tradition of reflection and research on marginality in relation to thought, sociologists have hardly begun the hard work of seriously investigating the family of hypotheses in the sociology of knowledge that derive from this conception of asymmetrical relations between diverse kinds of Insiders and Outsiders.

Outsider Doctrine and Perspectives

The strong version of the Insider doctrine, with its epistemological claim to a monopoly of certain kinds of knowledge, runs counter to a long history of thought. From the time of Francis Bacon, to reach no further back, students of the intellectual life have emphasized the corrupting influence of group loyalties upon the human understanding. Among Bacon's four Idols (or sources of false opinion), we need only recall the second, the Idol of the Cave. Drawing upon Plato's allegory of the cave in the *Republic,* Bacon undertakes to tell how the immediate social world in which we live seriously limits what we are prepared to perceive and how we perceive it. Dominated by the customs of our group, we maintain received opinions, distort our perceptions to have them accord with these opinions, and are thus held in ignorance and led into error which we parochially mistake for the truth. Only when we escape from the cave and extend our vision do we provide for access to authentic knowledge. By implication, it is through the iconoclasm that comes with changing or multiple group affiliations that we can destroy the Idol of the Cave, abandon delusory doctrines of our own group, and enlarge the prospects for reaching the truth. For Bacon, the dedicated Insider is peculiarly subject to the myopia of the cave.

In this conception, Bacon characteristically attends only to the dysfunctions of group affiliation for knowledge. Since for him access to authentic knowledge requires that one abandon superstition and prejudice, and since these stem from groups, it would not occur to Bacon to consider the possible functions of social locations in society as providing for observability and access to particular kinds of knowledge.

In far more subtle style, the founding fathers of sociology in effect also argued against the strong form of the Insider doctrine *without turning to the equal and opposite error of advocating the strong form of the Outsider doctrine* (which would hold that knowledge about groups, unprejudiced by membership in them, is accessible only to outsiders).

The ancient epistemological problem of subject and object was taken up in the discussion of historical *Verstehen.* Thus, first Simmel and then, repeatedly, Max Weber symptomatically adopted the memorable aphorism: "one need not be Caesar in order to understand Caesar."[41] In making this claim, they rejected the extreme Insider thesis which asserts in effect that one *must* be Caesar in order to understand him just as they rejected the extreme Outsider thesis that one must *not* be Caesar in order to understand him.

The observations of Simmel and Weber bear directly upon implications of the Insider doctrine that reach beyond its currently emphasized scope. The dedicated Insider argues that the authentic understanding of group life can be achieved only by those who are directly engaged as members in it. Taken seriously, the doctrine puts in question the validity of just about all historical writing, as Weber clearly saw.[42] If direct engagement in the life of a group is essential to understanding it, then the only authentic history is contemporary history, written in fragments by those most fully involved in making inevitably limited portions of it. Rather than constituting only the raw materials of history, the documents prepared by engaged Insiders become all there is to history. But once the historian

41. Thanks to Donald N. Levine (*Georg Simmel: On Individuality and Social Forms* [Chicago: University of Chicago Press, 1971], p. xxiii), I learn that in often attributing the aphorism, with its many implications for social epistemology, to Weber, I had inadvertently contributed to a palimpsestic syndrome: assigning a striking idea or formulation to the author who first introduced us to it when in fact that author had simply adopted or revived a formulation that he (and others versed in the same tradition) knew to have been created by another. As it happens, I first came upon the aphorism in Weber's basic paper on the categories of a *verstehende* sociology published in 1913. In that passage, he treats the aphorism as common usage which he picks up for his own analytical purposes: "Man muss, wie oft gesagt worden ist, 'nicht Cäsar sein, um Cäsar zu verstehen.' " Alerted by Levine's note, I now find that Weber made earlier use of the aphorism back in 1903–6 (*Gesammelte Aufsätze zur Wissenschaftslehre*, pp. 100–101) as he drew admiringly upon Simmel's *Probleme der Geschichtsphilosophie* to which he attributes the most thoroughly developed beginnings of a theory of *Verstehen.* Properly enough, Weber devotes a long, long note to the general implications of Simmel's use of the aphorism, quoting it just as we have seen but omitting the rest of Simmel's embellished version: "Und kein zweiter Luther, um Luther zu begreifen." In his later work, Weber incorporated the aphorism whenever he examined the problem of the "understandability" of the actions of others.

42. See *Gesammelte Aufsätze zur Wissenschaftslehre*, p. 428. Having quoted the Caesar aphorism, Weber goes on to draw the implication for historiography: "Sonst wäre alle Geschichtsschreibung sinnlos."

elects to write the history of a time other than his own, even the most dedicated Insider, of the national, sex, age, racial, ethnic, or religious variety, becomes the Outsider, condemned to ignorance and error. If the Insider is capable of knowing and understanding because he was actually there—in that place, in that time, and, above all else, in that active role—then all historians, black or white, old or young, men or women, are permanently estopped from writing history of the remote time or place.

Writing some twenty years ago in another connection, Claude Lévi-Strauss noted the parallelism between history and ethnography. Both subjects, he observed,

> are concerned with societies *other* than the one in which we live. Whether this *otherness* is due to remoteness in time (however slight) or to remoteness in space, or even to cultural heterogeneity, is of secondary importance compared to the basic similarity of perspective. All that the historian or ethnographer can do, and all that we can expect of either of them, is to enlarge a specific experience to the dimensions of a more general one, which thereby becomes accessible *as experience* to men of another country or another epoch. And in order to succeed, both historian and ethnographer must have the same qualities: skill, precision, a sympathetic approach and objectivity.[43]

Our question is, of course, whether the qualities required by the historian and ethnographer as well as other social scientists are confined to or largely concentrated among Insiders or Outsiders. Simmel, and after him, Schutz, and others have pondered the roles of that incarnation of the Outsider, the stranger who moves on.[44] In a fashion oddly reminiscent of the anything-but-subtle Baconian doctrine, Simmel develops the thesis that the stranger, not caught up in commitments to the group, can more readily acquire the strategic role of the relatively objective inquirer. "He is freer, practically and theoretically," notes Simmel, "he surveys conditions with less prejudice; his criteria for them are more general and more objective ideals; he is not tied down in his action by habit, piety, and precedent."[45] Above all, and here Simmel departs from the simple Baconian conception, the objectivity of the stranger "does not simply involve passivity and detachment; it is a particular structure composed of distance and nearness, indifference and involvement." It is the stranger, too, who finds what is

43. The essay from which this is drawn was first published in 1949 and is reprinted in Claude Levi-Strauss, *Structural Anthropology* (New York: Basic Books, 1963); see p. 16 for the quotation.

44. See Simmel, *Soziologie;* and Alfred Schutz, "The Stranger: An Essay in Social Psychology," *American Journal of Sociology* 49 (May 1944):499–507. It is symbolically appropriate that Simmel should have been attuned to the role of the stranger as outsider. For as Lewis Coser has shown, Simmel's style of sociological work was significantly influenced by his role as "the Stranger in the Academy"; see *Georg Simmel*, pp. 29–39.

45. Georg Simmel, *The Sociology of Georg Simmel,* translated, edited, and with an introduction by Kurt H. Wolff (New York: Free Press, 1950), pp. 404–5.

familiar to the group significantly unfamiliar and so is prompted to raise questions for inquiry less apt to be raised at all by Insiders.

As was so often the case with Simmel's seminal mind, he thus raised a variety of significant questions about the role of the stranger in acquiring sound and new knowledge, questions that especially in recent years have begun to be seriously investigated. A great variety of inquiries into the roles of anthropological and sociological fieldworkers have explored the advantages and limitations of the Outsider as observer.[46] Even now, it appears that the balance sheet for Outsider observers resembles that for Insider observers, both having their distinctive assets and liabilities.

Apart from the theoretical and empirical work examining the possibly distinctive role of the Outsider in social and historical inquiry, significant episodes in the development of such inquiry can be examined as "clinical cases" in point. Thus, it has been argued that in matters historical and sociological the prospects for achieving insights and understanding may actually be somewhat better for the Outsider. Soon after it appeared in 1835, Tocqueville's *Democracy in America* was acclaimed as a masterly work by "an accomplished foreigner." Tocqueville himself expressed the opinion that "there are certain truths which Americans can only learn from strangers." These included what he described as the tyranny of majority opinion and the particular system of stratification which even in that time involved a widespread preoccupation with relative status in the community that left "Americans so restless in the midst of their prosperity." (This *is* Tocqueville, not Galbraith, writing.) All the same,

46. Many of these inquiries explicitly take off from Simmel's imagery of the roles and functions of the stranger. From the large and fast-growing mass of publications on fieldwork in social science, I cite only a few that variously try to analyze the roles of the Outsider as observer and interpreter. From an earlier day dealing with "stranger value," see O. A. Oeser, "The Value of Team Work and Functional Penetration as Methods in Social Investigation," in *The Study of Society*, ed. F. C. Bartlett, M. Ginsberg, E. J. Lindgren, and R. H. Thouless (London: Kegan Paul, 1939); S. F. Nadel, "The Interview Technique in Social Anthropology," in ibid.; Robert K. Merton, "Selected Problems of Field Work in the Planned Community," *American Sociological Review* 12 (June 1947): 304–12; and Benjamin D. Paul, "Interview Techniques and Field Relationships," in *Anthropology Today,* ed. A. L. Kroeber (Chicago, Universitv of Chicago Press, 1953). For more recent work on the parameters of adaptation by strangers as observers, see especially the imaginative analysis by Dennison Nash ("The Ethnologist as Stranger: An Essay in the Sociology of Knowledge," *Southwestern Journal of Anthropology* 19 [1963]: 149–67) and the array of papers detailing how the sex role of women anthropologists affected their access to field data (e.g. Peggy Golde, ed., *Women in the Field* [Chicago: Aldine Press, 1970]). On comparable problems of the roles of Insiders and Outsiders in the understanding of complex public bureaucracies, see the short, general interpretation by Merton ("Role of the Intellectual in Public Bureaucracy," *Social Forces* 23 [May 1945]: 405–15) and the comprehensive, detailed one by Charles Frankel (*High on Foggy Bottom: An Outsider's Inside View of the Government* [New York: Harper & Row, 1969]).

this most perceptive Outsider did not manage to transcend some of the deep-seated racial beliefs and myths he encountered in this country.

Having condemned the Anglo-Americans whose "oppression has at one stroke deprived the descendants of the Africans of almost all the privileges of humanity"; having described slavery as mankind's greatest calamity and having argued that the abolition of slavery in the North was "not for the good of the Negroes, but for that of the whites"; having identified the marks of "oppression" upon both the oppressed Indians and blacks *and* upon their white oppressors; having noted "the tyranny of the laws" designed to suppress the "unhappy blacks" in the states that had abolished slavery; having approximately noted the operation of the self-fulfilling prophecy in the remark that "to induce the whites to abandon the opinion they have conceived of the moral and intellectual inferiority of their former slaves, the Negroes must change; but as long as this opinion subsists, to change is impossible"; having also approximated the idea of relative deprivation in the statement that "there exists a singular principle of relative justice which is very firmly implanted in the human heart. Men are much more forcibly struck by those inequalities which exist within the circle of the same class, than with those which may be remarked between different classes"; having made these observations and judgments, this talented Outsider nevertheless accepts the doctrine, relevant in his time, that racial inequalities "seem to be founded upon the immutable laws of nature herself"; and, to stop the list of particulars here, assumes, as an understandable and inevitable rather than disturbing fact that "the Negro, who earnestly desires to mingle his race with that of the European, cannot effect it."[47]

Without anachronistically asking, as a Whig historian might, for altogether prescient judgments from this Outsider who, after all, was recording his observations in the middle third of the last century, we can nevertheless note that the role of Outsider no more guarantees emancipation from the myths of a collectivity than the role of the Insider guarantees unfailing insight into its social life and belief-systems.

What was in the case of Tocqueville an unplanned circumstance has since often become a matter of decision. Outsiders are sought to observe

47. Alexis de Tocqueville, *Democracy in America* ([1835]; New York: Alfred A. Knopf, 1945), vol. 1, pp. 332, 360–61, chap. 18 passim, pp. 368, 358n, 373–74, 358–59, 335. Tocqueville also assumes that the "fatal oppression" has resulted in the enslaved blacks becoming "devoid of wants," and that "plunged in this abyss of evils, [he] scarcely feels his own calamitous situation," coming to believe that "even the power of thought . . . [is] a useless gift of Providence" (ibid., 1:333). Such observations on the dehumanizing consequences of oppression are remarkable for the time. As Oliver Cromwell Cox observes about part of this same passage, Tocqueville's point "still has a modicum of validity" (*Caste, Class and Race* [New York: Doubleday, 1948], p. 369n).

social institutions and cultures on the premise that they can do so with comparative detachment. In the first decade of this century, for example, the Carnegie Foundation for the Advancement of Teaching, in its search for someone to investigate the condition of medical schools, reached out to appoint Abraham Flexner, after he had confessed to never before having been inside a school of medicine. It was a matter of policy to select a total Outsider who, as it happened, produced the uncompromising Report that did much to transform the state of American medical education at the time.

Later, casting about for a scholar who might do a thoroughgoing study of the Negro in the United States, the Carnegie Corporation searched for an Outsider, preferably one, as they put it, drawn from a country of "high intellectual and scholarly standards but with no background or traditions of imperialism." These twin conditions of course swiftly narrowed the scope of the search. Switzerland and the Scandinavian countries alone seemed to qualify, the quest ending, as we know, with the selection of Gunnar Myrdal. In the preface to *An American Dilemma,* Myrdal reflected on his status as an Outsider who, in his words, "had never been subject to the strains involved in living in a black-white society" and who "as a stranger to the problem . . . has had perhaps a greater awareness of the extent to which human valuations everywhere enter into our scientific discussion of the Negro problem."[48]

Reviews of the book repeatedly alluded to the degree of detachment from entangling loyalties that seemed to come from Myrdal's being an outsider. J. S. Redding, for one, wrote that "as a European, Myrdal had no American sensibilities to protect. He hits hard with fact and interpretatiton." Robert S. Lynd, for another, saw it as a prime merit of this outsider that he was free to find out for himself "without any side glances as to what was politically expedient." And, for a third, Frank Tannenbaum noted that Myrdal brought "objectivity in regard to the special foibles and shortcomings in American life. As an outsider, he showed the kind of objectivity which would seem impossible for one reared within the American scene." Even later criticism of Myrdal's work—notably, the comprehensive critique by Cox—does not attribute imputed errors of interpretation to his having been an outsider.[49]

Two observations should be made on the Myrdal episode. First, in the judgment of critical minds, the Outsider, far from being excluded from the

48. Gunnar Myrdal, *An American Dilemma: The Negro Problem and Modern Democracy* (New York and London: Harper & Bros., 1944), pp. xviii–xix.

49. See J. S. Redding, "Review," *New Republic*, 20 March 1944, pp. 384–86; Robert S. Lynd, "Prison for Our Genius," *Saturday Review*, 22 April 1944, pp. 5–7, 27; Frank Tannenbaum, "An American Dilemma," *Political Science Quarterly* 59 (September 1944): 321–30; and Cox, *Caste, Class and Race*, chap. 23.

understanding of an alien society, was able to bring needed perspectives to it. And second, that Myrdal, wanting to have both Insider and Outsider perspectives, expressly drew into his circle of associates in the study such Negro and white insiders, engaged in the study of Negro life and culture and of race relations, as E. Franklin Frazier, Arnold Rose, Ralph Bunche, Melville Herskovits, Otto Klineberg, J. G. St. Clair Drake, Guy B. Johnson, and Doxey A. Wilkerson.

It should be noted in passing that other spheres of science, technology, and learning have accorded distinctive and often related roles to both the Insider and the Outsider (See chapter 22 in this volume, pp. 517–19). As long ago as the seventeenth century, for example, Thomas Sprat, the historian of the Royal Society, took it "as evident, that divers sorts of Manufactures have been given us by men who were not bred up in Trades that resembled those which they discover'd. I shall mention Three: that of Printing, [Gun] Powder, and the Bow-Dye." Sprat goes on to expand upon the advantages of the Outsider for invention, concluding with the less-than-science-based observation that "as in the Generation of Children, those are usually observ'd to be most sprightly, that are the stollen Fruits of an unlawful Bed; so in the Generations of the Brains, those are often the most vigorous, and witty, which men beget on other Arts, and not on their own."[50]

In our own time, Gilfillan reported that the "cardinal inventions are due to men outside the occupation affected, and the minor, perfective inventions to insiders."[51] And in a recent and more exacting inquiry, Joseph Ben-David[52] found that the professionalization of scientific research "does not in itself decrease the chances of innovation by outsiders to the various fields of science." For the special case of Outsiders to a particular discipline, Max Delbrück, himself a founding father of molecular biology, notes that although "nuclear physics was developed almost exclusively within the framework of academic institutes at universities, molecular biology, in contrast, is almost exclusively a product of outsiders, of chemists, physicists, medical microbiologists, mathematicians and engineers."[53]

Siegfried Kracauer, the deeply perceptive historian and sociologist, suggested in his posthumous book, *History: The Last Things Before the Last,* that it is the exile, that onetime Insider become Outsider, who "may look at his previous existence with the eyes of one 'who does not belong

50. Thomas Sprat, *History of the Royal Society*, ed. Jackson I. Cope and Harold W. Jones ([1667]; London: Routledge & Kegan Paul, 1959), pp. 391–93.

51. S. C. Gilfillan, *The Sociology of Invention* (Chicago: Follett, 1935), p. 88.

52. "Role and Innovations in Medicine," *American Journal of Sociology* 65 (May 1960): 557–68.

53. Max Delbrück, "Das Begriffsschema der Molekular-Genetik," *Nova Acta Leopoldina* 26 (1963): 9–16.

to the house.' " And he reminds us of Thucydides saying in so many words that his long exile enabled him " 'to see something of both sides—the Peloponnesian as well as the Athenian.' "[54]

The cumulative point of this variety of intellectual and institutional cases is not—and this needs to be repeated with all possible emphasis—is *not* a proposal to replace the extreme Insider doctrine by an extreme and equally vulnerable Outsider doctrine. The intent is, rather, to transform the original question altogether. We no longer ask whether it is the Insider or the Outsider who has monopolistic or privileged access to social knowledge; instead, we begin to consider their distinctive and interactive roles in the process of seeking truth.

Interchange, Trade-offs, and Syntheses

The actual intellectual interchange between Insiders and Outsiders—in which each adopts perspectives from the other—is often obscured by the rhetoric that commonly attends intergroup conflict. Listening only to that rhetoric, we may be brought to believe that there really is something like "black knowledge" and "white knowledge," "man's knowledge" and "woman's knowledge" which somehow manage to be both incommensurable and antithetical. Yet the boundaries between Insiders and Outsiders are far more permeable than this allows. Just as with the process of competition generally, so with the competition of ideas. Competing or conflicting groups take over ideas and procedures from one another, thereby denying in practice the rhetoric of total incompatibility. Even in the course of social polarization, conceptions with cognitive value are utilized all apart from their source. Concepts of power structure, cooptation, the dysfunctions of established institutions, and findings associated with these concepts are utilized by social scientists, irrespective of their social or political identities. Nathan Hare,[55] for example, who remains one of the most articulate exponents of the Insider doctrine, does not hesitate to use the notion of the self-fulfilling prophecy in trying to explain how it is, in this day and age, that organizations run by blacks find it hard to work out.[56] As he puts it, "White people thought that we could not have

54. Siegfried Kracauer, *History: The Last Things Before the Last* (New York: Oxford, 1969), p. 84. See also Peter Gay, *Weimar Culture: The Outsider as Insider* (New York: Harper & Row, 1968).

55. See "Interview with Nathan Hare," *U.S. News and World Report* 22 (May 1967): 64–68.

56. Elsewhere, Hare treats certain beliefs of "Negro dignitaries" as a self-fulfilling prophecy; see his *Black Anglo-Saxons* (London: Collier-Macmillan, 1970), p. 44. A recent work on women's liberation movements, both new and old, also observes: "Feminists argue further that there is a self-fulfilling prophecy component: when one group dominates another, the group with power is, at best, reluctant to relinquish its

any institutions which were basically black which were of good quality. This has the effect of a self-fulfilling prophecy, because if you think that black persons cannot possibly have a good bank, then you don't put your money in it. All the best professors leave black universities to go to white universities as soon as they get the chance. The blacks even do the same thing. And this makes your prediction, which wasn't true in the beginning, come out to be true."[57]

Black scholars and women scholars utilize the conception of the self-fulfilling prophecy as a matter of course whenever it seems to illuminate the condition they seek to understand. They do so without a backward glance at the functionally irrelevant circumstance that the conception was set forth and developed by scholars who happened to be neither black nor female. Correlatively, white sociologists, both male and female, utilize the conception of "status without substance" without pausing to consider that it was originated by the black sociologist, Franklin Frazier.[58]

Such diffusion of ideas across the boundaries of groups and statuses has long been noted. In one of his more astute analyses, Mannheim states the general case for the emergence and spread of knowledge that transcends even profound conflicts between groups:

> Syntheses owe their existence to the same social process that brings about polarization; groups take over the modes of thought and intellectual achievements of their adversaries under the simple law of 'competition on the basis of achievement.' . . . In the socially-differentiated thought process, even the opponent is ultimately forced to adopt those categories and forms of thought which are most appropriate in a given type of world order. In the economic sphere, one of the possible results of competition is that one competitor is compelled to catch up with the other's technological advances. In just the same way, whenever groups compete for having their interpretation of reality accepted as the correct one, it may happen that one of the groups takes over from the adversary some fruitful hypothesis or category—anything that promises cognitive gain. . . . [In due course, it becomes possible] to find a position from which both kinds of thought can be envisaged in their partial correctness, yet at the same time also interpreted as subordinate aspects of a higher synthesis.[59]

The essential point is that, with or without intent, the process of intellectual exchange takes place precisely because the conflicting groups are in interaction. The extreme Insider doctrine, for example, affects the thinking of sociologists, black and white, who reject its extravagant claims. Intellectual conflict sensitizes them to aspects of their subject that they have otherwise not taken into account.

---

control. Thus in order to keep woman in 'her place,' theories are propounded which presume that her place is defined by nature" (Judith Hole and Ellen Levine, *Rebirth of Feminism* [New York: Quadrangle Books, 1971], p. 193).

57. Hare, *Black Anglo-Saxons*, p. 65.

58. See *Black Bourgeoisie*.

59. Karl Mannheim, *Essays on the Sociology of Knowledge*, pp. 221–23.

Social sadism and sociological euphemism

As a case in point of this sort of sensitization through interaction, I take what can be described as a composite of social sadism and sociological euphemism. "Social sadism" is more than metaphor. The term refers to social structures which are so organized as to systematically inflict pain, humiliation, suffering, and deep frustration upon particular groups and strata. Social sadism need have nothing to do with the psychic propensities of individuals to find sexual pleasure in cruelty. It is an objective, socially organized, and recurrent set of situations that has these cruel consequences, however diverse its historical sources and whatever the social processes that maintain it.

The sadistic type of social structure is readily overlooked by the perspective that can be described as that of the sociological euphemism. This term does not refer to the obvious cases in which ideological support of the structure is simply couched in sociological language. Rather, it refers to the kind of conceptual apparatus that, once adopted, requires us to ignore such intense human experiences as pain, suffering, humiliation, and so on. In this context, analytically useful concepts such as social stratification, social exchange, reward system, dysfunction, symbolic interaction are altogether bland in the fairly precise sense of being unperturbing, suave, and soothing in effect. To say this is not to say that the conceptual repertoire of sociology (or of any other social science) must be purged of such impersonal concepts and filled with mawkish, sentiment-laden substitutes. But it should be noted that analytically useful as these impersonal concepts are for certain problems, they also serve to exclude from the attention of the social scientist the intense feelings of pain and suffering that are the experience of some people caught up in given patterns of social life. By screening out these profoundly human experiences, the impersonal concepts become sociological euphemisms.

Nor is there any easy solution to the problem of sociological euphemism. True, we have all been warned off the Whiteheadian fallacy of misplaced concreteness, the fallacy of assuming that the particular concepts we employ to examine the flow of events capture their entire content. No more than in other fields of inquiry are sociological concepts designed to depict the concrete entirety of the psychosocial reality to which they refer. But the methodological rationale for conceptual abstraction has yet to provide a way of assessing the cognitive costs as well as the cognitive gains of abstraction. As the biologist Paul Weiss has put the general issue: "How can we ever retrieve information about distinctive features once we have tossed it out?"[60]

Consider some outcomes of the established practice of employing bland

60. Paul Weiss, "One Plus One Does Not Equal Two," in *Within the Gates of Science and Beyond* (New York: Hafner, 1971), p. 213.

sociological concepts that systematically abstract from certain elements and aspects of the concreteness of social life. It is then only a short step to the further tacit assumption that the aspects of psychosocial reality which these concepts help us to understand *are the only ones worth trying to understand.* The ground is then prepared for the next seemingly small but altogether conclusive step. The social scientist sometimes comes to act as though the aspects of reality that are neglected in his analytical apparatus *do not even exist.* By that route, even the most conscientious of social scientists are often led to transform their concepts and models into scientific euphemisms.

All this involves the special irony that the more intellectually powerful a set of social science conceptions has proved to be, the less the incentive for trying to elaborate it in ways designed to catch up the humanly significant aspects of the psychosocial reality that it neglects.

It is this tendency toward sociological euphemism, I suggest, that some (principally but not exclusively black) social scientists are forcing upon the attention of (principally but not exclusively white) social scientists. No one I know has put this more pointedly than Kenneth Clark: "More privileged individuals may understandably need to shield themselves from the inevitable conflict and pain which would result from acceptance of the fact that they *are* accessories to profound injustice. The tendency to discuss disturbing social issues such as racial discrimination, segregation, and economic exploitation in detached, legal, political, socioeconomic, or psychological terms as if these persistent problems did not involve the suffering of actual human beings is so contrary to empirical evidence that it must be interpreted as a protective device."[61]

Perhaps enough has been said to indicate how Insider and Outsider perspectives can converge through reciprocal adoption of ideas and the developing of complementary and overlapping foci of attention in the formulation of scientific problems. But these intellectual potentials for synthesis are often curbed by the social processes dividing scientists that transform intellectual controversy into social conflict.[62]

When a reverse transition from social conflict to intellectual controversy is achieved, when the perspectives of each group are taken seriously enough to be carefully examined rather than rejected out of hand, there can develop trade-offs between the distinctive strengths and weaknesses of Insider and Outsider perspectives that enlarge the chances for a sound and relevant understanding of social life.

Insiders, Outsiders, and types of knowledge

If we indeed have distinctive contributions to make to social knowledge

61. Kenneth Clark, *Dark Ghetto* (New York: Harper & Row, 1965), p. 75.
62. For discussion of these processes, see chapter 3 of this volume, pp. 55–58.

in our roles as Insiders or Outsiders—and it should be repeated that all of us are Insiders in one context and Outsiders in another—then those contributions probably link up with a long-standing distinction between two major kinds of knowledge, a basic distinction that is blurred in the often ambiguous use of the word "understanding." In the language of William James, drawn out of John Grote, who was in turn preceded by Hegel, this is the distinction between "acquaintance with" and "knowledge about."[63] The one involves direct familiarity with phenomena that is expressed in depictive representations; the other involves more abstract formulations which do not at all "resemble" what has been directly experienced.[64] As Grote noted a century ago, contrasting pairs of words have embedded the distinction in various languages:

"Acquaintance with"	"Knowledge about"
noscere	scire
kennen	wissen
connaître	savoir

These distinct and connected kinds of understanding may turn out to be distributed, in varying mix, among Insiders and Outsiders. The introspective meanings of experience within a group may be more readily accessible, for all the seemingly evident reasons, to those who have shared part or all of that experience. But authentic awareness, even in the sense of acquaintance with, is not guaranteed by social affiliation, as the concept of false consciousness is designed to remind us. Determinants of social life—for an obvious example, ecological patterns and processes—are not necessarily evident to those directly engaged in it. In short, sociological understanding involves much more than the acquaintance-with of the Insider. It includes an empirically confirmable comprehension of the conditions and the often complex processes in which people are caught up without much awareness of what is going on. To analyze and understand these requires a theoretical and technical competence which, as such, transcends one's status as Insider or Outsider. The role of social scientist concerned with achieving knowledge about society requires enough detachment and trained capacity to know how to assemble and assess the evidence without regard for what the analysis seems to imply about the worth of one's group.

63. See William James, *The Meaning of Truth* ([1885]; New York: Longmans Green, 1932), pp. 11–13; John Grote, *Exploratio Philosophica* (Cambridge: Deighton, Bell & Co, 1865), p. 60; Georg Hegel, *The Phenomenology of Mind*, 2d rev. ed. ([1807]; New York: Macmillan, 1961). Hegel catches the distinction in his aphorism: "Das Bekannte überhaupt ist darum, weil est bekannt is, nicht erkannt." Polanyi has made a significant effort to synthesize these modes of understanding, principally in his conception of "tacit knowing"; see his *Study of Man* and *The Tacit Dimension*.

64. Merton, *Social Theory and Social Structure*, p. 545.

Other attributes of the domain of knowledge dampen the relevance of Insider and Outsider identities for the validity and worth of the intellectual product. It is the character of an intellectual *discipline* that its evolving rules of evidence are adopted *before* they are used in assessing a particular inquiry. These criteria of good and bad intellectual work may turn up to differing extent among Insiders and Outsiders as an artifact of immediate circumstance, and that is itself a difficult problem for investigation. But the margin of autonomy in the culture and institution of science means that the intellectual criteria, as distinct from the social ones, for judging the validity and worth of that work transcend extraneous group allegiances. The acceptance of criteria of craftsmanship and integrity in science and learning cuts across differences in the social affiliations and loyalties of scientists and scholars. Commitment to the intellectual values dampens group-induced pressures to advance the interests of groups at the expense of these values and of the intellectual product.

The consolidation of group-influenced perspectives and the autonomous values of scholarship is exemplified in observations by John Hope Franklin who, for more than a quarter-century, has been engaged in research on the history of American Negroes from their ancient African beginnings to the present.[65] In the first annual Martin Luther King, Jr., Memorial Lecture at the New School for Social Research, he observes in effect how great differences in social location of both authors and audiences can make for profound differences in scholarly motivation and orientation. Franklin notes that it was the Negro teacher of history, "outraged by the kind of distorted history that he was required to teach the children of his race," who took the initiative in the nineteenth century to undo what one of them described as "the sin of omission and commission on the part of white authors, most of whom seem to have written exclusively for white children."[66] The pioneering revisionist efforts of W. E. B. DuBois and others found organized expression in the founding in 1915 of the Association for the Study of Negro Life and History and, a year later, of the *Journal of Negro History* by Carter G. Woodson and his associates. This institutionalization of scholarship helped make for transfer and interchange of knowledge between Insiders and Outsiders, between black historians and white. In Franklin's words, the study of Negro history became "respectable. Before the middle of the twentieth century it would entice not only a large number of talented Negro scholars to join in the quest for a revised and more valid American history, but it would also bring into its fold a considerable number of the ablest white historians who could no longer tol-

65. Perhaps the best known of Franklin's many writings is *From Slavery to Freedom: A History of Negro Americans* (New York: Alfred A. Knopf, 1967), now in its third edition.

66. John Hope Franklin, *The Future of Negro American History* (New York: New School for Social Research, 1969), p. 4.

erate biased, one-sided American history. Thus, Vernon Wharton's *The Negro in Mississippi,* Kenneth Stampp's *The Peculiar Institution,* Louis Harlan's *Separate but Unequal* and Winthrop Jordan's *White over Black* —to mention only four—rank among the best of the efforts that any historians, white or black, have made to revise the history of their own country. In that role they, too, became revisionists of the history of Afro-Americans"[67]

These efforts only began to counter the "uninformed, arrogant, uncharitable, undemocratic, and racist history [which] . . . spawned and perpetuated an ignorant, self-seeking, superpatriotic, ethnocentric group of white Americans who can say, in this day and time, that they did not know that Negro Americans had a history."[68] But much needed counter-developments can induce other kinds of departure from scholarly standards. Franklin notes that the recent "great renaissance" of interest in the history of Negro Americans has found proliferated and commercialized expression. "Publishers are literally pouring out handbooks, anthologies, workbooks, almanacs, documentaries, and textbooks on the history of Negro Americans. . . . Soon, we shall have many more books than we can read; indeed, many more than we should read. Soon, we shall have more authorities on Negro history than we can listen to; indeed, many more than we should listen to."[69]

Franklin's application of exacting, autonomous, and universalistic standards culminates in a formulation that, once again, transcends the statuses of Insiders and Outsiders:

> Slavery, injustice, unspeakable barbarities, the selling of babies from their mothers, the breeding of slaves, lynchings, burnings at the stake, discrimination, segregation, these things too are a part of the history of this country. If the Patriots were more in love with slavery than freedom, if the Founding Fathers were more anxious to write slavery into the Constitution than they were to protect the rights of men, and if freedom was begrudgingly given and then effectively denied for another century, these things too are a part of the nation's history. It takes a person of stout heart, great courage, and uncompromising honesty to look the history of this country squarely in the face and tell it like it is. But nothing short of this will make possible a reassessment of American history and a revision of American history that will, in turn, permit the teaching of the history of Negro Americans. And when this approach prevails, the history of the United States and the history of the black man can be written and taught by any person, white, black, or otherwise. For there is nothing so irrelevant in telling the truth as the color of a man's skin.[70]

Differing profoundly on many theoretical issues and empirical claims, Cox and Frazier are agreed on the relative autonomy of the domain of

67. Ibid., pp. 5–6.
68. Ibid., p. 9.
69. Ibid., pp. 10–11.
70. Ibid., pp. 14–15.

knowledge and, specifically, that white scholars are scarcely barred from contributing to what Frazier described as a "grasp of the condition and fate of American Negroes."[71] Recognition of what has been called "the mark of oppression," Frazier notes, "was the work of two white scholars that first called attention to this fundamental aspect of the personality of the American Negro. Moreover, it was the work of another white scholar, Stanley M. Elkins, in his recent book on *Slavery*, who has shown the psychic trauma that Negroes suffered when they were enslaved, the pulverization of their social life through the destruction of their clan organization, and annihilation of their personality through the destruction of their cultural heritage."[72] And Cox, in his strong criticism of what he describes as "the black bourgeoisie school" deriving from Frazier's work, emphasizes the distorting effects of the implicitly black nationalist ideology of this school on the character of its work.[73]

It should now be evident that structural analysis applied to the domain of knowledge provides an ironically self-exemplifying pattern. For just as the unity of any other collectivity based on a single status—of Americans or of Nigerians, of blacks or of whites, of men or of women—is continuously subject to the potential of inner division owing to the other statuses of its members, so with the collectivities often described as the scientific community and the community of scholars. Their functional autonomy is also periodically subject to great stress, owing in part to the complex social differentiation of the population of scientists and scholars that weakens their response to external pressures. The circumstances and processes making for the fragility or resiliency of that autonomy constitute one of the great questions in the sociology of knowledge.

It is nevertheless that autonomy which still enables the pursuit of truth to transcend other loyalties, as Michael Polanyi, more than most of us, has long recognized: "People who have learned to respect the truth will feel entitled to uphold the truth against the very society which has taught them to respect it. They will indeed demand respect for themselves on the grounds of their own respect for the truth, and this will be accepted, even against their own inclinations, by those who share these basic convictions."[74]

A paper such as this one needs no peroration. Nevertheless, here is mine. Insiders and Outsiders, unite. You have nothing to lose but your claims. You have a world of understanding to win.

71. See Cox, *Caste, Class and Race*, and his introduction to Hare, *Black Anglo-Saxons*; see also Frazier's *Black Bourgeoisie* and his "Failure of the Negro Intellectual," in *E. Franklin Frazier on Race Relations*, ed. G. Franklin Edwards (Chicago: University of Chicago Press, 1968).

72. Frazier, "Failure of the Negro Intellectual," p. 272.

73. Cox, introduction to Hare, *Black Anglo-Saxons*, pp. 15–31.

74. Polanyi, *Study of Man*, pp. 61–62.

The Sociology of Scientific Knowledge

Part 2

Prefatory Note

A fundamental problem in the sociology of scientific knowledge is, of course, that of how science comes to develop in the first place. This is followed by the correlative, more specific question: Once science becomes culturally and institutionally established, what affects the rate and directions of its development? The five papers assembled in this section focus on the problematics of scientific knowledge and are arranged in order of increasing specificity with regard to particular bodies of knowledge, beginning with a critique of Sorokin's handling of the larger question and concluding with an analysis of why the sociology of science was itself not a focus of scholarly attention for so long a time.

In the first paper, Merton and Bernard Barber undertake a detailed consideration of Sorokin's "emanationist" theory of the development of science. Both authors studied under Sorokin at Harvard, Merton in the early 1930s and Barber in the late 1930s and after World War II. They are thoroughly familiar with his work, Merton having collaborated with Sorokin on several investigations. Sorokin's extensive reliance on the concept of "cultural mentalities" (idealistic, ideational, and sensate) as the basic sources and results of social change, and his dual use of science as both symptom and consequence of the "sensate" mentality, they argue, fail to provide either opportunities for empirical verification or sufficient grounds for explaining exceptions to theory-based expectations. The paper concludes with seven concise questions directed to Sorokin's theoretical intent and his methodological standards. Published in the same volume as the Merton-Barber paper, Sorokin's detailed and appreciative but unyielding reply—partly clarification, mostly rebuttal—rounds out an instructive dialogue in the developing sociology of scientific knowledge.

The second paper, "Social and Cultural Contexts of Science," appeared in 1970 as a new preface to the reprinting of Merton's dissertation, which was first printed as a monograph by *Osiris*. Here the author finds an appropriate level of objectivity vis-à-vis his own work of more than thirty years ago on the emergence of organized science in England, and uses the occasion to provide not only some details of its background but to consider various scholarly discussions of it. He remarks on the apparently excessive amount of attention in these discussions that has focused upon the section of the book that deals with the links between Puritanism and science, and responds to some methodological criticisms of the empirical basis for some of his hypotheses about these links.

It is interesting that Merton fails to examine these criticisms as data relevant to the sociology of scientific knowledge. Just as the substance of developing scientific knowledge is influenced by both intra- and extra-scientific forces, so too, presumably, are commentaries on scientific advances. A careful analysis of questions raised at different times since the appearance of the monograph in 1938 would probably have provided further insight into the relations between scientific knowledge, the scientific community, and the environing society in which science exists.

The third paper, originally chapter 3 in the same monograph, identifies a problem that has come to be recognized as basic to the sociology of scientific knowledge. This involves the question of the processes that make for changing distributions of intellectual interest among the various sciences, technology, and the other scholarly disciplines. It also involves the question of what makes for changing foci of interest within each science in the course of its development. Merton attacked the problem by the then unfamiliar use of quantitative measures of shifts of interest among the sciences in England during the last third of the seventeenth century. Here the units are scientific articles and enumerated scientific discoveries rather than individual scientists. His conclusion, based upon comparisons of measured shifts in interest with important substantive developments within sciences, is that the short-range ebb and flow of scientific attention can be attributed largely to causes internal to science (the sources of more long-range shifts are taken up later in the monograph, with the conclusion that they are much more affected by extrascientific changes).

"Interactions of Science and Military Technique" takes up the other side of this coin, the role of practical interests in determining foci of scientific attention. First published in 1935 and later elaborated in his dissertation monograph, the paper also supplements qualitative observations with a quantitative basis for the conclusions. Research reports are categorized in terms of their direct and indirect relevance to problems of military technique (largely gunpowder and ballistics) and a fairly stable 10 percent of all inquiries conducted in the Royal Society between 1661 and 1687 are found to fall into this category. Beyond this, the paper contains an early formulation of the autonomy-oriented character of scientific knowledge in its demonstration that aggregates of scientists will frequently pick up a topic because of its applied value and then transform it into a "pure" scientific problem that leads off in directions having less and less relevance to the originating occasion.

It would be exceedingly difficult to replicate this study today, given both the existence of security regulations and the complexities of categorizing research, but if research and development funds are both considered (and if it can be assumed that relevant and irrelevant projects cost roughly the same amount), the proportion of total scientific effort that bears upon military technique today should be considerably above 50 percent in many countries.

In the final paper of this section, Merton focuses specifically on the sociology of science as a case study of a neglected specialty. This was his foreword to Barber's *Science and the Social Order* (1952), and in it he takes up the question of why this subject had so long been neglected by sociologists. The context of this paper and some of its implications are treated in the introduction to this volume; suffice it to note here that this catalogue of the circumstances which might have accounted for the neglect of the sociology of science fairly cries out for empirical research, and that those sociologists currently studying the growth of scientific specialties would do well to take explicit account of these extra-scientific circumstances in the course of their own investigations.

N. W. S.

6

Sorokin's Formulations in the Sociology of Science

1963

[With Bernard Barber]

From the beginning we must abandon the attempt to put into short compass all the wide-ranging, diversified, and developing observations in the sociology of science set forth by Pitirim Sorokin. Any such effort would be the work of a sizable book, not of a short essay. For his contributions to the sociology of science engage almost every other major part of his empirically connected sociological theory. To try to trace out each component in his sociology of science—or, more generally, in his sociology of knowledge—would mean to touch upon every other aspect of his voluminous works and this, even were it within our powers, would lead to an intolerable and presumptuous duplication of much found elsewhere in the volume devoted to his work. In place of a systematic treatment of Sorokin's contributions to the sociology of science, therefore, we shall substitute some observations that bear upon their most significant and sometimes thorny aspects. In place of the many details that enter into his sociology of science, we shall put the more general formulations that encompass these details, knowing that this means the exclusion of issues that in a more thorough examination would have to be taken up substantially. And finally, in place of tracking down the development of Sorokin's ideas about the sociology of science as these emerged over the course of almost half a century, we shall deal primarily with his later ideas, particularly as these were set forth in his *Social and Cultural Dynamics*. In short, this is only an essay toward a critical understanding of Sorokin's work in the sociology of science; it is not a comprehensive and methodical analysis of that work.

As though this were not enough of a limitation, we must confess to another. We find it difficult, not to say impossible, to achieve a sufficient

Reprinted with permission from *Pitirim A. Sorokin in Review*, Philip J. Allen, ed., (Durham, N.C.: Duke University Press, 1963), pp. 332–68.

sense of historical distance from a scholar of our time and a scholar, moreover, who has been the teacher of us both. But here Sorokin himself has come to our rescue. And the respect in which he has done so is itself almost an essay in the microsociology of science, showing how the social structure of a university affects the relations between a professor and his students and so affects the transmission of knowledge. Sorokin began to make his imprint upon American sociology shortly after he took up his first post in the United States at the University of Minnesota. But it was especially after he was brought to Harvard to found the department of sociology there that this emphatic, straight-spoken, and *urgent* man began to influence the thinking of substantial numbers of students. In order to understand the character of this influence, we should try the thought-experiment of imagining that Sorokin had gone not to Harvard but to some other important university—in Europe. He would have held "the chair of sociology" in that university. He would have had a number of assistants as well as students who, in accord with the cultural expectations of docility, would have become his disciples, echoing his words and thoughts almost as though they were their own. But in the American scheme of things academic, and particularly in a university such as Harvard, the social structure and the culturally defined patterns of expectations were of course quite otherwise. The authority structure of the department was pluralistic rather than strongly centralized. There was not only *the* occupant of The Chair but other members of the faculty who had equal access to students. Nor was there an unchallenged norm of obedient agreement with the major professor. This structural situation meant that, at Harvard, many graduate students in sociology were apt to become anything but disciples. Moreover, Sorokin's own personality and role-behavior reinforced his tendency toward independence of mind among his students. They tended to adopt the same critical stance toward aspects of Sorokin's work as he, in the capacity of a role-model, was taking toward the work of others, both contemporary and bygone. All this meant that the structurally defined role of Sorokin was primarily that of alerting students to intellectual alternatives rather than that of imprinting the particulars of his own theory upon them. And all this, we conjecture, helps explain how it is that Sorokin's students have not hesitated to differ with him when, rightly or wrongly, they did not see matters just as he did. In adopting this sometimes timorous but socially supported position of criticism, they were helped in no small measure by the prototype of Sorokin's own behavior, when he was persuaded that other scholars had erred. This, then, is no occasion for exhibiting alumnal piety in public but rather an occasion for applying to Sorokin's work the same critical standards that he has applied to the work of others. It is primarily an essay in criticism rather than an essay in exposition.

Sorokin's Sociology of Science: The Central Position

Sorokin has explicitly adopted an idealistic and emanationist theory of the sociology of science. Unlike the theories of a Marx or a Mannheim, which seek primarily to account for the character and limits of knowledge obtaining in a particular society in terms of its social structure, Sorokin's theory tries to derive every aspect of knowledge from underlying "culture mentalities."[1] So prominent is this aspect of Sorokin's theory that it has been variously noted by every commentator upon his work. In somewhat restricted terms, for example, it has been observed by Maquet, a thoroughly sympathetic critic, that in Sorokin's theory the "independent variable is the intellectual position in regard to ultimate reality and ultimate value. . . . The three premises of culture are nothing else but philosophic positions."[2] This statement, as we shall soon see, places far too restrictive an interpretation upon the theory. By Sorokin's own testimony, much more is contained in his concept of types of culture than can be aptly described as the philosophic position basic to it. Nevertheless, Maquet does take hold of the essential fact that each of the types of culture discriminated by Sorokin has its distinctive ontological orientation. This has also been noted, to cite only one other commentator on Sorokin, by Stark, who observes that "It is essential for the understanding of the whole theory to realize that it considers the ontological convictions prevailing at a given time not so much as culture contents but rather as culture premises, from which the culture proceeds and emanates as a whole."[3]

What, then, is the character of the "culture mentalities" that are variously expressed in distinctive kinds of (claims to) knowledge and how do these mentalities differ? There are two "pure types" of these mentalities, which differ fundamentally in what is taken to be the nature of ultimate reality and value. The first is the ideational, which conceives of reality as "non-material, ever-lasting Being," which defines human needs as primarily spiritual and seeks the satisfaction of these needs through "self-imposed minimization or elimination of most physical needs."[4] The ideational culture adopts the "truth of faith." At the other extreme is the sensate mentality, which limits reality to what can be perceived through the senses. Concerned primarily with physical needs, this culture calls for satisfaction of these needs, not through modification of self, but through modification of the external world. It is oriented to the "truth of senses." Intermediate

1. Robert K. Merton, *Social Theory and Social Structure*, rev. ed. (Glencoe, Ill.: The Free Press, 1957), p. 466.

2. Jacques Maquet, *The Sociology of Knowledge* (Boston: Beacon Press, 1951), pp. 135, 187.

3. W. Stark, *The Sociology of Knowledge* (Glencoe, Ill.: The Free Press, 1958), p. 226.

4. Pitirim A. Sorokin, *Social and Cultural Dynamics*, 4 vols. (New York: American Book Co., 1937), 1:72–73.

to these two is a "mixed type" of culture mentality, the idealistic, which represents a kind of balance of the foregoing types. It is oriented toward a "truth of reason." From these three types of mentalities—the major premises of each kind of culture—Sorokin derives their distinctive systems of truth and knowledge.

We shall have something more to say about the meanings, both expressed and implicit, of the idea that these culture premises are basic to distinctive kinds of knowledge developing within each kind of sociocultural system. For the moment, however, we need note only that Sorokin takes a great array of somewhat more specific types of knowledge as *dependent* upon these cultural premises. These include such dependent "variables" as the fundamental categories of causality, time, space, and number; basic philosophical conceptions such as idealism-materialism, eternalism-temporalism, realism-conceptualism-nominalism; various conceptions of cosmic, biological, and sociocultural processes, as expressed, for example, in notions of mechanism or vitalism in biology; the rate of scientific advance; the prevailing kinds of moral philosophy; and to take just one other, the various kinds of criminal law.

Sorokin himself best summarizes the brooding omnipresence of the three principal types of culture:

> Each has its own mentality; its own system of truth and knowledge; its own philosophy and *Weltanschauung*; its own type of religion and standards of "holiness"; its own system of right and wrong; its own forms of art and literature; its own mores, laws, code of conduct; its own predominant forms of *social relationships*; its own economic and political organization; and, finally, its own type of *human personality*, with a peculiar mentality and conduct.[5]

In short, more than the forms of "knowledge" alone are dependent upon the premises underlying each type of sociocultural system. The forms of social structure and the kinds of prevailing personality also share this condition of dependence. Every sphere of culture, social structure, and personality is seen as emanating from the fundamental orientations characteristic of each of the three kinds of sociocultural systems.

As we shall also see in due course, Sorokin considers that particular theories of science as well as the rate of scientific advance are dependent upon these underlying cultural premises. Here we need note only the first of the puzzles presented in Sorokin's conception, which, so far as we can see, he does not solve for us. How can he escape from the self-contained emanationism of the theoretical position he adopts? For it would appear tautological to say, as Sorokin does, that "in a Sensate society and culture the Sensate system of truth based upon the testimony of the organs of senses has to be dominant."[6] For, sensate *mentality*—that abstraction

5. Ibid., 1:67.
6. Ibid., 2:5.

which Sorokin makes ontologically basic to the culture—has already been *defined* as one conceiving of "reality as only that which is presented to the sense organs."[7] In this case, as in other comparable cases, Sorokin seems to vacillate between treating his types of culture mentality as a defined concept or as an empirically testable hypothesis. This is the first of several questions which it would be useful to have Sorokin examine anew and clarify for future reference.

By way of introduction, then, we see that Sorokin's sociology of science, and by extension, his sociology of knowledge, is idealistic and emanationist. In this respect it differs fundamentally from the materialistic conception of Marx, which focuses on the intellectual perspectives generated by the position of thinkers in the class structure of their time, just as it differs from the quasi-Marxist conceptions of Mannheim, which extend Marx's notion of the structural bases of thought. It is this contrast, presumably, that led Maquet to conclude that "since Sorokin's independent variable is the intellectual position in regard to ultimate reality and ultimate value, his sociology will have a very idealistic character (in the current sense of the word . . . 'ideas rule the world')."[8]

This can be put somewhat differently without altering the basic point, but directing attention to some of its further implications. In one of his many books published after *Social and Cultural Dynamics*, Sorokin addressed himself to the tripartite distinction of culture, social structure, and personality as abstracted aspects of all human action. This can be seen from the title of the book: *Society, Culture and Personality*. Although he attends to all three aspects, even in his earlier work, it is plain that in his sociology of science Sorokin asserts the dominance of culture over the other two aspects of social structure and personality. It is being asserted, apparently, that deep-rooted cultural assumptions override any variation in social structure and in diverse types of personality, producing a basic uniformity of outlook that is characteristic of men living in a particular kind of culture. Cultural mentality is regarded as fundamental; social structure and personality as producing, at most, minor variations on culturally embedded themes.

There is much to be said for this position, *providing* that it is not allowed to become a barely disguised tautology. To the extent that men in a particular society do in fact share the same fundamental assumptions about a reality significant to them, to the extent that they share much the same values, they will indeed tend to express this in their behavior and in their works. But the extent to which they do share these orientations and values is of course a question of fact, rather than an assumption of concept. It is a matter for inquiry, not a matter of conviction, to find out how far this

7. Ibid., 1:73.
8. Maquet, *Sociology of Knowledge*, p. 135.

obtains in different cultures. In short, a major problem of inquiry is to find out how far there obtains a consensus of outlook and to explain the differences that are found to exist among men living in the "same" culture. Sorokin has addressed himself to this question, but only in part. His data, as we shall now see, demonstrate substantial variability *within* each particular type of culture, but Sorokin's theoretical commitments and preferences are such that he does not go on to examine some of the implications of this variability in order to extend and refine his theory of the sociology of science. In a word, it is being said that Sorokin has confined himself to a first approximation—to be sure, a vast and comprehensive one—but one which sets excessive limits to a sociology of science that must attend also to the bases of the considerable variability of scientific outlook within the same culture. The different conceptions set forth within the same culture, it is being suggested, need not be merely minor variations on a major theme; some of these variations are precisely those that lead to basic developments in thought and science. Committed to the first approximation—to the focus on dominant tendencies within a culture—Sorokin largely shuts himself out from *analyzing* those variations which often make for the advancement of knowledge, that which is recognized as authentic knowledge in cultures that otherwise differ in many respects.

This statement is a large claim. At the least, it must be elaborated if we are to assess the limitations as well as the contributions of Sorokin's sociology of knowledge in general, and of his sociology of science in particular.

Macro- and Microsociological Perspectives on Knowledge

Sorokin's theory of sociocultural systems is offered as a partial description and analysis of the whole vast sweep of Greco-Roman and Western societies—with something more than a casual orientation to Eastern societies—during the last three thousand years or so. The theory may fairly be described as a macrosociological perspective.[9] It attends to the gross rather than the microscopic features of each society and culture under view. The centuries-long periods of culture described as ideational, idealistic, or sensate, for example, are characterized in terms of their dominant traits. Discriminations within each culture are largely excluded by the very scope of the conception. This exclusion, we must realize, is imposed by the theoretical commitment, not by the external reality. Concretely, it means, for example, that for Sorokin the last four or five hundred years in the West comprise a single sociocultural type dominated by a sensate culture mentality. All the substantial variations of science and knowledge generally that are to be found in this period are, under Sorokin's compre-

9. For a comparison of the macrosociology and microsociology of knowledge, see Stark, *Sociology of Knowledge*, pp. 19–37; also Maquet, *Sociology of Knowledge*.

hensive scheme of analysis, regarded as expressions of one fundamental orientation toward reality.

Now, first approximations in approaches to social reality have a way of concealing, deep within them, basic commitments to values. For what is singled out as *fundamental* is what the observer takes to be that which "really matters." And in the same obvious sense, the variations that are excluded from notice by the observer's conception are thereby regarded as inconsequential for what the observer regards as significant. So it is that by characterizing the entire period as sensate, Sorokin does not direct his analytical attention to the differing kinds of scientific work that are to be found in that uniformly sensate period. This is a perspective which, precisely because it is macroscopic, throws together, for all pertinent purposes, the work of a Galileo, Kepler, and Newton, on the one hand, and the work of a Rutherford, Einstein, and, shall we say, Yang and Lee, on the other. It thus excludes from analysis the great differences that, for many human and intellectual purposes, are to be found in the science of the sixteenth or seventeenth century and the twentieth. It is a gross approximation that threatens to usurp the attention of those who have reason to regard the variability *within* the macroscopic sensate period as also fundamental.

These remarks may be enough to raise the second question that, in our view, needs to be dealt with by Sorokin in reexamining his sociology of knowledge and of science. What components of his theory will help us to account for the variability of thought within the societies and eras which Sorokin assigns to one or another of his sociocultural types? Has Sorokin imposed excessive limits upon his theory by a commitment to a kind of macroscopic analysis that excludes from detailed investigation the very questions which many would consider central?

In one restricted sense, Sorokin does address himself to this problem, as has been noted in chapter 1 of this volume. He observes, for example, that the failure of the sensate "system of truth" (empiricism) to monopolize our sensate culture testifies to the fact that the culture is not "fully integrated." But this would seem to surrender inquiry into the bases of those very differences of thought with which our contemporary world is concerned. On the Sorokinian theory, how does one account for these differences? The same question applies to other categories and principles of knowledge with which he deals on the plane of macrosociology. He finds, for example, that in present-day sensate culture, "materialism" is less prevalent than "idealism," and that "temporalism" and "eternalism" are almost equally current, as are "realism" and "nominalism," "singularism" and "universalism." And now we come again to the decisive (and we repeat, self-imposed) limitation of Sorokin's theory: since, by his own testimony, these diverse doctrines exist within the same culture, how can the general characterization of the culture as "sensate" help us to explain why some thinkers subscribe to one mode of thought, and others to another?

The essential point is that Sorokin's theory does not lead us to explore variations of thought *within* a society or culture, for he looks to the "dominant" themes of the culture and imputes these to the culture as a whole.[10] Quite apart from the *differences* of intellectual outlook of different classes and groups, contemporary society, for example, is regarded by Sorokin as an integral example of sensate culture. On its own major premises, Sorokin's theory is primarily suited to characterize cultures in the large, not to analyze the connections between various positions in the social structure and the styles and content of thought which are distinctive of them.[11]

That the macrosociological level of analysis excludes from attention problems which are of import in understanding varied developments within a culture has been noticed by several critics of Sorokin's work. The anthropologist, Alexander Goldenweiser, soon picked up the issue but stated it so extravagantly as to convert a sound observation into a self-defeating exaggeration, saying that "the meshes of his [Sorokin's] net are spread so wide that all [?] that counts in history slips right through it."[12] And essentially the same point is made by Maquet about differing social and political systems when he notes that "some differences which are significant from a microscopic point of view are neglected. Thus, communism, capitalism, fascism are subsumable under the same category of sensate culture . . . the use of conceptual tools like the three premises of culture will let a rather large number of differences very important in regard to a narrower frame of reference escape."[13]

Inspecting the course of science through Sorokin's macrosociological lens is apt to blur specific developments in science rather than to bring them into sharp focus. For example, in his account of how culture mentalities affect the foci of attention in science—a problem important in its own right—Sorokin observes: "the scientists of Ideational culture would be more *interested* in the study of spiritual, mental, and psychological *phenomena*. . . . Scientists of Sensate culture would probably be more *in-*

10. So far as we can see, on only one occasion does Sorokin relate the internal differentiation of a society to any aspect of the types of thought obtaining in that society. This he does tangentially when he contrasts the tendency of the "clergy and religious landed aristocracy to become the leading and organizing classes in the Ideational, and capitalistic bourgeoisie, intelligentsia, professionals, and secular officials in the Sensate culture" (*Social and Cultural Dynamics*, 3:250). See also his account of the diffusion of culture among the social classes in ibid., 4:221 ff.

11. Cf. Merton, *Social Theory*, pp. 466–67.

12. "Sociologos," *Journal of Social Philosophy*, July 1938, p. 353, cited in *Dynamics*, 4:291n.

13. Maquet, *Sociology of Knowledge*, p. 199.

terested in the purely material *phenomena*"[14] (italics inserted). The *comparative* degrees of interest in these two broad classes of phenomena at any one time need not be put in question. But it does divert us from considering the import of the fact that an immense interest has in recent generations developed in the sciences of human behavior which are concerned with "spiritual, mental, and psychological phenomena." Paraphrasing Derek Price's estimate of physical scientists, we have only to remember that more than 90 percent of all social and behavioral scientists that have ever lived are still alive. This great interest in the scientific investigation of man and his works is a historical fact that requires interpretation by the sociology of science, but it is not one readily explained by Sorokin's macrosociological conceptions.

In emphasizing this general point, we should prefer not to be misunderstood. It is not being said that Sorokin's macrosociology of science is *theoretically incompatible* with the more detailed analysis of varying developments of thought and science within each of his major types of culture. It is not a matter of theoretical inconsistency but rather a matter of the kinds of inquiry in the sociology of science that tend to be emphasized and those that tend to be neglected in the macrosociological perspective. That is what we mean by saying that, in this respect, Sorokin's theory is a first approximation. It can be and, we argue, should be complemented by intensive inquiries into the connections between types of scientific work by men variously located in the social structure of a particular society.

Much the same issue is involved in Sorokin's treatment of long-run and short-run fluctuations in the modes of thought prevailing in one or another sphere of culture. Sorokin is of course primarily interested in the long-run fluctuations of culture mentalities which he regards as fundamental to all the rest. But he does attend—for example, in chapter 12 of the second volume of the *Dynamics*—to short-run changes in such scientific theories as atomism, vitalism, and mechanism in biology, abiogenesis, and corpuscular and wave theories of light, going on to note that "across the ever-recurring alternation of these theories, short-time fluctuations may also be perceived."[15] But these short-run variations do not engage Sorokin's interest; he makes no effort to investigate their social and cultural sources. More specifically, he observes, "The situation in regard to mechanistic and vitalistic conceptions in the present century appears to be one of armed conflict. Both conceptions seem to be existing side by side and both seem to be flourishing."[16] Again, Sorokin does not consider it part of his theoretical commitment to examine the social and cultural conditions under which these opposed biological theories are found in a state of armed

14. Sorokin, *Social and Cultural Dynamics,* 2:13.
15. Ibid., 2:446.
16. Ibid., 2:454.

coexistence. Yet this would plainly be a major problem for the micro-sociology of science.

Cultural Determinism and the Relative Autonomy of Subsystems

Up to this point we have treated Sorokin's general theoretical position in its more extreme and emphatically reiterated form. This position holds that the three types of "culture mentalities" alone determine the form, substance, and development of knowledge in general and of science in particular. It is compactly expressed, for example, in Sorokin's assertion that "Scientific theory thus is but an opinion made 'creditable' and 'fashionable' by the type of the prevalent culture."[17] That theories in science which are not acknowledged as valid by a substantial part of the community of scientists form no significant part of the science of the time is of course the case. But this is a far cry from concluding that scientific theory is *nothing but* a matter of accreditation and fashion. If this were so, it would negate a principal fact about the history of science, the selective accumulation of certified knowledge, albeit an accumulation that proceeds at uneven rates. Whatever else may be disputed, we can scarcely deny that there exists a greater stock of scientific knowledge today than in the remote past. There is more here than a mere matter of belief and fashion.

In point of fact, Sorokin does not confine himself to the extreme position that holds the development of science to be wholly determined by the prevailing culture mentality. Instead, he introduces two qualifications, the one emphasized as an integral part of his theory and the other treated casually and only in passing. The first qualification assigns a margin of autonomy or independence to each subsystem in a culture, especially the subsystem of disciplined thought and science; the second briefly acknowledges that a differentiated social structure as well as the dominant culture mentality affects the development of knowledge. Both qualifications, and particularly the first of these, are essential to a sound reading of Sorokin's theory.

Possibly because Sorokin himself so often emphasizes the dependence of everything in a sociocultural system upon its "cultural premises," critics of his work understandably take him to subscribe to a doctrine of rigid cultural determinism. The dominant emphasis overshadows the basic restriction upon this doctrine expressly introduced in the first chapter of his *Dynamics*, where each subsystem of a sociocultural system is seen as having a degree of autonomy or independence. Put most generally,

> The autonomy of any system means . . . the existence of some margin of choice or selection on its part with regard to the infinitely great number of varying

17. Ibid., 2:455.

external agents and objects which may influence it. It will ingest some of these and not others. . . . [O]ne of the most important "determinators" of the functioning and course of any system lies within the system itself, is inherent in it. In this sense any inwardly integrated system is an autonomous self-regulating, self-directing, or, if one prefers, "equilibrated" unity. . . . This is one of the specific aspects of the larger principle which may be called "immanent self-regulation and self-direction."[18]

The problem then becomes one of developing a theory adequate to account for the different "margins" of autonomy possessed by various kinds of institutions and other subsystems. So far as we can see, this is another gap—in our accounting, the third gap—in Sorokin's theory as it now stands. Apart from the roughly ascertainable *fact* that particular institutions have a smaller or greater measure of independence of their social and cultural environments, there seems nothing *in the theory* to help us anticipate how this will turn out for various kinds of institutional spheres in various kinds of sociocultural systems.

With regard to this problem, Sorokin would seem to hold a position formally (not of course substantively) like that adopted by Marx and Engels. In view of Sorokin's well-known opposition to Marxist theory, this statement may at first seem to be implausible, not to say extravagant. Yet when theorists are confronted with the same problem, they not infrequently converge in their formal analysis of it, however much they may differ in their substantive conclusions. And this, it seems to us, is the case with Marx and Sorokin in their treatment of the relative autonomy of institutional spheres within society. Consider only these few parallelisms of formal analysis.

Just as Sorokin in the main makes his culture mentalities the effective determinant of what develops in a sociocultural system, so, of course, Marx makes the "relations of production" the "real foundation" which "determines the general character of the social, political and intellectual processes of life."[19] Substantively, Marx and Sorokin could not be farther apart: Marx adopts a "materialistic" position in the sense of the social relations of production largely determining the superstructure of ideas;[20] Sorokin adopts an "idealistic" position in which the underlying premises and cultural mentality largely determine the general character of the society and culture, including its social relations. But both agree on the formal position of positing *primary* social or cultural determinants that nevertheless leave

18. Ibid., 1:50–51.

19. Karl Marx, *A Contribution to the Critique of Political Economy* (Chicago: C. H. Kerr, 1904), pp. 11–12.

20. As a reminder of his strongest formulation, we have only to read the sentence that follows the passage quoted from Marx in the above text "It is not the consciousness of men that determines their [social] existence, but on the contrary, their social existence determines their consciousness."

room for some degree of independence in the spheres of thought and knowledge.

Just as Sorokin postulates some measure of autonomy for social and cultural subsystems, so does the alter ego of Marx in attributing substantial autonomy to law.[21] And, as Engels goes on to say, what holds for the legal sphere holds for other spheres in the superstructure, such as science and religion, which interact with the economic base rather than remain wholly determined by it.[22]

As we have seen, Sorokin puts all this more generally in his concept of the autonomy of logically or functionally unified systems. And he applies the concept, among other cases, specifically to the institutional sphere of science in these words:

> It is not claimed that all scientific theories show, or must show such a connection [with the underlying cultural premises]; many of them can fluctuate independently of our main variables, within their limited sphere of autonomy and the immediate mental atmosphere of their compartment. . . . Due to the *Principle of Autonomy* of any really integrated system, each of the integrated currents of culture mentality studied should be expected to have some margin of this autonomy.[23]

Thus, just as Marx-Engels regard the *pressure toward internal consistency* within each institutional sphere as a source of its comparative independence of the social relations of production which are the "ultimate determinant, so Sorokin regards the *integration* of a subsystem as a source of its comparative independence of culture mentality as the "ultimate" determinant. The difference of theory is substantive rather than formal.

This brings us back, then, to the third problem of Sorokin's theory to which we have alluded: How does the theory deal with the comparative degrees of autonomy characteristic of different institutional subsystems in a society? Is the measure of autonomy the same for them all—for religion and law, for science and philosophy? Or is there a theoretical basis for assuming that the degree of autonomy characteristically differs for these subsystems? To raise the question is one thing; to supply a satisfactory answer is quite another. When we suggest that the Sorokinian theory seems to provide no answer, we consider this rather as an identifiable and instructive gap than as an observation that undercuts the basis of the theory. The following loose formulation by Engels provides only a suggested clue to the solution, rather than the solution itself: "The further the particular sphere which we are investigating is removed from the eco-

21. Friedrich Engels to Conrad Schmidt, 17 October 1890, in Karl Marx, *Selected Works*, 2 vols. (Moscow: Cooperative Publishing Co., 1936), 1: 385. (For the full quotation, see chapter 1 of this volume, p.19.—Ed.)

22. Ibid., 1: 386; Engels to Heinz Starkenburg, 25 January 1894, in ibid., 1: 392.

23. Sorokin, *Social and Cultural Dynamics*, 2:474–75.

nomic sphere and approaches that of pure abstract ideology, the more shall we find it exhibiting accidents [that is, deviations from "the expected"] in its development, the more will its curve run in zig-zag."[24] This suggestion still leaves open the difficult question of how to find out the "distance" of each institutional sphere from the economic sphere. But if Engels left the problem unresolved, so, too, it seems, does Sorokin.

That this is so is further suggested by Sorokin's passing observations on the independent functions of science in any sociocultural system. After pointing out that a society like the United States has a "highly integrated and differentiated system of science," while primitive societies have "little developed" systems of science,[25] he notes: "But in some form science will be found as a system in any culture area, because [note the functional assumption] any social group, as long as it lives, must have and does have a minimum of knowledge of the world that surrounds it, of the phenomena and objects that are important for its survival and existence. No group entirely devoid of any knowledge can exist and survive for any length of time."[26] It is not the functional assumption of some indispensable minimum of authentic knowledge that concerns us here; rather, it is that this assumption ascribes an independent function to "science" in every society, so that we see Sorokin once again implying that science is not merely the reflection of the culture mentality but has its own functional basis as well.

After this extended discussion of the principle of autonomy of subsystems in Sorokin's theory and of the unfilled gap in that theory, we may turn for a moment to the second of his restrictions on the cultural determination of science. This must be brief, not because we consider it unimportant, but because, as we have intimated in the foregoing section, Sorokin has elected to give it only fleeting attention in his own work. This restriction upon the determination of thought by the general culture mentality deals with the connections between the internal differentiation of the social structure and the character of the diversified thought that obtains in the society. It deals, in the language of the foregoing section, with problems in the microsociology of science rather than its macrosociology.

Symptomatically enough, Sorokin touches upon this only in a long footnote. Moreover, the note is not in his *Dynamics* where he most fully develops his sociology of knowledge and science but in his later general introduction to sociology, *Society, Culture and Personality*. Only there,

24. Engels to Starkenburg, 25 January 1894, in Marx, *Selected Works*, 1: 393.

25. As Sorokin informs us, he uses the term "science" as a shorthand expression for science-and-technology. In this passage he is evidently concerned with the low technology of everyday life in nonliterate societies rather than with science, strictly speaking.

26. *Social and Cultural Dynamics*, 4:111.

in discussing what he describes as the "non-logicity" of ideas, does Sorokin remark of Mannheim's analysis:

> He rightly looks for the cause in the group affiliations of a person; but . . . his theory remains vague, and in many respects incorrect. Meanwhile the real reasons for non-logicity are at hand. They are the nature of one's group affiliations and one's cultural affiliations. . . . Unfortunately how our social affiliations influence our logic and judgments is still but little known. The so-called "sociology of knowledge" has hardly reached a clear formulation of this problem.[27]

In short, Sorokin here acknowledges the saliency of the problem of how, within the same culture, differences in social status and group affiliations affect the nature of nonlogical sentiments, of logical thought, and, presumably, of scientific inquiry. No good purpose would be served in discussing Sorokin's appraisal of the current state of the sociology of knowledge, particularly since we are agreed that singularly little empirical inquiry has been developed in this field of enduring intellectual interest. What is pertinent is, not so much the appraisal of work left undone, but the *theoretical* issue that such inquiry, advocated by Sorokin, presupposes the probability of distinct lines of thought that will differ according to the social status and group affiliations of men of science and of the intellect generally. For this implies a conception of the sociology of knowledge which allows for significant variability in the ideas and knowledge developed *within* a particular culture—variability that results from social differentiation—and so supports our interpretation that Sorokin's macrosociology of knowledge does not deny in principle the pertinence of socially differentiated sources of knowledge. This, then, indicates once again the theoretical receptivity on the part of Sorokin, although he has chosen not to pursue this tack for himself, to the notion that the broad cultural mentality does not fully determine the character of knowledge but allows for significant and socially patterned variations in that knowledge.

Thus, if we take account of Sorokin's two basic qualifications to the determination of science and other knowledge by the prevailing culture mentality, we find that his theoretical position is not as far removed as it would seem from that adopted by other sociologists of science. His theory sets us the empirical task of trying to ferret out the ways in which culture and social structure affect the development of knowledge, allowing some measure of independence to the requirements internal to each branch of knowledge and science. The appearance of the volume of papers edited by Philip Allen affords an opportunity for Sorokin to set out his present

27. Sorokin, *Society, Culture and Personality*, (New York: Harper & Bro., 1947), pp. 352–53n.

thinking on this question central to the sociology of knowledge. It will then be possible to decide whether Maquet is correct, or merely vague, in his conclusion that "For Sorokin, it is certain that the existential factor [that is, the social structure] is the least important. . . . He considers that the premises of culture are really the most important factors [sic] for the determination of mental productions. . . . We can say that in reality [for Sorokin] the cultural premise truly exercises a predominant influence."[28] Whatever else can legitimately be said of Sorokin's theory, it cannot be described as a theory of "factors," of great, middling, or slight "importance." When Sorokin undertakes to translate Maquet's fuzzy expressions —such as "least important" and "most important factors"—into ideas that are definite enough to bear inspection, we shall be the better able to appreciate his current position on the place, in his theory, of the socially patterned distribution of types of knowledge that is found within each kind of culture.

Empirical Research: Quantitative Indicators in the Sociology of Science

Thus far we have attended to certain components of Sorokin's macrosociological theory of science, singling out those which give rise to theoretical issues that would profit from further clarification. In doing so, we have raised three questions about puzzles that persist in Sorokin's theory: first, how the theory escapes from an emanationist position which postulates underlying culture mentalities that seem to include, in their definition, what is later said to be an expression of these mentalities; second, and to our mind, basically, how the theory accounts for the socially patterned distributions, within a particular type of culture, of diverse modes of thought that do not correspond to the prevailing tendencies; third and correlatively, how the theory deals with the comparative degrees of autonomy characteristic of various subsystems within a sociocultural system, so that it can treat the extent of observed autonomy not simply as an empirical given but as theoretically explainable.

But since Sorokin's is an empirically connected theory, rather than one presented as a set of abstractions remote from systematically assembled data, we have now to turn to selected aspects of his empirical inquiries in the sociology of science. And here the most striking feature of Sorokin's work is the creation of massive accumulations of social and cultural statistics, designed to serve as basic empirical indicators of underlying changes in the rate and character of social and cultural changes. Surely more than any other single scholar dealing with problems in the sociology of knowledge, Sorokin has in effect heeded the maxim of the French social historian, Georges Lefebvre, "Il faut compter."

28. Maquet, *Sociology of Knowledge*, p. 202.

As one of us has had occasion to note before, "Studies in historical sociology have only begun to quarry the rich ore available in comprehensive collections of biographies and other historical evidence. Although statistical analysis of such materials cannot stand in place of detailed qualitative analysis of the historical evidence, they afford a *systematic* basis for new findings and, often, for correction of received assumptions. . . . The most extensive use of such statistical analysis is found in Sorokin's *Dynamics*."[29]

When we speak of Sorokin's "creation" of these social and cultural statistics, we do so advisedly. For, unlike the operations of governmental bureaus of the census, there are few kinds of social bookkeeping that systematically record evidence on the kinds of intellectual developments with which Sorokin's theory requires him to concern himself. (Statistics of patents for inventions in the modern period and data on the numbers of books published in various fields practically exhaust all that is readily available on these subjects.) And so, in spite of Sorokin's remarkably ambivalent attitude toward the use of sociological statistics, he found himself required, by the implications of his own theory, to assemble statistics that would testify to the degree of integration empirically found in each of his theoretically constructed types of culture.

By assembling these statistics, Sorokin boldly confronted the problem of how to find out the *extent* to which cultures are in fact integrated. Despite his vitriolic comments on the statisticians of our sensate age, he recognized that to deal with the extent of integration implies some statistical measure. Accordingly, for the field of knowledge, he developed numerical indexes of writings and authors in each time and place, had these coded and classified in appropriate categories, and thus assessed the comparative frequency and inferred influence of various systems of thought.

In the sociology of science, for example, the data cover the period from 3500 B.C. to the twentieth century, being based upon counts from such standard sources as Darmstädter's *Handbuch zur Geschichte der Naturwissenschaften und der Technik,* F. H. Garrison's *Introduction to the History of Medicine*, and the ninth edition of the *Encyclopaedia Britannica.* Counts such as these provide the basis for empirical confirmation of the theoretically derived proposition that "The rate of scientific development tends to become slow, stationary, even regressive in Ideational cultures . . . becoming rapid and growing apace in Sensate cultures.[30]

29. Merton, *Social Theory*, p. 599n. On a far less extensive scale, the use of quantitative indicators in the sociology of science will be found in R. K. Merton, *Science, Technology and Society in Seventeenth-Century England* (New York: Howard Fertig, 1970 [1938]), and in Nicholas Hans, *New Trends in Education in the Eighteenth Century* (London: Routledge, 1951).

30. Sorokin, *Social and Cultural Dynamics*, 2:125. All of chap. 3 is devoted to the presentation of such evidence.

There is neither need nor space to report the limitations of his quantitative indicators as these are set out by Sorokin.[31] In any event, he concludes that, whatever their limitations, the indicators provide a valid and reliable measure of fluctuations in the rate of scientific discovery and technological invention as well as of other intellectual and artistic expressions of the culture. That is why he is prepared to assert that "Not only do the first principles and categories of human thought fluctuate, but also most of the scientific theories of a more or less general nature."[32] Plainly, he bases his empirical conclusions very largely upon these cultural statistics.

In view of the basic part played by these statistics in his sociology of knowledge, Sorokin adopts a curiously ambivalent attitude toward them. This can be seen in his approval of the remark by Robert E. Park that his statistics are merely a concession to the prevailing sensate mentality and that "if they want 'em, let 'em have 'em."[33] Park's facetious remark was not intended to obscure the symptomatic nature of Sorokin's fundamental ambivalence toward criteria of scientific validity, an ambivalence deriving from his effort to cope with quite disparate "systems of truth." In view of the vast effort that went into compiling the cultural statistics that underlie Sorokin's work in the sociology of science, it seems safe to say that these systematic data were designed as more than mere trappings considered necessary to "convince the vulgar." The fact is that Sorokin's empirical descriptions are very largely based on these statistics. They are essential to his argument. To remove them would not be to remove a façade, leaving the essential structure of his theory intact; it would be to undercut his macrosociological theory of science and to leave it suspended in the thin air of unrestrained speculation.

This, then, leads to another, the fourth, question and this one two-pronged. In view of Sorokin's ambivalence toward social and cultural statistics, about which we shall have more to say, we must ask: What is his current and perhaps consolidated position with regard to the place of such statistics in sociological inquiry, primarily in the sociology of science and, by implication, in other branches of sociology as well? Further, how does his discussion of this question help clarify his position on the criteria of scientific truth which he adopts: does he regard systematic evidence of the kind caught up in his statistics as merely a mode of communication to his scientific compeers in a sensate culture or as a substantial basis both for confirming and developing his theory?

It is these cultural statistics, moreover, that serve to highlight once again two of the principal questions that we consider still unresolved in

31. Ibid., pp. 125–31.
32. Ibid., p. 439.
33. Sorokin, *Sociocultural Causality, Space, Time* (New York: Russell and Russell, 1964), p. 95n. See also chapter 1 in this volume.—Ed.

Sorokin's sociology of knowledge: the question of how the theory accounts for observed variations in the modes of knowledge within a culture and the question of accounting for the distribution of these differences among various groups and strata in the social structure. Take just one case in point. Sorokin describes empiricism as "the typically sensate system of truth." The last five centuries, and more particularly the last century, represent "sensate culture *par excellence!*"[34] Yet even in this flood tide of sensate culture, Sorokin's statistical indices show only some 53 percent of influential writings to be characterized by "empiricism." Furthermore, in the earlier centuries of this sensate culture, from the late sixteenth to the mid-eighteenth, the indices of empiricism are consistently *lower* than the indices for rationalism (which, in the theory, is associated with an idealistic rather than a sensate culture). The statistical indicators, then, show that the notion of a "prevailing" system of truth needs to be greatly qualified, if it is to cover both the situations in which it represents a bare statistical "majority" and even a statistically indicated minority in the writings of a period.

Even more is implied by Sorokin's statistics. For the main purpose of our observations is not to raise the question of the extent to which Sorokin's conclusions coincide with his statistical data: it is not to ask why the sixteenth and seventeenth centuries are said to have a predominantly "sensate system of truth" in the light of these data. Rather, the purpose is to suggest that, even on Sorokin's own premises, the general characterizations of historical cultures as sensate, idealistic, or ideational constitute only a first step in the analysis, a step which must be followed by further detailed analyses of deviations from the central tendencies of the culture. Once Sorokin has properly introduced the notion of the *extent* to which historical cultures are in fact integrated, he cannot, in all theoretical conscience, treat the existence of types of knowledge which differ from the dominant tendencies as evidence of a mere "congeries" or as a merely accidental fact. It is as much a problem of the sociology of science to account for these substantial "deviations" from the central tendency as to account for these tendencies themselves. And for this, we suggest yet again, it is necessary to develop a theory of the sociostructural bases of thought in a fashion that a cultural-emanationist theory does not permit.

Apart from these theoretical implications, Sorokin's statistics presented in his *Dynamics* afford an occasion for exploring further the intellectual grounds of his ambivalence toward social and cultural statistics altogether. As is well known, Sorokin devotes a considerable portion of his book *Fads and Foibles in Modern Sociology* to an attack on "quantophrenia" or an uncritical devotion to faulty statistics. That quantitative methods in

34. Sorokin, *Social and Cultural Dynamics,* 2:51.

sociology can be, and have been, abused is surely not in question, any more than that qualitative methods, based on ill-devised and ill-confirmed impressions, can be and have been abused. And surely, no sober man will declare himself in favor of faulty craftsmanship, unsound assumptions, and mistaken inferences. The question is therefore not one of identifying this or that case of a fallacy in quantitative analysis in sociology but, rather, one of setting out the criteria and limits of sound quantitative analysis. And since so much of Sorokin's work in the sociology of science is pervaded by empirically grounded statistics, this question becomes thoroughly germane to our discussion.

What, then, are Sorokin's criteria for the appropriate use of social and cultural statistics? We find it decidedly easier to raise the question than to answer it. Indeed, we raise the question in the hope that Sorokin will seize upon the dialogue in the Allen volume devoted to his work as an occasion for giving his pointed and definite answer to it. This becomes all the more pertinent when we find that some authors are prepared to adopt an even more extreme perspective on social and cultural statistics than Sorokin's own. Werner Stark, for example, says of Sorokin's *Dynamics* that

> our criticism . . . is one of principle. His whole procedure assumes *a radice* the possibility of quantifying what is qualitative, and this is almost like supposing it is possible to square the circle. A book, or a work of art, is all quality [n.b.], because it is all spirit. . . . It is to be feared that the sociology of knowledge will never be able to get much assistance from statistical techniques. Much as we may regret the fact, it will always have to rely heavily on the more cumbersome monographic and descriptive methods.[35]

The issue is even more stark than Stark apparently supposes. For everyone in his senses would agree that what is "inherently" and "exclusively qualitative" cannot, by definition, be quantified. It is really asking too much to ask us to reject a strict tautology. But when we get down to cases, the crucial question, of course, is begged by such an affirmation; the question is precisely one of establishing criteria of what is irrefragably qualitative and of what, in some aspect and degree, can be reasonably and usefully quantified. And since, in our opinion, Sorokin has wisely and justifiably counted *aspects* of complex works of science and art, it would be helpful to have him clarify the sense in which he found these to be quantifiable.

Sorokin's restatement of his position on the issue of such quantification would be particularly instructive in view of what he has said about the issue in his *Fads and Foibles*. At one place in that work, for example, he declares that

35. Stark, *Sociology of Knowledge*, p. 280.

only through direct empathy, co-living and intuition of the psychosocial states can one grasp the essential nature and differences . . . of religious, scientific, aesthetic, ethical, legal, economic, technological, and other cultural value-systems and their subsystems. Without the direct living experience of these cultural values, they will remain *terra incognita* for our outside observer and statistical analyst. . . . These methods are useless in understanding the nature and difference between, say, Plato's and Kant's systems of philosophy, between the ethics of the Sermon on the Mount and the ethics of hate, between Euclidean and Lobachevskian geometry and between different systems of ideas generally. Only after successfully accomplishing the mysterious inner act of "understanding" each system of ideas or values, can one classify them into adequate classes, putting into one class all the identical ideas, and putting into different classes different ideas or values. Only after that, can one count them, if they are countable, and perform other operations of a mathematical or statistical nature, if they are possible. Otherwise, all observations and statistical operations are doomed to be meaningless, fruitless, and fallacious simulacra of real knowledge.[36]

It would no doubt be generally agreed that a proper understanding of cultural content is required for it to be validly classified so that specimens in each class can then be counted. The vast compilations and counts of such data in the *Dynamics* testify that Sorokin also thinks this can be done. But is it not too stringent a criterion to require a "direct living experience of the cultural values"[37] in order for them to be classified and counted? Some substantial knowledge about the materials in hand is of course necessary but this would seem to fall far short of the extreme requirement exacted by Sorokin. We cannot assume that all of Sorokin's research associates and assistants had a "direct living experience" of the many thousands of scientific discoveries, technological inventions, philosophical doctrines, and art objects which they classified and counted in order to provide an empirical test of Sorokin's ideas. It is certain that one of his research assistants, R. K. Merton, had no such demanding experience of the almost thirteen thousand discoveries and inventions he computed on the basis of the Darmstädter *Handbuch*, just as it is probable that J. W. Boldyreff, another of his assistants, had no such experience of the thousands of scholars, scientists, artists, statesmen, and so on, mentioned in the ninth edition of the *Encyclopaedia Britannica*, who were classified and assigned weights on the basis of the amount of space devoted to them in the *Encyclopaedia*.

Nevertheless, there is internal evidence that these counts were not vitiated by limited knowledge (though, we suggest, knowledge enough for

36. Sorokin, *Fads and Foibles in Modern Sociology* (Chicago: Henry Regnery, 1956), pp. 160–61.

37. On the general issue of social epistemology involved in Sorokin's statement, see "The Perspectives of Insiders and Outsiders," chapter 5 in this volume.—Ed.

the purpose in hand). For independent classifications and counts of different but theoretically related materials produced much the same empirical results. As Sorokin reports for one such case dealing with data on the "empirical system of truth (of senses)" and data on the rate of scientific discovery:

> The items and the sources were entirely different and the computations were made by different persons who were not aware of the work of the other computers. (Professors Lossky and Lapshin had no knowledge of my study, and Dr. Merton, who made the computation of the scientific discoveries, was unaware not only of my study but also of the computations made by Professors Lossky and Lapshin.) Under the circumstances, the agreement between the curve of the scientific discoveries and inventions and the curve of the fluctuations of the influence of the system of truth of senses is particularly strong evidence that the results obtained in both cases are neither incidental nor misleading.[38]

In a word, the quantification of cultural contents cannot, need not, and is not intended to reproduce the entire complex whole of each item entering into the computation. Only selected aspects and attributes are classified and counted. And for this purpose, full, detailed, and empathic understanding of each cultural item is not, apparently, required. It would therefore be instructive to have Sorokin redirect his attention to the seeming discrepancy between the actual practice employed in quantifying cultural items in the *Dynamics* and the far more demanding criteria for such quantification proposed in the *Fads and Foibles*. What Sorokin actually *does* in the one case seems to us more compelling than what he *says* in the other. In making this observation, we only adopt and adapt the sage advice of Albert Einstein: "If you want to find out anything from the theoretical physicists about the methods they use, I advise you to stick closely to one principle: don't listen to their words, fix your attention on their deeds."[39]

All this allows us to note that not the least advance in sociology during the last century or so is reflected in the growing recognition that even crude quantitative data can serve the intellectual purpose of enabling the sociologist to reject or to modify his initial hypotheses when they are in fact defective. To see this change in outlook we have only to contrast the encyclopedic efforts of a Comte with those of a Sorokin. Comte handles scattered facts gingerly and infrequently, as though they were unfamiliar and even dangerous things; he does not think of so assembling systematic arrays of data that they could, in principle, put his intuitive or reasoned guesses to the test of empirical reality. Sorokin drenches us in quantitative facts—for example, in the *Dynamics*, but not only there—and thus provides both himself and his readers with the occasion for matching theoret-

38. *Social and Cultural Dynamics*, 2:20.
39. *The World as I See It* (New York: Philosophical Library, 1934), p. 30.

ical expectations and empirical data. This practice would seem particularly required when scholars turn to the sociological drama of large-scale changes in the cultures and social structures that make up the framework of world history. For entirely qualitative claims to facts prove to be excessively pliable, easily bent to fit the requirements of a comprehensive theory. But if it is to be more than a dogma, a theory must state the empirical observations that will be taken to disprove it or, at least, to require its substantial revision. Independently collected, systematic and quantitative data supply the most demanding test called for by such an empirically-connected theory. And that Sorokin also thinks this to be the case seems implied by the way in which he has gone about his task of conducting empirical inquiries in the sociology of science.

Relativism and the Criteria of Scientific Truth

We have alluded, once or twice, to the problem confronted by Sorokin of locating his own work in one or another of the "systems of truth" which he makes distinctive of each of his three major types of culture. What criteria of truth does he employ in setting about his own work? Is he a thoroughgoing relativist, regarding scientific truth as *nothing but* a matter of satisfying the different criteria that obtain in each particular type of culture? Does he consider each system of truth just as compelling (or as arbitrary) as the next? Does he see himself as a creature of contemporary sensate culture, subject to contemporary criteria of scientific truth, or has he found an Archimedean point to stand upon, which enables him to move beyond these criteria? If so, what is this point and how does he assure himself and his prevalently sensate readers that it is an effective and justifiable one? In short, how does Sorokin try to escape the relativistic impasse?

This barrage of questions—at bottom, they of course comprise only one question—is something more than a matter of rhetoric. We are genuinely puzzled and unable to identify, with any assurance, the position taken by Sorokin on this matter. Our confusion is further confounded by what seems to be Sorokin's indecisive and possibly changing conception of science in today's sensate culture. We recall his statement that "In a Sensate society and culture, the Sensate system of truth based upon the testimony of the organs of senses has to be dominant."[40] But, it turns out, this statement is only a gross approximation. For reason enters into the system as well. The sensate method of validation requires "Mainly the reference to the testimony of the organs of senses . . . , supplemented by logical reasoning, especially in the form of mathematical reasoning. But

40. *Social and Cultural Dynamics,* 2:5.

even the well-reasoned theory remains in the stage of pure hypothesis, unproved until it is tested by the sensory facts; and it is unhesitatingly rejected if these 'facts' contradict it."[41] And, for our immediate purposes, finally, he writes that the sensate system of scientific truth "possesses some of the elements of the rationalistic system of truth in various forms; in the forms of the laws of logic which are obligatory for scientists and which are hardly mere results of the sensory experience; in that of deductions, which are incorporated in the queen of these sciences, mathematics; of many conceptual elements in the form of the fundamental concepts and principles of the sciences; and in several other forms."[42]

With this statement Sorokin seems to have returned, from a distant point of departure, close to the position which, except for turns of language, is that generally adopted by working scientists in our time. Intuition, hunch, and guess may, and often do, originate ideas, but they do not provide a sufficient basis for choosing among ideas. Logical analysis and abstract reasoning interlock with empirical inquiry and it is only when the results of these two prove consistent that contemporary scientists consider them to be an authentic part of validated scientific knowledge. However much Sorokin may on occasion seem to take joy in the system of truth described as characteristic of an idealistic culture, he nevertheless *practices* under the rules of a sensate system. That, we suppose, is what lies behind his footnoted remark: "however surprised a contemporary partisan of scientism may be at my impartiality in 'observing and ascribing' the existence of various systems of truth . . . , he has to countenance it because they are empirical facts witnessed by the testimony of our organs of senses, as will be demonstrated further. In other words, in my study I shall intentionally follow the 'empirical system of truth' which must be convincing to such a partisan of 'scientism.' "[43]

Here Sorokin says that he adopts as criteria for his own work that complex of rational discourse and empirical data which is characteristic of a sensate science. But he implies that he does so only as a *façon de parler*. Yet this reply-in-advance to our question seems facile rather than adequate. Does it mean that Sorokin as a social scientist is truly prepared to abandon empirical tests of his ideas? that he is ready to propose characterizations of historical societies and cultures which are at odds with the empirical evidence he has assembled? We suspect not. The composite of reason and ordered experience seems to us precisely what Sorokin in fact employs as a guide to his own inquiry and as a measure of the acceptability of the results of the inquiry. Intuition, scriptures, chance experiences, dreams, or whatever may be the psychological source of an

41. Ibid., p. 9.
42. Ibid., p. 11n.
43. Ibid., pp. 11–12n.

idea. (Remember only Kekulé's dream and intuited imagery of the benzene ring which converted the idea of the mere number of atoms in a molecule into the structural idea of their being arranged in a pattern resulting from the valences of different kinds of atoms.) But whatever the source, the idea itself must be explored in terms of its implications and these implications then examined in terms of how far they hold empirically.

To put the issue directly and so to afford Sorokin an occasion in this dialogue for further clarifying his position, we suggest that, whatever asides may be tucked away in footnotes, Sorokin adopts, in the course of his inquiries in the sociology of science, a thoroughgoing commitment to the combined criteria of internal consistency and empirical observation that are the mark of scientific work in our sensate age.

Sorokin's image of sensate science notwithstanding, the fact is that concepts and rules of reasoning are no mere props in modern science. They are as indispensable as the testimony of the senses. We call only one witness, although many more are waiting in the corridors of today's science:

> Our experience hitherto justifies us in believing that nature is the realization of the simplest conceivable mathematical ideas. I am convinced that we can discover by means of purely mathematical constructions the concepts and the laws connecting them with each other, which furnish the key to the understanding of natural phenomenena. Experience may suggest the appropriate mathematical concepts, but they most certainly cannot be deduced from it. Experience remains, of course, the sole criterion of the physical utility of a mathematical construction. But the creative principle resides in mathematics. In a certain sense, therefore, I hold it true that pure thought can grasp reality, as the ancients dreamed.

This is not the voice of the thirteenth-century Robert Grosseteste speaking; it is the voice of the decidedly twentieth-century Albert Einstein.[44]

Moreover, as the history of science during the last centuries testifies, not only can empirical data challenge established concepts and theories, but concepts and theories often challenge the superficial testimony of the senses. It is a familiar part of everyday practice in science to reject misleading empirical impressions when these run counter to theories that have themselves been firmly embedded in scientific thought. Any sharp separation of reason and empirical data in contemporary science must therefore distort much of the operative reality. Work in the scientific laboratory rests upon both, with one or the other raising questions that must be resolved by a congruence between them. Only then is there a reasonable prospect that an idea or a finding will enter permanently into the repertory of science. And this sensate conception of science, we suggest, is basic to Sorokin's own work, his incidental disclaimers notwithstanding. This, at

44. Einstein, *The World as I See It*, pp. 36–37.

least, is a sixth puzzle which Sorokin might helpfully unravel in his part of the dialogue.

The Selective Cumulation of Scientific Knowledge

The issue we have just identified leads us directly to still another question about Sorokin's theory of social and cultural dynamics, this one cutting deeply enough to isolate, for a moment, his sociology of science from the rest of his theory. The fact that it is an issue hoary with age does not make it any the less in point. We refer, as the caption of this section implies, to the particular sense in which science, as distinct from other spheres of culture, tends to be accumulative. In our view, this raises a question, deeply embedded in Sorokin's theory of culture change, that requires him to consolidate his role as sociological historian and as sociological theorist.

As sociological theorist, and on his own accounting, Sorokin has identified two full cycles of ideational-idealistic-and-sensate phases in Greco-Roman and Western cultures. He sees a third sensate phase beginning roughly in the fifteenth century. In his vocabulary of abstract types of culture, one ideational phase in history is much like the other; one idealistic phase is much like the next; and one sensate phase is much like the rest. For these are described and analyzed in terms of general categories and criteria, in the light of which they seem to be "of the same kind."

As sociological historian, however, Sorokin must reckon with quite another question. Whatever his theory may identify as similar *kinds* of culture phases, there remains the historical question of the extent to which prior cultural products accumulate and become the possession of men living in a later period of the same or differing cultural type. The cycles of cultural change do not start anew. Particularly with regard to science, each succeeding historical phase makes use of antecedent knowledge on which it builds. In this more nearly concrete, historical sense, the sensate phase of the last centuries is *not*, of course, identical with the sensate phase of the preceding cycles. The phases are *alike*, in terms of the abstract categories employed by Sorokin, else they would not be classified as sensate. But they differ—and science remains our test case of this—in that some of the cultural products of the past are available to those living in the later phase. Science did not start anew in the sensate phase, said to begin early in the sixteenth century; as historians of science periodically remind us, it built upon the selective accumulation of what had gone before.

All this seems evident enough. Yet, possibly because Sorokin is adamantean in his rejection of a unilinear doctrine of cultural change, he tends to neglect the *implications* of selective cultural accumulation[45] for his theory. It is this accumulation and its consequences that distinguish the sensate culture of the twentieth century from the sensate culture of, say, the Hellenistic period. To describe both periods as sensate is justified only abstractly but not historically. For the accumulation of scientific and technological knowledge makes a difference that can make a very great difference to men living in the later of these sensate phases. To say that the thesis of a unilinear accumulation of knowledge cannot qualify as historical truth is one thing. But to ignore selective accumulation of knowledge is quite another. (To put it vulgarly, the Hellenistic Greeks did not have a body of knowledge about quantum mechanics or a technology of spacecraft; or to scramble legend and history, Icarus really cannot be equated with the astronauts Gagarin and Glenn.)

What we have been saying raises two related questions about Sorokin's macrosociology of science. These are questions about what he is prepared to take as significant similarities and what as significant differences in the scientific knowledge found in historical eras of the same abstract type.

The first question comes to the fore when we examine again his observations on short-run fluctuations of particular scientific theories in various periods.[46] He summarizes his judgment in these words: "as far as mere oscillation is concerned, there probably has been no scientific theory which has not undergone it, and, like a fashion, now has been heralded as the last word of science, and now has fallen into disrepute."[47] This judgment leads us to ask in what sense recurrent sets of ideas constitute one and the same theory that now finds general acceptance and later, rejection, only to be accepted again, still later. To consider one of the instances cited by Sorokin as a case of fluctuations in a theory, in what sense is present-day "atomistic theory" to be taken as the same as "atomistic theory" in ancient Greece? Similarities are there, of course, but also, obvious and significant differences. And it is these accumulative differences in what is on the surface the same kind of scientific theory that constitute an advance in science. To attend only to the formal similarity is to jettison the historically significant differences that enable present-day atomic theory to deal with problems in science that could not, of course, even be dreamt of by the Greeks. Or take the case of fluctuations in the long

45. That is to say, Sorokin amply recognizes the fact but does not draw the possible implications of the fact for his theory. Thus: "The trend for the last four centuries has been for empiricism to rise steadily until, at the beginning of the twentieth century, it reached a unique, unprecedented [n. b.] level. . . . There was also a unique and unprecedented multiplication . . . of important discoveries and inventions in the sciences. Thus we truly live in the age of truth of senses, of a magnitude, depth, and brilliancy hardly witnessed in other cultures and periods" (*Social and Cultural Dynamics*, 2:113).

46. Ibid., vol. 2, chap. 12.

47. Ibid., 2:467.

history of "the theory" of biological evolution which has been so often traced. Darwin's was not just another version of evolution; it differed from what had gone before by beginning to specify the processes through which the evolution of species took place. Again, to identify Darwin's or later evolutionary theory with ancient versions is to ignore that aspect of the development of science that leads to an enlarged scientific knowledge: selective accumulation. To attend only to similarities between early and later versions of a theory is to become subject to adumbrationism, "the practice of claiming to find dim anticipations of current scientific discoveries in older, and preferably ancient, work by the expedient of excessively liberal interpretations of what is being said now and of what was said then."[48] This is a practice that can only stir up anew the obsolete quarrel between the ancients and the moderns.

The second question raised by Sorokin's relative neglect of the selective accumulation of scientific knowledge has to do with his diagnosis of the present condition of science and his forecast for its immediate future. Knowing Sorokin's sentiments about sensate culture, we can anticipate that his picture of the present state of science will be a gloomy one, and he does not disappoint us:

> One can turn to any field of science now and find first of all a multitude of different theories and sometimes even opposite hypotheses fighting one another for "recognition" as true theory. Such as opulence of contradictions and mutual criticism does not permit any certitude, especially concerning the most important principles, and therefore fosters more and more uncertainty. . . . If such a situation continues—and empiricism, as long as it is dominant, cannot help continuing it—the incertitude will increase. . . . The boundary lines between knowledge and nonknowledge thus are bound to become less and less clear. . . . In such circumstances the truth of senses can easily give way to a truth of faith. In other words, neither doubt, nor uncertainty, nor changeability of the scientific theories can be pushed too far without destroying science itself and its truth. Contemporary science has already possibly gone too far in that direction and therefore is already exposed to danger.[49]

This is strong language. It is the prophetic utterance of a sociological Jeremiah. But, on *that* account, it is not to be lightly dismissed. No one who sat in Sorokin's classes during the 1930s is apt to forget his annual impassioned lecture announcing that one day men of science would create the possibility of destroying all that lives on the earth and that when that day comes, some of these men will be curious to see what really happens when the button is pressed.[50]

48. See "Singletons and Multiples in Science," chapter 16 of this volume.
49. *Social and Cultural Dynamics*, 2:119–20.
50. In less passionate prose than he employed in his lecture, Sorokin wrote in 1937: "Suppose someone should discover a simple but terrific explosive which could easily destroy a considerable part of our planet. Scientifically, it would be the greatest

From this apocalyptic vision, we return to Sorokin's forecast of a decline in science. This raises again the question of how his theory takes account of the cross-cultural accumulation of scientific knowledge. When Western society largely turned its back on science—in Sorokin's overview, from the third to the eleventh centuries—it turned from a comparatively small stock of accumulated knowledge.[51] It is vastly different in our own sensate age. We are the legatees and initiators of an incomparably greater body of scientific knowledge and of an associated technology not as easily put to one side. To base a prediction on the two preceding cycles of culture would seem hazardous at best; to predict the decline of science in our world means to discount the immensely greater store of science that has accumulated since the last sensate phase and so to treat it as though it were of a piece with the limited scientific knowledge of ancient Rome.

Nor is it evident that confidence in science as a source of knowledge shows signs of diminishing. True, we find expressions of hostility toward science, largely because of the social consequences of some of the technology it has made possible. Science is seen as originating those engines of human destruction which may plunge our civilization into everlasting night and confusion. But there is little of that alienation from science which Sorokin believes to be immanent in the very development of contemporary science. The tonicity of scientists themselves seems more aptly expressed by C. P. Snow when, speaking of what happened in science during two decades at Cambridge University, he says: "I was privileged to have a ringside view of one of the most wonderful creative periods in all physics."[52] He then goes on to describe "a much louder voice, that of another archetypal figure, Rutherford, trumpeting: 'This is the heroic age of science! This is the Elizabethan age!' "[53] And finally, he expresses his conviction that this is a revolutionary time for science, a time in which to take joy in science:

> About two years ago, he writes, one of the most astonishing experiments in the whole history of science was brought off. . . . I mean the experiment at Columbia by Yang and Lee. It is an experiment of the greatest beauty and originality, but the result is so startling that one forgets how beautiful the experiment is.

---

discovery, but socially the most dangerous for the very existence of mankind, because out of 1,800,000,000 human beings there certainly would be few individuals who, being 'scientifically minded,' would like to test the explosive and as a result would destroy our planet. Such an explosion would be a great triumph of science. . . . This half-fantastic example shows that there must be limitations of science imposed by the reasons which are outside it, and these reasons usually come from the truth of faith and that of reason" (*Social and Cultural Dynamics*, 2:20).

51. A. C. Crombie, *Augustine to Galileo: The History of Science A.D. 400–1650* (London, 1952), chaps. 1–4.

52. C. P. Snow, *The Two Cultures and the Scientific Revolution* (London: Cambridge University Rede Lectures, 1959), p. 1.

53. Ibid., p. 4.

It makes us think once again about some of the physical world. Intuition, common sense—they are neatly stood on their heads. The result is usually known as the contradiction of parity.[54]

That each such advance in science enlarges our awareness of how little is still known is a judgment that has been endlessly reiterated by scientists, particularly the greatest among them.[55] But this does seem far removed from the portrait of uncertainty and confusion among scientists painted by Sorokin. In any case, it would be instructive to have him restate the place occupied by the fact of the accumulation of scientific knowledge in his macrosociological theory of science.

Themes of the Dialogue

And here we must stop putting questions in detail. Not that other questions fail to make their appearance as we continue to study Sorokin's macrosociology of science. We encounter the question, for example, of Sorokin's appraisal of the current condition of social science. Is it in a thoroughly parlous state, as he suggests in his *Dynamics*,[56] or is it, as he suggests later in *Society, Culture and Personality*, "entering the stage of a new synthesis and a further clarification of its logical structure"?[57] Or again, we meet the question: How does Sorokin see the relations between science and other social institutions, in particular, the institution of religion? Are the relations between the two confined, as he suggests, to those of active combat or of absorption of one by the other, with "rarely, if ever, close cooperation between them"? The matter appears more complex than that. At least, the work of Whitehead and others finds that some religions have inadvertently lent support to the pursuit of science and that, apart from the times of conspicuous conflict between them, the institutions of science and religion have not infrequently been mutually supporting.

And so we might continue to raise further questions for Sorokin to reexamine. But like those we have just mentioned, these would touch only the surface of Sorokin's sociological theory of science. It might therefore be more useful to wind up our discussion by recapitulating the puzzles and questions that seem to us unresolved in that theory. This would afford

54. Ibid., p. 15.

55. On the norm of humility in science, see "Priorities in Scientific Discovery," chapter 14 in this volume, especially pp. 303–5. The most famous expression of this norm by Newton can perhaps bear still another repetition: "I do not know what I may appear to the world, but to myself I seem to have been only like a boy playing on the seashore, and diverting myself in now and then finding a smoother pebble or a prettier shell than ordinary, whilst the great ocean of truth lay all undiscovered before me."

56. *Social and Cultural Dynamics*, 2:304.

57. *Society, Culture and Personality*, p. 30.

Professor Sorokin an occasion, in his comment on our paper, to give us the benefit of his current thinking on these issues and even, perhaps, to divest himself of ideas which once had a definite place in his evolving theory but which now, in the light of further inquiry and reflection, he no longer sees any need for retaining.

In short summary, then, these are the major questions that puzzled us as we reworked our way through Sorokin's sociology of science.

1. Does the theory really adopt an emanationist position which assumes that the principal features of science and knowledge in a particular culture merely emanate from the culture mentality that underlies it? And since the culture mentality seems to include, in its definition, what are later said to be expressions of that mentality, must we not take this as rather an implied definition than an empirically testable hypothesis?

2. What is there in the theory to account for the variability of thought and science *within* each of the societies and cultures that are generally characterized as being of one or another type: as ideational, idealistic, or sensate?

3. Since integrated subsystems in a culture are said to have a margin of autonomy, a degree of independence of their social and cultural environment, how does the theory account for the margins of autonomy exhibited by the various subsystems? By way of example, does the theory lead us to expect the same or differing margins of autonomy for science and politics, for religion, law, and the economy?

4. What place is assigned, in this theory, to the connections between social differentiation and knowledge? How does the theory deal with the possibility that differences in the social location of men of knowledge, in their statuses and group affiliations, affect the character of what they take as authentic knowledge and what they produce as new claims to knowledge?

5. In view of his variously expressed ambivalence toward social and cultural statistics, what is Sorokin's current position on the place of such statistics in the sociology of science and, by implication, in other branches of sociology as well? Does he consider cultural statistics of the kind employed in the *Dynamics* as only a means of communicating with his sensate compeers or as also a basis for testing and developing his theory? And since the very stringent criteria of sound quantitative analysis he advocates in *Fads and Foibles* do not seem fully met even by his own cultural statistics in the *Dynamics*, would we do better to take his precepts or his practice as guidelines to quantitative inquiries in sociology?

6. Which criteria of scientific truth are utilized in Sorokin's own theory? How does it escape from the relativistic impasse of making scientific truth only a matter of taste, in which each type of culture prescribes its own criteria? Does Sorokin consider each system of truth just as compelling or just as arbitrary as the next?

7. How does the theory take account of the selective accumulation of scientific knowledge? Does this accumulation make our sensate period different from the sensate cultures that have gone before?

If we are right in supposing that the foregoing questions direct us to gaps in this macrosociological theory of science, then perhaps the gaps will be bridged as we listen to the rest of this dialogue.[58]

58. Sorokin's detailed replies to our questions are set forth on pp. 474–95 of the same volume in which this critique first appeared; See Allen, *Pitirim A. Sorokin in Review*.

7 Social and Cultural Contexts of Science

1970

Begun as a doctoral dissertation in 1933 and completed two years later, this monograph was first published in 1938, appearing in *Osiris*[1] at the invitation of its founder-editor and my teacher, George Sarton. That was a time when the sociology of science lay dormant. In contrast, American sociologists in particular were wide awake to the problems of urban life, family, and community, of racial and ethnic groups, of poverty, crime, and delinquency, and all the rest of that manifold of human problems in an industrial civilization brought into prominence by the Great Depression. Preoccupied with these conspicuous problems, sociologists easily managed to avoid all study of the behavior patterns of scientists and of science as an evolving social institution. An abundance of monographs dealt with the juvenile delinquent, the hobo and saleslady, the professional thief and the professional beggar, but not one dealt with the professional scientist.

I wish it could be said that appearance of this monograph promptly repaired that condition of remarkable neglect. It cannot. A decade after its publication, that astute observer of the state of American sociology, Edward Shils, could still count "science and scientific institutions" among the undeveloped areas of sociological inquiry, proving the rule by citing this monograph as a lone "exception" to it. Nothing much had changed by 1952 when, in a foreword to Bernard Barber's *Science and the Social Order*, I puzzled over this state of continued neglect and concluded that sociologists would turn seriously to the systematic study of interaction

Originally published as "Preface: 1970," in Robert K. Merton, *Science, Technology and Society in Seventeenth-Century England* (New York: Howard Fertig, Inc., and Harper & Row, 1970), pp. vii–xxix; reprinted with permission. The writing of this essay was supported by a grant from the National Science Foundation.

1. See *Osiris* 4, no. 2 (1938): 360–632.

between science and society only when science itself came to be widely regarded as something of a social problem or as a prolific source of social problems. That crude prognosis has since been borne out by developments of the last decade or so.

This fairly recent emergence of a distinct interest in the sociology of science is the principal reason for my agreeing, not without qualms, to the issuing in book form of a monograph first published more than thirty years ago. My misgivings have been somewhat allayed by the circumstance that both historians and sociologists of science continue to discuss and criticize it. In his comprehensive article on the history of science in the new *International Encyclopedia of the Social Sciences*, for one instance, Thomas S. Kuhn argued that the monograph sets forth a conception of how "the larger culture impinges on scientific development" that needs to be incorporated in the new "direction in which the history of science must now develop." And in his vigorous and conscientious critique of the monograph, A. Rupert Hall[2] went on to express the hope that it "might be speedily reprinted." Promptly suiting my action to these words only six years later, here it is.

Looking back at this *Jugendwerk*, hopefully without the condescension that age so often adopts toward youth, I must say, in all candor, that the cadence of its prose does not give me much pleasure. But if its style is stiff and self-conscious, the exposition is, to give the author his due, reasonably clear. I can find few murky passages in it, although a fair number of turgid ones.

When it comes to the substance of the monograph, I do not consider it at all evident that I am now more competent to gauge its merits and flaws. True, I am the beneficiary of the scholarship that has since examined the questions it treats and, more particularly, of the articles and books that have been addressed to the monograph itself. But the author was far more continuously immersed in the prime facts of the historical case as he sifted and organized them in the course of intensive inquiry than I can possibly be. It is possible, however, to compensate for this deficiency in two respects. I can draw upon the advantage of hindsight to examine briefly the problems and ideas advanced in the study that have enduring interest for us today, partly as a result of the intervening scholarship of historians and sociologists of science, partly as a result of recent conspicuous changes in the relations of science to its social and cultural contexts. And nothing helps one quite so much to acquire a degree of detachment from one's own work as having survived its first appearance by several decades.

In their most general aspect, the principal questions raised in this study

2. "Merton Revisited, Or, Science and Society in the Seventeenth Century," *History of Science: An Annual Review* 2 (1963): 1–16.

are with us still. What are the modes of interplay between society, culture, and science? Do these vary in kind and extent in differing historical contexts? What makes for those sizable shifts in recruitment to the intellectual disciplines—the various sciences and humanities—that lead to great variations in their development? Among those engaged in the work of science, what makes for shifts in the foci of inquiry: from one science to another and, within each of the sciences, from one set of problems to another? Under which conditions are changes in the foci of attention the planned results of deliberate policy, and under which the largely unanticipated consequences of value commitments among scientists and those controlling the support of science? How did these matters stand while science was being institutionalized and how do they stand since its thoroughgoing institutionalization? And once science has evolved forms of internal organization, how do patterns and rates of social interaction among scientists affect the development of scientific ideas? How does a cultural emphasis upon social utility as a prime, let alone an exclusive, criterion for scientific work variously affect the rate and direction of advance in science?

These are plainly questions of enough generality to be addressed to every society and historical epoch where an appreciable number of people are at work in science. What the author of this monograph undertook, with all the uninhibited innocence of youth, was to pose these general questions for the historically specific case of seventeenth-century England, all unknowing that these questions would have continuing relevance for an understanding of the place of science in society and for its internal workings. The theoretical mode in which he attacked these questions still holds a certain interest.

A principal sociological idea governing this empirical inquiry holds that the socially patterned interests, motivations, and behavior established in one institutional sphere—say, that of religion or economy—are interdependent with the socially patterned interests, motivations, and behavior obtaining in other institutional spheres—say, that of science. There are various kinds of such interdependence, but we need touch upon only one of these here. The same individuals have multiple social statuses and roles: scientific and religious and economic and political. This fundamental linkage in social structure in itself makes for some interplay between otherwise distinct institutional spheres even when they are segregated into seemingly autonomous departments of life. Beyond that, the social, intellectual, and value consequences of what is done in one institutional domain ramify into other institutions, eventually making for anticipatory and subsequent concern with the interconnections of institutions. Separate institutional spheres are only partially autonomous, not completely so. It is only after a typically prolonged development that social institutions, including the institutions of science, acquire a significant degree of autonomy.

In its bare bones, this conception of the interdependence of social institutions is plainly not a new idea—no more when this study was first carried out than now. All the same, it is an idea that has still not been thoroughly worked out in its many implications. Even now, there are scholars who would argue that science goes its own way, unaffected by changes in the environing social structure. Moreover, this is an idea that has often been distorted into a doctrine of "factors" in social development: of social, economic, religious, political, military, technological, and scientific factors in this or that historical society. It is an idea that has also been stretched into doctrines of universally dominant factors resulting in claims to "the economic determination of historical change," or its "technological determination" or "political determination," as the case may be.

It is now quite evident to me and I hope will become evident to its readers that this inquiry into the interdependence of science and other institutional spheres in seventeenth-century England neither adopts a factor theory nor supposes that the character of interchanges between institutional spheres that occurred in that period is much the same in other cultures and other times. Rather, it states in so many words that the nature and extent of these interchanges differ in various societies, depending on the state of their science and of their institutional systems of economy, politics, religion, military, and so on. This should not come as a strange idea. After all, the relations between science, economy, and government in England of the seventeenth century, when modern science and its technological offshoots were only in their beginning, differ palpably from their relations in the twentieth-century United States or Soviet Union, where science has long been institutionalized, where scientific research requires vast support, and where it has acquired new magnitudes of consequence for technologies of production and destruction. The recent highly publicized discovery of the industrial-military-scientific complex only brings to our notice tendencies toward the interdependence of science and other social institutions that have, to a degree, been present all along. That, at least, is the import of the chapters in this monograph that examine the relations of science and technology with economic development and military technique.

Another aspect of this case study in the historical sociology of science should be noted here if only because the author did not emphasize it enough to ensure its being brought to the reader's notice. By inquiring into the *reciprocal* relations between science, as an ongoing intellectual activity, and the environing social and cultural structure, the monograph managed to bypass the then current tendency—one still marked today in some quarters of historical and sociological scholarship—of giving uneven attention to the distinct directions of that reciprocity, with the impact of science (and of science-based technology) upon society eliciting much attention, and the impact of society upon science very little. The study takes seriously the notion of institutional *inter*change as it abandons the easy assumption of exclusively one-sided impact.

In this short preface, I should like, whenever possible, to give the author the benefit of the doubt. That is why I do not outline the book in detail, assuming that the structure and substance of the argument have been set out clearly enough for the reader to grasp them without great difficulty. Nevertheless, the published responses of scholars to the book through the years occasionally lead me to doubt this assumption. These responses seldom attend to the total structure of the inquiry. I should estimate that some nine of every ten discussions of the book (listed in the bibliography of commentaries and continuities in this volume) have centered on just one part of it, the one dealing with the interrelations between Puritanism and the institutionalization of science. This concentration of attention creates something of a puzzle, as can be seen when I employ the device of quantitative content analysis which George Sarton was so fond of. In his unique *Introduction to the History of Science*, whose three volumes in five books and 4,243 pages take us from the ninth century B.C. to the end of the fourteenth century A.D., Sarton would repeatedly assay the structure of a work by indicating the amount of space devoted to each of its constituent parts, just as he would quantify its citations to previous works as one way of establishing its intellectual heritage. (He could not be expected to foresee that computerized citation indices would become an important tool for the sociological analysis of contemporary scientific development.) Adapting the Sartonian procedure of quantitative content analysis to the structure of the monograph, we arrive at this arithmetical distribution:

Subject	Number of Pages	Percentage of Content
Recruitment to various occupational fields and shifts of interest among the sciences (chaps. 2, 3)	52	20
Puritanism and science hypothesis (chaps. 4–6)	82	32
Economic and military influences on spectrum of scientific investigation (chaps. 7–10 and Appendix A)	91	36
Population, social interaction, and science (chap. 11)	30	12
	255	100

This Sartonian arithmetic tells us that, if anything, somewhat more space in the monograph has been devoted to the hypotheses about economic and military influences on the range of scientific inquiry than to the hypotheses that link up Puritanism with recruitment and commitment to work in science. And yet, as I have told, the trio of chapters on the second subject has received all manner of attention in scholarly print while the quartet of chapters on the first subject has received remarkably little.

Now there is no reason, of course, for readers to attend evenly to every part of a book. Some chapters may have even less to say than others; the intellectual issues they treat may not be uniformly interesting; arid pieces of exposition which passeth all understanding may enlist the interest only of those decoders of scholarship who find themselves most challenged by thoroughgoing obscurity. Still, none of these differentiations seems to apply here. The chapters dealing with economic and military influences on science seem no more enigmatic than the ones on Puritanism and science. Perhaps the sharply uneven distribution of scholarly response strikes me as odd only because I respond to the book in quite the other way.

I find myself more partial to the section dealing with economic and military influences on the spectrum of scientific work, and for a variety of reasons. To begin with, it exhibits a little more acumen than the preceding section both in the formulation of theoretical ideas and in the method of investigation. For one thing, it distinguishes throughout, with reasonable clarity, between science and technology, an essential distinction that, as I recall, was not uniformly made back in the days when the monograph was written and one that is often blurred even today. Further, it does not for a moment adopt the simplistic choice between arguing that the selection of problems for investigation was either entirely governed by economic and military concerns or was not at all influenced by such concerns. It rejects, in other words, the mock choice between a vulgar Marxism and an equally vulgar purism. The need for this evenhanded rejection of both simplisms is widely recognized today. But when the monograph was being written—during the Great Depression, it will be remembered—vulgar Marxism was just about the only variety of Marxism that was being expounded on the periphery of American academic circles.[3]

Another conceptual distinction drawn in the analysis of economic and military influences on the development of seventeenth-century science remains basic though still often unregarded. This is the distinction between motivational and institutional levels of analysis. It is theoretically naïve to assume that such influences operate *only* through the *motivations* of scientists as they deliberately select their agenda for investigation in an explicit effort to solve concrete practical problems called to their attention. This type of case involves what the monograph describes (in chapter 10) as scientific research that is "directly related" to economic, social, or military demands. But other investigations are only indirectly related to such "practical" interests, with the investigators having no explicit concern with them. In the behavior of scientists, then as now, subjective intent and objective consequence are analytically distinct, coinciding at times and

3. I resist the temptation to note the reversion to this practice among some American academic youth today, for that would take me away from the preface to this book.

differing at others. Just as inquiries aimed at fundamental knowledge repeatedly turned up previously unsuspected applications, so inquiries aimed at application occasionally turned up new understandings of uniformities in nature. By examining several hundred investigations reported to the Royal Society in those days, the author was able to indicate the extent to which its working members devoted themselves to "pure science" as compared to science that was directly or indirectly related to contemporary economic and military concerns. The significance of this type of analysis, it now seems to me, rests less in the particular estimates of such extrinsic influences upon the spectrum of problems for investigation—these are inevitably crude—than in the mode of reasoning that escapes from the premises which assume either a fully autonomous development of science or one fully determined by extrinsic forces. The author seems to have gotten hold here of a procedure which, however coarse-grained, may not be inappropriate for analyzing the spectrum of scientific work today.

This brings us directly to another characteristic of the monograph: whenever possible, it develops corrigible quantitative data to generate and to test its principal conclusions. Thus, as we have seen, in place of merely asserting that the foci of scientific problems were entirely or not at all affected by practical tasks of the time—the kind of statement encouraged by broad qualitative observations—it provides sobering statistical evidence of estimatable magnitudes of such relationships. That is one reason why, in looking back, I cannot find it in me to say that the monograph arrived at hasty conclusions. Imperfect, perhaps, but not hasty. The statistical evidence is laboriously assembled and sometimes labored. Witness the inventory of some 6,000 biographies in the *Dictionary of National Biography*, which were sorted without benefit of the then rare IBM equipment (which in any case, in those distant days, would scarcely have been made available to a mere graduate student). And again, witness the classification of the 2,000 papers published in the *Philosophical Transactions* and hundreds of reports in the Minutes of the Royal Society as transcribed in Birch's *History*. The statistics so laboriously and, as it still seems, conscientiously assembled are employed as an impersonal check on various hypotheses about the development of science in that time and place.

Such statistical data are, of course, merely approximations to what the author should have ideally wanted. But it must have been as agreeable as it was surprising to that apprentice to find both the scientist, Joseph Needham, just then beginning work on his magisterial *Science and Civilization in China*, and the humanist, Marjorie Nicolson, even then clearly destined to become dean of the history of literary and scientific ideas, commenting favorably on this addition to the historian's art. The quantitative orientation is designed, so far as possible, to put interpretative ideas on trial by facing them with suitable compilations of statistical data, rather than relying wholly on selected bits and scraps of evidence that too often catch the scholar's eye simply because they are consistent with his ideas. This refers, in particular, to statistics put together by the historical sociologist, and rather less to the statistics generated by the social, political, or economic system of the time which are then reproduced and put to analytical use by the investigator.

Beyond its use of historical quantification, this monograph has not exactly suffered from inattention. Yet, in spite of all the excellent reasons for critically considering its other thematics, the scholars who turned their attention to it at all generally preferred, as noted earlier, to center on the hypotheses of linkages between Puritanism and science. Had educated and articulate Puritans of the seventeenth century been social scientists, they would have found this focus of interest passing strange. For they took it almost as self-evident that science made not for the dethronement of God but rather provided a means of celebrating His wisdom and the tidiness of the universe He had created.

What, then, has converted the commonplace of that time into the paradox of this time? Here is an adventitious clue. It happens that this dissertation was written in a university which, if I may put it so, enjoys a distinctly Puritan heritage, although it was written at a time when that heritage was no longer omnipresent and controlling. Indeed, the section of the dissertation dealing with Puritanism focused on what then seemed to many an improbable, not to say absurd, relation between religion and science. At least among those who had been reared on such positivistic works as John W. Draper's *History of the Conflict between Religion and Science* and A. D. White's *History of the Warfare of Science with Theology*, it was widely believed, as some still believe, that the prime historical relation between religion and science is bound to be one of conflict. An abundance of historical evidence testified that history is chock full of conflict between the two: witness only the heretical shades of Giordano Bruno and Michael Servetus. In good positivistic style, then, it was only a short leap from such empirical episodes of conflict to a belief in the logical and historical necessity for conflict. Since science was engaged in assaulting the dogmatic assumptions about reality that were tucked away in theology and associated religious beliefs and practices, or was, at least, busily nibbling away at these assumptions, a state of war between the two was constrained to be continuous and inevitable. The only concession made by scholars holding this view was that, on occasion, the inherently opposed forces of science and religion were accommodated each to the other in an effort to dampen the intensity of that warfare.

Quite another kind of interplay between seventeenth-century ascetic Protestantism and the contemporary science was conceived of in this study. It was proposed that Puritanism inadvertently contributed to the legitimacy of science as an emerging social institution. It may be of cursory interest that the author did not begin with that hypothesis. Rather, it happened in quite another way. The inquiry began as he was rummaging about in seventeenth-century England, trying to make some sense of the remarkable efflorescence of science at that time and place, being directed in the search by a general sociological orientation. The orientation was simple enough: various institutions in the society are variously interdependent so that what happens in the economic or religious realm is apt to have some perceptible connections with some of what happens in the realm of science, and conversely. In the course of reading the letters, diaries, memoirs, and papers of seventeenth-century men of science, the author slowly noted the frequent religious commitments of scientists in this time, and even more, what seemed to be their Puritan orientation. Only then, and almost as though he had not been put through his paces during the course of graduate study, was he belatedly put in mind of that intellectual tradition, established by Max Weber, Troeltsch, Tawney, and others, which centered on the interaction between the Protestant ethic and the emergence of modern capitalism. Swiftly making amends for this temporary amnesia, the author turned to a line-by-line reading of Weber's work to see whether he had anything at all to say about the relation of Puritanism to science and technology. Of course, he had. It turns out that Weber concluded his classic essay by describing one of the "next tasks" as that of searching out "the significance of ascetic rationalism, which has only been touched in the foregoing sketch, for a variety of cultural and social developments," among them "the development of philosophical and scientific empiricism . . . technical development and . . . spiritual ideals." Once identified, Weber's recommendation became a mandate.

I do not intend to review once again the accumulating details of evidence that pointed toward a significant interaction between the ethos of Puritanism and the emerging social institution of science, for that would amount to only a slightly revised version of what will be found in the book. Nor do I intend to discuss in detail the body of confirming and critical literature that has grown up about this hypothesis.[4] Instead, I shall examine the texture of the argument—its theory, if you will—with the aim of undoing some of the critical *and* appreciative misunderstandings of it that have found their way into print.

I might just as well begin with the misunderstanding that invites a muddle from which there is no easy escape. It would have been fatuous for the author to maintain, as some swift-reading commentators upon the

4. Much of this body of literature is listed in the bibliography of commentaries and continuities included in this volume. Some of it has also been discussed in recent editions of Merton's *Social Theory and Social Structure*; see, for example, the 1968 edition, pp. 649–60.—ED.

book would have him maintain, that, without Puritanism, there could have been no concentrated development of modern science in seventeenth-century England. Such an imputation betrays a basic failure to understand the logic of analysis and interpretation in historical sociology. In such analysis, a particular *concrete* historical development cannot be properly taken as indispensable to other concurrent or subsequent developments. In the case in hand, it is certainly not the case that Puritanism was indispensable in the sense that if it had not found historical expression at that time, modern science would not then have emerged. The historically concrete movement of Puritanism is not being put forward as a prerequisite to the substantial thrust of English science in that time; other functionally equivalent ideological movements could have served to provide the emerging science with widely acknowledged claims to legitimacy. The interpretation in this study assumes the functional requirement of providing socially and culturally patterned support for a not yet institutionalized science; it does not presuppose that only Puritanism could have served that function. *As it happened*, Puritanism provided major (not exclusive) support in that historical time and place. But that does not make it indispensable. However, and this requires emphasis, neither does this functional conception convert Puritanism into something epiphenomenal and inconsequential. It, rather than conceivable functional alternatives, happened to advance the institutionalization of science by providing a substantial basis for its legitimacy. But the imputed drastic simplification that would make Puritanism historically indispensable only affords a splendid specimen of the fallacy of misplaced abstraction (rather than concreteness). It would mistakenly have the author undertake an exercise in historical prophecy (to adopt the convenient term that Karl Popper uses to describe efforts at concrete historical forecasts and retrodictions), even though the much less assuming author himself has only tried his hand at an analytical interpretation in the historical sociology of science.

This brings us directly to a principal assumption underlying the entire book. The substantial and persistent development of science occurs only in societies of a certain kind, which provide both cultural and material conditions for that development. This becomes particularly evident in the early days of modern science before it was established as a major institution with its own, presumably manifest, value. Before it became widely accepted as a value in its own right, science was required to justify itself to men in terms of values other than that of knowledge itself. This underlying idea unites the several themes of the monograph, the one dealing with the role of Puritanism and the other with the role of economic and military utilities in the institutionalization of science. When the author, who now has my belated sympathy, brought these two themes together, they seemed strange bedfellows indeed. Commentators who concern themselves with segregating theoretical perspectives in historical sociology found them not merely oddly assorted but, on their very face, altogether contradictory. The theme of Puritanism-and-science seemed to exemplify the "idealistic" interpretation of history in which values and ideologies expressing those values are assigned a significant role in historical development. The theme of the economic-military-scientific interplay seemed to exemplify the "materialistic" interpretation of history in which the economic substructure determines the superstructure of which science is a part. And, as everyone knows, "idealistic" and "materialistic" interpretations are forever alien to one another, condemned to ceaseless contradiction and intellectual warfare.

Still, what everyone should know from the history of thought is that what everyone knows often turns out not to be so at all. The model of interpretation advanced in this study does provide for the mutual support and independent contribution to the legitimatizing of science of both the value orientation supplied by Puritanism and the pervasive belief in, perhaps more than the occasional fact of, scientific solutions to pressing economic, military, and technological problems. In that remote time of the seventeenth century—which somehow manages to be more than a little reminiscent of today's far different social structure and the place of science in it—the emerging men of modern science were evolving the social role of natural philosopher (scientist) and the social organization of science. For this, they required legitimation and support in all manner of ways. Not least, as the evidence recorded in this book testifies, they were at times seeking to justify the ways of science to themselves. And to make science go, they required more by way of resources and facilities than was being provided. As emblematic of this continuing need, there is the plaintive postscript by Henry Oldenburg, the secretary of the Royal Society, in a letter to Robert Boyle: "How large and useful a Philosophicall trade could I drive, had I but any competent assistance."[5]

As science grows, then, resources must grow in order that science may continue to grow. It was in this connection that religion and economy coalesced to provide arguments for the "utility" of science (often to that degree of exaggeration that reminds us of the claims put forward by some scientists and laymen today). To the implicit and sometimes explicit question raised by the doubting Thomases of the time—why practice science and why support it?—natural philosophers, clergymen, merchants, mineowners, soldiers, and civilian officials developed an impressive inventory of the diverse "utilities" of science:

5. See Henry Oldenburg, *Correspondence of Henry Oldenburg*, ed. and trans. A. Rupert Hall and Marie Boas Hall, 6 vols. (Madison: University of Wisconsin Press, 1965–69), 3:613.

1. the religious utility of exhibiting the wisdom of the divine handiwork;
2. the economic and technological utility of enabling mines to be workable at increasing depths;
3. the economic and technological utility of helping mariners to sail safely to ever more far-off places, in quest of adventure and trade;
4. the military utility of providing for ever more efficient and inexpensive ways of killing the enemy;
5. the self-development utility of providing a form of mental discipline (much as the study of Latin or even mathematics sometimes continues to be justified today); and
6. the nationalistic utility of enlarging and deepening the collective self-esteem of Englishmen as they advanced their claims to priority of discovery and invention.

The religious, economic, technological, military, and even the self-development utilities may seem to provide an extrinsic rationale for the support and cultivation of science that requires no further elaboration. But what of the nationalistic utility of extending the collective self-esteem of England? Until a few years ago, this might have seemed to many a flimsy and improbable pretext for giving support to science, but that, of course, would only have been before Sputnik. Today, to appreciate the power of the drive for ethnocentric esteem, we need only paraphrase that epitaph to Christopher Wren: *"Si exemplum requiris, circumspice,"* as we contemplate the billions of dollars happily expended by the United States to win the race to the moon.

The Americans contesting with the Russians for scientific preeminence today represent only a highly elaborated version of the English contesting with the French or Germans in the seventeenth century. The difference is one merely of magnitude of resources allocated to the ethnocentric competition, not one of kind. To give an example, Wallis, writing to Oldenburg about the new transfusion of blood between animals, voices the type of ethnocentric concern with national priority of discovery that turns up time and again among both men of science and laymen in the seventeenth century: "Onely I could wish that those of our own Nation; were a little more forward than I find them generally to bee (especially the most considerable) in timely publishing their own Discoveries, & not let strangers reape ye glory of what those amongst ourselves are ye Authors."[6]

To reap the glory of being first in scientific discovery was not, of course, only a matter of ethnocentric pride; competition in science was an intensely personal matter as well. Tucked away in a footnote in this book is an observation of the many and intense personal disputes over priority of discovery in seventeenth-century England, coupled with the suggestion

6. Oldenburg, *Correspondence*, 3:373.

that the race for priority might constitute a strategic subject for study, providing clues to the ways in which the institution of science shapes the motives, passions, and social relations of scientists. So far as I can tell, the youthful author of that footnote proved to be its only reader. At any rate, no one, not even he, heeded the muted clarion call. Slowly it dawned upon him that although it was well enough for others to ignore the plain wisdom of this prescription, he, at least, was obliged to take his own medicine. Some ten years ago I first tried to make amends for this lapse of two decades and have since examined the import of priority races for an understanding of both the institution of science and the behavior of scientists.

Ethnocentric glory, then, as well as the other diverse expressions of utilitarianism, provided a substantial basis for legitimatizing early modern science. In those days, little thought was given to the possibility that a major accent on the utility of science might eventually confine the free play of the scientific imagination. But perhaps it is just as well that men do not commonly attend to the historically distant, accumulative consequences of current commitments, or human action might be altogether paralyzed. Before science had acquired a substantial autonomy as an institution, it needed those extraneous sources of legitimation. It was only later that this dependence of science upon other institutionalized values began slowly to change. Science gradually acquired an increasing degree of autonomy, claiming legitimacy as something good in its own right, just as much so as literature and the other arts, as the quest for physical well-being or for personal salvation. The autonomous case for pure science evolved out of the derivative case for applied science. The new attitude is reflected in Ben Franklin's reply to the question put to him about the use of a new discovery: "What good is a newborn baby?"—a reply echoed by Pasteur and Faraday in the century to come, but not one readily brought to mind in the century before. The new attitude expresses a double confidence: that fundamental scientific knowledge is a self-contained good and that, as a surplus value, it will in due course lead to all manner of practical consequences serving the other varied interests of man.

The emphases upon the intrinsic and the utilitarian rationales for basic science have since varied as changing social circumstances invited differing strategy and tactics for maintaining legitimacy and enlisting support. The changing visible social consequences of science have served to shift these emphases, as we can observe in today's world. The pressing claims for the social utility of science—or to use the catchword, its "relevance"—perhaps foreshadow a new epoch limiting the spectrum of scientific inquiry. But in the seventeenth century, the sometimes excessive claims for the utilities of science were mainly prelude to its institutionalization. Once science was established with a degree of functional autonomy, the doctrine of basic scientific knowledge as a value in its own right became an integral part of the creed of scientists. Even so, the demand for "practical payoffs" continued to be periodically expressed, in greater or less degree.

Having identified the spiritual and material benefits of science, as these were expressed in seventeenth-century England, the author raises the question about the kind of social structure in which these doctrines of the utility of science are apt to emerge and prove consequential. (In one form or another, they had of course been around for a long time before that eventful century.) In rough approximation, he answers that question. Emerging new social strata, with their great expectations, found both kinds of utility congenial to them. This does not mean that the development in science was *confined* to ascetic Protestant strata or societies. But then, except in the frivolous all-or-none imputations of a few historians and sociologists, who could bring himself to allege such an *exclusive* concern with science? In this book, just as in the interpretations of economic systems set out by Weber, Troeltsch, Tawney, and all the numerous tribe that followed after them, there is no unthinking suggestion that interest in science is uniquely confined to certain groups. As with all patterns of behavior and attitudes in society, it is, of course and most significantly, a matter of degree. Within the social and cultural complex of the time, certain social strata developed, for the reasons set out, a disproportionately marked interest in science, both among the supporting population and the comparative few who actually entered upon scientific work. That is where the social arithmetic of the ascetic Protestant orientations of scientists becomes germane. It confirms this differential, not exclusive, tendency for those coming out of this religious subculture to educate for science, to provide support for science, to work at science.

But social arithmetic applied to history has its pitfalls, as we shall soon see. Some time ago, Dorothy Stimson, Raymond Stearns, and the author of this monograph arrived at much the same results (as one might hope would be the case with independent computations of substantially the same data) in finding a distinct Puritan cast among English scientists in the latter part of the seventeenth century.[7] Lately, a remarkably different set of arithmetical results has been reported. Lewis S. Feuer[8] tells us in effect

7. Reviewing a wider sweep of evidence, Ben-David notes: "In Europe, as a whole, men of Protestant origin were more numerous among scientists than were Catholics. Although the data on the religious background of scientists are not entirely reliable and it is difficult to estimate precisely the size of the religious communities in the countries from which they came, all the existing computations (including those of authors intent on disproving the hypothesis about the relationship between Protestantism and science) show that Protestants were disproportionately highly represented among scientists from the sixteenth to the end of the eighteenth century" ("The Scientific Role: The Conditions of Its Establishment in Europe," *Minerva* 4 [Autumn 1965]:46). What holds generally for Protestants in Europe holds specifically for ascetic Protestants in England.

8. *The Scientific Intellectual* (New York: Basic Books, 1963).

that these previous counts are hopelessly awry: that, in truth, of the one hundred nineteen members of the Royal Society in June 1663, "adherents to the Puritan ethics [sic]" can be counted, quite literally, on the fingers of one hand. For the rest, about whom there is enough information to judge, the great majority—forty-three or fifty or fifty-four of them (the figures vary a bit on page 421 of the Feuer book)—were what he describes as "hedonist-libertarians." This report naturally invites the nagging suspicion that something has gone wrong, badly wrong, somewhere. How can this new arithmetical result be so wholly at odds with the earlier ones? A detailed answer would run beyond the limits of this preface. It would require a report, case by case, of Feuer's sortings and countings. But we can, in passing, take note of how he reaches his extraordinary results.

On the road to his concluding statistics, Feuer alerts us by engaging in straightforward misquotation, with this monograph as the victim. To exhibit this particular art, we adopt the age-old practice—it was already prevalent in the seventeenth century—of reproducing the text which Mr. Feuer ostensibly quotes and his distinctively private version of it.

Merton Text	*Feuer "Quotation"*
It is hardly a fortuitous circumstance that the leading figures of this[9] nuclear group of the Royal Society were divines and eminently religious men, though it is not quite accurate to maintain, as did Dr. Richardson, that the beginnings of the Society were amongst a small group of learned men in which Puritan *divines* predominated. But quite clearly it is true that the originative spirits of the Society were markedly influenced by Puritan conceptions. (1970 ed., pp. 113–14)	"It is hardly a fortuitous circumstance," writes Merton, "that the leading figures of the [n.b.] nuclear group were divines and eminently religious men . . ." [the ellipsis and deletion are wholly Mr. Feuer's]. The largest active professional group in the Royal Society, on the contrary, was that of the physicians . . . [the ellipsis here is entirely mine]. How many divines, on the other hand, were there among the founders of the Royal Society? (pp. 68–69)

The ellipsis and consequent miscounting testify to a powerful (though perhaps slightly willful) imagination. To begin with, "this" nuclear group which refers to the group of 1645 is transmuted into "the" nuclear group of the society located some time in the 1660s. Then, "divines and eminently religious men" become truncated, in Feuer's actual count, into "divines" alone. And finally, to guarantee his statistical outcome beyond all doubt, Feuer simply deletes the author's own rejection of the view that "divines" rather than religious men predominated in this group, deploying a con-

9. As the reader can see for himself (on p. 112 of the 1970 edition of *Science, Technology and Society*), the antecedent of "this" nuclear group is specifically designated as the group "found in the occasional meetings of devotees of science in 1645 and immediately thereafter," including "John Wilkins, John Wallis, Jonathan Goddard, and soon afterwards, Robert Boyle and Sir William Petty, upon all of whom religious forces seem to have had a singularly strong influence."

venient ellipsis to excise from the original text the very conclusion at which he is to arrive himself. After this skillful bit of surgery, nothing more is needed than to proceed with the count showing that divines did not in truth predominate. Like most of us who are a little partial to our own ideas, the author of the monograph would doubtless want to know why Feuer responds to cogent reasoning with logic-chopping.

All this is foretaste to the agreeable elasticity of social arithmetic when one arranges it so. Another feature of Feuer's brand of nose-counting is that it takes all noses as being of a kind. He arrives at his dramatic results by taking all Fellows of the Royal Society in 1663 as having equal weight for the hypothesis under review. His list includes all courtiers—whether or not they had given any evidence of an interest, let alone an understanding of science—who could enter at will into that royally sponsored fellowship. Historians of science have ordinarily been a little more discriminating. Charles C. Gillispie (for one) has noted that "For various obscure legal reasons, the 115 names listed in the Royal Society's second charter, granted on May 20, 1663, constitute the officially recognized original Fellows." But as an historian rather than a legalist, he goes on to note that the list included substantial numbers who could be admitted at once to the fellowship simply because "they were of, or above, the degree of barons." Unlike the rest, these were not first scrutinized for evidence of scientific competence or even of interest in science.[10] Once having elected to include these courtly royalists in the relevant population, Feuer could proceed with confidence to arrive at the guaranteed but not altogether relevant results.

Still, all this is only preliminary to the final expedient that leads inescapably to the finding that "the dominant ethic of the membership as a whole of the Royal Society on May 20, 1663, was not that of the Puritan virtues; it was hedonist-libertarian." The criteria of hedonist-libertarianism might seem elastic and loose enough to serve the purpose. Even so, Feuer

10. Christopher Hill, ("The Intellectual Origins of the Royal Society—London or Oxford," *Notes and Records of the Royal Society of London* 23 [December 1968]: 144–56) has lately described the functions of this practice: "Wilkins had presented Oxford's congratulations to Oliver Cromwell when he became Lord Protector, and himself married Cromwell's sister. We can see why it was necessary that 'our company at Gresham College,' in Wallis's words, should be 'much again increased by the accession of divers eminent and noble persons upon his Majesty's return.' The first President of the Society was a peer who had some scientific pretensions but would scarcely have obtained the office for that reason alone. Peers were admitted to Fellowships without scrutiny, and the door was opened wide to gentlemanly amateurs. The ultimate results of this social dilution were unhappy; but in the short run it won Charles II's patronage for the Royal Society, which continued to be run by John Wilkins, Cromwell's brother-in-law, and Henry Oldenburg, Cromwell's admirer." In Feuer's accounting, of course, Oldenburg proudly emerges as a royalist and hedonist-libertarian, while Wilkins appears as merely a nonpartisan hedonist.

apparently does not regard this elasticity as enough to ensure the result. For he goes on to include among the criteria of "hedonistic-libertarianism" any expression of joy in scientific work. Once he has achieved this act of definition in which the pleasures of scientific inquiry become just another expression of "hedonism," he is plainly destined to arrive at the conclusion that scientists in that time (and, one must add, in every other) are surely committed to hedonism. The historian Donald Fleming says all this more crisply:

> It is only fair to emphasize that Feuer's quarrel with Calvinism is merely incidental to his vindication of "hedonism" as the mainspring of science. In this cause every scientist who ever drank a glass of beer or looked at a woman, preferably both but one will do, becomes a "hedonist." Even by these standards, Newton remains a problem; but Feuer endeavors to distract the reader's attention by a "liaison" between Newton's niece and a "notorious libertine." Where everything else fails, Feuer counts as a hedonist anybody who ever said he enjoyed doing science.[11]

Nevertheless, Feuer's use of these generously enveloping criteria does produce one puzzle beyond the reach of his predestined conclusions: in terms of these criteria, how was it possible for even five members of the Royal Society to be tagged as Puritan in their value orientations? (Legitimate curiosity should be indulged. Here are the oddly assorted five who miraculously retain their Puritan identity after this grim ordeal by classificatory fiat: "the Earl of Crawford and Lindsay, Haake [probably Theodore Haak], perhaps Hill, Viscount Massarene, and Vermuyden.") Once joy in scientific work becomes a sign of hedonism, there should be no great difficulty in establishing the intensely Puritan scientist John Ray or the Puritan physician Thomas Sydenham or the pious Robert Boyle as incorrigible hedonists. And going on from there to what was yet to come, to count in his happy company of hedonist-libertarians the altogether committed Nonconformist Joseph Priestly; the devout Sandemanian Michael Faraday, whose "religious feelings," as his biographer L. Pearce Williams has observed, "were deep and permanent"; the pious devotee of Baxter, Clerk Maxwell; the Calvinistic Sir William Ramsay; and that even more deeply committed Calvinist son of a Calvinist pastor, Leonhard Euler.

This excessively detailed and somewhat jaundiced digression into the Feuerian statistics is excusable only on one count—and not, perhaps, even on that one. It may serve to induce in the reader of this book a proper skepticism about the use of social arithmetic designed to capture the value orientations of seventeenth-century scientists. Feuer's example persuades us, if any of us needed persuasion, that, in the end, such figures can be no better than the procedures adopted to generate them.

11. Donald Fleming, "*The Scientific Intellectual*: A Review," *Isis* 56 (1965): 369–70.

Of quite another kind is the difficulty of resolving a basic issue in the scholarly debate that has centered on the question of the interaction of Puritanism and science. Deeply involved in that debate are such students of the historical case as Christopher Hill, Theodore K. Rabb, R. Hooykaas, Hugh F. Kearney, Joseph Needham and Lawrence Stone. Here, I do not take up matters of detail but focus only on the central issue. To what extent did the old Puritans turn their attention to science (and, for that matter, to commercial and industrial activity) because this interest was generated by their ethos, and to what extent was it rather the other way, with those having entered upon a career in science (or commerce and industry), for whatever other reasons, subsequently finding the values of Puritanism congenial to them? It would be satisfying to be able to answer the question, to estimate the proportions of cases which can be ascribed to the Puritanism → science sequence and to the science → Puritanism sequence. But it seems no more possible today to proceed to such a rigorous analysis than when this investigation was first begun. The needed data are simply not at hand. Yet it is something more than a banality to conjecture that both processes were at work to unknown extent. From evidence available from other spheres of activity for our own time, we know that the relations between institutional values, interests, and affiliations are typically those of reciprocal interplay. Certain religious orientations tend to make for entrance into particular economic, political, and occupational categories, just as, conversely, some who have for other reasons entered into these categories tend to adopt religious orientations which they find congenial to their ways of life. For each historical society, it is, in the end, a question of the preponderance of one or the other direction of reciprocal influence. A conclusion of this sort cannot provide the satisfaction of theoretical closure. But it serves no good purpose, particularly as scholarly polemics push toward partisan convictions, to adopt a fixed conclusion in advance of the needed evidence.

Another remark on this general point. Having lately reread this book, I must admit that only in a dozen or so places did the author take note of the partly immanent, partly accommodative, changes in the character of religious belief and commitment to science. But unschooled in the art of scholarly exposition, the author evidently did not realize that these periodic statements would be lost to view unless they were caught up in the emphatic general formulation that, interrelated as they were, both science and religion also developed under their own steam.

Even now, I am led to subscribe to the subdued concluding sentence of this aged but perhaps not yet obsolete essay.[12]

12. That sentence reads as follows: "On the basis of the foregoing study, it may not be too much to conclude that the cultural soil of seventeenth-century England was peculiarly fertile for the growth and spread of science."—Ed.

8 Changing Foci of Interests in the Sciences and Technology

1938

An unusually large number of seventeenth-century Englishmen who made their mark in science and technology first turned to these fields during the fifth and sixth decades of the century. Before exploring the causes of this apparent expansion of interest in science, it would seem wise to find out whether the fact itself is securely established. Does additional evidence sustain the impression derived from the statistical analysis of the *Dictionary of National Biography* memoirs? In any event, which of the sciences were accorded the greatest measure of attention? Did any of them maintain an unchallenged primacy throughout this period, or were there continually shifting foci of scientific interest?

Method of Study

An independent source of data which may afford a basis for answering the first of these questions is the compilation by Ludwig Darmstädter and twenty-six associated scholars: *Handbuch zur Geschichte der Naturwissenschaften und der Technik*.[1] This chronological list of important scientific and technologic discoveries and inventions is far from exhaustive, but it is the fullest list available. It contains occasional errors in the attribution and dating of discoveries,[2] but these do not vitiate our use of this material,

Originally published, in a slightly different form, as "Foci and Shifts of Interest in the Sciences and Technology," chapter 3 of Robert K. Merton, *Science, Technology and Society in Seventeenth-Century England* (Bruges, Belgium: Saint Catherine Press, Ltd., 1938); and with a new introduction (New York: Howard Fertig, Inc., and Harper and Row, 1970) pp. 38–54. Reprinted with permission.

1. Berlin: J. Springer, 1908.

2. Cf. George Sarton, *Introduction to the History of Science* (Baltimore: Williams and Wilkins, 1927–47), vol. 1 (1927), p. 144. Boris Weinberg, "Les lois d'évolution des découvertes de l'humanité," *Revue générale des sciences* 37 (1926): 43–44.

since we are not concerned with the attribution of discoveries to particular individuals. Furthermore, since the Darmstädter entries pertaining to seventeenth-century England have been verified, there is little basis for significant residual error.

The difficulties inherent in the statistical use of lists of discoveries have not been fully overcome in this study, but they have been reduced by the nature of the inquiry. Our primary objective is not so much to estimate rates of scientific advance, as to estimate the relative degrees of scientific interest reflected in output. There is consequently no need to establish a precise one-to-one correspondence between individual discoveries and the "units" of the tabulation. Each scientific increment, irrespective of its significance for scientific development, indicates interest in that field; certain strictures concerning the comparability and additiveness of heterogeneous units can be put aside for the purpose of the tabulations.[3] Each increment recorded by Darmstädter may not reflect an equal amount of interest in the corresponding science, but in the absence of any contrary evidence, it seems admissible to assume that the variations are not cumulative, that they do not, in short, lead to systematically biased results.

The procedure involved in such tabulations depends upon certain assumptions. As we have said, it is assumed that the number of scientific discoveries mentioned in Darmstädter's *Handbuch* is, in general, a function of the extent of contemporary interest in science. Manifestly, this situation does not always obtain; possibly not in the century of scientific genius here under consideration. A few brilliant scientists—Newton, Boyle, Halley—may produce more noteworthy discoveries than say a hundred times as many pedestrian investigators.[4] As an approximate index of interest in science this list taken alone will not suffice, but taken in conjunction with the data derived from the *Dictionary of National Biography* as well as with

3. For a critical discussion of the difficulties involved in statistical analysis of lists of inventions, see Floyd H. Allport and Dale A. Hartmann, "The Prediction of Cultural Change," in *Methods in Social Science,* ed. S. A. Rice (Chicago: University of Chicago Press, 1931), pp. 307–52. But contrast T. J. Rainoff, "Wave-like Fluctuations of Creative Productivity in the Development of West-European Physics," *Isis* 12 (1929): 287–88; P. A. Sorokin and R. K. Merton, "The Course of Arabian Intellectual Development: A Study in Method," *Isis* 22 (1935): 516–24; Robert K. Merton, "Fluctuations in the Rate of Industrial Invention," *The Quarterly Journal of Economics* 44 (1935): 454–74.

4. However, broadly speaking, there seems to be an appreciable correlation between the *number* of workers and discoveries in any given field and the *importance* of both the scientists and discoveries. Cf. Sorokin and Merton, "Arabian Intellectual Development," pp. 522–24; also Professor Sorokin's *Social and Cultural Dynamics* (New York: American Book Co., 1937), esp. vol. 2, chaps. 1–4, which seem to establish this point. The correlation is probably due to the fact that outstanding scientists frequently attract a cluster of less talented followers, so that periods with an unusually large number of brilliant scientists are also periods of wide interest in science. Moreover, conspicuous success in any given field, as we shall see, is apt to attract the attention of a number of mediocre as well as able investigators.

additional data garnered from independent sources, its reliability can be variously monitored.

Scientific Productivity

The tabulation based on the *Handbuch*—by counting each discovery or invention as one "unit"—is presented in the table 1. Because of the small number of items for any one science during each decade, the discoveries are not classified separately by scientific field.

TABLE 1
Number of Important Discoveries and Inventions, England, 1601–1700

Years	Number	Years	Number
1601–10	10	1651–60	13
1611–20	13	1661–70	44
1621–30	7	1671–80	29
1631–40	12	1681–90	32
1641–50	3	1691–1700	17

Compiled from Darmstädter, *Handbuch zur Geschichte der Naturwissenschaften.*

The contrast between the productivity of the two halves of the century is marked: there are three times as many discoveries in the second half as in the first. This coincides with the frequent observation by historians of science that the advance of science in England became especially notable during the latter half of the century. After a period of trendless fluctuation, there is a pronounced increase in the number of important discoveries during the decade 1661–70, after which productivity slackens considerably. The low point in scientific output is understandably reached during the period when the Civil Wars were rife. Wallis, Boyle, and others of their company frequently remarked on the distracting influences of the wars. The same pattern occurred during the internal disturbances engendered by William's entry into England in 1688, as was observed by the editor of the *Philosophical Transactions* when its interrupted publication was resumed: "The Publication of these Transactions [has] for some time past been suspended, chiefly by reason that the unsettled posture of Publick Affairs did divert the thoughts of the curious towards Matters of more immediate Concern than are Physical and Mathematical Enquiries."[5]

Internal conflict evidently retarded the acceleration of scientific research during the two middle decades of the century. The following decade, marked by the undisturbed and enlarged interaction of scientists attendant upon the inauguration of the Royal Society, was one of great scientific

5. *The Philosophical Transactions of the Royal Society of London* 17 (1693): 452.

activity. In fact, the cessation of the civil war as well as the great increase of interest in science during the decades immediately preceding[6] probably account for the marked and "sudden" appearance of fundamental discoveries in the sixties. The scientific movement had been gathering momentum for some time previous, but had been repressed by the uncertainties and disorders of the time of strife.[7]

The reliability of this tabulation of Darmstädter data can be gauged by comparing it with data concerning scientific interests derived from the *D.N.B.* memoirs, set down in table 2.

TABLE 2
Number of Initial Interests in Science and Technology, England, 1601–1700

Years	Number	Years	Number
1601–10	17	1651–60	46
1611–20	18	1661–70	41
1621–30	23	1671–80	43
1631–40	39	1681–90	38
1641–50	46	1691–1700	35

Compiled from *Dictionary of National Biography.*

It will be noted that with a lag of about a decade there is correlation between indices of initial scientific interests and of scientific productivity. Although interest in science was consistently increasing during the first half of the century, it was, as far as productivity is concerned, a period of incubation. Interaction between scientists was slight as compared with the intensive contact and discussion that attended the emergence of the "Invisible College" and the outgrowing Royal Society. Another of the discrepancies may be accounted for, as previously suggested, by the Civil Wars.[8] Moreover, the interaction between the degree of interest manifested in science and the rate of scientific output is complex. Enlarged interest in a particular field may lead with a lag to greater productivity, and this increase in the number of discoveries may in turn elicit still greater interest.

6. As we have seen from the data based on the *D.N.B.*

7. G. N. Clark adopts this same explanation for the "burgeoning" of science from 1660 on. See his excellent work, *The Later Stuarts* (Oxford: Clarendon Press, 1934), pp. 28–29.

8. Professor Sorokin has indicated that "an increase of men of genius [including scientists and inventors] during and after revolutions and wars has been manifested many times." Cf. his *Social Mobility* (New York and London: Harper & Brothers, 1927), p. 513. This was true during the public troubles of 1642–60, at which time, according to our *D. N. B.* data, the peak of notable seventeenth-century scientists in England was reached. This correlation has also been noted by R. T. Gunther, *Early Science in Oxford*, 10 vols. (Oxford: Printed for the Author, 1935), 3:330, and by the mathematician, S. Brodetsky, *Sir Isaac Newton* (London: Methuen & Co., 1927), p. 13.

But, as will be indicated presently, this process occurs only when there are no shifts of interest from science to other fields of activity.

The foregoing data suggest that scientific development in England became especially marked about the middle of the seventeenth century. As the late Martha Ornstein indicated in her exemplary study,[9] a dividing line may be drawn at this point, for the forces in Western Europe that produced the science of the early part of the century were different from those that produced the science of the later part. This is true, a fortiori, for England. The first period included Gilbert and Harvey, as well as that peer of scientific propagandists, Francis Bacon, but science in the form of a social movement did not develop until later. Beyond the coterie of great names at this time was the *popularity* of science. The new experimental philosophy became fashionable; aristocrats began to dabble in nature's arcana.[10] The new-found popularity did not result in notable scientific achievements, but it helped legitimatize the place of science.

Indices of Interest in the Sciences

Which shifts of attention occurred within the context of this enhanced interest in science; what relative variations of interest were manifested in the several sciences and technology? The best accessible indicators of such shifts and variations can be derived from the only scientific journal published in seventeenth-century England: *The Philosophical Transactions of the Royal Society of London.* But the *Philosophical Transactions* cannot furnish indices of changing scientific interests during the first part of the century since it began publication only in 1665, three years after the official inauguration of the Royal Society. For only the latter part of the century, then, will it be possible to gauge the reliability of these indicators of shifting foci of scientific interest by comparing the data derived from Darmstädter's *Handbuch* with the data derived from the *Philosophical Transactions.*

The classification of sciences employed in the tabulation of articles in the *Transactions* has been adapted from the "Systematic Classification" developed by the editors of *Isis.*[11] Each article was taken as one unit and classified in that field of science to which it chiefly pertained.[12] Reviews

9. *The Rôle of Scientific Societies in the Seventeenth Century* (Chicago: University of Chicago Press, 1938 [first printed privately in 1913]), chap. 2.

10. Thomas Sprat described the change in these words: "It [science] has begun to keep the best company, refine its fashion and appearance and become the employment of the rich and great, instead of being the subject of men's scorn" (*The History of the Royal-Society of London* [London, 1667], p. 149).

11. Cf. *Isis: International Review Devoted to the History of Science and Civilization,* passim (for example, 19 [1933]: 431 ff).

12. Compare the methods used in Hornell Hart, "Changing Social Attitudes and Interests," *Recent Social Trends* (New York: McGraw-Hill, 1933), 1:384 ff.;

of books, which usually involved lengthy discussions, were treated in the same way. The "indices of interest" in the various sciences simply represent the percentage of articles in each period devoted to each field of science. The indices are obviously crude and so minor fluctuations can be neglected. For example, the fact that there were no articles in mathematics for the year 1677 obviously does not signify a cessation of interest in the subject. The chief use of the indices is to identify *trends*, not annual variability.

Shifts of Interest between the Sciences

The results arranged by triennia for the period 1665–1702 appear in tables 3 and 4.[13] The number of articles and indices of attention are grouped by three-year periods since we want to detect trends rather than random year-to-year oscillations.

The indices identify several well-defined movements. Thus, in the formal sciences—logic, epistemology, and chiefly mathematics—there are three clearly marked "cycles." The first reaches its peak in the years 1668–70, when Newton, Wallis, and James Gregory, and to a lesser degree, John Collins, Christopher Wren, J. J. Ferguson, and Viscount Brouncker were advancing mathematics. The *Logarithmotechnia* of Nicolaus Mercator was published in 1668 and aroused interest. It was at this time (in 1669), moreover, that Newton communicated to Barrow his conception of the method of fluxions and quadrature of curves. This tract, *De Analysi per Aequationes Numero Terminorum Infinitas*, sent in turn to Collins, stimulated further interest in the subject. From that time, a slight decline continues until a new cycle of rising interest reaches its peak in the years 1681–83. It was about then that interest was revived in John Pell's *Idea of Mathematics*, first written in 1639 and sent by him to Mersenne and Descartes. Hooke included their comments in the reprinting of Pell's tract in the *Transactions*.[14] The peak of interest for the entire period was reached

Howard Becker, "Distribution of Space in the American Journal of Sociology," *American Journal of Sociology* 36 (1930): 461–66, and 38 (1932): 71–78.

13. It will be noted that there were two lapses in the publication of the *Transactions,* the first beginning in 1678 and lasting three years. In 1681, Robert Hooke began to publish the *Philosophical Collections,* which have always been considered as constituting a part of the *Transactions.* (Cf. Thomas Thomson, *History of the Royal Society* [London, 1812], p. 7 ff.) *The Collections* were published until 1683, when the secretary of the Royal Society, Robert Plot, revived the original publication, which continued until 1687, when another three-year intermission occurred. In 1691, Richard Waller and more especially Edmond Halley edited the *Transactions* once again, and since that time the publication of this journal has continued unbroken.

14. Pell never lived up to the early expectations of his friends and associates in mathematics. John Collins regarded him as "a very learned man, more knowing in algebra, in some respects, than any other." "But," he went on to say, "to incite him

TABLE 3

Classifications of Articles in Philosophical Transactions, 1665–1702

Fields of Interest	1665–67	1668–70	1671–73	1674–76	1677–78	1681–83	1684–87	1691–93	1694–96	1697–66	1700–1702	Total
A. Philosophy	2	4	6	3	1	—	1	—	—	—	—	17
B. Formal Sciences	4	20	22	8	1	7	9	2	11	13	4	101
1. Logic and Epistemology	—	1	—	1	—	—	—	—	—	—	—	
2. Mathematics	4	19	22	7	1	7	9	2	11	13	4	
C. Physical Sciences	87	79	102	94	39	50	110	29	16	59	22	686
3. Astronomy	36	20	29	39	11	27	29	6	3	15	4	
4. Physics	22	32	39	22	5	7	34	7	6	23	5	
5. Chemistry	6	14	13	16	5	10	16	7	2	12	7	
6. Technology	23	13	21	17	17	6	31	9	5	9	6	
D. Biological Sciences	34	39	53	42	17	20	26	29	17	55	34	366
7. Biology	17	14	10	24	6	4	7	8	7	16	11	
8. Botany	7	15	24	13	7	8	5	14	5	20	16	
9. Zoology	10	10	19	5	4	8	14	7	5	19	7	
E. Sciences of the Earth	24	22	20	14	5	9	21	15	14	34	8	186
10. Geodesy	5	2	1	4	—	1	3	—	—	2	1	
11. Geography and Oceanography	7	3	5	4	1	2	3	2	4	8	3	
12. Geology, Mineralogy, Paleontology	11	14	10	2	3	6	10	11	6	9	1	
13. Meteorology, Climatology	1	3	4	4	1	—	5	2	4	15	3	
F. Anthropological Sciences (physical)	26	24	23	18	11	22	33	7	14	28	20	226
14. Anatomy	13	17	16	9	8	14	24	5	8	19	9	
15. Physiology	13	7	7	9	3	8	9	2	6	9	11	
G. Anthropological Sciences (cultural)	—	3	2	11	3	6	19	8	7	12	13	84
16. History and Archaeology	—	1	1	4	2	6	15	3	7	7	5	
17. Economics	—	—	—	7	—	—	—	—	—	—	—	
18. Philology	—	2	1	—	1	—	4	2	—	4	2	
19. Political Arithmetic	—	—	—	—	—	—	—	3	—	1	6	
H. Medical Sciences	22	28	33	16	8	11	30	12	23	43	31	257
20. Pharmacy, Pharmacology	—	1	8	2	3	—	—	2	—	6	4	
21. Medicine	22	27	25	14	5	11	30	10	23	37	27	
I. Alia	14	12	12	8	3	3	7	5	12	16	15	107
Total	213	231	273	214	87	128	256	107	114	260	147	2030

TABLE 4

Indices of Interest in the Several Sciences, Philosophical Transactions, 1665–1702

Fields of Interest	1665–67	1668–70	1671–73	1674–76	1677–78	1681–83	1684–87	1691–93	1694–96	1697–99	1700–1702	Arithmetic Mean
A. Philosophy	.9	1.7	2.2	1.4	1.1	—	.4	—	—	—	—	.9
B. Formal Sciences	1.9	8.6	8.0	3.7	1.1	5.5	3.5	1.9	9.7	5.0	2.7	5.0
1. Logic and Epistemology	—	.4	—	.5	—	—	—	—	—	—	—	—
2. Mathematics	1.9	8.2	8.0	3.3	1.1	5.5	3.5	1.9	9.7	5.0	2.7	
C. Physical Sciences	40.8	34.1	37.4	43.9	43.5	39.1	42.8	27.0	14.1	22.7	15.0	33.8
3. Astronomy	16.9	8.6	10.6	18.2	12.6	21.1	11.3	5.6	2.6	5.8	2.7	
4. Physics	10.3	13.8	14.3	10.3	5.7	5.5	13.2	6.5	5.3	8.8	3.4	
5. Chemistry	2.8	6.1	4.8	7.5	5.7	7.8	6.2	6.5	1.8	4.6	4.8	
6. Technology	10.8	5.6	7.7	7.9	19.5	4.7	12.1	8.4	4.4	3.5	4.1	
D. Biological Sciences	16.0	16.9	19.5	20.0	19.5	15.7	10.1	31.1	14.9	21.2	23.2	18.0
7. Biology	8.0	6.1	3.7	11.2	6.9	3.1	2.7	7.5	6.1	6.2	7.5	
8. Botany	3.3	6.5	8.8	6.5	8.0	6.3	1.9	17.1	4.4	7.7	10.9	
9. Zoology	4.7	4.3	7.0	2.3	4.6	6.3	5.5	6.5	4.4	7.3	4.8	
E. Sciences of the Earth	11.3	9.6	7.4	6.6	5.6	7.1	8.2	14.1	12.3	13.2	5.4	9.2
10. Geodesy	2.3	.9	.4	1.9	—	.8	1.2	—	—	.8	.7	
11. Geography and Oceanography	3.3	1.3	1.8	1.9	1.1	1.6	1.2	1.9	3.5	3.1	2.0	
12. Geology, Mineralogy, Paleontology	5.2	6.1	3.7	.9	3.4	4.7	3.9	10.3	5.3	3.5	.7	
13. Meteorology, Climatology	.5	1.3	1.5	1.9	1.1	—	1.9	1.9	3.5	5.8	2.0	
F. Anthropological Sciences (physical)	12.2	10.4	8.5	8.4	12.6	17.2	12.9	6.6	12.3	10.8	13.6	11.1
14. Anatomy	6.1	7.4	5.9	4.2	9.2	10.9	9.4	4.7	7.0	7.3	6.1	
15. Physiology	6.1	3.0	2.6	4.2	3.4	6.3	3.5	1.9	5.3	3.5	7.5	
G. Anthropological Sciences (cultural)	—	1.3	.8	5.2	3.4	4.7	7.5	7.5	6.1	4.6	8.9	4.1
16. History and Archaeology	—	.4	.4	1.9	2.3	4.7	5.9	2.8	6.1	2.7	3.4	
17. Economics	—	—	—	3.3	—	—	—	—	—	—	—	
18. Philology	—	.9	.4	—	1.1	—	1.6	1.9	—	1.5	1.4	
19. Political Arithmetic	—	—	—	—	—	—	—	2.8	—	.4	4.1	
H. Medical Sciences	10.3	11.7	12.1	7.4	9.1	8.6	11.7	11.2	20.2	16.5	21.1	12.6
20. Pharmacy, Pharmacology	—	4	2.9	.9	3.4	—	—	1.9	—	2.3	2.7	
21. Medicine	10.3	11.3	9.2	6.5	5.7	8.6	11.7	9.3	20.2	14.2	18.4	
I. Alia	6.6	5.2	4.4	3.7	3.4	2.3	2.7	4.7	10.5	6.2	10.2	5.3
Total	100.0	100.0	100.0	100.0	100.0	100.0	100.0	100.0	100.0	100.0	100.0	100.0

in 1694–96, with the publication of the complete mathematical works of John Wallis. Interest centered on the problems with which Wallis had been primarily concerned: the method of finding the areas of curves by use of the infinite series (which constituted an approach to the calculus) and the fluxionary calculus. Although Newton had employed the principles of the calculus in the *Principia* (1687), its peculiar English notation did not appear until the publication of Wallis's *Algebra* in 1693. These events evoked considerable interest, as can be seen from the articles on quadrature problems and the like by Halley, Abraham DeMoivre, and Wallis himself.

Apparently, short-term fluctuations of interest in particular fields derive from developments *internal* to the discipline. That is, publications recognized as highly significant make for foci of interest in the range of problems that have been brought to light and give promise of solution.

The impression that minor fluctuations in foci of interest are primarily determined by the internal history of the science is confirmed by other kinds of data. Thus, we know from the explicit statements by their authors, that many works were brought into being as a result of Gilbert's researches on magnetism.[15] Harvey's achievements especially stimulated English scientists to engage in research in both anatomy and physiology.[16] English anatomists, such as Glisson and Wharton, turned their attention to those organs concerned with the preparation and movement of the blood: the liver and the heart.[17] The small number of investigators all told meant that the attention paid to these fields sometimes deflected interest from related subjects. The shift of interest to physiology and microscopic anatomy led to a decline of interest in surgery.[18] On occasion, the contributions of

---

to publish anything seems to be as vain an endeavour as to think of grasping the Italian Alps in order to their removal. He hath been a man accounted incommunicable." All that is remembered of Pell today is his invention of the sign ÷ for division. For Collins's observations, see S. J. Rigaud, *Correspondence of Scientific Men of the Seventeenth Century*, 2 vols. (Oxford, 1841), 1:195–97; also see the *D.N.B. s.v.* John Pell.

15. For example, works by Edward Wright, Thomas Blundeville, Marke Ridley, William Barlowe, Nathaniel Carpenter, George Hakewell, Henry Gellibrand, Henry Bond, Sir Kenelm Digby, and Sir Thomas Browne. Cf. P. F. Mottelay, *Bibliographical History of Electricity and Magnetism* (London: Charles Griffin & Co., 1922), passim.

16. Erik Nordenskiöld, *The History of Biology* (New York: A. A. Knopf, 1932), p. 147; cf. J. L. Pagel who observes: "Unmittelbar nach dieser Publikation [Harvey's *De motu cordis*] und in direkter Folge derselben wurde nun die Lösung einer ganzen Reihe wichtiger physiologischer Fragen angebahnt, und die grossen Physiologen schossen wie Pilzen aus der Erde hervor, lediglich angeregt durch Harveys Entdeckung." *Einführung in die Geschichte der Medizin,* revised and adapted by Karl Sudhoff (Berlin: S. Karger, 1915), pp. 262–63.

17. Heinrich Haeser, *Lehrbuch der Geschichte der Medizin und der epidemischen Krankheiten,* 2 vols. (Jena, 1887), 2:287–88.

18. Ibid., 2:430; Pagel, *Einführung in die Geschichte der Medizin,* p. 289; Edward

supreme scientific genius inhibited further work. The prestige accorded the work of Newton, for example, prevented his immediate successors from daring any advance upon his discoveries. Thus, in the field of hydrodynamics (even though, as Lagrange said, Newton's work was least satisfactory in this subject) no significant advance was made in England until the time of Thomas Young.[19]

Although the *general types of problems* confronting the scientist which in turn suggest a host of derivative problems may be suggested by extra-scientific factors—as we shall see later—it is the development of the derivative problems uncovered through sustained scientific study that largely accounts for the foci and shifts of attention in given sciences over the short run in this period. In this limited sense, the study of short-time fluctuations would seem the province of the historian of science rather than that of the sociologist.

This conclusion seems again warranted if we turn to the trends of interest in the physical sciences—astronomy, physics, and chemistry—during the latter seventeenth century. Up to this point the physical sciences claim by far the greater part of scientific attention but the approach of the "Age of Enlightenment" with its predominate interest in man rather than matter is heralded by the gradual decline of articles in these fields. The decline is even more marked by contrast with the increase of interest in the cultural sciences—history, archaeology, economics, philology, and political arithmetic—revealed by the indices of attention.[20]

Setting of a Problem

The data point to various trends in scientific interests during this period which can be briefly summarized. In mathematics, as we have seen, there are three short-time cycles. In the physical sciences there is a more or less continuously sustained high degree of interest until 1684–87, when a perceptible decline sets in. Irvine Masson, the historian of chemistry, notes a relapse in chemical investigation after Boyle's death;[21] a short-term trend which is reflected in our data. The biological sciences—natural history,

---

Withington, *Medical History* (London, 1894), p. 329: "The surgery of the seventeenth century is much less important that that which came before or after it, for the wonderful progress of physiology seems to have attracted the ablest minds to the study of medical problems."

19. Christopher Wordsworth, *Scholae Academicae,* p. 66 ff.; William Whewell, *History of the Inductive Sciences,* 2 vols. (New York, 1858), 1:349–50.

20. Partial corroboration of this trend is furnished by David Ogg who notes the increased popular interest in history beginning with the late seventies. Cf. his *England in the Reign of Charles II,* 2 vols. (Oxford: Clarendon Press, 1934), 2:714–15.

21. Irvine Masson, *Three Centuries of Chemistry* (London: E. Benn, 1925), p. 100.

botany, and zoology—maintain a roughly constant share of scientific attention.

The sciences of the earth—geodesy, geography and oceanography, geology, mineralogy, paleontology, meteorology, and climatology, to use the modern terminology—exhibit no marked changes. The sciences of anatomy and physiology show a slight decrease from 1665 to 1676, followed by a greater increase to 1683, followed by another small cycle. The cultural sciences—history, archaeology, philology, and political arithmetic—show an almost continuous increase, from 1668 to 1702. Finally the medical sciences—pharmacy, pharmacology, and medicine—also show an increase during the period from 1665 to 1702.

The gross trends can be further identified by grouping the sciences into two general categories: those dealing with inorganic nature, and those dealing with organic nature. In the first group are physical sciences to which we add the formal disciplines; in the latter, the biological, anthropological, and medical sciences. Sciences of the earth are omitted since they imply a knowledge of both organic and inorganic nature. The cultural sciences are also omitted. Table 5 indicates the changing emphasis on the two types of sciences: interest in the inorganic remains consistently higher than in the organic until the late eighties, when there is an increasingly pronounced shift to interest in the latter group of sciences.

These trends may only reflect a changing policy of the editors of the *Transactions* rather than actual changes in the foci of scientific interests. This is of course possible but several considerations put it in doubt. First,

TABLE 5
Shifts of Attention between Groups of Sciences

Years	Physical and Formal Sciences (B + C) Index of Attention	Sciences of Organic Life (D + F + H) Index of Attention
1665–67	42.7	38.5
1668–70	42.7	39.0
1671–73	45.4	40.1
1674–76	47.6	35.8
1677–78	44.6	41.2
1681–83	44.6	41.5
1684–87	46.3	34.7
1691–93	28.9	48.9
1694–96	23.8	47.4
1697–99	27.7	48.5
1700–1702	17.7	57.9

The letters B, C, and so on, refer to the categories in the preceding tables. The figures are percentages of the total number of articles published in the *Transactions*. The necessity for exercising care in the interpretation of these figures because of the relative base is apparent.

the editors of the *Transactions* were members of the Royal Society who were in direct touch with the leading scientists. Secondly, the articles were contributed by the comparatively few investigators of the time and so, in the nature of the case, would reflect their interests. Finally, much the same results appear from independent sources.

A comparison of the data derived from the *Philosophical Transactions* and from Darmstädter's *Handbuch* must be restricted to the rank order of the various groups of sciences since the Darmstädter material does not include enough English discoveries for this period to warrant comparison either by specific sciences or by yearly periods. However, since it is the reliability of the generalized picture of shifts of scientific interests which is in question, this is not a cramping restriction. The comparison is shown in Table 6.[22]

TABLE 6
Relative Interest in the Various Sciences, England, 1665–1702

	Darmstädter			Philosophical Transactions		
	Absolute number	Percent	Rank order	Absolute number	Percent	Rank order
Formal Sciences	5	5.2	6	101	5.4	6
Physical Sciences	54	55.6	1	686	37.6	1
Biologic Sciences	13	13.4	2	366	20.0	2
Sciences of the Earth ..	6	6.2	5	186	10.2	5
Anthropological Sciences	10	10.3	3	226	12.4	4
Medical Sciences	9	9.3	4	257	14.4	3
Total	97	100.0		1822	100.0	

The agreement between indices derived from two wholly independent sources argues for the reliability of the evidence. The rank-correlation between the two sets of data is well-nigh perfect. The one reversal, between ranks 3 and 4, represents a small difference in percentage. It is of further interest that a competent observer of scientific development during this period ranks the sciences in about the same order. Martha Ornstein[23] notes that "The greatest progress was evinced in physics, astronomy,

22. In the tabulation of the *Philosophical Transactions,* the philosophy category (17 items), the cultural anthropological category (84 items), and the *alia* category (107 items) are omitted, since there are no comparable items in the Darmstädter compilation. This reduces the *Transactions* total from 2,030 to 1,822 items. The items mentioned in Darmstädter's compilation refer, of course, only to the period 1665–1702, in order that an adequate comparison may be effected.

23. *The Role of Scientific Societies,* p. 19.

medicine and mathematics . . . ; considerable progress was shown in botany, zoology, and chemistry; least in geology and paleontology."

To be sure, the Darmstädter data cannot be used to confirm the *trends* obtained from the analysis of the *Philosophical Transactions*, since they include too few cases for trend-analysis. But the finding of the same rank-order in the two sets of material suggests that the trend data may also be sound.

We have noted that changing foci of scientific interest can be interpreted as the result of developments internal to the various sciences. But it would be misleading to assume that this is entirely so. As Rickert and Max Weber forcefully indicated through the concept of *Wertbeziehung*, scientists often choose problems for investigation that are vitally linked with major values and interests of the time.[24] Much of this study will examine some of the extra-scientific elements which significantly influenced, if they did not wholly determine, the foci of scientific interest.

24. See the profound discussion of the use of this concept in Alexander von Schelting, *Max Webers Wissenschaftslehre* (Tübingen: J. C. B. Mohr, 1934), esp. pp. 235 ff.

9 Interactions of Science and Military Technique

1935

The foci of scientific interest are determined by social forces as well as by the immanent development of science. We must therefore examine extra-scientific influences in order to comprehend more fully the reasons why scientists have applied themselves to one field of investigation rather than another. With the view of tracing such a connection between science and society, I shall indicate the ways in which military exigencies have encouraged the growth of one branch of science, especially in seventeenth-century England.[1]

The seventeenth century in England was one in which war and revolution were rife. Moreover, the dominance of firearms (both muskets and artillery) over sidearms first became marked at this time—swords and pike disappeared almost completely as weapons (except as they were incorporated in the removable bayonet about 1680). Especially notable was the use of heavy artillery, for in this field occurred a change of scale which raised new technical problems. Ever since the early fourteenth century, cannon, or "firepots," had been used in warfare, but it was not until three centuries later that they played an important role in military technology.

Leonardo was one of the first to combine military engineering and scientific prowess, as is evidenced by his polygonal fortress, steam cannon, breech-loading cannon, rifled firearms, and wheel-lock pistol. Other scientists also employed themselves with such matters. Niccolo Tartaglia, in his *Nuova scienza* (1537), dealt with the theory and practice of gunnery. Georg Hartmann invented a scale of calibers which provided a standard

Originally published as "Science and Military Technique," *Scientific Monthly* 41, no. 6 (December 1935): 542–45.

1. Compare the paper by B. Hessen in *Science at the Cross-Roads* (London: Kniga, 1932), pp. 151–212.

for the production of guns and advanced the empirical laws of firing. Galileo, in his *Dialoghi*, suggested that (ignoring air resistance) the trajectory of a projectile described a parabola; while Torricelli concerned himself in great detail with the problems of the trajectory, range, and fire zone of projectiles. Leibniz, as is evidenced by his posthumous writings, was greatly concerned with such aspects of military problems as "military medicine," "military mathematics," and "military mechanics." He also worked on a "new air-pressure gun," as did Otto von Guericke and Denis Papin. Isaac Newton, in his *Principia* (bk. 2, sect. 1–4) attempted to calculate the effect of air resistance upon the trajectory of a projectile. Johannes Bernoulli, who also studied the expansion of gunpowder gases, pointed out Newton's error, with the result that it (bk. 2, prop. 37) was eliminated in the second edition of the *Principia*. Euler continued the theory that the parabola best approximates the actual trajectory of a projectile; a subject dealt with minutely by Maupertuis.

But all this simply indicates that outstanding scientists have at times been directly and perhaps tenuously concerned with matters of military technique. In order to see the ways in which such practical exigencies stimulated research in certain fields of "pure" science, it is necessary to make a detailed study of another sort.

The technical and scientific problems set by the development of artillery in the seventeenth century were these. Interior ballistics is concerned with the formation, temperature, and volume of the gases into which the powder charge is converted by combustion, and the work performed by the expansion of these gases upon the gun, carriage, and projectile. Formulae for the velocity imparted to a projectile by the gases of given weights of gunpowder and for their reaction upon the gun and carriage must be computed to determine the correct relation of the weight of charge to the projectile's weight and length of bore, the velocity of recoil, and the like.

Not only were such nineteenth-century scientists as Gay-Lussac, Chevreul, Graham, Piobert, Cavalli, Mayevski, Otto, Neumann, Noble, and Abel concerned with these problems, but also many investigators before them. Of obvious fundamental importance for interior ballistics is the relation between pressure and volume of gases. That the volume of any gas varies inversely as its pressure was stated by Boyle in 1662 and verified independently by Mariotte some fourteen years later. Apparently Boyle was not unaware of the relation between his discovery and problems of interior ballistics, for he proposed to the Royal Society "that it might be examined what is really the expansion of gunpowder, when fired."[2] This same problem was investigated in detail by Leeuwenhoek, who, although

2. Thomas Birch, *The History of the Royal Society of London*, 4 vols. (London, 1756), 1:455.

he resided in Holland, may be considered in the stream of "English science," by virtue of the 375 papers that he sent to the Royal Society, of which he was a member. His experiments in interior ballistics, published in the *Transactions*, aroused sufficient interest to be repeated before the society by Papin.

At one of the early meetings of the society, both Boyle and Lord Brouncker, the latter being especially interested in ballistics, suggested experiments on air pressure and the expansion of gases. One of the proposed experiments dealt with the inflammation and combustion of a charge of powder—a problem basic to the noted memoirs in interior ballistics centuries later by Noble and Abel, read before the Royal Society in 1874 and 1879, respectively.

At one of the earliest meetings of the society "the lord viscount Brouncker was desired to prosecute the experiment of the recoiling of guns, and to bring it in at the next meeting."[3] These experiments were repeated and followed with great interest by the other members of the society.

Exterior ballistics is concerned with the motion of a projectile after it leaves the gun: it treats of the trajectory and the relation between the velocity of a projectile and the resistance of the air. The most notable experiments in exterior ballistics in the eighteenth and nineteenth centuries were those by Robins, Hutton, Didion, Poisson, Helie, Bashforth, Mayevski, and Siacci, but these were largely based upon scientific work of the preceding period.

As is well known, Galileo introduced his *Discorsi*, in which he dealt with the trajectory of a projectile, with an acknowledgment of the assistance rendered him by the Florentine arsenal. Moreover, as Whewell noted, the practical military applications of the doctrine of projectiles doubtlessly helped to establish the truth of Galileo's views.

The study of the free fall of bodies—so essential to exterior ballistics in its initial stages—was continued by Robert Hooke in his experiments with the fall of "steel" bullets. He followed this investigation with some experiments designed to determine the resistance of air to projectiles. This resistance, he maintained, could be tested by "shooting horizontally from the top of some high tree." Hooke went further and constructed an engine "for determining the force of gunpowder by weight"—an experiment that proved of sufficient interest to be repeated at two subsequent meetings of the society.

Christopher Wren, concerned with the invention of "offensive and defensive engines," was, with Wallis and Huygens, the first to state correctly the motion of bodies in their direct impact. This law, together with the first two laws of motion, affords a basis for an approximation to the trajectory

3. Ibid., 1:8.

of a projectile.[4] Roger Cotes, Newton's disciple who edited the second edition of the *Principia*, similarly occupied himself in his *Harmonia Mensuarum* with the motion of projectiles.

The problem of the trajectory was attacked by Halley, who demonstrated the utility of Newton's analysis in the *Principia* for this purpose. Halley even referred to the economic, as well as technical, advantages to be derived from his mathematico-mechanical formulation of the approximate trajectory of projectiles, indicating that his "rule may be of good use to all Bombardiers and Gunners, not only that they may use no more Powder than is necessary to cast their Bombs into the place assigned, but that they may shoot with much more certainty."[5]

Sir Robert Moray introduced to the Royal Society Prince Rupert's gunpowder, "in strength far exceeding the best English powder," as well as a new gun invented by the prince. Moray also suggested a series of experiments in gunnery, which were broadcast in the *Transactions*. These trials aimed to determine the relation between the quantity of powder, caliber of gun, and the carrying distance of the shot. Similar experiments were conducted by the sometime Savilian professor of astronomy, John Greaves.

It appears evident that contemporary scientists were interested in problems and investigations which pertained directly to military technology, but, what is perhaps less evident and more significant, many researches devoted to "pure science" were also related to such problems. For example, the study of the free fall of bodies, which since Galileo occupied such a prominent place in physical research, is necessary, if the trajectory and velocity of a projectile are to be determined. The connection between these problems was made explicit when an experiment was frequently performed before the Royal Society "for finding the velocity of a bullet by means of the instrument for measuring the time of falling bodies."[6]

In an effort to determine the approximate extent to which military technology tended to focus the attention of scientists upon certain problems, I have computed the number of inquiries and experiments carried on by the Royal Society which are related, directly and indirectly, to such military needs. By tabulating the number of experiments listed in the minutes of the society during four years in the later seventeenth century and classifying them in terms of the fields to which they pertained, it is possible to secure an approximate comparison of this kind. Each experiment or inquiry listed in the society's minutes (as published in Birch's *History*) is counted as one "unit" and classified in the field to which it is most closely allied. The field of military technology was classified in the following fashion:

4. See the *Principia*, "Scholium to Laws of Motion."
5. *Philosophical Transactions* 16 (1686): 3–20.
6. Birch, *History of the Royal Society*, 1:461, 474, passim.

Military Technology

A. *Directly related research:*

1. Study of trajectory and velocity of projectiles.
2. Processes of casting and improvement of arms.
3. Studies of the relation of the length of gun-barrels to the range of bullets.
4. Study of the recoil of guns.
5. Experiments with gunpowder.

B. *Indirectly related research:*

1. Compression and expansion of gases: relations between volume and pressure in the gun.
2. Strength, durability, and elasticity of metals: elastic strength of guns.
3. Free fall of a body and conjunction of its progressive movement with its free fall: determination of the trajectory of projectiles.
4. Movement of bodies through a resistant medium: approximations to the trajectory of projectiles as influenced by air resistance.

In the year 1661, the Royal Society pursued 191 separate scientific investigations. Of these, 18 (15 directly related, 3 indirectly related) or 9.4 percent were related to military technology. In 1662, of 203 projects, 23 (4 direct, 19 indirect) or 11.3 percent; in 1686, of 241 projects, 32 (22 direct, 10 indirect) or 13.3 percent; and in 1687, of 171 projects, 14 (8 direct, 6 indirect) or 8.2 percent. Thus it appears that on the average about 10 percent of the research carried on by the foremost scientific body in seventeenth-century England was devoted to some aspect of military technology. To an appreciable extent, then, this one extrascientific concern tended to focus scientific attention upon a given body of scientific problems.

The relations between military demands and the scientific development of the time were primarily of two sorts. The first involved conscious, deliberate efforts of the contemporary scientists directly to solve problems of military technology. This is what is meant by a *direct* relation. The second type is less apparent, for it concerns scientific attention to problems which, although either imposed or emphasized by military needs, seemed to the scientists as of purely scientific interest. This was called the *indirect* or derivative relation. Concern with the expansion of gases, combustibility of powder and the consequent pressure of the products of combustion, and the durability and resistance of materials may in individual instances involve no consciously felt connection with military technology, although the original impetus for such study may have come from military requirements. These derivative problems may enter into scientific tradition as problems of purely scientific interest. This is to say that scientific research often proceeds along lines largely independent of social forces once the initial problems have become evident, and in this way, much research may be related only in tenuous degree to military or economic developments. Thus science develops an autonomous corpus of investigation which has its origin in strictly scientific, not utilitarian, considerations. It is these developments (which probably constitute the greater part of science) arising from the relative autonomy of scientific work that seem to have little or no connection with social forces.

10 The Neglect of the Sociology of Science

1952

Although the interaction between science and society has been a subject of occasional interest to scholars for more than a century, there has been little effort to provide a systematic organization of the facts and ideas which comprise that subject—the sociology of science. Numerous works, particularly in recent years, have variously dealt with one or another part of the subject—for example, the writings of Bernal, Crowther, and Farrington, of Lilley, Pledge, and Hogben. But these, with the important exception of Lilley's "Social Aspects of the History of Science,"[1] have not examined the linkage between science and social structure by means of a conceptual framework that has proved effective in other branches of sociology. It is the special distinction of this book that it puts into provisional order an accumulation of otherwise fragmentary and uncoordinated materials on the interplay of science and society.

When a book is clearly and closely organized, it becomes redundant to sketch out its design in a foreword. That is certainly the case with Mr. Barber's book. There is no need to enumerate its major themes here, for he does that himself, lucidly and succinctly. But there may be some value in attempting to place his book, and what it represents, in its social setting, to consider why it is that we have had to wait so long for a book which essays the task Mr. Barber has set himself: "to get a better understanding of science by applying to it the kind of sociological analysis that has proved fruitful when directed to many other kinds of social activities." How does it happen that the sociology of science is still a largely unfulfilled promise

Originally published as foreword to Bernard Barber, *Science and the Social Order* (New York: The Free Press, 1952), pp. xi–xxiii; reprinted with permission.

1. See *Archives Internationales d'Histoire des Sciences* 28 (1949): 376–443.

rather than a highly developed special field of knowledge, cultivated jointly by social, physical, and biological scientists? What are its present resources and prospects?

It is nothing new to observe that this field has long remained in a condition of remarkable neglect. In his recent diagnosis of "the present state of American sociology," for example, Edward Shils counts the study "of science and scientific institutions" among the major undeveloped areas of sociological inquiry. The evidence for such a judgment is varied but consistent. Consider for a moment the sphere of teaching: of the several thousand classes devoted to one or another branch of sociology in American colleges and universities, fewer than a handful are devoted to the sociology of science. Textbooks which, after an appreciable lag, ordinarily mirror the foci of attention in a discipline, also bear out this impression of neglect. Among current introductory textbooks in sociology, typically designed to acquaint students with the specialized spheres of interest in the field, all deal at length with the institutions of family, state, and economy, many with the institution of religion, but very few indeed with science as a major institution in modern society. Incidental discussions of the "important role" of science in society turn up in abundance, but there is little by way of a systematic analysis of that role.

Or consider the evidence in the sphere of research. It is true, of course, that comparatively little research is conducted in sociology altogether. In contrast to the tens of thousands of inquiries reported annually in physics, chemistry, and biology or, for that matter, of the thousands in history and English literature, there are only hundreds reported for the entire field of sociology. Among these, scores of studies deal with the sociology of marriage and the family, with population and with crime, and an appreciable number deal even with the sociology of religion, but the sociology of science has not yet enlisted sufficient interest to merit separate notice in the annual catalogues of sociological research.

Another telling sign of this studied neglect is found in the social organization of social research. Particularly, though not exclusively, in social science, institutes for specialized research are typically established in response to social, economic, and political needs, as these are defined by influential groups in the society. Each "social problem" seems to generate its own complement of research centers. Thus, as public alarm is voiced at the alleged instability of the family and the rising rate of divorce, universities establish institutes specializing in research on the family; as the course of world affairs focuses attention on Russia, the Near East, or the Far East, universities establish institutes devoted to social research on these regions. Yet among these scores of research centers in the social sciences, not one is devoted on any substantial scale to the sociology of science.

The inventory of neglect need not be continued. These varied evidences all reflect the central fact that the sociology of science claims the undivided attention of only a negligible number of specialists, most of these in England and elsewhere in Europe. Among the several thousand American sociologists, not even a dozen report this as their field of primary interest. Indeed, the sociology of science has been nudged into being, not so much by sociologists as by occasional physical and biological scientists who occupy their leisure hours by working on this subject. That it is these scientists who have contributed most to what is presently known in the field can be seen from Mr. Barber's critically selected bibliography. Of the many books and articles on which he has drawn, roughly half were written by practicing physical and life scientists or by scientists who have turned to administration; more than a quarter, by historians and philosophers of science; and only the remaining fraction by sociologists. Granted that these numbers are rough approximations and that they may reflect the bent of the author who, for one reason or another, may have leaned toward writings by natural scientists. Yet other, more inclusive and less exacting, bibliographies of the sociology of science have much the same character: not many persons cultivate the field altogether and those who do are for the most part physical and biological, rather than social, scientists.

All this has left its mark on the nature of existing materials in the sociology of science. Since many of those interested in the relations between science and society are primarily engaged in other fields of inquiry, they have usually not been able to express this interest in the form of time-consuming disciplined research. They have, instead, written speculative books and articles, making use of the historical evidence lying close to hand. In these works, therefore, the historical anecdote often stands in place of systematic data and the opinion, in place of documented inference. Generalizations spring easily from a few selected particulars. Thus, Newton is noted to have been a bachelor and so, celibacy manifestly contributes to a wholehearted devotion to science; in his invasion of Egypt, Napoleon was accompanied by nearly two hundred astronomers, archaeologists, chemists, geometers, and mineralogists, from which it appears that war generally facilitates scientific advance. These works also draw, again and again, upon the same small number of empirical studies, although these are insufficient to bear the weight of the numerous conclusions which are precariously based upon them.

The circumstance that much of the material in this field has been fashioned by physical and biological scientists for whom this is an avocation rather than a major concern has left another mark, which Mr. Barber has tried to erase. Unlike the pattern in solidly established disciplines, in the sociology of science facts are typically divorced from systematic theory. Empirical observation and hypothesis do not provide mutual assistance.

Not having that direct bearing on a body of theory which makes for cumulative knowledge, the empirical studies that have been made, from time to time, by natural scientists have resulted in a thin scattering of unconnected findings rather than a chain of closely linked findings.

As a consequence of all this, the sociology of science has long been in a disordered condition: on the one hand, it is unduly speculative, with few established facts altogether and, on the other, it suffers from an excess of empiricism, since these facts are ordinarily not cast in the mold of theory. Largely absent in this field are the productive patterns of inquiry in which, as has been said, men pursue facts until they uncover ideas or pursue ideas until they uncover facts.

It is by no means inherent in the subject matter that the years and decades should have slipped by with relatively few accretions to our knowledge of the interconnections of science and society. Rather, this is the natural outcome of continued neglect. Since the social scientists having the needed grounding in theory have not ordinarily taken up empirical studies in the sociology of science and since the physical and biological scientists who have conducted such studies ordinarily lack the needed theory, it is no wonder that the growth of the field has been stunted. Not that mere numbers of students devoted to a special branch of knowledge ensure its rapid growth—some problems remain refractory to quick solution—but the converse is a truism: a field of knowledge does not prosper if it is neglected.

The slow, uncertain, and sporadic development of the sociology of science has meant that its leading ideas have grown worn with repetition. As one example among many of this, consider the history of the inferences that have been drawn from the multiple and independent appearance of the same scientific discovery, or of the same invention. It may not be too much to say that the implications about the cultural context of innovation which have been drawn from this strategic fact are among the more significant conceptions found in the sociology of science. These conceptions are, properly enough, associated with the sociologists, William F. Ogburn and Dorothy S. Thomas, who drew up a list of almost 150 independently duplicated scientific discoveries and technological inventions, pointing out that these innovations became virtually inevitable as certain types of knowledge accumulated in the cultural heritage and as social needs directed attention to particular problems.

In two respects, the history of this idea illustrates the slow pace of development in the sociology of science: first, this idea has been little elaborated or extended since it was emphasized by Ogburn and Thomas a generation ago and second, essentially the same idea regarding the sociological significance of multiple independent discoveries had been repeatedly formulated, particularly throughout the century before. As early as 1828, Macaulay, in his essay on Dryden, had noted that the independent invention of the calculus by Newton and Leibniz belonged to a larger class of instances in which the same discoveries and inventions had been made by scientists working apart from one another. This coincidence he attributed to an accumulated common stock of knowledge and to a common focus of attention. As he put it, in terms now grown old with repetition, "Mathematical science, indeed, had then reached such a point, that if neither [Leibniz nor Newton] had ever existed, the principle must inevitably have occurred to some person within a few years." Nor was Macaulay alone in this idea, which turned up in the most diverse quarters of English society. In spite of Carlyle's doctrine of culture heroes, this unheroic idea was regarded as a useful commonplace by the Victorian manufacturers who testified before Royal Commissions that, after all, inventions only constituted small and inevitable increments in existing technology as could be seen from repeated instances of the virtually simultaneous and independent appearance of the same invention. Not much later, a prominent writer who detested his own position as a Manchester manufacturer expressed the same thesis when he wrote of his partner in ideas that "while Marx discovered the materialist conception of history, Thierry, Mignet, Guizot and all the English historians up to 1850 are the proof that it was being striven for, and the discovery of the same conception by Morgan proves that the time was ripe for it and that indeed it *had* to be discovered." Meanwhile, the same conception, based on the same kind of evidence, was being promulgated in the United States. In 1885, William H. Babcock and P. B. Pierce were informing their colleagues in the Anthropological Society of Washington that the "synchronism of inventions" testifies that "the progress of a certain art has reached a point where a given step becomes inevitable" and that "this shows that the individual man is of less importance in invention than his environment." Shortly thereafter in France, Gabriel Tarde, in 1902, and Abel Rey, in 1922, were arriving at the same conclusion, observing that the simultaneity of discoveries and inventions was sufficient evidence of the crucial part played by cultural accumulation.

The point of this is not, of course, that Macaulay or the Victorian manufacturers, that Engels or the American anthropologists said it first. Nor that this multiple and, in some instances, independent rediscovery of the same idea may be regarded as an hypothesis which is confirmed by its own history. Nor again, is it to detract from the genuine contribution of Ogburn and Thomas who did so much to establish this hypothesis in sociological thinking. The point is, rather, that periodic rediscovery of the same hypothesis characteristically results from the neglect, by sociologists, of the sociology of science and technology so that this special field, in contrast to other specialized branches of sociology, has developed little that is new through the years. Very few, for example, have followed up the implications of the hypothesis to determine, through actual empirical study, the extent to which, as the hypothesis supposes, the same constituents were indeed comparably developed in the different cultures where the same discovery or invention appeared. Consequently, this hypothesis, like other hypotheses in the sociology of science, has remained substantially unextended for more than a century.[2]

It is no easy matter to say why it is that the sociology of science has for so long remained in a state of comparative lethargy. This condition is particularly anomalous, since it seems widely agreed that science constitutes one of the major dynamic forces in modern society. Possibly there are social and institutional circumstances which, largely unnoticed, combine to divert the attention of scholars and scientists from a subject which one would expect to be of central interest in a world where science looms large.

The relative neglect of this subject by physical and biological scientists perhaps requires little explanation. After all, specialization in science calls for devoted concentration of effort, and the sociology of science is not their *métier*. Hard at work on research in their own science, they are scarcely in a position to take up yet another life as sociologists. Furthermore, current practices and assumptions in the world of natural science may militate against their developing even a casual interest in the linkages of science and social structure. For example, there may prevail, among these scientists, the assumption that the history of science is comprised by a succession of great minds—an assumption with an easy plausibility since turning points in the history of science are indeed commonly associated with great scientists. Standing on such an assumption, scientists may readily lose sight of the less visible social processes which play their indispensable part. In paying its homage to these great minds, society may inadvertently reinforce that assumption. Eponymy, the practice of affixing a scientist's name to his discovery, as with Boyle's law or Planck's constant; Nobel prizes and lesser testimonials; nationalistic claims to scientific preeminence which lead to a focus on the contributions of one's own nationals; the virtual anonymity of the lesser breed of scientists whose work may be indispensable for the accumulation of scientific knowledge—these and similar circumstances may all reinforce an emphasis on the great men of science and a neglect of the social and cultural contexts which have significantly aided or hindered their achievements.

Physical and biological scientists may be reluctant to consider the bearing of social environment upon science for quite another set of reasons. They may be apprehensive that the dignity or integrity of their work might be damaged were they to recognize the implications of the fact that, as Mr. Barber points out, science is an organized social activity, that it

2. See chapters 14–17 in this volume, which treat multiple independent discoveries as data strategic for investigating the character of the reward system in science.

presupposes support by society, that the measure of this support and the types of scientific work for which it is given differ in different social structures, and that the directions of scientific advance may be appreciably affected by all this. Or perhaps their reluctance comes from the widespread and mistaken belief that to trace such connections between science and society is to impugn the motives of scientists. But, as Mr. Barber and others have shown, this belief involves a confusion between the motives of scientists and the social environment which affects the course of science. It assumes also that scientists are consistently aware of the social influences which affect their behavior, and this is by no means a self-evident truth. To consider how and how far various social structures canalize the directions of scientific research is not to arraign scientists for their motives. Nor, as Mr. Barber reminds us by emphasizing the relative autonomy of science, is it to make the institution of science the mere appendage of political, economic, and other social institutions.

Whether these are, or are not, the reasons why physical and biological scientists neglect the sociology of science, they can scarcely be the reasons why sociologists have given this field such little attention. The mythopoeic view of history has had little standing among them for generations—if anything, they are more likely to underestimate the distinctive role of great men in social change. Nor do sociologists generally assume that to study the social patterning of human behavior is to condemn the motives of those acting out those patterns—they are more likely to adopt the relativistic opinion that to understand is to excuse, that the conception of individual responsibility is alien to social determinism. It would seem, therefore, that the absence of concerted research interest in this field among sociologists must have another explanation.

Although there is little evidence on which to base an explanation, the fact itself is at once so conspicuous, and strange, that it invites conjecture. It may be that the connections between science and society constitute a subject matter which has become tarnished for academic sociologists who know that it is close to the heart of Marxist sociology. Such an attitude need not stem from a fear of guilt by association with politically condemned ideas, though this, too, may play a part.[3] Like attitudes toward most revolutionary ideas, attitudes toward Marxism have long been polarized: they have typically called for total acceptance or for total rejection. Sociologists who have come to reject the Marxist conceptions out of hand have not uncommonly rejected also the subject matters to which they pertained: American sociologists do not much study the conflicts between social classes just as they do not much study the relations between science

3. Written in 1952, this interpretation evidently alludes to the fears rampant in academic circles during the McCarthy period. Cf. Paul F. Lazarsfeld and Wagner Thielens, Jr., *The Academic Mind* (New York: Free Press, 1958).

and society. At the other pole, those who regard themselves as disciples of Marxist theory seem to act as disciples merely, content to reiterate what the masters have said or to illustrate old conclusions with newly selected examples, rather than to consider these conclusions as hypotheses which they are to test, extend, or otherwise modify through actual empirical inquiry. At both polar extremes, the sociology of science suffers, either by inattention or by preconception.

In part, also, the field is the victim of existing programs of higher education. Physical and biological scientists have typically had their rigorous training confined to the specialized skills and knowledge of their field, and few have had more than a slight acquaintance with social science. Social scientists, similarly, have typically had little training in one or another branch of the more exact sciences or even in the history of science, and consequently feel reluctant to take up a specialization for which they see themselves as unprepared. In the meantime, the sociology of science falls unnoticed between these two academic stools.

Yet to emphasize the relative neglect of this field is not to say that it is wholly barren or condemned to slow growth. Mr. Barber's book would belie any such rash claim. Actually, there are many signs that this condition of neglect is drawing to an end and that the prospects for growth are greater than ever before.

Various social tendencies, not entirely new but now more conspicuous and compelling, are forcing attention to the relations between science and its environing social structure. The politicalizing of science in Nazi Germany and in Soviet Russia, for instance, has aroused the interest of many in identifying the particular kinds of social contexts in which science thrives, a problem central to the sociology of science and one which Mr. Barber treats more systematically than has been done before. In liberal societies as well, recent changes have subjected scientists to abrupt conflicts between their several social roles and between their deep-seated values. Early in their apprenticeship, scientists have commonly acquired certain values which, as a result of changed social conditions, they are asked to unlearn and abandon at a later point in their career. The value, for example, which calls for making new-won knowledge part of the commonwealth of science now clashes with the demands made upon them, in their role as citizens, to keep some of this knowledge secret. Men previously unaware of the social contexts of their attitudes and values are apt to become acutely aware of them when they are frustrated in their aims by strains and stresses which are manifestly social in origin. Even the most artless and singleminded of scientists, living out their work-lives within the confines of the laboratory, must now know, to adapt a remark by Butterfield, that they are "not autonomous god-like creatures acting in a world of unconditioned freedom."

More particularly, these historical developments have given rise, among scientists themselves, to polemics and controversies about the "social control of science"—a hot conflict which Mr. Barber judiciously analyzes in chapter 10 of his book. However unproductive they may be in settling the points at issue, these warring opinions have had the collateral result of exciting and maintaining interest in the social relations of science at a higher pitch than ever before.

Not only scientists, but a wider public, have had their attention drawn to the social implications of science by recent events. The explosion over Hiroshima, and other experimental atomic explosions, have had the incidental consequence of awakening a dormant public concern with science. Many people who had simply taken Science for granted, except when they occasionally marveled at the Wonders of Science, have become alarmed and dismayed by these demonstrations of human destructiveness. Science has become a "social problem," like war, or the perennial decline of the family, or the periodic event of economic depressions.

Now, as we have noted, when something is widely defined as a social problem in modern Western society, it becomes a proper object for study. Particularly in American sociology, new special branches have developed in response to new sets of problems. A few generations ago, the great influx of immigrants evoked deep sociological interest in the processes of assimilation and acculturation, just as changes in the status of Negroes in American society have intensified the specialized study of race relations. So, too, the more conspicuous problems of city life so nearly usurped the attention of sociologists that, for a time, their research sites were typically in urban slums, the better to observe the behavior of juvenile delinquents, adult criminals, and other presumed aberrants. With the great diffusion of the motion picture and the appearance of radio, there began a new phase of concerted research on mass communications and public opinion, thus reviving another sociological specialty on a scale previously unknown. In more recent years, the effective organization of trade unions in this country and the attendant organization of conflict between workers and employers belatedly brought in their wake the special field of industrial sociology.

There are indications that the sociology of science, as a distinct field of specialized research, is now in much the same situation as was industrial sociology a scant twenty years ago. Previously amorphous and sporadic interest in the subject is becoming crystallized and continued. There is, however, a basic difference in the social contexts of the two fields which may make for a different outcome: industrial sociology concerned itself largely with problems involving the economic interests of industry—with problems of worker morale, with the connections between informal group structure and productivity, with relations between management and labor.

As with technological research, when it promised rich yields, so with sociological research: industry was prepared to support these studies because it was good business to do so. Profit-making organizations are constrained to make their decisions in terms of expectable profit, and, in this narrowly economic sense, sociological studies of science and the scientist hold little promise. It is from the agencies not organized for economic gain that support must come.

Out of the complex of recent historical developments—among them, the attempts to subordinate science to political control, the deepening conflicts between men's roles as scientist and as citizen, the events leading science to be widely regarded as a source of social problems—there has come a renascence of interest in the sociology of science. Thus, in cooperation with the American Academy of Arts and Sciences, Philipp Frank has lately gathered together a group of scholars to carry forward empirical and theoretical studies in this field. Another group has been formed, under the auspices of the American Council of Learned Societies, to study the humanistic, including the social, aspects of science. L'Union International d'Histoire des Sciences has broadened its scope to include a commission for the history of the social relations of science and its important first report, prepared by S. Lilley, gives ample proof of its sociological orientation. Some of the relatively few university departments of the history of science have also begun to attend to sociological considerations, and this may be expected to develop all the more rapidly as suitable research materials accumulate.

Another kind of academic development promises, in due course, to provide some of these research materials. For more than a decade, sociologists have exhibited increasing interest in the structure, role, and functions of the professions in society—medicine and the law, the ministry and engineering, among others, are being studied in their social implications. This may well carry over to comparable studies of science and scientists. Should this happen, it would have the further advantage of making for a synthesis of historical materials and of materials based on firsthand fieldwork. Until now, the great bulk of studies in the sociology of science have been based almost entirely on historical data—the documents left behind by scientists, autobiographies, diaries, and reports of scientific societies. This is of course indispensable material, but it is not sufficient. Scientists, like others, are apt to be so deeply immersed in their own work that they cannot take cognizance of the multitude of social actions and interactions which presumably take place in the laboratory, as in the factory, and which are, in significant degree, below the threshold of awareness of those involved in them. There already exists, of course, a vast literature on "scientific method" and, by inference, on the "attitudes" and "values" of scientists. But this literature is concerned with what the social scientists would call ideal patterns, that is, with ways in which scientists *ought* to think, feel, and act. It does not necessarily describe, in needed detail, the ways in which scientists actually do think, feel, and act. Of these actual patterns, there has been little systematic study—the psychological examination of biologists and physicists by Roe representing a rare exception. It is at least possible that if social scientists were to begin observations in the laboratories and field stations of physical and biological scientists, more might be learned, in a comparatively few years, about the psychology and sociology of science than in all the years that have gone before.

From all this, it seems that this book could scarcely have appeared at a more fitting time. For at a time of renewed interest in a field, even a single book, which provides a tentative systematic overview of that field, can have a disproportionately great effect. It is not unlikely that Mr. Barber's book, together with others that will probably follow in its path, will do much to encourage the establishment of university courses introducing students to the sociology of science. It may well be that some of the students electing these courses, perhaps because the recent course of history has aroused their curiosity about the social environment of science, will develop an abiding interest in the subject. These would then be the new and substantially the first generation of recruits, trained both in social science and in one or another of the physical and biological sciences, who, as they mature into independent scholars, could establish the sociology of science as a specialized field of disciplined knowledge. Mr. Barber's book takes a long step in that direction.

The Normative Structure of Science

Part

3

Prefatory Note

The papers making up part 3 constitute the core of Merton's work on the ethos of science—"the emotionally toned complex of rules, prescriptions, mores, beliefs, values and presuppositions that are held to be binding upon the scientist." They are concerned explicitly with science as a social institution rather than as a type of knowledge, and the fact that his writings on this topic have appeared periodically since the 1930s, just as have his writings in the sociology of knowledge, attests his continuing active engagement in both levels of analysis.

The papers illustrate the transition from treating science as a "strategic research site" for the sociology of knowledge to treating it as a subject worth investigation in its own right. The path of logical development in these successive papers in clear: from the societal values that encouraged science, through the concept of an ethos that makes science a viable institution distinct from (and sometimes pitted against) other parts of society, to an analysis of the component institutional imperatives that make up the core of the ethos. The importance of these papers to the development of the sociology of science can hardly be overestimated.

The first paper provides the qualitative foundation for Merton's argument that the growth of science in England was considerably aided by the remarkable parallel between its values and perspectives and those of the Puritan ethic. (Quantitative data supporting this hypothesis, not reprinted here, are reported in Merton's *Science, Technology and Society in Seventeenth-Century England.*) One will not find here the naïve claim that science was "caused" by Puritanism, nor that *only* Puritans became scientists in seventeenth-century England; the argument is more subtle and more persuasive. Merton is dealing with a case study of the fledgling institution of science groping for legitimacy; he suggests that science not only found firm support in the values of Puritanism, but that Puritans found in the pursuit of science an activity that embodied to a notable degree the kinds of activity enjoined by Puritan teachings.

Parenthetically, it is startling to see, in the paragraphs that most clearly delineate the background of this argument—those opening the part of this paper entitled "To the 'Glory of the Great Author of Nature' "—such a lucid statement about the interplay of values and social institutions along lines that have appeared more recently in phenomenological analyses of society (for example, Peter L. Berger and Thomas Luckmann's *The Social Construction of Reality,*[1] and H. Taylor Buckner's *Deviance, Reality and Change*[2]).

At this point, Merton was still some distance from conceptualizing the ethos of science as a more-or-less coherent, self-sufficient complex of norms and values. Indeed, his interest was primarily in the basic cultural conditions that promote or curb the growth of science—the overarching values prescribing how one is to approach the world—rather than in the rules governing the social relationships through which this approach is to be implemented. Rationality and empiricism he found common to both science and Puritanism, with the subsidiary emphases on promoting the good of mankind and glorifying God through explicating His works being more directly relevant at the time to the motivational bases of scientific activity.

It is in the second of these papers, "Science and the Social Order" (1938) that Merton moves to a more focused concern with the ethos of science itself. Here he develops the problematic relationships between science as an institution and the society in which it exists. This strategy of analysis presupposes that science is an identifiably distinctive part of society, an assumption that requires sociological justification. With this emphasis on societal responses to science it becomes necessary to identify the nature of that which stimulates these responses. Here, then, is Merton's first direct consideration of the ethos of science, signaling a marked

1. New York: Doubleday and Co., 1966.
2. New York: Random House, 1971.

shift in interest from values to norms, or from how people define the world and their role in it to the rules by which they organize their interactions in playing out this larger role.

In laying out the general character of the ethos of science, however, Merton does not yet specify the central norms and their interrelationships. But the potential is obviously there. He mentions, for instance, "the sentiments embodied in the ethos of science—characterized by such terms as intellectual honesty, integrity, organized scepticism, disinterestedness, impersonality." The idea of universalism is implicit in much of his discussion of the problems raised by the imposition of irrelevant criteria for evaluating scientific knowledge (as in the Nazi claim that only Aryans can do valid science). His treatment of organized scepticism, however, reminds one more of Barber's later definition of rationality as a norm of science ("the critical approach to all the phenomena of human existence in the attempt to reduce them to ever more consistent, orderly, and generalized forms of understanding") than of his own subsequent definition (1942) of organized scepticism as emphasizing primarily an institutionally enjoined critical attitude toward the work of fellow scientists.

Still, one sees here the emerging outlines of a more concise depiction of the ethos of science. In addition Merton makes ingenious use of a concept employed in the first paper to aid his analysis of what we would today call societal "backlash" against science. In explaining how it was that religion could both encourage science and also find itself threatened by science, he invokes the distinction between a religious ethos and an explicit theology and points out that as long as action reflects "proper" motivations (that is, so long as it is congruent with the religious ethos), there is little concern for its concrete historical consequences until after they have appeared. In the same way, he suggests that scientists have often acted in accord with the motivations encouraged by their ethos but have failed to consider the impact that the outcome of their actions will have upon society. The lesson that had only been implied in Merton's earlier paper, "The Unanticipated Consequences of Purposive Social Action," is now made explicit: the institutional imperatives of science must be construed within their wider social context if scientists are to avoid unanticipated (and unwanted) consequences of their behavior. Premonitory antipositivistic passages such as this one, drawn from a paper written in 1938, take on added interest in the light of current worldwide attacks mounted against science and scientists:

> Stress upon the purity of science has had other consequences that threaten rather than preserve the social esteem of science. It is repeatedly urged that scientists should in their research ignore all considerations other than the advance of knowledge. Attention is to be focused exclusively on the scientific significance of their work with no concern for the practical uses to which it

may be put or for its social repercussions generally. . . . The objective consequences of this attitude have furnished a further basis of revolt against science; an incipient revolt that is found in virtually every society where science has reached a high stage of development. Since the scientist does not or can not control the direction in which his discoveries are applied, he becomes the subject of reproach and of more violent reactions insofar as these applications are disapproved. . . . The antipathy toward the technological products is projected toward science itself. Thus, when newly discovered gases or explosives are applied as military instruments, chemistry as a whole is censured by those whose humanitarian sentiments are outraged. Science is held largely responsible for endowing those engines of human destruction which, it is said, may plunge our civilization into everlasting night and confusion.

The third paper, perhaps the least "complete" (in the traditional Aristotelian sense of being a well-rounded essay) of all Merton's papers in this volume, is yet one of the most significant in the history of the sociology of science. It was originally written at the request of Georges Gurvitch, then a refugee from Nazi-occupied France, for the first issue of his ill-fated *Journal of Legal and Political Sociology,* which expired soon after it was born. To fit the theme of that first issue of the *Journal*, the paper was saddled with the rather misleading title, "A Note on Science and Democracy." While it provided an opportune occasion for Merton to formulate a concise outline of the normative structure of science, the necessarily brief paper did not cover the subject in depth. Possibly as a result of this, the paper has provided a convenient target for critics of the Mertonian paradigm despite the fact that his other writings show that much of this criticism is misplaced.

The paper is essentially a definition of the four major norms, or institutional imperatives, that comprise the ethos of science and a statement of their interdependence as well as their functional relationships to the formal goal of scientific work: "the extension of certified knowledge."

Merton gives less emphasis here than in the 1938 paper to the nonlogical aspects of the ethos, and less reference to the historical origins and transmutations of these norms. In a footnote in the earlier paper, for instance, Merton writes: "Some phases of this complex [of rules, etc.] may be methodologically desirable, but observance of the rules is not dictated solely by methodological considerations"; while in this paper he states more bluntly, "The institutional imperatives (mores) derive from the goal and the methods. . . . [They] possess a methodologic rationale but they are binding, not only because they are procedurally efficient, but because they are believed right and good." The difference between the two formulations is slight, but apparently significant, with the latter seeming to imply (despite Merton's explicit rejection of this view in his theoretical work in functional analysis) that functional needs somehow create the structures necessary to implement them.

Later work in the sociology of science[3] has indicated how the four norms are functional with respect to the goal of science, but perhaps there would have been less occasion for controversy if Merton had drawn out the striking parallels between these norms and components of the seventeenth-century Protestant ethic. The norm of disinterestedness, for example, could have been reinforced by its similarity to the idea of stewardship, or "the calling"; the evangelical implications of the desire to glorify God through uncovering and publicizing the plan of His works could certainly have legitimated the scientific norm of communism. Particularly in the Calvinist stress on the equality of souls before the Almighty, there is strong support for the norm of universalism, and one might even suppose that organized scepticism could have received substantial impetus from the mutual suspicion that existed among those who could never be sure which of their family and friends were "saved" and which "damned."

Despite its brevity, the paper also contains strong hints of at least two items which were to come much later from Merton's typewriter. In the footnote discussing the origins of Newton's aphorism, "If I have seen farther, it is by standing on the shoulders of giants," is to be found the first indication of an abiding interest that would result finally in the book *On the Shoulders of Giants* (1965). And in Merton's brief reference to the importance of eponymy and to "institutional emphasis upon recognition and esteem as the sole property right of the scientist in his discoveries," as well as to controversies over priority, he comes tantalizingly close to identifying the principal components in the reward system of science. We cannot now guess what the immediate history of the sociology of science might have been had the Mertonian paradigm been established in 1942 instead of 1957, but this near-miss only goes to underline the critical importance of the paper.

N. W. S.

3. Recent work utilizing Merton's analysis of the social norms of science includes: Warren O. Hagstrom, *The Scientific Community* (New York: Basic Books, 1965); Norman W. Storer, *The Social System of Science* (New York: Holt, Rinehart and Winston, 1966); Norman W. Storer and Talcott Parsons, "The Disciplines as a Differentiating Force," in Edward B. Montgomery, ed., *The Foundations of Access to Knowledge* (Syracuse, N.Y.: Syracuse University Press, 1968), pp. 101–21; André F. Cournand and Harriet Zuckerman, "The Code of Science," *Studium Generale* 23 (October 1970): 942–61; Stephen Cotgrove and Steven Box, *Science, Industry and Society: Studies in the Sociology of Science* (London: George Allen and Unwin, Ltd., 1970), chapter 2; Maurice N. Richter, Jr., *Science as a Cultural Process* (Cambridge: Schenkman Publishing Co., 1972), chapter 6; Marlan Blissett, *Politics in Science* (Boston: Little, Brown, 1972), chapter 3; Seymour Martin Lipset and Richard R. Dobson, "The Intellectual as Critic and Rebel: With Special Reference to the United States and the Soviet Union," *Daedalus* 101 (Summer 1972): 137–98, esp. 159 ff.; Seymour Martin Lipset, "Academia and Politics in America," in T. J. Nossiter, ed., *Imagination and Precision in the Social Sciences* (London: Faber, 1972), pp. 211–89.

11 The Puritan Spur to Science

1938

What we call the Protestant ethic was at once a direct expression of dominant values and an independent source of new motivation. It not only led men into particular paths of activity; it exerted a constant pressure for unswerving devotion to this activity. Its ascetic imperatives established a broad base for scientific inquiry, dignifying, exalting, consecrating such inquiry. If the scientist had hitherto found the search for truth its own reward, he now had further grounds for disinterested zeal in this pursuit. And those once dubious of the merits of men who devoted themselves to investigation of the "petty, insignificant details of a boundless Nature" now confronted a developing rationale for such inquiry.

The capital elements of the Puritan ethic were related to the general climate of sentiment and belief. In a sense, these tenets and convictions have been accentuated through a biased selection, but this sort of bias is common to all scholarly inquiries. Theories which attempt to account for certain phenomena require facts, but not all facts are equally pertinent to the problem in hand. "Selection," determined by the limits of the problem, is necessary. Among the cultural variables which invariably influence the development of science are the dominant values and sentiments. At least, this is our working hypothesis. In this particular period, religion in large part made articulate much of the prevailing value-complex. For this reason, we must consider the scope and bearing of the contemporary religious convictions, since these may have been related, in one way or another, to the upsurge of science. But not all of these convictions were relevant. A

Originally published as "Motive Forces of the New Science," chapter 5 in Robert K. Merton, *Science, Technology and Society in Seventeenth-Century England* (Bruges, Belgium: Saint Catherine Press, Ltd., 1938; with a new preface, New York: Howard Fertig, Inc., and Harper & Row, 1970). Reprinted with permission.

certain degree of selection is therefore necessary for the purpose of abstracting those elements which had such a perceivable relation.

Puritanism attests to the theorem that nonlogical notions with a transcendental reference may nevertheless exercise a considerable influence upon practical behavior. If the fancies of an inscrutable deity do not lend themselves to scientific investigation, human action predicated upon a particular conception of this deity does. It was precisely Puritanism which built a new bridge between the transcendental and human action, thus supplying a motive force for the new science. To be sure, Puritan doctrines rested ultimately upon an esoteric theological base but these were translated into the familiar and cogent language of the laity.

Puritan principles undoubtedly represent to some extent an accommodation to the current scientific and intellectual advance. Puritans had to find some meaningful place for these activities within their view of life. But to dismiss the relationship between Puritanism and science with this formula would be superficial. Clearly, the psychological implications of the Puritan system of values independently conduced to an espousal of science, and we would grossly simplify the facts to accord with a pre-established thesis if we failed to note the convergence of these two movements. Moreover, the changing class structure of the time reinforced the Puritan sentiments favoring science since a large proportion of Puritans came from the rising class of bourgeoisie, of merchants.[1] They manifested their increasing power in at least three ways. First, in their positive regard for both science and technology which reflected and promised to enhance this power. Second was their increasingly fervent belief in progress, a profession of faith that stemmed from their growing social and economic importance. A third manifestation was their hostility toward the existing class structure which limited and hampered their participation in political control, an antagonism which found its climax in the Revolution.

Yet we cannot readily assume that the bourgeoisie were Puritans solely because the Puritan ethic appealed to bourgeois sentiments. The converse was perhaps even more important, as Weber has shown. Puritan sentiments and beliefs prompting rational, tireless industry were such as to aid economic success. The same considerations apply equally to the close connection between Puritanism and science: the religious movement partly "adapted" itself to the growing prestige of science but it initially involved deep-seated sentiments which inspired its followers to a profound and consistent interest in the pursuit of science.

1. Cf. Ernst Troeltsch, *The Social Teachings of the Christian Churches*, 2 vols. (New York: Macmillan, 1931), 2:681; Roland Usher, *The Reconstruction of the English Church*, 2 vols. (New York: Appleton, 1910), especially vol. 2, which contains a statistical study of the social origins of Puritan ministers.

The Puritan doctrines were nothing if not lucid. If they provided motivation for the contemporary scientists, this should be evident from their words and deeds. Not that scientists, any more than other mortals, are necessarily aware of the sentiments which invest with meaning their way of life. Nonetheless, the observer may often, though not readily, uncover these tacit valuations and bring them to light. Such a procedure should enable us to determine whether the putative consequences of the Puritan ethic truly proved effective. Moreover, it will disclose the extent to which all this was perceived by the persons whom it most concerned. Accordingly, we shall examine the works of the natural philosopher who "undoubtedly did more than any one of his time to make Science a part of the intellectual equipment of educated men," Robert Boyle.[2] His investigations in physics, chemistry, and physiology, to mention only the chief fields of achievement of this omnifarious experimentalist, were epochal. Add to this the fact that he was one of the individuals who attempted explicitly to establish the place of science in the scale of cultural values and his importance for our particular problem becomes manifest. But Boyle was not alone. Equally significant for our purpose were John Ray, whom Haller termed, a bit effusively, the greatest botanist in the history of man; Francis Willughby, who was perhaps as eminent in zoology as was Ray in botany; John Wilkins, one of the leading spirits in the "invisible college" which developed into the Royal Society; Oughtred, Barrow, Grew, Wallis, Newton; but a complete list would comprise a Scientific Register of the time. Further materials for our purpose are provided by the Royal Society which, instituted just after the middle of the century, stimulated scientific advance. In this instance we are fortunate to have a contemporary account, written under the constant supervision of the members of the Society in order that it might be representative of the motives and aims of that group. This is Thomas Sprat's widely read *History of the Royal-Society of London,* published in 1667, after it had been examined by Wilkins and other representatives of the Society.[3] From these works, then, and from the writings of other scientists of the period, we may glean the chief motive forces of the new science.

2. J. F. Fulton, "Robert Boyle and His Influence on Thought in the Seventeenth Century," *Isis* 18 (1932): 77–102. The range of Boyle's prolific writings is shown in Professor Fulton's exemplary bibliography.

3. Cf. Charles L. Sonnichsen, "The Life and Works of Thomas Sprat" (doctoral dissertation, Harvard University, 1931), p. 131 ff., where substantial evidence of the fact that the *History of the Royal-Society* was representative of the views of the Society is presented. As we shall see, the statements in Sprat's book concerning the aims of the Society bear distinct similarity on every score to Boyle's characterizations of the motives and aims of scientists in general. See ibid., p. 167. This similarity is evidence of the dominance of the ethos which included these attitudes.

To the "Glory of the Great Author of Nature"

Once science has become firmly institutionalized, its attractions, quite apart from any economic benefits it may bestow, are those of all established and elaborated social activities. These attractions are essentially twofold: generally prized opportunities of engaging in socially approved patterns of association with one's fellows and the consequent creation of cultural products which are esteemed by the group. Such group-sanctioned conduct tends to continue unchallenged, with little questioning of its reason for being. Institutionalized values are conceived as self-evident and require no vindication.

But all this is changed in periods of sharp transition. New patterns of conduct must be justified if they are to take hold and become the foci of social sentiments. A new social order presupposes a new scheme of values. And so it was with the new science. Unaided by forces which had already gripped man's will, science could claim only a bare modicum of attention and loyalty. But in partnership with a powerful social movement which induced an intense devotion to the active exercise of designated functions, science was launched in full career.

A clear manifestation of this process is not wanting. The Protestant ethic had pervaded the realm of science and had left its indelible stamp upon the attitude of scientists toward their work. Expressing his motives, anticipating possible objections, facing actual censure, the scientist found motive, sanction, and authority alike in the Puritan teachings. Such a dominant force as religion in those days was not and perhaps could not be compartmentalized and delimited. Thus in Boyle's highly-commended apologia of science, we read:

> It will be no venture to suppose that at least in the Creating of the Sublunary World, and the more conspicuous Stars, two of God's principal ends were, the Manifestation of His own Glory, and the Good of Men.[4]

> It will not be perhaps difficult for you [Pyrophilus]: to discern, that those who labour to deter men from sedulous Enquiries into Nature, do (though I grant, designlessly) take a Course which tends to defeat God of both those mention'd Ends.[5]

4. Robert Boyle, *Some Considerations Touching the Usefulness of Experimental Natural Philosophy*, 2nd ed. (Oxford, 1664), p. 22.

5. Ibid., p. 27. This allusion to contemporary opposition to science refers to that of some zealous divines. Generally speaking, strictures on science arose from four primary sources. First, there were disgruntled individuals such as Robert Crosse, an upholder of pseudo-Aristotelianism, who held that the Royal Society was a Jesuitical conspiracy against society and religion, and Henry Stubbe, a professional literary bravo and Galenical physician who entered the fray for reasons of personal and professional aggrandizement. These exaggerated gestures of antagonism had little influence and are not at all indicative of the place held by science and men of science

This is the motif that recurs in constant measure in the very writings which often contain considerable scientific contributions: these worldly activities and scientific achievements manifest the Glory of God and enhance the Good of Man. The juxtaposition of the spiritual and the material is characteristic and significant. This culture rested securely on a substratum of utilitarian norms which identified the useful and the true. Puritanism itself had imputed a threefold utility to science. Natural philosophy was instrumental first, in establishing practical proofs of the scientist's state of grace; second, in enlarging control of nature; and third, in glorifying God. Science was enlisted in the service of individual, society, and deity. That these were adequate grounds could not be denied. They comprised not merely a claim to legitimacy, they afforded incentives which cannot be readily overestimated. One need only look through the personal correspondence of seventeenth-century scientists to realize this.[6]

John Wilkins proclaimed the experimental study of Nature to be a most effective means of begetting in men a veneration for God.[7] Francis Willughby, probably the most eminent zoologist of the time, was prevailed upon to publish his works—which his excessive modesty led him to deem unworthy of publication—only when Ray insisted that it was a means of

---

in the latter part of the century. See Arthur E. Shipley, "The Revival of Science in the Seventeenth Century," in *Vanuxem Lectures* (Princeton, N.J.: Princeton University Press, 1914); F. Greenslet, *Joseph Glanvill* (New York: Columbia University Press, 1900), p. 78. The second source of opposition was literary. For example, Shadwell, in his comedy "The Virtuoso" (1676), and Butler, in his "Elephant in the Moon" and "Hudibras," ridiculed the pursuits of certain "scientists," but these literary satires were criticism of exaggerated scientism and dilettantism rather than the significant scientific works of the day. Cf. *The Record of the Royal Society* (Oxford: Oxford University Press, 1912), pp. 45 ff. A third source of opposition, and by far the most important, was found among those churchmen who felt that the theological foundations of their beliefs were being undermined by scientific investigations. But theology and religion must not be confused. Orthodox, dogmatic theologians then, as ever, opposed any activity which might lead to the contravention of their dogmas. But the implications of religion, particularly of the religious ethic, were quite contrariwise. It is this ethic, following with equal ineluctability from diverse theological bases, which in its consequences was of far greater social significance than the abstruse doctrines which rarely penetrated to the life of the people. Professor R. F. Jones suggests a fourth source. After the restoration, ardent royalists impugned science, and particularly the Royal Society, because of the close connection between these, Baconianism, and Puritanism. This suggests that contemporaries recognized the strong Puritan espousal of the new experimental science, as indeed they did. See Jones's excellent study, *Ancients and Moderns: A Study of the Background of the Battle of the Books* (St. Louis: Washington University Studies, 1936), pp. 191–92, 224.

6. See, for example, the letters of William Oughtred in *Correspondence of Scientific Men of the Seventeenth Century,* ed. S. J. Rigaud, pp. xxxiv, and passim. Or see the letters of John Ray in the *Correspondence of John Ray,* ed. Edwin Lankester (London, 1848), pp. 389, 395, 402, and passim.

7. *Principles and Duties of Natural Religion,* 6th ed. (London, 1710), pp. 236 ff.

glorifying God.[8] And Ray's panegyric of those who honor Him by studying His works was so well received that five large editions were issued in some twenty years.[9]

Many emancipated souls of the present day, accustomed to a radical cleavage between religion and science and convinced of the relative social unimportance of religion for the modern Western world, are apt to generalize this state of affairs. To them, these recurrent pious phrases signify Machiavellian tactics or calculating hypocrisy or at best merely customary usage, but nothing of deep-rooted motivating convictions. The evidence of extreme piety invites the charge that *qui nimium probat nihil probat*. But such an interpretation is possible only upon the basis of an unwarranted extrapolation of twentieth-century beliefs and attitudes to seventeenth-century society. Though it always serves to inflate the ego of the iconoclast and sometimes to extol the social images of his own day, "debunking" can supplant truth with error. As a case in point, it is difficult to believe that Boyle, who manifested his piety by expending considerable sums to have the Bible translated into foreign tongues as well as in less material ways, was simply rendering lip service to Protestant beliefs. As Professor G. N. Clark properly notes in this connection:

> There is . . . always a difficulty in estimating the degree to which what we call religion enters into anything which was said in the seventeenth century in religious language. It is not solved by discounting all theological terms and treating them merely as common form. On the contrary, it is more often necessary to remind ourselves that these words were then seldom used without their accompaniment of meaning, and that their use did generally imply a heightened intensity of feeling. This sense of the closeness of God and the Devil to every act and fact of daily life is an integral part of the character of the century.[10]

In various ways, then, general religious ideas were translated into concrete policy. This was no mere intellectual exercise. Puritanism transfused ascetic vigor into activities which, in their own right, could not as yet achieve self-sufficiency. It so redefined the relations between the divine and the mundane as to move science to the front rank of social values. As it happened, this was at the immediate expense of literary, and ultimately,

8. See Edwin Lankester, ed., *Memorials of John Ray* (London, 1846), p. 14n.

9. John Ray, *Wisdom of God* (London: 1691), pp. 126–29, and passim. Striking illustrations of the extent to which Ray had assimilated the Puritan sentiments are to be found throughout his correspondence. For example, he writes in a letter to James Petiver (4 April 1701): "I am glad your business increases so as to require more attendance, and take up more of your time, which cannot be better employed than in the works of your proper callings. What time you have to spare you will do well to spend, as you are doing, in the inquisition and contemplation of the works of God and nature" (*Correspondence of John Ray*, p. 390).

10. *The Seventeenth Century* (Oxford: University Press, 1929), p. 323.

of religious pursuits. For if the Calvinist God is irrational in the sense that He cannot be directly grasped by the cultivated intellect, He can yet be glorified by a clear-sighted, meticulous study of His natural works.[11] Nor was this simply a compromise with science. Puritanism differed from Catholicism, which had gradually come to tolerate science, in demanding, not merely condoning, its pursuit. An elastic concept,[12] the Catholic and Protestant definitions of which differed so fundamentally as to produce entirely opposed consequences, the "glorification of God" thus came to be, in Puritan hands, the "fructification of science."

"Comfort of Mankind"

Protestantism afforded further grounds for the cultivation of science. The second dominant tenet in the Puritan ethos, it will be remembered, designated social welfare, the good of the many, as a goal ever to be held in mind. Here again, the contemporary scientists adopted an objective that carried with it, in addition to its own obvious merits, a cluster of religious sentiments. Science was to be fostered and nurtured as leading to the improvement of man's lot on earth by facilitating technological invention. The Royal Society, we are told by its worthy historian, "does not intend to stop at some particular benefit, but goes to the root of all noble inventions."[13] Further, the experiments that do not bring with them immediate gain are not to be condemned, for as the noble Bacon had declared, experiments of Light ultimately conduce to a whole troop of inventions useful to the life and state of man.[14] This power of science to better the material condition of man, he continues, is, apart from its purely mundane value, a good in the light of the Evangelical Doctrine of Salvation by Jesus Christ.

Boyle, in his last will and testament, echoes the same attitude, petitioning the Fellows of the Society in this wise: "Wishing them also a happy success in their laudable Attempts, to discover the true Nature of the Works of God; and praying that they and all other Searchers into Physical Truths, may Cordially refer their Attainments to the Glory of the Great

11. Cf. Troeltsch, *Social Teachings,* 2:585.

12. The changing definitions of nominally identical concepts comprise a fruitful field for sociological research. Such students of the sociology of knowledge as Mannheim and historians of ideas (Lovejoy, Boas, Crane) have contributed significant studies of such developments.

13. Thomas Sprat, *The History of the Royal-Society of London* (London, 1667), pp. 78–79.

14. Ibid., pp. 245, 351 ff.

Author of Nature, and to the Comfort of Mankind."[15] "Experimental science was to Boyle, as to Bacon, itself a religious task."[16]

Earlier in the century, this keynote had been sounded in the resonant eloquence of that "veritable apostle of the learned societies," Francis Bacon. Himself the initiator of no scientific discoveries; unable to appreciate the importance of his great contemporaries, Gilbert, Kepler, and Galileo; naïvely believing in the possibility of a scientific method that "places all wits and understandings nearly on a level"; a radical empiricist holding mathematics to be of no use in science; he was, nevertheless, highly successful in being one of the principal propagandists for positive social evaluation of science and disclaim of "sterile scholasticism." As one would expect from the son of a "learned, eloquent and religious woman, full of puritanic fervor" who was admittedly influenced by his mother's attitudes,[17] he speaks in the *Advancement of Learning* of the true end of scientific activity as the "glory of the Creator and the relief of man's estate."[18] Since, as is quite clear from many official and private documents, the Baconian teachings constituted the basic principles on which the Royal Society was patterned, it is not strange that in the charter of the Society, the same sentiment is expressed.[19] Thomas Sydenham, the zealous Puritan,[20] likewise had a profound admiration for Bacon. And, like Bacon, he was prone to exaggerate the importance of empiricism to the very point of

15. Quoted by Gilbert Burnet, *A Funeral Sermon Preached at the Funeral of the Honourable Robert Boyle* (London, 1692), p. 25.

16. Edwin A. Burtt, *The Metaphysical Foundations of Modern Physical Science* (New York: Harcourt, Brace & Co., 1927), p. 188.

17. Cf. Mary Sturt, *Francis Bacon* (London: K. Paul, Trench, Trubner & Co., 1932), pp. 6 ff. It is true, as Professor M. M. Knappen has pointed out to the writer, that Bacon supplied James with his legal arguments against the Puritans. But this should not be confused with Bacon's tacit acceptance of many of the *nonpolitical* phases of Puritanism. For the congeniality of Bacon's philosophy, see R. F. Jones's *Ancients and Moderns,* p. 92 ff.

18. In the *Novum Organum,* book 1, aphorism 89, science is characterized as the handmaid of religion since it serves to display God's power. This is not, of course, a novel contention.

19. In the second Charter, which passed the Great Seal on 22 April 1663, and by which the Society is governed to this day, we read that the studies of its Fellows "are to be applied to further promoting by the authority of experiments the sciences of natural things and of useful arts, to the glory of God the Creator, and the advantage of the human race" (*Record of the Royal Society,* p. 15). Note the increased emphasis upon utilitarianism.

20. See Joseph F. Payne, *Thomas Sydenham* (New York: Longmans, Green & Co., 1900), pp. 7–8, passim, where abundant evidence of Sydenham's sternly Puritan background is presented. "We cannot appreciate his whole character and career without remembering that he was imbued with the intense earnestness of the Puritans, and was quite prepared, in opposition to authority of any kind, to be called, if necessary, a rebel."

excluding theoretical interpretation entirely. "Pure intellectual curiosity . . . seemed to him, perhaps partly owing to the Puritan strain in his character, of little importance. He valued knowledge only either for its ethical value, as showing forth the glory of the Creator or for its practical value, as promoting the welfare of man."[21] Empiricism characteristically dominated Sydenham's approach to medicine which set above all the value of clinical observation, the "repeated, constant observation of particulars." It is of some interest that the greatest clinical observers of this century, Mayerne and Sydenham, were of Puritan stock.

Throughout there was the same point-to-point correlation between the principles of Puritanism, and the avowed attributes, goals, and results of scientific investigation. Such, at least, was the contention of the protagonists of science at that time. If Puritanism demands systematic, methodic labor, constant diligence in one's calling, what, asks Sprat, more active and industrious and systematic than the Art of Experiment, which "can never be finish'd by the perpetual labours of any one man, nay, scarce by the successive force of the greatest Assembly?"[22] Here is employment enough for the most indefatigable industry since even those hidden treasures of Nature which are farthest from view may be uncovered by pains and patience.[23]

Does the Puritan eschew idleness because it conduces to sinful thoughts (or interferes with the pursuit of one's vocation)? "What room can there be for low, and little things in a mind so *usefully* and successfully employ'd [as in natural philosophy]?"[24] Are plays and playbooks pernicious and flesh-pleasing (and subversive of more serious pursuits)?[25] Then it is the "fittest season for Experiments to arise, to teach us a Wisdom, which springs from the depths of Knowledge, to shake off the shadows, and to scatter the mists [of the spiritual distractions brought on by the Theater]."[26] And finally, is a life of earnest activity within the world to be preferred to monastic asceticism? Then recognize the fact that the study of natural philosophy "fits us not so well for the secrecy of a Closet: It makes us serviceable to the World."[27] In short, science embodies patterns of behavior

21. Ibid., p. 234.
22. Ibid., pp. 341–42.
23. Ray, *Wisdom of God*, p. 125.
24. Sprat, *History of the Royal-Society*, pp. 344–45.
25. Cf. Richard Baxter, *A Christian Directory: or, A Body of Practical Divinity*, 5 vols. (London, 1825 [1664–65]), 1:152, 2:167. Cf. also Robert Barclay, the Quaker apologist, who specifically suggests "geometrical and mathematical experiments" as innocent divertisements to be sought instead of pernicious plays; see *An Apology for the True Christian Divinity* (Philadelphia, 1805 [1675]), pp. 554–55.
26. Sprat, *History of the Royal-Society*, p. 362.
27. Ibid., pp. 365–66.

that are congenial to Puritan tastes. Above all, it embraces two highly prized values: utilitarianism and empiricism.[28]

In a sense this explicit coincidence between Puritan tenets and the eminently desirable qualities of science as a calling which was suggested by the historian of the Royal Society is casuistry. No doubt it is partly an express attempt to fit the scientist qua pious layman into the framework of the prevailing moral and social values. Since both the constitutional position and the personal authority of the clergy were far more important then than now, it probably constituted a bid for religious and social sanction. Science, no less than literature and politics, was still, to some extent, subject to approval by the clergy.[29]

But this is not the entire explanation. Present-day discussions of "rationalization" and "derivations" have been wont to becloud certain fundamental issues. It is true that the "reasons" adduced to justify one's actions often do not account satisfactorily for this behavior. It is also an acceptable hypothesis that ideologies seldom *give rise* to action and that both the ideology and action are rather the product of common sentiments and values upon which they in turn react. But these ideas cannot be ignored for two reasons. They provide clues for detecting the basic values that motivate conduct. Such signposts cannot be profitably neglected. Of even greater importance is the role of ideas in directing action into *particular* channels. It is the dominating system of ideas that determines the choice between alternative modes of action which are equally compatible with the underlying sentiments. Without such guidance and direction, nonlogical action would become, within the limits of the value-system, random.[30]

In the seventeenth century, the frequent recourse of scientists to religious vindication suggests first of all that religion was a sufficiently powerful

28. Sprat perspicaciously suggests that monastic ascetism induced by religious scruples was partially responsible for the lack of empiricism of the Schoolmen: "But what sorry kinds of Philosophy must they needs produce, when it was a part of their religion, to separate themselves, as much as they could, from the converse of mankind? when they were so farr from being able to discover the secrets of Nature, that they scarce had opportunity, to behold enough of its common works" (ibid., p. 19).

29. H. H. Henson, *Studies in English Religion in the Seventeenth Century* (London: J. Murray, 1903), p. 29.

30. Operationally, there is often a thin, uncertain line between "derivations" and "residues" (Pareto). Constant elements in the speech reactions associated with action manifest deep-rooted, effective sentiments. Speaking elliptically, these constant elements may be held to provide motivations for behavior, whereas the variable elements are simply *post factum* justifications. But, in practice, it is at times exceedingly difficult to discriminate between the two. Once aware of the strong emotional charge which certain religious convictions carried at the time, we may find it justifiable to treat these as residues rather than derivations.

social force to be invoked in support of an activity that was intrinsically less acceptable at the time. It also leads the observer to the peculiarly effective religious orientation that could invest scientific pursuits with all manner of values and could thus serve to direct the interests of believers into the channels of science.

The efforts of Sprat, Wilkins, Boyle, or Ray to justify their interest in science do not represent simply opportunistic obsequiousness, but rather an earnest attempt to justify the ways of science to God. The Reformation had transferred the burden of individual salvation from the Church to the individual, and it is this "overwhelming and crushing sense of the responsibility for his own soul" that accounts in part for both the acute longing for religious justification[31] and the intense pursuit of one's calling. If science were not demonstrably a "lawful" and desirable calling, it dare not claim the attention of those who felt themselves "ever in the Great Taskmaster's eye." It is to this intensity of feeling that such apologias were due.

Rationalism and Empiricism

The exaltation of the faculty of reason in the Puritan ethos—based partly on the conception of rationality as a curbing device of the passions—inevitably led to a sympathetic attitude toward those activities that demand the constant application of rigorous reasoning.[32] But again, in contrast to medieval rationalism, reason is deemed subservient and auxiliary to empiricism. Sprat is quick to indicate the preeminent adequacy of science in this respect.[33] It is on this point probably that Puritanism and the scientific temper are in most salient agreement, for the combination of rationalism and empiricism that is so pronounced in the Puritan ethic forms the essence of the spirit of modern science. Puritanism was suffused with the

31. Usher, *Reconstruction of the English Church*, 1:15.

32. It must be remembered that the use of reason was lauded by the Puritans partly because it served to differentiate man from beast. The extent to which this idea seeped into the thought of contemporary scientists may be indicated by a statement made by Boyle. "So much admirable workmanship as God hath displayed in the universe, was never meant for eyes that wilfully close themselves, and affront it with the not judging it with the speculating, *Beasts inhabit and enjoy the world, man, if he will do more, must study & spiritualize it*" (Boyle, *Works*, ed. Thomas Birch, 3:62).

33. "Who ought to be esteem'd the most carnally minded? The Enthusiast, that pollutes his Religion, with his own passions? or the Experimenter, that will not use it [reason] to flatter and obey his own desires, but to subdue them?" (Sprat, *History of the Royal-Society*, p. 361). Baxter, it will be remembered, in a fashion representative of the Puritans, had decried the invasion of "enthusiasm" into religion. Reason "must maintain its authority in the command and government of your thoughts" (*C. D.*, 2:199, and passim). In like spirit, those who, at Wilkins's lodgings, laid foundations of the Royal Society "were invincibly arm'd against all the inchantments of Enthusiasm" (Sprat, *History of the Royal-Society*, p. 53).

rationalism of neo-Platonism, derived largely through an appropriate modification of Augustine's teachings. But it did not stop there. Associated with the designated necessity of dealing successfully with the practical affairs of life within this world—a derivation from the peculiar twist afforded largely by the Calvinist doctrine of predestination and *certitudo salutis* through successful worldly activity—was an emphasis upon empiricism. These two currents brought to converge through the logic of an internally consistent system of theology were so associated with the other attitudes of the time as to prepare the way for the acceptance of a similar coalescence in natural science.

The Puritan insistence upon empiricism, upon the experimental approach, was intimately connected with the identification of contemplation with idleness, of the expenditure of physical energy and the handling of material objects with industry.[34] Experiment was the scientific expression of the practical, active, and methodical bents of the Puritan. This is not to say, of course, that experiment was derived in any sense from Puritanism. But it serves to account for the ardent support of the new experimental science by those who had their eyes turned toward the other world and their feet firmly planted on this. Moreover, as Troeltsch has suggested, Calvinism which abolished the absolute goodness of the Godhead tended to an emphasis on the individual and the empirical, the practically untrammeled and utilitarian judgment of all things. He finds in the influence of this spirit a most important factor of the empirical and positivist tendencies of Anglo-Saxon thought.[35]

A blunt Puritan, Noah Biggs, evidences this attitude in his sharp attack on the universities of his day.

> Wherein do they [universities] contribute to the promotion or discovery of truth? . . . Where have we any thing to do with Mechanicall Chymistrie the hand maid of Nature, that hath outstript the other Sects of Philosophy, by her multiplied real experiences? Where is there an examination and consecution of Experiments? encouragements to a new world of Knowledge, promoting, completing, and actuating some new Inventions? where have we constant reading upon either quick or dead Anatomies, or an ocular demonstration of Herbs? Where a Review of the old Experiments and Traditions, and casting out the rubbish that has pestered the Temple of Knowledge?[36]

34. This observation constitutes one of the many contributions of Professor Jones's valuable book, *Ancients and Moderns.* Cf. chap. 5, esp. pp. 112–13. The derivation of this emphasis upon empiricism was not sufficiently clarified in my paper on "Puritanism, Pietism and Science," *The* [*English*] *Sociological Review* 28 (1936): 1–30.

35. Ernst Troeltsch, *Die Bedeutung des Protestantismus für die Entstehung der modernen Welt* (Munich: R. Oldenbourg, 1911), pp. 80–81.

36. I am indebted to Professor Jones (*Ancients and Moderns,* p. 104) for this quotation from Noah Biggs's *Mataeotechnia Medicine Praxeos* (London, 1651), dedicated to the Reformist Parliament of the time.

It was a common practice for Puritans to couple their intense scorn for a "jejeune Peripatetick Philosophy" with extravagant admiration for "mechanicall knowledge," which substituted fact for fantasy. From every direction, elements of the Puritan ethic converged to reinforce this set of attitudes. Active experimentation embodied all the select virtues and precluded all the baneful vices. It represented a revolt against that Aristotelianism which was traditionally bound up with Catholicism; it supplanted passive contemplation with active manipulation; it promised practical utilities instead of sterile figments; it established in indubitable fashion the glories of His creation. Small wonder that the Puritan transvaluation of values carried with it the consistent endorsement of experimentalism.[37]

Empiricism and rationalism were canonized, beatified, so to speak. It may very well be that the Puritan ethos did not directly influence the method of science and that this was simply a parallel development in the internal history of science, but it becomes evident that, through the psychological sanction of certain modes of thought and conduct, this complex of attitudes made an empirically founded science commendable rather than, as in the medieval period, reprehensible or at best acceptable on sufferance. In short, Puritanism altered social orientations. It led to the setting up of a new vocational hierarchy, based on criteria that bestowed prestige upon the natural philosopher. As Professor Speier has well said, "There are no activities which are honorable in themselves and are held excellent in all social structures."[38] And one of the consequences of Puritanism was the reshaping of the social structure in such fashion as to bring esteem to science. This must have influenced the direction of some talents into scientific fields who, in other social contexts, would have turned to other pursuits.

The Shift to Science

As the full import of the Puritan ethic manifested itself—even after the political failure of the revolution which should not be erroneously identified with the collapse of Puritan influence upon social attitudes—the sciences became foci of social interest. Their new fashionableness contrasts with

37. Dury, in a *Lettre du sieur Jean Dury touchant l'état présent de la religion en Angleterre* (London, 1658), writes of the Independents, "Ils ne croient que ce qu'ils voient." Quoted by Georges Ascoli, *La Grande-Bretagne devant l'opinion française au XVIIe siècle* (Paris: Librairie Universitaire J. Gamber, 1930), 1:407. "I am an enemy of their philosophy that vilify sense!" wrote Baxter. And, on the practical side, John Wilkins affirms that "our best and most Divine Knowledge is intended for Action; and those may justly be counted barren studies, which do not conduce to Practice as their proper End" (*Mathematical Magick* [London, 1648], p. 2).

38. Hans Speier, "Honor and Social Structure," *Social Research* 2 (1935): 79.

their previous condition.[39] This was not without its effects. Some, who before might have turned to theology or rhetoric or philology, were directed, through the subtle, largely unperceived and newly arisen predisposition of society, into scientific channels. Thus, Thomas Willis, whose *Cerebri Anatome* was probably the most complete and accurate account of the nervous system up to that time and whose name is immortalized in the "circle of Willis," "was originally destined to theology, but in consequence of the unfavorable conditions at that age for theological science, he turned his attention to medicine."[40]

No less indicative of a shift of interest is the lament of Isaac Barrow, when he was professor of Greek at Cambridge: "I sit lonesome as an Attic owl, who has been thrust out of the companionship of all other birds; while classes in Natural Philosophy are full."[41] Evidently, Barrow's loneliness proved too much for him, for, as is well known, in 1663 he left this chair to accept the newly established Lucasian Professorship of Mathematics, in which he was Newton's predecessor.

The science-loving amateur, so prominent a feature of the latter part of the century, also reflects this new attitude. Nobles and wealthy commoners turned to science not as a means of livelihood but as an object of devoted interest. Particularly for them direct economic benefits were a negligible consideration. Science afforded them the opportunity to devote their energies to an honored task; an obligation as the comforts of unrelieved idleness vanished from the new scale of values.[42]

In the history of science the most famous of these amateurs is of course Robert Boyle, but perhaps the best index of their importance is to be found in their role in the formation of the Royal Society.[43] Of those who, in that "wonderful pacifick year," 1660, constituted themselves into a definite association, a considerable number—among them Lord Brouncker, Boyle, Lord Bruce, Sir Robert Moray, Dr. Wilkins, Dr. Petty, and Abraham Hill

39. Cf. Sprat, *History of the Royal-Society*, p. 403.

40. Johann H. Baas, *Outlines of the History of Medicine and the Medical Profession* (New York, 1889).

41. Quoted by Hermann Hettner, *Geschichte der englischen Literatur* (Brunswick, 1894), pp. 16–17.

42. This is clearly brought out by William Derham's estimate of the virtuoso and zoologist, Willughby. "He prosecuted his design with as great application as if he had to get his bread thereby; all of which I mention . . . for an example to persons of great estate and quality that they may be excited to answer the ends for which God gives them estates, leisure, parts and gifts, or a good genius; which was not to exercise themselves in vain or sinful follies, but to be employed for the glory and in the service of the infinite Creator, and in doing good offices in the world" (*Memorials of John Ray, Consisting of His Life by Dr. Derham*, ed. Edwin Lankester [London, 1846], pp. 34–35).

43. Martha Ornstein, *The Rôle of the Scientific Societies in the Seventeenth Century*, p. 91 ff.

—were amateurs of this type. Hardly less assiduous were the efforts of such virtuosi as Lord Willughby, John Evelyn, Samuel Hartlib, Francis Potter, and William Molineux.

The social emphasis on science had a peculiarly fruitful effect, probably because of the general state of scientific development. The methods and objects of investigation were frequently not at many removes from daily experience, and could be understood not only by the especially equipped but by those with comparatively little technical education.[44] To be sure, dilettantish interest in science seldom enriched its fruits directly, but it served to establish it more firmly as a socially estimable pursuit. And this same function was performed no less by Puritanism. The fact that science today is almost wholly divorced from religious sanctions is itself of interest as an example of the process of secularization. Having grown away from its religious moorings, science has in turn become a dominant social value to which other values are subordinated. Today it is much more common in the Western world to subject the most diverse beliefs to the sanctions presumably afforded by science than to those yielded by religion; the increasing reference to scientific authority in contemporary advertisements and the long-standing eulogistic connotation of the very word "scientific" diversely reflect the social standing of science.[45]

The Process of Secularization

The beginnings of such secularization, faintly perceptible in the latter Middle Ages,[46] were, in one sense, emerging more fully in the Puritan

44. Ibid., p. 53.

45. As Professor C. Bouglé remarks, "Science has decidedly advanced to the first rank in the table of values" (*The Evolution of Values* [New York: Henry Holt & Co., 1926], p. 201).

46. G. R. Owst, in his exemplary study based upon new documentary evidence, *Literature and Pulpit in Medieval England* (Cambridge: Cambridge University Press, 1933), pp. 188–89 ff; pp. 554–57, and passim, presents a painstaking analysis of medieval homilies, then so effective for the determination of the outlook of the folk, and notes this adumbrated tendency. As we have indicated it was the spokesmen of the medieval Church themselves who bade men to consider the work of God's hand in the multifarious appearances of Nature, and this was indeed a powerful justificatory principle for scientific pursuits. Associated with this, however, was the *odium theologicum* of secularized knowledge, but it was too much to expect the permanent abeyance of concerted efforts because of the peremptory prohibitions of the theologians. It was in part to combat this threatened secularization of knowledge, which became alarmingly noticeable with the great University movement of the twelfth century, that the Mendicant Orders were established. But "the very Mendicant preaching originally designed to steer a safe middle course in the moral and mental instruction of lay-folk was itself helping unconsciously to create a fresh crisis, in which such secularization would become at last inevitable" (p. 189). It was with the advent of the post-Reformation religious ethic, which burst the last bonds of inhibitive control of natural philosophy and which created for it a role,

ethos. But the Puritan was not simply the last of the medievalists or the first of the moderns. He was both. It was in the system of Puritan values, as we have seen, that reason and experience began to be considered as independent means for ascertaining even religious truth. Faith that is unquestioning and not "rationally weighed," proclaimed Baxter, is not faith, but a dream or fancy or opinion. In effect this grants to science a power that may ultimately limit that of religion. This unhesitant assignment of a virtual hegemony to science is based on the explicit assumption of the unity of knowledge, experiential and supersensuous, so that the testimony of science must perforce corroborate religious beliefs.[47]

This conviction of the mutually confirmatory nature of reason and revelation afforded a further basis for the favorable attitude toward experimental studies, which, it is assumed, will simply reinforce basic theological dogmas. However, the active pursuit of science, thus freely sanctioned by unsuspecting religionists created a new tone and habit of thought—to use Lecky's phrase—that is the "supreme arbiter of the opinions of successive periods."[48] As a consequence of this change, ecclesiastics, no longer able to appeal to commonly accepted teachings of science which seem rather to contravene various theological doctrines, are likely once again to substitute authority for reason in an effort to emerge victorious from the conflict.

In one direction, then, Puritanism led inevitably to the elimination of religious restrictions on scientific work. This was the distinctly modern element of Puritan beliefs. But this did *not* involve the relaxation of religious discipline over conduct; quite the contrary. Compromise with the

---

then acceptable to scientists and religionists alike, of subserviency to ultimate religious goals and of autonomy within the scope of its investigations, that secularization became as explicit and pronounced as it had hitherto been implicit and subdued. The Reformist tenets did not arise full-blown; they did not in their implications represent a radical break with the past, but, through a shift and intensification of emphasis, helped effect a change which, though prepared by a long history of antecedent tendencies, seemed saltatory. As Owst suggests, the Lollard teachings remind us "of the honoured place which Work has continued to hold in Protestant faith and practice. Its subsequent achievements, alike in science and industry, . . . when 'meritory works' are finally discountenanced, prove once again our kinship with the past. The gulf of the Reformation is thus bridged once more and the spiritual continuity of our history maintained in the face of all such inevitable changes" (p. 557).

47. There is so admirable an accord and correspondency between the findings of natural science and supernatural divinity, says Baxter, that the former "greatly advantageth us" in the belief of the latter. *C. D.*, 1:172–74. This illustrates the incipient tendency of theology to become in a sense the handmaid of science since religious concepts become dependent upon the type of universe which man can know. Cf. Paul R. Anderson, *Science in Defense of Liberal Religion* (New York: Putnam, 1933), p. 191 ff.

48. William E. H. Lecky, *History of the Rise and Influence of the Spirit of Rationalism in Europe*, 2 vols. (London, 1865), 1:7. See also A. C. McGriffert, *The Rise of Modern Religious Ideas*, p. 18, and passim.

world was intolerable. It must be conquered and controlled through direct action and this ascetic compulsion was to be exercised in every sphere of life. It is, therefore, an error to portray the Puritan espousal of science as simply an "accommodation" to the intellectual environment of the age.[49] Such secularized elements there were, especially with the passage of time, but these were far less significant than the unyielding constraint for devotion to the thrice-blessed calling of natural philosopher.

Paradoxically but persistently, then, this religious ethic, based on rigid theological foundations, furthered the development of the very scientific disciplines that later seem to confute orthodox theology.

The articulation of these several ideas, each the focus of strong sentiments, into a system that was all the more forceful precisely because it was psychologically rather than logically coherent, led to a long chain of consequences not least of which was the substantial destruction of the system of ideas itself. Though the corresponding religious *ethic,* as we shall see, does not necessarily lose its effectiveness as a social force immediately upon the undermining of its theological foundations, it tends to do so in time. This sketch of the influence of science in the processes of secularization should serve to make intelligible the diverse, quite opposed roles that religion and theology may play in their relations to science.

A religion—understood here, as throughout this inquiry, as the ethical and moral beliefs and practices which constitute a system of faith and worship, that is, as a religious ethic—may indirectly promote the cultivation of science, although specific scientific discoveries are at the same time vehemently attacked by theologians, who suspect their possibly subversive nature. Precisely because this pattern of interlocking and contradictory forces is so often unanalyzed, we must distinguish clearly between the intentions and aims of religious leaders and the (frequently unforeseen) consequences of their teachings.[50] Once this pattern is clearly understood, it is not surprising or inconsistent that Luther particularly, and Melanchthon less strongly, execrated the cosmology of Copernicus. In

49. This assumption is the one fundamental shortcoming of Olive Griffith's otherwise excellent monograph, *Religion and Learning: A Study in Presbyterian Thought from 1662 to the Foundation of the Unitarian Movement* (Cambridge: Cambridge University Press, 1935). Her treatment unwarrantedly presupposes throughout that religious convictions are intrinsically static and change only through external pressures, whereas it is the contention of the present analysis that such changes are, in great part, the outcome of inherent tendencies which are gradually realized in the course of time. See my review of Dr. Griffith's work in *Isis* 26 (1936): 237–39.

50. As Mr. Tawney put it, "So little do those who shoot the arrows of the spirit know where they will light" (*Religion and the Rise of Capitalism*, [New York: Harcourt, Brace, 1926] p. 277). But Calvin could also say: "Dieu a ressuscité les sciences humaines qui sont propres et utiles à la conduite de nostre vie, et, servant à nostre utilité, peuvent aussi servir à sa gloire." Cf. Otto Rodewald, *Johannes Calvins Gedanken über Erziehung Bünde* (Westphalia: Ziegemeyer & Co., 1911), pp. 37 ff.

magisterial mood, Luther berates the Copernican theory: "Der Narr will die ganze Kunst Astronomiae umkehren. Aber wie die heilige Schrift anzeigt, so hiess Josua die Sonne still stehen, und nicht das Erdreich."[51] Calvin also frowned upon the acceptance of numerous scientific discoveries of his day, whereas the religious ethic which stemmed from him inspired the pursuit of natural science.[52]

This failure of the Reformers to foresee some of the most fundamental social effects of their teachings was not simply the result of ignorance. It was rather an outcome of that type of nonlogical thought which deals primarily with the motives rather than the probable results of behavior. Righteousness of motive is the basic concern; other considerations, including that of the probability of attaining the end, are precluded. Action enjoined by a dominant set of values *must* be performed. But, with the complex interaction of forces in society, the effects of action ramify. They are not restricted to the specific sphere in which the values were originally centered, occurring in interrelated fields ignored at the outset. Yet it is precisely because these spheres of society are interrelated that the further consequences in adjacent areas react upon the basic system of values. It is this usually unlooked-for reaction which constitutes a most important factor in the process of secularization, of the transformation or breakdown of value-systems. This is the essential paradox of social action—the "realization" of values may lead to their renunciation. We may thus para-

51. Quoted by Dorothy Stimson, *The Gradual Acceptance of the Copernican Theory* (New York, 1917), p. 39. As Dean Stimson suggests (p. 99), such denunciations were less influential than those of the Catholic clergy, largely because of the Protestant doctrine of the right to individual interpretation. This was one effective source of secularization.

52. In view of this analysis, it is surprising to note the statement *credited* to Max Weber, that the opposition of the Protestant Reformers is sufficient reason for not linking Protestantism with scientific advance. See *Wirtschaftsgeschichte* (Munich: Duncker & Humblot, 1924), p. 314. This remark is especially unanticipated since it does not at all accord with Weber's discussion of the same point in his other works. Cf. *Gesammelte Aufsätze zur Religionssoziologie* (Tübingen: J. C. B. Mohr, 1922), 1:564, 141; *Wissenschaft als Beruf* (Munich: Duncker & Humblot, 1921), pp. 19–20. The explanation may be that the first is not Weber's statement, since the *Wirtschaftsgeschichte* was compiled from scraps of Weber's notes by two of his students: "a bundle of sheets of notes that were little more than catchwords, put down in a handwriting hardly legible even to those familiar with it." It is unlikely that Weber would have made the elementary error of confusing the opposition to scientific discoveries of the Reformers with the unforeseen consequences of the Protestant ethic, particularly since he expressly warns against the failure of such discrimination in his *Religionssoziologie*. Nor would he have been apt to identify the attitudes of the Reformers themselves with those of their followers as the Protestant movement developed. See further the comment of Troeltsch (*Social Teachings*, 2:879–80), to the effect that although Calvin was himself antagonistic to some scientific discoveries, the consequence of his doctrine was to provide a ferment of opinion directly favorable to the espousal of science.

phrase Goethe and speak of "Die Kraft, die stets das Gute will, und stets das Böse schafft."[53]

Insofar as the attitudes of the theologians dominate over the, in effect, subversive religious ethic—as did Calvin's authority largely in Geneva until the first part of the eighteenth century—scientific development may be greatly impeded. For this reason, it is important to discriminate between the early and late periods of Calvinism. The implications of its dogmas found expression only with the passage of time. But upon the relaxation of this hostile influence and with the influx of an ethic, stemming from it and yet progressively differing from it, science takes on new life, as indeed was the case in Geneva from about the middle of the eighteenth century.[54] This development was particularly retarded in Geneva because there the authority resting in Calvin himself, rather than in the implications of his religious system, was not soon dissipated.

The Integration of Religion and Science

It is thus to the religious ethos, not the theology, that we must turn if we are to understand the integration of science and religion in seventeenth-century England.

Perhaps the most directly effective belief in this ethos for the sanction of natural science held that the study of nature enables a fuller appreciation of His works and thus leads us to admire and praise the Power, Wisdom, and Goodness of God manifested in His creation. Though this conception was not unknown to medieval thinkers, the consequences deduced from it were entirely different. For example, Arnaldus of Villanova, in studying the products of the Divine Workshop, adheres strictly to the medieval scholastic ideal of determining the properties of phenomena from *tables* (in which, according to the canons of logic, all combinations of chance were set forth).[55] But in the seventeenth century, the contemporary emphasis upon empiricism led to the investigation of nature primarily through experience.[56] This difference in interpretation of substantially the same doctrine can only be understood in the light of the different values permeating the two cultures. Cloistered contemplation was forsaken; active experimentation was introduced.

53. See Robert K. Merton, "The Unanticipated Consequences of Purposive Social Action," *American Sociological Review* 1 (1936): 894–904.

54. See Alphonse de Candolle, *Histoire des sciences et des savants depuis deux siècles* (Geneva-Basel: H. Georg, 1885), pp. 335–36.

55. Walter Pagel, "Religious Motives in the Medical Biology of the XVIIth Century," *Bulletin of the Institute of the History of Medicine* 3 (1935): 112.

56. Ibid., pp. 214–15. It is not maintained, of course, that this empiricist bent derived solely from Puritanism. As we shall see, at least one other source was economic and technological. But Puritanism did contribute an added force to this development which has often been overlooked.

The Royal Society was of inestimable importance, both in the propagation of this new point of view and in its actual application. Its achievements gained added stature by contrast with the lethargy of the English universities. As is well known, the universities were the seats of conservatism and virtual neglect of science, rather than the nurseries of the new philosophy. It was the learned society that chiefly effected the association and social interaction of scientists with such signal results. The *Philosophical Transactions* and later journals greatly expanded upon the previously prevailing and unsatisfactory mode of communicating scientific ideas through personal correspondence. Associated with the popularity of science was the new tendency to write even scientific works in the vernacular—so especial a characteristic of Boyle—or, in any case, to have English translations of the esoteric Latin and Greek. It was this type of cumulative interaction between science and society that was destined to mold a climate of opinion in which science stood high in public esteem, long after its religious justification had been forgotten.

But in the seventeenth century, this justification was of sterling importance, not only in preparing the social atmosphere for a welcome acceptance of scientific contributions, but also in providing an ultimate aim for many scientists of the period. For a Barrow, Boyle, or Wilkins, a Ray or Nehemiah Grew, science found its rationale in the end and all of existence —His glorification and the Good of Man. Thus, from Boyle: "The knowledge of the Works of God proportions our Admiration of them, they participating and disclosing so much of the inexhausted Perfections of their Author, that the further we contemplate them, the more Footsteps and Impressions we discover of the Perfections of their Creator; and our utmost Science can but give us a juster veneration of his Omniscience."[57]

Ray carries this conception to its logical conclusion, for if Nature is the manifestation of His power, then nothing in Nature is too mean for scien-

57. *Usefulness of Experimental Natural Philosophy*, pp. 51–52. Boyle continues in this vein. "God loving, as he deserves, to be honour'd in all our Faculties, and consequently to be glorified and acknowledged by the acts of Reason, as well as by those of Faith, there must be sure a great Disparity betwixt that general, confus'd and lazy Idea we commonly have of his Power and Wisdom, and the distinct, rational and affecting notions of those Attributes which are form'd by an attentive inspection of those Creatures in which they are most legible, and which were made chiefly for that very end" (p. 53). Cf. Ray, *Wisdom of God*, p. 132; Wilkins, *Natural Religion*, p. 236 ff; Isaac Barrow, *Opuscula*, 4:88 ff. Cf. Nehemiah Grew, *Cosmologia Sacra* (London, 1701), who points out that God is "the Original, and Ultimate End" and that "we are *bound* to study His works" (pp. 64, 124). Sprat, speaking for the Royal Society, explicitly defines the place of science in the means-end schema of life. "It cannot be deny'd, but it lies in the Natural Philosophers hands, best to advance that part of Divinity [knowledge]: which though it fills not the mind, with such tender, and powerful contemplations, as that which shews us Man's Redemption by a Mediator, yet it is by no means to be pass'd by unregarded: *but is an excellent ground to establish the other*" (*History of the Royal-Society*, p. 83; also pp. 132–33, and passim).

tific study.[58] The universe and the insect, the macrocosm and microcosm alike, are indications of "divine Reason, running like a Golden Vein, through the whole Leaden Mine of Brutal Nature."

On such bases, then, religion was invoked as a sanctioning power of science. But it is necessary to place this and the similar connections previously noted in proper perspective. For it might seem that I take religion as the independent and science as the dependent variable during this period, although, as was remarked at the outset, this is not in the least my intention.

The integration of the Puritan ethic with the accelerated development of science appears evident, but this is simply to maintain that they were elements of a culture which was largely centered about the values of utilitarianism and empiricism.[59] It is perhaps not too much to say, with Lecky, that the acceptance of every great change of belief depends less upon the intrinsic force of its doctrines or the personal capabilities of its proponents than upon the previous social changes which are seen—a posteriori, it is true—to have brought the new doctrines into congruence with the dominant conditions and values of the period. The reanimation of ancient learning; the hesitant, but perceptibly defined, instauration of science; the groping, yet persistent, intensification of economic tendencies; the revolt against scholasticism; all helped bring to a focus the social situation in which the Protestant beliefs and scientific interests both found acceptance.[60] But to realize this is simply to recognize that both Puritanism and science were components of a complicated system of mutually dependent variables. If some comprehensible order is to be attained, a simplified picture of this complex situation must be substituted for the whole; a defensible procedure only if the provisional formulation is not taken for a "complete" explanation.

The integration of religious values and many of the values basic to the contemporary scientists' activity is not fully evidenced by the fact that

58. *Wisdom of God*, p. 130. "If Man ought to reflect upon his Creator the glory of all his Works, then ought he to take notice of them all, and not to think anything unworthy of his Cognizance. And truly the Wisdom, Art and Power of Almighty God, shines forth as visibly in the Structure of the Body of the minutest Insect, as in that of a Horse or Elephant. . . . Let us not then esteem any thing contemptible or inconsiderable, or below our notice taking; for this is to derogate from the Wisdom and Art of the Creator, and to confess our selves unworthy of those Endowments of Knowledge and Understanding which he hath bestowed upon us." Max Weber remarks this same attitude of Swammerdam, whom he quotes as saying: "ich bringe Ihnen hier den Nachweis der Vorsehung Gottes in der Anatomie einer Laus" (*Wissenschaft als Beruf*, p. 19). This constant tendency of leading scientists themselves to relate their studies to dominantly religious ideas gives proof that religion as a social force was considerable and that its high estimation of any activity was of moment.

59. See Troeltsch, *Die Bedeutung des Protestantismus*, p. 80 ff., for a lucid expostion of this point.

60. Lecky, *History*, 1:6.

many of the leading scientists and mathematicians of the day—for example, Oughtred, Barrow, Wilkins, Ward, Ray, Grew, and so on—were also clerics. Such service in the church may have been—though other evidence leads us to doubt it in these instances—a matter of economic consideration since the clerical life provided a fairly adequate income and ample leisure for the pursuit of science. It must be remembered, moreover, that every person appointed to a college fellowship had to be in holy orders. Hence such "external" considerations are less significant than those disclosed by study of the lives of outstanding scientists. Boyle, though he never took orders, was deeply religious: not only did he devote large sums for the translation of the Bible and establish the Boyle lectures in theology, but he learned Greek, Hebrew, Syriac, and Chaldee that he might read the Scriptures in the original![61] For much the same reason, as he states in his *Cosmologia Sacra,* Nehemiah Grew, the estimable botanist, studied Hebrew. Napier and Newton assiduously pursued theological studies and, for the latter, science was in part highly valued because it revealed the divine power.[62]

Religion, then, was a prime consideration and as such its teachings were endowed with a power that emerges with striking emphasis. There is no

61. Boyle was governor of the Corporation for the Propagation of the Gospel in New England, established by Parliament in 1649. On the deep and sincere religiosity of Boyle, cf. Gilbert Burnet, *Lives and Characters* (London, 1833), pp. 351–60, for an account by a contemporary and friend; William Whewell, *Bridgewater Treatise* (London, 1852), p. 273. See H. T. Buckle, *History of Civilization in England* (New York: Boni, 1925), p. 210.

62. Cf. Louis T. More, *Isaac Newton: A Biography* (New York: Charles Scribner's Sons, 1934), p. 134; Burtt, *The Metaphysical Foundations,* pp. 281–83. Of some interest is the attitude towards Newton's work in theology displayed by Pareto and Lombroso. The former states that it appears incredible, though true, that the great Newton could have written a book on the Apocalypse. See *Traité de sociologie générale,* 1:354. Cesare Lombroso is much more extreme. "Newton himself can scarcely be said to have been sane when he demeaned his intellect to the interpretation of the Apocalypse" (*The Man of Genius* [London: Scott, 1891], p. 324). Pareto and Lombroso might have caviled similarly concerning John Napier, the inventor of logarithms, who likewise deemed the writing of a book on the Apocalypse of greater importance than his work in mathematics. Cf. Arthur Schuster and Arthur E. Shipley, *Britain's Heritage of Science,* pp. 6 ff. These declarations of astonishment and dismay over the "inconsistencies" of the seventeenth-century scientists neglect both the nonlogical linkages in human conduct and the particular value-context of the age. Once these are taken into account, Newton's diverse interests appear quite compatible within the given social context. It is significant that the influence of Puritan attitudes can be seen in the instances of Barrow and his successor Newton. Cf. Barrow's *Of Industry,* pp. 2 ff., where, in typically Puritan terms, he exalts the serious and steady application of mind in the prosecution of reasonable designs for the accomplishment of some considerable good. Time must be employed usefully, and games, gaming, theatergoing, poetry, and so on, must be eschewed. Newton likewise had a contempt for the "merely beautiful" and preferred the strictly "useful." His library represents an "almost puritanical selection"—there are no books of humor and practically none of literature, while in poetry there is represented only the Puritan bard, Milton. Cf. R. de Villamil, *Newton: The Man* (London: G. D. Knox, 1931), pp. 10–16.

need to enter into the motivations of individual scientists in order to trace this influence. It is only maintained that the religious ethic, considered as a social force, so consecrated science as to make it a highly respected and laudable focus of attention.

It is this *social* animus that facilitated the development of science by removing the incubus of derogatory social attitudes and instilling favorable ones instead. And it is precisely this social influence that would not necessarily be noticed by the individual scientists upon whom it impinged.[63] We note that religion directly exalted science, that religion was a dominant social force, that science was held in higher social esteem during the latter part of the century, and on the basis of much corroborative evidence we conclude that religion played an important part in this development.

Community of Tacit Assumptions in Science and Puritanism

Up to this point we have been concerned with the directly felt sanction of science by the Protestant ethic. Now, while this was of great importance, there was another relationship which, subtle and difficult of apprehension, was perhaps of equal significance. Puritanism was one element in the preparation of a set of largely implicit assumptions that made for the ready acceptance of the characteristic scientific temper of the seventeenth and subsequent centuries. It is not simply that Protestantism promoted free inquiry, *libre examen,* or decried monastic ascetism. These oft-mentioned characteristics touch only the bare surface of the relationship.

It has become manifest that in each age there is a system of science that rests upon a set of assumptions, usually implicit and seldom questioned by most of the scientific workers of the time.[64] The basic assumption in modern science, that is, in the type of scientific work which becoming pronounced in the seventeenth century has since continued, "is a widespread, instinctive conviction in the existence of an *Order of Things,* and, in particular, of an Order of Nature."[65] This belief, this faith, for at least since Hume it must be recognized as such, is simply "impervious to the demand for a consistent rationality."[66]

63. The difficulty of an individual clearly perceiving the influence of social forces upon his own behavior is expounded by Eduard Spranger. "It is possible to understand . . . a historical character better than he does himself, partly because he has not made himself the object of theoretic reflection . . . and because he is unaware of all the facts which are necessary to the understanding of oneself" (*Types of Men,* p. 367).

64. Alfred North Whitehead, *Science and the Modern World* (New York: Macmillan, 1925), chap. 1; A. E. Heath in *Isaac Newton: A Memorial Volume*, edited for the Mathematical Association by W. J. Greenstreet (London: G. Bell, 1929), p. 133; Burtt, *The Metaphysical Foundations.*

65. Whitehead, *Science and the Modern World*, p. 5.

66. Ibid., p. 6.

In the systems of scientific thought of Galileo, of Newton, and of their successors, the testimony of experiment is a basic criterion of truth, but, as has been suggested, the very notion of experiment is ruled out without the prior *assumption* that Nature constitutes an intelligible order, that when appropriate questions are asked, she will answer, so to speak. Hence it is this assumption that is final and absolute.[67] Now, as Professor Whitehead has indicated, this "faith in the possibility of science, generated antecedently to the development of modern scientific theory, is an unconscious derivative from medieval theology."[68] But this conviction, prerequisite to modern science though it is, was not sufficient to induce its development. What was needed was a constant interest in searching for this order of nature in an empirical and rational fashion, that is, an *active interest* in this world and in its occurrences plus a specifically empirical and methodical approach. With Protestantism, religion provided this interest—it actually imposed obligations of intense concentration on secular activity with an emphasis on experience and reason as bases for action and belief. The conception of good works that provided conviction of grace for the Calvinist-influenced sects is not to be confused with the Catholic conception of "good works." In the Puritan case it involved the notion of a transcendental God and an orientation to the "other world," it is true, but it also demanded a mastery over this world through study of its processes; while in the Catholic instance, it demanded absorption, save for an unbanishable minimum, in the supersensuous, in an intuitive love of God.

It is just at this point that the Protestant emphasis upon reason and experience is of prime importance. In the Protestant system of religion, there is the unchallenged axiom, *gloria Dei,* and, as we have seen, the scheme of behavior which was nonlogically linked with this principle assumes a utilitarian tinge. Virtually all conceptions other than this are subject to, even demand, the examination of reason and experience. Even the Bible as final and complete authority was subject to the interpretation of the individual upon these bases, for though the Bible is infallible, the "meaning" of its content must be sought, as will be remembered from Baxter's discussion of the point. The similarity between the approach and intellectual attitude implicit in the religious and scientific systems is of more than passing interest. This religious point of view molded an attitude of looking at the world of sensuous phenomena that was highly conducive to the willing acceptance of, and preparation for, the same attitude in

67. Cf. Edwin A. Burtt in *Isaac Newton: A Memorial Volume*, p. 139. For a classic exposition of this scientific faith, see Isaac Newton's Rules of Reasoning in Philosophy, in the *Principia*, trans. Andrew Motte (London, 1803), 2:160 ff.

68. See Whitehead, *Science and the Modern World*, p. 19 and preceding, for a discussion of this development.

science. A similarity of this sort is noted by a recent commentator on Calvin's theology. "Die Gedanken werden objektiviert und zu einem objektiven Lehrsystem aufgebaut. Es bekommt geradezu ein naturwissenschaftliches Gepräge; es ist klar, leicht fassbar und formulierbar, wie alles, was der äusseren Welt angehört, klarer zu gestalten ist als das, was im Tiefsten sich abspielt."[69]

The conviction in immutable law is as pronounced in the doctrine of predestination as in scientific investigation: "the immutable law is there and must be acknowledged" (das unabänderlich Gesetz ist da und muss anerkannt werden).[70] The similarity between this concept and the scientific approach is also clearly drawn by Hermann Weber: "Die Lehre von der Prädestination in ihrem tiefsten Kerne getroffen zu sein, wenn mann sie als Faktum im Sinne eines naturwissenschaftlichen Faktums begreift, nur dass das oberste Prinzip, das auch jedem naturwissenschaftlichen Erscheinungskomplex zugrunde liegt, die im tiefsten erlebte gloria dei ist."[71]

The commitment of the Protestant leaders to have reason and experience "test" all religious beliefs, except the basic assumption, which, just as in science, is simply accepted as a matter of faith, is in part grounded upon the previously mentioned conviction of the inherent consistency, congruence, and mutually confirmatory nature of all knowledge, sensory and supersensory. It would seem, then, that there is, to some extent, a community of assumptions in ascetic Protestantism and science: in both there is the unquestioned basic assumption upon which the entire system is built by the utilization of reason and experience. Within each context there is rationality, though the bases are nonrational.[72] The significance of this fundamental similarity is profound though it could hardly have been consciously recognized by those whom it influenced: religion had, for

69. Hermann Weber, *Die Theologie Calvins* (Berlin: Elsner, 1930), p. 23.

70. Ibid., p. 29. The significance of the doctrine of God's foreknowledge for the reinforcement of the belief in natural law is remarked by Buckle, *History of Civilization*, p. 482. It is significant that the first writer who maintained that even lotteries are governed by purely natural laws was a Puritan minister, Thomas Gataker, in his curious little book, *On the Nature and Use of Different Kinds of Lots* (London, 1619). This assumption, which ran over the barriers of religious differences, is not unrelated to the later development of "political arithmetic" by Graunt, Petty, and Halley.

71. Weber, *Die Theologie Calvins*, p. 31. See Whitehead, *Science and the Modern World*, chap. 1, for a statement of similar characteristics of modern science.

72. A modern logican has aptly remarked that the social sciences must locate the irrational (rather, nonlogical) sources of both rational and irrational thought. Cf. Rudolf Carnap, "Logic," in *Factors Determining Human Behavior* (Cambridge, Mass.: Harvard Tercentenary Publications, 1937), p. 118. Certainly Puritanism was not "the source" of modern science, but apparently it acted to stimulate such thought. Cf. Walter Pagel's similar comparison of the "irrationality and empiricism" of seventeenth-century religion and science. ("Religious Motives," p. 112.)

whatever reasons, adopted a cast of thought which was essentially that of science and so reinforced the typically scientific attitudes of the period. The society was permeated with attitudes toward natural phenomena that, derived from both science and religion, unwittingly helped maintain conceptions characteristic of the new science.

12 Science and the Social Order

1938

About the turn of the century, Max Weber observed that "the belief in the value of scientific truth is not derived from nature but is a product of definite cultures."[1] We may now add: and this belief is readily transmuted into doubt or disbelief. The persistent development of science occurs only in societies of a certain order, subject to a peculiar complex of tacit presuppositions and institutional constraints. What is for us a normal phenomenon which demands no explanation and secures many self-evident cultural values, has been in other times and still is in many places abnormal and infrequent. The continuity of science requires the active participation of interested and capable persons in scientific pursuits. But this support of science is assured only by appropriate cultural conditions. It is, then, important to examine the controls that motivate scientific careers, that select and give prestige to certain scientific disciplines and reject or blur others. It will become evident that changes in institutional structure may curtail, modify, or possibly prevent the pursuit of science.[2]

Sources of Hostility toward Science

Hostility toward science may arise under at least two sets of conditions, although the concrete systems of values—humanitarian, economic, politi-

Reprinted, with permission, from *Philosophy of Science* 5 (1938): 321–37. Paper first read at the American Sociological Society, December 1937. The writer is indebted to Read Bain, Talcott Parsons, E. Y. Hartshorne, and E. P. Hutchinson for their helpful suggestions.

1. Max Weber, *Gesammelte Aufsätze zur Wissenschaftslehre* (Tubingen: J. C. B. Mohr, 1922), p. 213; cf. Pitirim A. Sorokin, *Social and Cultural Dynamics*, 4 vols. (New York: American Book Company, 1937), esp. vol. 2, chap. 2.

2. Cf. Robert K. Merton, *Science, Technology and Society in Seventeenth-Century England*, chap. 11.

cal, religious—upon which it is based may vary considerably. The first involves the logical, though not necessarily empirically sound, conclusion that the results or methods of science are inimical to the satisfaction of important values. The second consists largely of nonlogical elements. It rests upon the feeling of incompatibility between the sentiments embodied in the scientific ethos and those found in other institutions. Whenever this feeling is challenged, it is rationalized. Both sets of conditions underlie, in varying degrees, current revolts against science. It might be added that such logical and affective responses are also involved in the social approval of science. But in these instances science is thought to facilitate the achievement of approved ends and basic cultural values are felt to be congruent with those of science rather than emotionally inconsistent with them. The position of science in the modern world may be analyzed, then, as a resultant of two sets of conflicting forces, approving and opposing science as a large-scale social activity.

We restrict our examination to a few conspicuous instances of hostile revaluation of the social role of science, without implying that the anti-science movement is in any sense thus localized. Much of what is said here can probably be applied to the cases of other times and places.[3]

The situation in Nazi Germany since 1933 illustrates the ways in which logical and nonlogical processes converge to modify or curtail scientific activity. In part, the hampering of science is an unintended by-product of changes in political structure and nationalistic credo. In accordance with the dogma of race purity, practically all persons who do not meet the politically imposed criteria of "Aryan" ancestry and of avowed sympathy with Nazi aims have been eliminated from universities and scientific institutes.[4] Since these outcasts include a considerable number of eminent scientists, one indirect consequence of the racialist purge is the weakening of science in Germany.

Implicit in this racialism is a belief in race defilement through actual or symbolic contact.[5] Scientific research by those of unimpeachable "Aryan" ancestry who collaborate with non-Aryans or who even accept their scientific theories is either restricted or proscribed. A new racial-political category has been introduced to include these incorrigible scientists who were once declared to be *echt-arisch*: the category of "White Jews." A

3. The premature death of E. Y. Hartshorne halted a proposed study of science in the modern world in terms of the analysis introduced in this chapter.

4. See chap. 3 of E. Y. Hartshorne, *The German Universities and National Socialism* (Cambridge, Mass.: Harvard University Press, 1937), on the purge of the universities; cf. *Volk und Werden* 5 (1937): 320–21, which refers to some of the new requirements for the doctorate.

5. This is one of the many phases of the introduction of a caste system in Germany. As R. M. MacIver has observed, "The idea of defilement is common in every caste system" (*Society* [New York: Farrar & Rinehart, 1937], p. 172).

prominent member of this new race is the Nobel Prize physicist, Werner Heisenberg, who has persisted in his declaration that Einstein's theory of relativity constitutes an "obvious basis for further research."[6]

In these instances, the sentiments of national and racial purity have prevailed over utilitarian rationality. The application of such criteria has led to a greater proportionate loss to the natural science and medical faculties in German universities than to the theological and juristic faculties, as E. Y. Hartshorne has found.[7] In contrast, utilitarian considerations are foremost when it comes to official policies concerning the directions to be followed by scientific research. Scientific work which promises direct practical benefit to the Nazi party or the Third Reich is to be fostered above all, and research funds are to be reallocated in accordance with this policy.[8] The rector of Heidelberg University announces that "the question of the scientific significance [*Wissenschaftlichkeit*] of any knowledge is of quite secondary importance when compared with the question of its utility."[9]

The general tone of anti-intellectualism, with its depreciation of the theorist and its glorification of the man of action,[10] may have long-run

6. Cf. the official organ of the SS, the *Schwarze Korps*, 15 July 1937, p. 2. In this issue Johannes Stark, the president of the Physikalisch-Technischen Reichsanstalt, urges elimination of such collaborations which still continue and protests the appointment of three university professors who have been "disciples" of non-Aryans. See also Hartshorne, *The German Universities*, pp 112–13; Alfred Rosenberg, *Wesen, Grundsätze und Ziele der Nationalsozialistischen Deutschen Arbeiterpartei* (Munich: E. Boepple, 1933), p. 45 ff.; J. Stark, "Philipp Lenard als deutscher Naturforscher," *Nationalsozialistische Monatshefte* 71 (1936): 106–11, where Heisenberg, Schrödinger, von Laue, and Planck are castigated for not having divorced themselves from the "Jewish physics" of Einstein. See also chapter 5 of this volume.

7. The data upon which this statement is based are from an unpublished study by E. Y. Hartshorne.

8. Cf. *Wissenschaft und Vierjahresplan*, Reden anlässlich der Kundgebung des NSD-Dozentenbundes, 18 January 1937; Hartshorne, *The German Universities*, p. 110 ff.; E. R. Jaensch, *Zur Neugestaltung des deutschen Studententums und der Hochschule* (Leipzig: J. A. Bart, 1937), esp. p. 57 ff. In the field of history, for example, Walter Frank, the director of the Reichsinstitut für Geschichte des neuen Deutschlands, "the first German scientific organization which has been created by the spirit of the national-socialistic revolution," testifies that he is the last person to forgo sympathy for the study of ancient history, "even that of foreign peoples," but also points out that the funds previously granted the Archaeological Institute must be reallocated to this new historical body which will "have the honor of writing the history of the National Socialist Revolution." See his *Zukunft und Nation* (Hamburg: Hanseatische Verlagsanstalt, 1935), esp. pp. 30 ff.

9. Ernst Krieck, *Nationalpolitische Erziehung* (Leipzig: Armanen Verlag, 1935), p. 8.

10. The Nazi theoretician, Alfred Baeumler, writes: "Wenn ein Student heute es ablehnt, sich der politischen Norm zu unterstellen, es z. B ablehnt, an einem Arbeits- oder Wehrsportlager teilzunehmen, weil er damit Zeit für sein Studium versäume, dann zeigt er damit, dass er nichts von dem begriffen hat, was um ihn geschieht. Seine Zeit kann er nur bei einem abstrakten, richtungslosen Studium versäumen" (*Männerbund und Wissenschaft* [Berlin: Junker & Dünnhaupt, 1934], p. 153).

rather than immediate bearing upon the place of science in Germany. For should these attitudes become fixed, the most gifted elements of the population may be expected to shun those intellectual disciplines which have become disreputable. By the late thirties, effects of this anti-theoretical attitude could be detected in the allocation of academic interests in the German universities.[11]

It would be misleading to suggest that the Nazi government has completely repudiated science and intellect. The official attitudes toward science are clearly ambivalent and unstable. (For this reason, any statements concerning science in Nazi Germany are made under correction.) On the one hand, the challenging skepticism of science interferes with the imposition of a new set of values which demand an unquestioning acquiescence. But the new dictatorships must recognize, as did Hobbes who also argued that the State must be all or nothing, that science is power. For military, economic, and political reasons, theoretical science—to say nothing of its more respectable sibling, technology—cannot be safely discarded. Experience has shown that the most esoteric researches have found important applications. Unless utility and rationality are dismissed beyond recall, it cannot be forgotten that Clerk Maxwell's speculations on the ether led Hertz to the discovery that culminated in the wireless. And indeed one Nazi spokesman remarks: "As the practice of today rests on the science of yesterday, so is the research of today the practice of tomorrow."[12] Emphasis on utility requires an unbanishable minimum of interest in science which can be enlisted in the service of the State and industry.[13] At the same time, this emphasis leads to a limitation of research in pure science.

Social Pressures on Autonomy of Science

An analysis of the role of science in the Nazi state uncovers the following elements and processes. The spread of domination by one segment of the social structure—the State—involves a demand for primary loyalty to it. Scientists, as well as all others, are called upon to relinquish adherence to

11. Hartshorne, *The German Universities*, p. 106 ff.; cf. *Wissenschaft und Vierjahresplan*, pp. 25–26, where it is stated that the present "breathing-spell in scientific productivity" is partly due to the fact that a considerable number of those who might have received scientific training have been recruited by the army. Although this is a dubious explanation of that particular situation, a prolonged deflection of interest from theoretical science will probably produce a decline in scientific achievements.

12. Professor Thiessen in *Wissenschaft und Vierjahresplan*, p. 12.

13. For example, chemistry is highly prized because of its practical importance. As Hitler put it, "we will carry on because we have the fanatic will to help ourselves and because in Germany we have the chemists and inventors who will fulfil our needs." Quoted in *Wissenschaft und Vierjahresplan*, p. 6, and passim.

all institutional norms that, in the opinion of political authorities, conflict with those of the State.[14] The norms of the scientific ethos must be sacrificed, insofar as they demand a repudiation of the politically imposed criteria of scientific validity or of scientific worth. The expansion of political control thus introduces conflicting loyalties. In this respect, the reactions of devout Catholics who resist the efforts of the political authority to redefine the social structure, to encroach upon the preserves which are traditionally those of religion, are of the same order as the resistance of the scientist. From the sociological point of view, the place of science in the totalitarian world is largely the same as that of all other institutions except the newly dominant State. The basic change consists in placing science in a new social context where it appears to compete at times with loyalty to the state. Thus, cooperation with non-Aryans is redefined as a symbol of political disloyalty. In a liberal order, the limitation of science does not arise in this fashion. For in such structures, a substantial sphere of autonomy—varying in extent, to be sure—is enjoyed by nonpolitical institutions.

The conflict between the totalitarian state and the scientist derives in part, then, from an incompatibility between the ethic of science and the new political code which is imposed upon all, irrespective of occupational creed. The ethos of science[15] involves the functionally necessary demand that theories or generalizations be evaluated in terms of their logical consistency and consonance with facts. The political ethic would introduce the hitherto irrelevant criteria of the race or political creed of the theorist.[16]

14. This is clearly put by Reichswissenschaftsminister Bernhard Rust, *Das Nationalsozialistische Deutschland und die Wissenschaft* (Hamburg: Hanseatische Verlagsanstalt, 1936), pp. 1–22, esp. p. 21.

15. The ethos of science refers to an emotionally toned complex of rules, prescriptions, mores, beliefs, values, and presuppositions which are held to be binding upon the scientist. Some phases of this complex may be methodologically desirable, but observance of the rules is not dictated solely by methodological considerations. This ethos, as social codes generally, is sustained by the sentiments of those to whom it applies. Transgression is curbed by internalized prohibitions and by disapproving emotional reactions which are mobilized by the supporters of the ethos. Once given an effective ethos of this type, resentment, scorn, and other attitudes of antipathy operate almost automatically to stabilize the existing structure. This may be seen in the current resistance of some scientists in Germany to marked modifications in the content of this ethos. The ethos may be thought of as the "cultural" as distinct from the "civilizational" component of science. Cf. R. K. Merton, "Civilization and Culture," *Sociology and Social Research* 21 (1936): 103–13.

16. Cf. Baeumler, *Männerbund und Wissenschaft*, p. 145. Also Krieck, *Nationalpolitische Erziehung*, who states: "Nicht alles, was den Anspruch auf Wissenschaftlichkeit erheben darf, liegt auf der gleichen Rang- und Wertebene; protestantische und katholische, französische und deutsche, germanische and jüdische, humanistische oder rassische Wissenschaft sind zunächst nur Möglichkeiten, noch nicht erfüllte oder gar gleichrangige Werte. Die Entscheidung über den Wert der Wissenschaft fällt aus ihrer 'Gegenwärtigkeit,' aus dem Grad ihrer Fruchtbarkeit, ihrer geschichtsbildenden Kraft."

Modern science has considered the personal equation as a potential source of error and has evolved impersonal criteria for checking such error. It is now called upon to assert that certain scientists, because of their extra-scientific affiliations, are a priori incapable of anything but spurious and false theories. In some instances, scientists are required to accept the judgments of scientifically incompetent political leaders concerning *matters of science*. But such politically advisable tactics run counter to the institutionalized norms of science. These, however, are dismissed by the totalitarian state as "liberalistic" or "cosmopolitan" or "bourgeois" prejudices,[17] inasmuch as they cannot be readily integrated with the campaign for an unquestioned political creed.

From a broader perspective, the conflict is a phase of institutional dynamics. Science, which has acquired a considerable degree of autonomy and has evolved an institutional complex that engages the allegiance of scientists, now has both its traditional autonomy and its rules of the game —its ethos, in short—challenged by an external authority. The sentiments embodied in the ethos of science—characterized by such terms as intellectual honesty, integrity, organized skepticism, disinterestedness, impersonality—are outraged by the set of new sentiments that the State would impose in the sphere of scientific research. With a shift from the previous structure where limited loci of power are vested in the several fields of human activity to a structure where there is one centralized locus of authority over all phases of behavior, the representatives of each sphere act to resist such changes and to preserve the original structure of pluralistic authority. Although it is customary to think of the scientist as a dispassionate, impersonal individual, it must be remembered that the scientist, in company with all other professional workers, has a large emotional investment in his way of life, defined by the institutional norms which govern his activity. In terms of that ethos, the social stability of science can be ensured only if adequate defenses are set up against changes imposed from outside the scientific fraternity itself.

This process of preserving institutional integrity and resisting new definitions of social structure which may interfere with the autonomy of science finds expression in yet another direction. It is a basic assumption of modern science that scientific propositions "are invariant with respect to the individual" and group.[18] But in a completely politicized society—where as

17. Thus, says Ernst Krieck: "In the future, one will no more adopt the fiction of an enfeebled neutrality in science than in law, economy, the State or public life generally. The method of science is indeed only a reflection of the method of government" (*Nationalpolitische Erziehung*, p. 6). Cf. Baeumler, *Männerbund und Wissenschaft*, p. 152; Walter Frank, *Zukunft und Nation*, p. 10; and contrast with Max Weber's "prejudice" that "Politik gehört nicht in den Hörsaal."

18. H. Levy, *The Universe of Science* (New York: Century Co., 1933), p. 189.

one Nazi theorist put it, "the universal meaning of the political is recognized"[19]—this assumption is impugned. Scientific findings are held to be merely the expression of race or class or nation.[20] As such doctrines percolate to the laity, they invite a general distrust of science and a depreciation of the prestige of the scientist, whose discoveries appear arbitrary and fickle. This variety of anti-intellectualism which threatens his social position is characteristically enough resisted by the scientist. On the ideological front as well, totalitarianism entails a conflict with the traditional assumptions of modern Western science.

Functions of Norms of Pure Science

One sentiment which is assimilated by the scientist from the very outset of his training pertains to the purity of science. Science must not suffer itself to become the handmaiden of theology or economy or state. The function of this sentiment is to preserve the autonomy of science. For if such extrascientific criteria of the value of science as presumable consonance with religious doctrines or economic utility or political appropriateness are adopted, science becomes acceptable only insofar as it meets these criteria. In other words, as the pure science sentiment is eliminated, science becomes subject to the direct control of other institutional agencies and its place in society becomes increasingly uncertain. The persistent repudiation by scientists of the application of utilitarian norms to their work has as its chief function the avoidance of this danger, which is particularly marked at the present time. A tacit recognition of this function may be the source of that possibly apocryphal toast at a dinner for scientists in Cambridge: To pure mathematics, and may it never be of any use to anybody!

The exaltation of pure science is thus seen to be a defense against the invasion of norms that limit directions of potential advance and threaten the stability and continuance of scientific research as a valued social activity. Of course, the technological criterion of scientific achievement also has a social function for science. The increasing comforts and conveniences deriving from technology and ultimately from science invite the social support of scientific research. They also testify to the integrity of the scien-

19. Baeumler, *Männerbund und Wissenschaft*, p. 152.

20. It is of considerable interest that totalitarian theorists have adopted the radical relativistic doctrines of *Wissenssoziologie* as a political expedient for discrediting "liberal" or "bourgeois" or "non-Aryan" science. An exit from this cul-de-sac is provided by positing an Archimedean point: the infallibility of *der Führer* and his *Volk*. Cf. General Hermann Goering, *Germany Reborn* (London: Matthews & Marrot, 1934), p. 79. Politically effective variations of the "relationism" of Karl Mannheim (for example, *Ideology and Utopia*) have been used for propagandistic purposes by such Nazi theorists as Walter Frank, Krieck, Rust, and Rosenberg.

tist, since abstract and difficult theories which cannot be understood or evaluated by the laity are presumably proved in a fashion which can be understood by all, that is, through their technological applications. Readiness to accept the authority of science rests, to a considerable extent, upon its daily demonstration of power. Were it not for such indirect demonstrations, the continued social support of that science which is intellectually incomprehensible to the public would hardly be nourished on faith alone.

At the same time, this stress upon the purity of science has had other consequences that threaten rather than preserve the social esteem of science. It is repeatedly urged that scientists should in their research ignore all considerations other than the advance of knowledge.[21] Attention is to be focused exclusively on the scientific significance of their work with no concern for the practical uses to which it may be put or for its social repercussions generally. The customary justification of this tenet—which is partly rooted in circumstance[22] and which, in any event, has definite social functions, as we have just seen—holds that failure to adhere to this injunction will encumber research by increasing the possibility of bias and error. But this *methodological* view overlooks the *social* results of such an attitude. The objective consequences of this attitude have furnished a further basis of revolt against science; an incipient revolt that is found in virtually every society where science has reached a high stage of development. Since the scientist does not or cannot control the direction in which his discoveries are applied, he becomes the subject of reproach and of more violent reactions insofar as these applications are disapproved by the agents of authority or by pressure groups. The antipathy toward the technological products is projected toward science itself. Thus, when newly discovered gases or explosives are applied as military instruments, chemistry as a whole is censured by those whose humanitarian sentiments are outraged. Science is held largely responsible for endowing those engines of human

21. For example, Pareto writes: "The quest for experimental uniformities is an end in itself." See a typical statement by George A. Lundberg. "It is not the business of a chemist who invents a high explosive to be influenced in his task by considerations as to whether his product will be used to blow up cathedrals or to build tunnels through the mountains. Nor is it the business of the social scientist in arriving at laws of group behavior to permit himself to be influenced by considerations of how his conclusions will coincide with existing notions, or what the effect of his findings on the social order will be" (*Trends in American Sociology*, ed. G. A. Lundberg, R. Bain, and N. Anderson [New York: Harper, 1929], pp. 404–5). Compare the remarks of Read Bain on the "Scientist as Citizen," *Social Forces* 11 (1933): 412–15.

22. A neurological justification of this view is to be found in E. D. Adrian's essay in *Factors Determining Human Behavior* (Cambridge, Mass.: Harvard Tercentenary Publications, 1937), p. 9. "For discriminative behavior . . . there must be some interest: yet if there is too much the behavior will cease to be discriminative. Under intense emotional stress the behavior tends to conform to one of several stereotyped patterns."

destruction which, it is said, may plunge our civilization into everlasting night and confusion. Or to take another prominent instance, the rapid development of science and related technology has led to an implicitly antiscience movement by vested interests and by those whose sense of economic justice is offended. The eminent Sir Josiah Stamp and a host of less illustrious folk have proposed a moratorium on invention and discovery,[23] in order that man may have a breathing spell in which to adjust his social and economic structure to the constantly changing environment with which he is presented by the "embarrassing fecundity of technology." These proposals have received wide publicity in the press and have been urged with unslackened insistence before scientific bodies and governmental agencies.[24] The opposition comes particularly from those representatives of labor who fear the loss of investment in skills that become obsolete before the flood of new technologies. Although these proposals probably will not be translated into action within the immediate future, they constitute one possible nucleus about which a revolt against science in general may materialize. It is largely immaterial whether these opinions which make science ultimately responsible for undesirable situations are valid or not. W. I. Thomas's sociological theorem—"If men define situations as real, they are real in their consequences"—is much in point here.

23. Of course, this does not constitute a movement opposed to science as such. Moreover, the destruction of machinery by labor and the suppression of inventions by capital have also occurred in the past. Cf. R. K. Merton, "Fluctuations in the Rate of Industrial Inventions," *Quarterly Journal of Economics* 49 (1935): 464 ff. But this movement mobilizes the opinion that science is to be held strictly accountable for its social effects. Sir Josiah Stamp's suggestion may be found in his address to the British Association for the Advancement of Science, Aberdeen, 6 September 1934. Such moratoria have also been proposed by M. Caillaux (cf. John Strachey, *The Coming Struggle for Power* [New York, 1935], p. 183), by H. W. Summers in the U. S. House of Representatives, and by many others. In terms of current humanitarian, social, and economic criteria, some of the products of science are more pernicious than beneficial. This evaluation may destroy the rationale of scientific work. As one scientist pathetically put it: if the man of science must be apologetic for his work, I have wasted my life. Cf. *The Frustration of Science*, ed. F. Soddy (New York: Norton, 1935), p. 42, and passim.

24. English scientists have especially reacted against the "prostitution of scientific effort to war purposes." Presidential addresses at annual meetings of the British Association for the Advancement of Science and frequent editorials and letters in *Nature* attest to this movement for "a new awareness of social responsibility among the rising generation of scientific workers." Sir Frederick Gowland Hopkins, Sir John Orr, Professor F. Soddy, Sir Daniel Hall, Dr. Julian Huxley, J. B. S. Haldane, and Professor L. Hogben are among the leaders of the movement. See, for example, the letter signed by twenty-two scientists of Cambridge University urging a program for dissociating science from warfare (*Nature* 137 [1936]: 829). These attempts for concerted action by English scientists contrast sharply with the apathy of scientists in this country toward these questions. [This observation holds for the period prior to the development of atomic weapons.] The basis of this contrast might profitably be investigated.

In short, this basis for the revaluation of science derives from what I have called elsewhere the "imperious immediacy of interest."[25] Concern with the primary goal, the furtherance of knowledge, is coupled with a disregard of the consequences that lie outside the area of immediate interest, but these social results react so as to interfere with the original pursuits. Such behavior may be rational in the sense that it may be expected to lead to the satisfaction of the immediate interest. But it is irrational in the sense that it defeats other values which are not, at the moment, paramount but which are nonetheless an integral part of the social scale of values. Precisely because scientific research is not conducted in a social vacuum, its effects ramify into other spheres of value and interest. Insofar as these effects are deemed socially undesirable, science is charged with responsibility. The goods of science are no longer considered an unqualified blessing. Examined from this perspective, the tenet of pure science and disinterestedness has helped to prepare its own epitaph.

Battle lines are drawn in terms of the question: can a good tree bring forth evil fruit? Those who would cut down or stunt the tree of knowledge because of its accursed fruit are met with the claim that the evil fruit has been grafted on the good tree by the agents of state and economy. It may salve the conscience of the individual man of science to hold that an inadequate social structure has led to the perversion of his discoveries. But this will hardly satisfy an embittered opposition. Just as the *motives* of scientists may range from a passionate desire in the furtherance of knowledge to a profound interest in achieving personal fame and just as the *functions* of scientific research may vary from providing prestige-laden rationalizations of the existing order to enlarging our control of nature, so may other social *effects* of science be considered pernicious to society or result in the modification of the scientific ethos itself. There is a tendency for scientists to assume that the social effects of science *must* be beneficial in the long run. This article of faith performs the function of providing a rationale for scientific research, but it is manifestly not a statement of fact. It involves the confusion of truth and social utility which is characteristically found in the nonlogical penumbra of science.

Esoteric Science as Popular Mysticism

Another relevant phase of the connections between science and the social order has seldom been recognized. With the increasing complexity of scientific research, a long program of rigorous training is necessary to test or even to understand the new scientific findings. The modern scientist has

25. R. K. Merton, "The Unanticipated Consequences of Purposive Social Action," *American Sociological Review* 1 (1936): 894–904.

necessarily subscribed to a cult of unintelligibility. There results an increasing gap between the scientist and the laity. The layman must take on faith the publicized statements about relativity or quanta or other such esoteric subjects. This he has readily done inasmuch as he has been repeatedly assured that the technologic achievements from which he has presumably benefited ultimately derive from such research. Nonetheless, he retains a certain suspicion of these bizarre theories. Popularized and frequently garbled versions of the new science stress the theories that seem to run counter to common sense. To the public mind, science and esoteric terminology become indissolubly linked. The presumably scientific pronouncements of totalitarian spokesmen on race or economy or history are for the uninstructed laity of the same order as announcements concerning an expanding universe or wave mechanics. In both instances, the laity is in no position to understand these conceptions or to check their scientific validity and in both instances they may not be consistent with common sense. If anything, the myths of totalitarian theorists will seem more plausible and are certainly more comprehensible to the general public than accredited scientific theories, since they are closer to common-sense experience and cultural bias. Partly as a result of scientific advance, therefore, the population at large has become ripe for new mysticisms clothed in apparently scientific jargon. This promotes the success of propaganda generally. The borrowed authority of science becomes a powerful prestige symbol for unscientific doctrines.

Public Hostility toward Organized Skepticism

Another feature of the scientific attitude is organized skepticism, which becomes, often enough, iconoclasm.[26] Science may seem to challenge the "comfortable power assumptions" of other institutions,[27] simply by subjecting them to detached scrutiny. Organized skepticism involves a latent questioning of certain bases of established routine, authority, vested procedures, and the realm of the "sacred" generally. It is true that, *logically*, to establish the empirical genesis of beliefs and values is not to deny their validity, but this is often the psychological effect on the naïve mind. Institutionalized symbols and values demand attitudes of loyalty, adherence, and respect. Science, which asks questions of fact concerning every phase of nature and society, comes into psychological, not logical, conflict with

26. Frank H. Knight, "Economic Psychology and the Value Problem," *Quarterly Journal of Economics* 39 (1925): 372–409. The unsophisticated scientist, forgetting that skepticism is primarily a methodological canon, permits his skepticism to spill over into the area of value generally. The social functions of symbols are ignored and they are impugned as "untrue." Social utility and truth are once again confused.

27. Charles E. Merriam, *Political Power* (New York: Whittlesey House, 1934), pp. 82–83.

other attitudes toward these same data which have been crystallized and frequently ritualized by other institutions. Most institutions demand unqualified faith; but the institution of science makes skepticism a virtue. Every institution involves, in this sense, a sacred area that is resistant to profane examination in terms of scientific observation and logic. The institution of science itself involves emotional adherence to certain values. But whether it be the sacred sphere of political convictions or religious faith or economic rights, the scientific investigator does not conduct himself in the prescribed uncritical and ritualistic fashion. He does not preserve the cleavage between the sacred and the profane, between that which requires uncritical respect and that which can be objectively analyzed.[28]

It is this which in part lies at the root of revolts against the so-called intrusion of science into other spheres. In the past, this resistance has come for the most part from the church which restrains the scientific examination of sanctified doctrines. Textual criticism of the Bible is still suspect. This resistance on the part of organized religion has become less significant as the locus of social power has shifted to economic and political institutions which in their turn evidence an undisguised antagonism toward that generalized skepticism which is felt to challenge the bases of institutional stability. This opposition may exist quite apart from the introduction of scientific discoveries that appear to invalidate particular dogmas of church, economy, and state. It is rather a diffuse, frequently vague, recognition that skepticism threatens the status quo. It must be emphasized again that there is no logical necessity for a conflict between skepticism within the sphere of science and the emotional adherences demanded by other institutions. But as a psychological derivative, this conflict invariably appears whenever science extends its research to new fields toward which there are institutionalized attitudes or whenever other institutions extend their area of control. In the totalitarian society, the centralization of institutional control is the major source of opposition to science; in other structures, the extension of scientific research is of greater importance. Dictatorship organizes, centralizes, and hence intensifies sources of revolt against science that in a liberal structure remain unorganized, diffuse, and often latent.

In a liberal society, integration derives primarily from the body of cultural norms toward which human activity is oriented. In a dictatorial structure, integration is effected primarily by formal organization and centralization of social control. Readiness to accept this control is instilled by speeding up the process of infusing the body politic with new cultural values, by substituting high-pressure propaganda for the slower process of the diffuse inculcation of social standards. The differences in the mecha-

28. For a general discussion of the sacred in these terms, see Emile Durkheim, *The Elementary Forms of the Religious Life*, pp. 37 ff.

nisms through which integration is typically effected permit a greater latitude for self-determination and autonomy to various institutions, including science, in the liberal than in the totalitarian structure. Through such rigorous organization, the dictatorial state so intensifies its control over nonpolitical institutions as to lead to a situation that is different in kind as well as degree. For example, reprisals against science can more easily find expression in the Nazi state than in America, where interests are not so organized as to enforce limitations upon science, when these are deemed necessary. Incompatible sentiments must be insulated from one another or integrated with each other if there is to be social stability. But such insulation becomes virtually impossible when there exists centralized control under the aegis of any one sector of social life which imposes, and attempts to enforce, the obligation of adherence to its values and sentiments as a condition of continued existence. In liberal structures the absence of such centralization permits the necessary degree of insulation by guaranteeing to each sphere restricted rights of autonomy and thus enables the gradual integration of temporarily inconsistent elements.

Conclusions

The main conclusions of this paper can be briefly summarized. There exists a latent and active hostility toward science in many societies, although the extent of this antagonism cannot yet be established. The prestige which science has acquired within the last three centuries is so great that actions curtailing its scope or repudiating it in part are usually coupled with affirmation of the undisturbed integrity of science or "the rebirth of true science." These verbal respects to the pro-science sentiment are frequently at variance with the behavior of those who pay them. In part, the anti-science movement derives from the conflict between the ethos of science and of other social institutions. A corollary of this proposition is that contemporary revolts against science are *formally* similar to previous revolts, although the *concrete* sources are different. Conflict arises when the social effects of applying scientific knowledge are deemed undesirable, when the scientist's skepticism is directed toward the basic values of other institutions, when the expansion of political or religious or economic authority limits the autonomy of the scientist, when anti-intellectualism questions the value and integrity of science and when nonscientific criteria of eligibility for scientific research are introduced.

This paper does not present a program for action in order to withstand threats to the development and autonomy of science. It may be suggested, however, that as long as the locus of social power resides in any one institution other than science and as long as scientists themselves are uncertain of their primary loyalty, their position becomes tenuous and uncertain.

13 The Normative Structure of Science

1942

Science, like any other activity involving social collaboration, is subject to shifting fortunes. Difficult as the notion may appear to those reared in a culture that grants science a prominent if not a commanding place in the scheme of things, it is evident that science is not immune from attack, restraint, and repression. Writing a little while ago, Veblen could observe that the faith of western culture in science was unbounded, unquestioned, unrivaled. The revolt from science which then appeared so improbable as to concern only the timid academician who would ponder all contingencies, however remote, has now been forced upon the attention of scientist and layman alike. Local contagions of anti-intellectualism threaten to become epidemic.

Science and Society

Incipient and actual attacks upon the integrity of science have led *scientists to recognize their dependence on particular types of social structure.* Manifestos and pronouncements by associations of scientists are devoted to the relations of science and society. An institution under attack must reexamine its foundations, restate its objectives, seek out its rationale. Crisis invites self-appraisal. Now that they have been confronted with challenges to their way of life, scientists have been jarred into a state of acute self-consciousness: consciousness of self as an integral element of society with corre-

Originally published as "Science and Technology in a Democratic Order," *Journal of Legal and Political Sociology* 1 (1942): 115–26; later published as "Science and Democratic Social Structure," in Robert K. Merton, *Social Structure and Social Theory*. Reprinted with permission.

sponding obligations and interests.[1] A tower of ivory becomes untenable when its walls are under prolonged assault. After a long period of relative security, during which the pursuit and diffusion of knowledge had risen to a leading place if indeed not to the first rank in the scale of cultural values, scientists are compelled to vindicate the ways of science to man. Thus they have come full circle to the point of the reemergence of science in the modern world. Three centuries ago, when the institution of science could claim little independent warrant for social support, natural philosophers were likewise led to justify science as a means to the culturally validated ends of economic utility and the glorification of God. The pursuit of science was then no self-evident value. With the unending flow of achievement, however, the instrumental was transformed into the terminal, the means into the end. Thus fortified, the scientist came to regard himself as independent of society and to consider science as a self-validating enterprise which was in society but not of it. A frontal assault on the autonomy of science was required to convert this sanguine isolationism into realistic participation in the revolutionary conflict of cultures. The joining of the issue has led to a clarification and reaffirmation of the ethos of modern science.

Science is a deceptively inclusive word which refers to a variety of distinct though interrelated items. It is commonly used to denote (1) a set of characteristic methods by means of which knowledge is certified; (2) a stock of accumulated knowledge stemming from the application of these methods; (3) a set of cultural values and mores governing the activities termed scientific; or (4) any combination of the foregoing. We are here concerned in a preliminary fashion with the cultural structure of science, that is, with one limited aspect of science as an institution. Thus, we shall consider, not the methods of science, but the mores with which they are hedged about. To be sure, methodological canons are often both technical expedients and moral compulsives, but it is solely the latter which is our concern here. This is an essay in the sociology of science, not an excursion in methodology. Similarly, we shall not deal with the substantive findings of sciences (hypotheses, uniformities, laws), except as these are pertinent to standardized social sentiments toward science. This is not an adventure in polymathy.

The Ethos of Science

The ethos of science is that affectively toned complex of values and

1. Since this was written in 1942, it is evident that the explosion at Hiroshima has jarred many more scientists into an awareness of the social consequences of their work.

norms which is held to be binding on the man of science.[2] The norms are expressed in the form of prescriptions, proscriptions, preferences, and permissions. They are legitimatized in terms of institutional values. These imperatives, transmitted by precept and example and reenforced by sanctions are in varying degrees internalized by the scientist, thus fashioning his scientific conscience or, if one prefers the latter-day phrase, his superego. Although the ethos of science has not been codified,[3] it can be inferred from the moral consensus of scientists as expressed in use and wont, in countless writings on the scientific spirit and in moral indignation directed toward contraventions of the ethos.

An examination of the ethos of modern science is only a limited introduction to a larger problem: the comparative study of the institutional structure of science. Although detailed monographs assembling the needed comparative materials are few and scattered, they provide some basis for the provisional assumption that "science is afforded opportunity for development in a democratic order which is integrated with the ethos of science." This is not to say that the pursuit of science is confined to democracies.[4] The most diverse social structures have provided some measure of support to science. We have only to remember that the Accademia del Cimento was sponsored by two Medicis; that Charles II claims historical attention for his grant of a charter to the Royal Society of London and his sponsorship of the Greenwich Observatory; that the Académie des Sciences was founded under the auspices of Louis XIV, on the advice of Colbert; that urged into acquiescence by Leibniz, Frederick I endowed the Berlin Academy, and that the St. Petersburg Academy of Sciences was instituted by Peter the Great (to refute the view that Russians are barbarians). But such historical facts do not imply a random association of science and social structure. There is the further question of the ratio of scientific achievement to scientific potentialities. Science develops in various social

2. On the concept of ethos, see William Graham Sumner, *Folkways* (Boston: Ginn, 1906), pp. 36 ff.; Hans Speier, "The Social Determination of Ideas," *Social Research* 5 (1938): 196 ff.; Max Scheler, *Schriften aus dem Nachlass* (Berlin, 1933), 1:225–62. Albert Bayet, in his book on the subject, soon abandons description and analysis for homily; see his *La morale de la science* (Paris, 1931).

3. As Bayet remarks: "Cette morale [de la science] n'a pas eu ses theoriciens, mais elle a eu ses artisans. Elle n'a pas exprimé son idéal, mais elle l'a servi: il est impliqué dans l'existence même de la science" (*La morale de la science*, p. 43).

4. Tocqueville went further: "The future will prove whether these passions [for science], at once so rare and so productive, come into being and into growth as easily in the midst of democratic as in aristocratic communities. For myself, I confess that I am slow to believe it" (*Democracy in America* [New York, 1898], 2: 51). See another reading of the evidence: "It is impossible to establish a simple causal relationship between democracy and science and to state that democratic society alone can furnish the soil suited for the development of science. It cannot be a mere coincidence, however, that science actually has flourished in democratic periods" (Henry E. Sigerist, "Science and Democracy," *Science and Society* 2 [1938]: 291).

structures, to be sure, but which provide an institutional context for the fullest measure of development?

The institutional goal of science is the extension of certified knowledge. The technical methods employed toward this end provide the relevant definition of knowledge: empirically confirmed and logically consistent statements of regularities (which are, in effect, predictions). The institutional imperatives (mores) derive from the goal and the methods. The entire structure of technical and moral norms implements the final objective. The technical norm of empirical evidence, adequate and reliable, is a prerequisite for sustained true prediction; the technical norm of logical consistency, a prerequisite for systematic and valid prediction. The mores of science possess a methodologic rationale but they are binding, not only because they are procedurally efficient, but because they are believed right and good. They are moral as well as technical prescriptions.

Four sets of institutional imperatives—universalism, communism, disinterestedness, organized skepticism—are taken to comprise the ethos of modern science.

Universalism

Universalism[5] finds immediate expression in the canon that truth-claims, whatever their source, are to be subjected to *preestablished impersonal criteria*: consonant with observation and with previously confirmed knowledge. The acceptance or rejection of claims entering the lists of science is not to depend on the personal or social attributes of their protagonist; his race, nationality, religion, class, and personal qualities are as such irrelevant. Objectivity precludes particularism. The circumstance that scientifically verified formulations refer in that specific sense to objective sequences and correlations militates against all efforts to impose particularistic criteria of validity. The Haber process cannot be invalidated by a Nuremberg decree nor can an Anglophobe repeal the law of gravitation. The chauvinist may expunge the names of alien scientists from historical textbooks but their formulations remain indispensable to science and technology. However *echt-deutsch* or hundred-percent American the final increment, some aliens are accessories before the fact of every new scientific advance. The imperative of universalism is rooted deep in the impersonal character of science.

5. For a basic analysis of universalism in social relations, see Talcott Parsons, *The Social System* (New York: Free Press, 1951). For an expression of the belief that "science is wholly independent of national boundaries and races and creeds," see the resolution of the Council of the American Association for the Advancement of Science, *Science* 87 (1938): 10; also, "The Advancement of Science and Society: Proposed World Association," *Nature* 141 (1938): 169.

However, the institution of science is part of a larger social structure with which it is not always integrated. When the larger culture opposes universalism, the ethos of science is subjected to serious strain. Ethnocentrism is not compatible with universalism. Particularly in times of international conflict, when the dominant definition of the situation is such as to emphasize national loyalties, the man of science is subjected to the conflicting imperatives of scientific universalism and of ethnocentric particularism.[6] The structure of the situation in which he finds himself determines the social role that is called into play. The man of science may be converted into a man of war—and act accordingly. Thus, in 1914 the manifesto of ninety-three German scientists and scholars—among them, Baeyer, Brentano, Ehrlich, Haber, Eduard Meyer, Ostwald, Planck, Schmoller, and Wassermann—unloosed a polemic in which German, French, and English men arrayed their political selves in the garb of scientists. Dispassionate scientists impugned "enemy" contributions, charging nationalistic bias, log-rolling, intellectual dishonesty, incompetence, and lack of creative capacity.[7] Yet this very deviation from the norm of universalism actually presupposed the legitimacy of the norm. For nationalistic bias is opprobrious only if judged in terms of the standard of universalism; within another institutional context, it is redefined as a virtue, patriotism. Thus in the process of condemning their violation, the mores are reaffirmed.

6. This stands as written in 1942. By 1948, the political leaders of Soviet Russia strengthened their emphasis on Russian nationalism and began to insist on the "national" character of science. Thus, in an editorial, "Against the Bourgeois Ideology of Cosmopolitanism," *Voprosy filosofii*, no. 2 (1948), as translated in the *Current Digest of the Soviet Press* 1, no. 1 (1 February 1949): 9: "Only a cosmopolitan without a homeland, profoundly insensible to the actual fortunes of science, could deny with contemptuous indifference the existence of the many-hued national forms in which science lives and develops. In place of the actual history of science and the concrete paths of its development, the cosmopolitan substitutes fabricated concepts of a kind of supernational, classless science, deprived, as it were, of all the wealth of national coloration, deprived of the living brilliance and specific character of a people's creative work, and transformed into a sort of disembodied spirit . . . Marxism-Leninism shatters into bits the cosmopolitan fictions concerning supraclass, non-national, 'universal' science, and definitely proves that science, like all culture in modern society, is national in form and class in content." This view confuses two distinct issues: first, the cultural context in any given nation or society may predispose scientists to focus on certain problems, to be sensitive to some and not other problems on the frontiers of science. This has long since been observed. But this is basically different from the second issue: the criteria of validity of claims to scientific knowledge are not matters of national taste and culture. Sooner or later, competing claims to validity are settled by universalistic criteria.

7. For an instructive collection of such documents, see Gabriel Pettit and Maurice Leudet, *Les allemands et la science* (Paris, 1916). Félix de Dantec, for example, discovers that both Ehrlich and Weismann have perpetrated typical German frauds upon the world of science. ("Le bluff de la science allemande.") Pierre Duhem concludes that the "geometric spirit" of German science stifled the "spirit of finesse": *La science allemande* (Paris 1915). Hermann Kellermann, *Der Krieg der Geister* (Weimar, 1915) is a spirited counterpart. The conflict persisted into the postwar period; see Karl Kherkhof, *Der Krieg gegen die Deutsche Wissenschaft* (Halle, 1933).

Even under counter-pressure, scientists of all nationalities adhered to the universalistic standard in more direct terms. The international, impersonal, virtually anonymous character of science was reaffirmed.[8] (Pasteur: "Le savant a une patrie, la science n'en a pas.") Denial of the norm was conceived as a breach of faith.

Universalism finds further expression in the demand that careers be open to talents. The rationale is provided by the institutional goal. To restrict scientific careers on grounds other than lack of competence is to prejudice the furtherance of knowledge. Free access to scientific pursuits is a functional imperative. Expediency and morality coincide. Hence the anomaly of a Charles II invoking the mores of science to reprove the Royal Society for their would-be exclusion of John Graunt, the political arithmetician, and his instructions that "if they found any more such tradesmen, they should be sure to admit them without further ado."

Here again the ethos of science may not be consistent with that of the larger society. Scientists may assimilate caste-standards and close their ranks to those of inferior status, irrespective of capacity or achievement. But this provokes an unstable situation. Elaborate ideologies are called forth to obscure the incompatibility of caste-mores and the institutional goal of science. Caste-inferiors must be shown to be inherently incapable of scientific work, or, at the very least, their contributions must be systematically devaluated. "It can be adduced from the history of science that the founders of research in physics, and the great discoverers from Galileo and Newton to the physical pioneers of our own time, were almost exclusively Aryans, predominantly of the Nordic race." The modifying phrase, "almost exclusively," is recognized as an insufficient basis for denying outcastes all claims to scientific achievement. Hence the ideology is rounded out by a conception of "good" and "bad" science: the realistic, pragmatic science of the Aryan is opposed to the dogmatic, formal science of the non-Aryan.[9] Or, grounds for exclusion are sought in the extrascientific capacity of men of science as enemies of the state or church.[10] Thus, the

8. See the profession of faith by Professor E. Gley (in Pettit and Leudet, *Les allemands et la science*, p. 181): "il ne peut y avoir une vérité allemande, anglaise, italienne ou japonaise pas plus qu'une française. Et parler de science allemande, anglaise ou française, c'est énoncer une proposition contradictoire à l'idée même de science." See also the affirmations of Grasset and Richet, ibid.

9. Johannes Stark, *Nature* 141 (1938): 772; "Philipp Lenard als deutscher Naturforscher," *Nationalsozialistische Monatshefte* 7 (1936): 106–12. This bears comparison with Duhem's contrast between "German" and "French" science.

10. "Wir haben sie ['marxistischen Leugner'] nicht entfernt als Vertreter der Wissenschaft, sondern als Parteigaenger einer politischen Lehre, die den Umsturz aller Ordnungen auf ihre Fahne geschrieben hatte. Und wir mussten hier um so entschlossener zugreifen, als ihnen die herrschende Ideologie einer wertfreien und voraussetzungslosen Wissenschaft ein willkommener Schutz fuer die Fortfuehrung ihrer Plaene zu sein schien. Nicht wir haben uns an der Wuerde der freien Wissenschaft vergangen. . ." Bernhard Rust, *Das nationalsozialistische Deutschland und die Wissenschaft* (Hamburg: Hanseatische Verlagsanstalt, 1936), p. 13.

exponents of a culture which abjures universalistic standards in general feel constrained to pay lip service to this value in the realm of science. Universalism is deviously affirmed in theory and suppressed in practice.

However inadequately it may be put into practice, the ethos of democracy includes universalism as a dominant guiding principle. Democratization is tantamount to the progressive elimination of restraints upon the exercise and development of socially valued capacities. Impersonal criteria of accomplishment and not fixation of status characterize the open democratic society. Insofar as such restraints do persist, they are viewed as obstacles in the path of full democratization. Thus, insofar as laissez-faire democracy permits the accumulation of differential advantages for certain segments of the population, differentials that are not bound up with demonstrated differences in capacity, the democratic process leads to increasing regulation by political authority. Under changing conditions, new technical forms of organization must be introduced to preserve and extend equality of opportunity. The political apparatus may be required to put democratic values into practice and to maintain universalistic standards.

"Communism"

"Communism," in the nontechnical and extended sense of common ownership of goods, is a second integral element of the scientific ethos. The substantive findings of science are a product of social collaboration and are assigned to the community. They constitute a common heritage in which the equity of the individual producer is severely limited. An eponymous law or theory does not enter into the exclusive possession of the discoverer and his heirs, nor do the mores bestow upon them special rights of use and disposition. Property rights in science are whittled down to a bare minimum by the rationale of the scientific ethic. The scientist's claim to "his" intellectual "property" is limited to that of recognition and esteem which, if the institution functions with a modicum of efficiency, is roughly commensurate with the significance of the increments brought to the common fund of knowledge. Eponymy—for example, the Copernican system, Boyle's law—is thus at once a mnemonic and a commemorative device.

Given such institutional emphasis upon recognition and esteem as the sole property right of the scientist in his discoveries, the concern with scientific priority becomes a "normal" response. Those controversies over priority which punctuate the history of modern science are generated by the institutional accent on originality.[11] There issues a competitive coopera-

11. Newton spoke from hard-won experience when he remarked that "[natural] philosophy is such an impertinently litigious Lady, that a man had as good be engaged in lawsuits, as have to do with her." Robert Hooke, a socially mobile individual whose rise in status rested solely on his scientific achievements, was notably "litigious."

tion. The products of competition are communized,[12] and esteem accrues to the producer. Nations take up claims to priority, and fresh entries into the commonwealth of science are tagged with the names of nationals: witness the controversy raging over the rival claims of Newton and Leibniz to the differential calculus. But all this does not challenge the status of scientific knowledge as common property.

The institutional conception of science as part of the public domain is linked with the imperative for communication of findings. Secrecy is the antithesis of this norm; full and open communication its enactment.[13] The pressure for diffusion of results is reenforced by the institutional goal of advancing the boundaries of knowledge and by the incentive of recognition which is, of course, contingent upon publication. A scientist who does not communicate his important discoveries to the scientific fraternity—thus, a Henry Cavendish—becomes the target for ambivalent responses. He is esteemed for his talent and, perhaps, for his modesty. But, institutionally considered, his modesty is seriously misplaced, in view of the moral compulsive for sharing the wealth of science. Layman though he is, Aldous Huxley's comment on Cavendish is illuminating in this connection: "Our admiration of his genius is tempered by a certain disapproval; we feel that such a man is selfish and anti-social." The epithets are particularly instructive for they imply the violation of a definite institutional imperative. Even though it serves no ulterior motive, the suppression of scientific discovery is condemned.

The communal character of science is further reflected in the recognition by scientists of their dependence upon a cultural heritage to which they lay no differential claims. Newton's remark—"If I have seen farther it is

12. Marked by the commercialism of the wider society though it may be, a profession such as medicine accepts scientific knowledge as common property. See R. H. Shryock, "Freedom and Interference in Medicine," *The Annals* 200 (1938): 45. "The medical profession . . . has usually frowned upon patents taken out by medical men. . . . The regular profession has . . . maintained this stand against private monopolies ever since the advent of patent law in the seventeenth century." There arises an ambiguous situation in which the socialization of medical practice is rejected in circles where the socialization of knowledge goes unchallenged.

13. Cf. Bernal, who observes: "The growth of modern science coincided with a definite rejection of the ideal of secrecy." Bernal quotes a remarkable passage from Réaumur (*L'Art de convertir le forgé en acier*) in which the moral compulsion for publishing one's researches is explicitly related to other elements in the ethos of science. For example, "il y eût gens qui trouvèrent étrange que j'eusse publié des secrets, qui ne devoient pas etre revelés . . . est-il bien sur que nos découvertes soient si fort à nous que le Public n'y ait pas droit, qu'elles ne lui appartiennent pas en quelque sorte? . . . resterait il bien des circonstances, où nous soions absolument Maîtres de nos découvertes? . . . Nous nous devons premiérement à notre Patrie, mais nous nous devons aussi au rest du monde; ceux qui travaillent pour perfectionner les Sciences et les Arts, doivent même se regarder commes les citoyens du monde entier" (J. D. Bernal, *The Social Function of Science* [New York: Macmillan, 1939] pp. 150–51).

by standing on the shoulders of giants"—expresses at once a sense of indebtedness to the common heritage and a recognition of the essentially cooperative and selectively cumulative quality of scientific achievement.[14] The humility of scientific genius is not simply culturally appropriate but results from the realization that scientific advance involves the collaboration of past and present generations. It was Carlyle, not Maxwell, who indulged in a mythopoeic conception of history.

The communism of the scientific ethos is incompatible with the definition of technology as "private property" in a capitalistic economy. Current writings on the "frustration of science" reflect this conflict. Patents proclaim exclusive rights of use and, often, nonuse. The suppression of invention denies the rationale of scientific production and diffusion, as may be seen from the court's decision in the case of *U.S.* v. *American Bell Telephone Co.*: "The inventor is one who has discovered something of value. It is his absolute property. He may withhold the knowledge of it from the public."[15] Responses to this conflict-situation have varied. As a defensive measure, some scientists have come to patent their work to ensure its being made available for public use. Einstein, Millikan, Compton, Langmuir have taken out patents.[16] Scientists have been urged to become promoters of new economic enterprises.[17] Others seek to resolve the conflict by advocating socialism.[18] These proposals—both those which demand economic returns for scientific discoveries and those which demand a change in the social system to let science get on with the job—reflect discrepancies in the conception of intellectual property.

Disinterestedness

Science, as is the case with the professions in general, includes disinterestedness as a basic institutional element. Disinterestedness is not to be equated with altruism nor interested action with egoism. Such equivalences

14. It is of some interest that Newton's aphorism is a standardized phrase which had found repeated expression from at least the twelfth century. It would appear that the dependence of discovery and invention on the existing cultural base had been noted some time before the formulations of modern sociologists. See *Isis* 24 (1935): 107–9; 25 (1938): 451–52.

15. 167 U. S. 224 (1897), cited by B. J. Stern, "Restraints upon the Utilization of Inventions," *The Annals* 200 (1938): 21. For an extended discussion, cf. Stern's further studies cited therein, also Walton Hamilton, *Patents and Free Enterprise*, Temporary National Economic Committee Monograph no. 31 (1941).

16. Hamilton, *Patents and Free Enterprise*, p. 154; J. Robin, *L'oeuvre scientifique: sa protection-juridique* (Paris, 1928).

17. Vannevar Bush, "Trends in Engineering Research," *Sigma Xi Quarterly* 22 (1934): 49.

18. Bernal, *The Social Function of Science*, pp. 155 ff.

confuse institutional and motivational levels of analysis.[19] A passion for knowledge, idle curiosity, altruistic concern with the benefit to humanity, and a host of other special motives have been attributed to the scientist. The quest for distinctive motives appears to have been misdirected. It is rather a distinctive pattern of institutional control of a wide range of motives which characterizes the behavior of scientists. For once the institution enjoins disinterested activity, it is to the interest of scientists to conform on pain of sanctions and, insofar as the norm has been internalized, on pain of psychological conflict.

The virtual absence of fraud in the annals of science, which appears exceptional when compared with the record of other spheres of activity, has at times been attributed to the personal qualities of scientists. By implication, scientists are recruited from the ranks of those who exhibit an unusual degree of moral integrity. There is, in fact, no satisfactory evidence that such is the case; a more plausible explanation may be found in certain distinctive characteristics of science itself. Involving as it does the verifiability of results, scientific research is under the exacting scrutiny of fellow experts. Otherwise put—and doubtless the observation can be interpreted as lese majesty—the activities of scientists are subject to rigorous policing, to a degree perhaps unparalleled in any other field of activity. The demand for disinterestedness has a firm basis in the public and testable character of science and this circumstance, it may be supposed, has contributed to the integrity of men of science. There is competition in the realm of science, competition that is intensified by the emphasis on priority as a criterion of achievement, and under competitive conditions there may well be generated incentives for eclipsing rivals by illicit means. But such impulses can find scant opportunity for expression in the field of scientific research. Cultism, informal cliques, prolific but trivial publications—these and other techniques may be used for self-aggrandizement.[20] But, in general, spurious claims appear to be negligible and ineffective. The translation of the norm of disinterestedness into practice is effectively supported by the ultimate accountability of scientists to their compeers. The dictates of socialized sentiment and of expediency largely coincide, a situation conducive to institutional stability.

In this connection, the field of science differs somewhat from that of other professions. The scientist does not stand vis-à-vis a lay clientele in the same fashion as do the physician and lawyer, for example. The possi-

19. Talcott Parsons, "The Professions and Social Structure," *Social Forces* 17 (1939): 458–59; cf. George Sarton, *The History of Science and the New Humanism* (New York, 1931), p. 130 ff. The distinction between institutional compulsives and motives is a key, though largely implicit, conception of Marxist sociology.

20. See the account of Logan Wilson, *The Academic Man* (New York: Oxford University Press, 1941), p. 201 ff.

bility of exploiting the credulity, ignorance, and dependence of the layman is thus considerably reduced. Fraud, chicane, and irresponsible claims (quackery) are even less likely than among the "service" professions. To the extent that the scientist-layman relation does become paramount, there develop incentives for evading the mores of science. The abuse of expert authority and the creation of pseudo-sciences are called into play when the structure of control exercised by qualified compeers is rendered ineffectual.[21]

It is probable that the reputability of science and its lofty ethical status in the estimate of the layman is in no small measure due to technological achievements.[22] Every new technology bears witness to the integrity of the scientist. Science realizes its claims. However, its authority can be and is appropriated for interested purposes, precisely because the laity is often in no position to distinguish spurious from genuine claims to such authority. The presumably scientific pronouncements of totalitarian spokesmen on race or economy or history are for the uninstructed laity of the same order as newspaper reports of an expanding universe or wave mechanics. In both instances, they cannot be checked by the man-in-the-street and in both instances, they may run counter to common sense. If anything, the myths will seem more plausible and are certainly more comprehensible to the general public than accredited scientific theories, since they are closer to common-sense experience and to cultural bias. Partly as a result of scientific achievements, therefore, the population at large becomes susceptible to new mysticisms expressed in apparently scientific terms. The borrowed authority of science bestows prestige on the unscientific doctrine.

Organized Skepticism

As we have seen in the preceding chapter, organized skepticism is variously interrelated with the other elements of the scientific ethos. It is both a methodological and an institutional mandate. The temporary suspension of judgment and the detached scrutiny of beliefs in terms of empirical and logical criteria have periodically involved science in conflict with other institutions. Science which asks questions of fact, including potentialities, concerning every aspect of nature and society may come into conflict with other attitudes toward these same data which have been crystallized and often ritualized by other institutions. The scientific investigator does not preserve the cleavage between the sacred and the profane, between that

21. Cf. R. A. Brady, *The Sprit and Structure of German Fascism* (New York: Viking, 1937), chap. 2; Martin Gardner, *In the Name of Science* (New York: Putnam's, 1953).

22. Francis Bacon set forth one of the early and most succinct statements of this popular pragmatism: "Now these two directions—the one active, the other contemplative—are one and the same thing; and what in operation is most useful, that in knowledge is most true" (*Novum Organum*, book 2, aphorism 4).

which requires uncritical respect and that which can be objectively analyzed.

As we have noted, this appears to be the source of revolts against the so-called intrusion of science into other spheres. Such resistance on the part of organized religion has become less significant as compared with that of economic and political groups. The opposition may exist quite apart from the introduction of specific scientific discoveries which appear to invalidate particular dogmas of church, economy, or state. It is rather a diffuse, frequently vague, apprehension that skepticism threatens the current distribution of power. Conflict becomes accentuated whenever science extends its research to new areas toward which there are institutionalized attitudes or whenever other institutions extend their control over science. In modern totalitarian society, anti-rationalism and the centralization of institutional control both serve to limit the scope provided for scientific activity.

The Reward System of Science

Part 4

Prefatory Note

With the papers in this section we come to the heart of the Mertonian paradigm—the powerful juxtaposition of the normative structure of science with its institutionally distinctive reward system—as it provides a simplified but basic model of the structure and dynamics of the scientific community. Published within a period of seven years, just at the time when other American sociologists were beginning to see in the field an attractive focus for research, these papers demonstrate how basic theoretical formulations can open the doors to a quick succession of investigations by others who capitalize on the opening wedge. In this section, of course, the papers are all by Merton, but the principle holds, and the record amply supports it. Within two years of the 1957 paper both Warren O. Hagstrom and this writer were at work on doctoral dissertations that drew heavily on the newly developed paradigm, and in the next decade or so more had been contributed by Crane, Zuckerman, Mullins, Stephen Cole, Jonathan Cole, Gaston, and a good many others than had been published in the forty years before.[1]

The central topic of these papers is cognitive achievement as it is enmeshed in the social matrix of science. Matters of priority, of the ubiquity of multiple discovery, and of scientists' attitudes towards the circumstances and consequences of achievement are dealt with in extensive detail, with historical materials arrayed against a clear-cut theoretical background to flesh out and extend the analyses. Not only is the essential character of the reward system set forth, but in coupling it with the normative system Merton is able to make sense of the problems which develop out of the incommensurate imperatives of these two components of the institution.

Several social pathologies of science are pinpointed as the *results of specific discontinuities* between the normative and the reward systems so that their investigation may contribute to a cumulating body of knowledge about science. It would not have been enough simply to show that the institution of science has its own systematic problems, just as the political and economic institutions do; without an appropriate conceptual framework, those problems and observed departures of scientists from the institutionalized norms stand simply as uncomfortable anomalies. It is thus the conception of variable relationships between the normative system and the reward system that gives the paradigm its capability of generating meaningful research questions about *both* conformity and deviance in the domain of science.

The particularly felicitous style of the first three papers, incidentally, may be due to their having been prepared for oral presentation—as the presidential address to the American Sociological Society in 1957, as the Phi Beta Kappa–Sigma Xi Address in 1968, and as a lecture commemorating the 400th anniversary of Francis Bacon's birth—for in addition to the usual high quality of Merton's scholarship they show a bit more humor than is ordinarily found in his work. This editor was present at the first occasion (his own first national meeting of sociologists) and still remembers the waves of response that greeted the recital of paternity of the sciences, the various echelons of eponymies, and the varieties of misbehavior of scientists. (Given the direct bearing of the subject on many of the listeners' deep personal concern with their roles in science, the laughter may also have provided a vent for the anxieties that were building up as Merton explored some rules of the game and some of their consequences.)

The initial paper in part 4, "Priorities in Scientific Discovery," is Merton's first major statement of the nature of the reward system in science. His earlier writings show that for twenty years he had sensed the potential theoretical significance of contests over priority and of eponymic

1. Michael J. Mulkay, a conscientious critic of Merton's paradigm, makes a similar observation, describing "Priorities in Scientific Discovery" as "a most important article from which stems a whole tradition of research on the reward system of science" (*The Social Process of Innovation: A Study in the Sociology of Science* [London: Macmillan, 1972], p. 62). I am suggesting that the article has had this impact not so much in its own right as because it helped round out the developing paradigm.

practices, but not until this paper did the elements of the paradigm all fall into place. Neither the abstract statement of the norms of science nor the conception of professional recognition as the institutionally central reward for scientific achievement could separately point to the sources of various forms of deviant behavior in science. But in combination, as here, they add dimension and organization to what had previously been little more than a congeries of unconnected incidents involving the "unfortunate" misbehavior of particular scientists. The basic idea of interaction between the normative structure and the reward structure of science provides a solid foundation for the understanding of science as a social institution.

From one point of view, the paper is an extension of the thesis Merton propounded in "The Unanticipated Consequences of Purposive Social Action" (1936) and greatly developed soon after in "Social Structure and Anomie" (1938). This is the idea that extreme institutional emphasis upon recognized achievement, greatly encouraged by the reward system, can be as dysfunctional for human purposes and social institutions as the systematic flouting of norms. Moreover, the one leads to the other. From another point of view, the paper copes with the general sociological problem, identified by Merton in his paper "Problem-Finding in Sociology,"[2] of "how to account for regularities of social behavior that are not prescribed by cultural norms or that are even at odds with those norms. It casts doubt on the familiar assumption that uniformities of social behavior necessarily represent conformity to norms calling for that behavior."

The theoretical approach adopted in his "Priorities" paper generates a systematic awareness of the varieties of pathogenic situations that occur within the scientific community and raises meaningful questions about a broader range of subjects: the consequences of inequalities in the productivity of scientists, the degree of equity in the system through which professional recognition is allocated, and even some of the myths to which observers of the work of science and some working scientists themselves subscribe.

One such myth, recently gaining wide currency, is that the competition among scientists for priority is something new—that it results entirely from the greatly increased size, affluence, and prominence of science in the modern era. This belief has shown up frequently in the reviews of *The Double Helix,*[3] James D. Watson's recollections, published in 1968, of the discovery of the structure of the DNA molecule which won the Nobel Prize for him and Francis Crick. Appearing more than a decade

2. *Sociology Today*, 1957, p. xxiii.

3. For a critical overview by a molecular biologist of six reviews of *The Double Helix*, including the one by Merton, see Gunther S. Stent, "What Are They Saying about Honest Jim?" *The Quarterly Journal of Biology* 43 (June 1968): 179–84.

after Merton's paper on priorities, the revelations of *The Double Helix* came as no surprise to the sociologists of science.

In the next paper, "Behavior Patterns of Scientists," Merton takes the Watson book as his text for an updating of the ideas advanced in the "Priorities" paper. As printed here, this later selection is actually an amalgamation of two papers: the first, under this title, appeared in 1969, and the second, "The Race for Priority," written with Richard Lewis, appeared in 1971. They are put together here because the later piece complements the preceding one by extending the analysis of the conditions under which competition in science becomes intensified. It is shown that scientists' behavior with regard to priority has *not* greatly changed over the past centuries, except possibly in the direction of greater civility in contests over priority. What variations there are in both personal perceptions of these conflicts and their actual frequency, Merton suggests, can be traced to a combination of the structural characteristics of scientific communities and of their bodies of specialized knowledge.

Another myth, the belief that multiple discoveries of like kind are rare, is addressed in the third paper in this section, "Singletons and Multiples in Scientific Discovery." Published in 1961, it attacks the problem of priorities in wider perspective to clarify alternative ideas about the process of scientific development. Suggesting that independent multiple discoveries occur far more frequently in science than is generally assumed, especially since announcements of discoveries often forestall others from completing their work along the same lines, Merton goes on to show the speciousness of the argument that scientific progress must stem *either* from the work of the great scientists *or* that it is a virtually inevitable outgrowth of the state of accumulated scientific knowledge. It is precisely the great scientists, he points out, who prove to be most capable of exploiting the current state of the art—they are involved in many more multiples than others—so that the two theoretical perspectives actually complement each other.

The discerning reader will also find here a hint of quite a different idea that would be developed some years later as "The Matthew Effect" (reprinted in part V of this volume). This appears in the statement toward the end of the paper: "it required a Freud to focus the attention of many on ideas which might otherwise not have come to their notice." Placed within the framework of the basic paradigm, this tentative observation gave rise to the problem, later examined in detail, of the functions and dysfunctions for the communication system of science and for individual scientists of great differences in the standing accorded scientists.

The next two papers were originally parts of one long paper which has been divided here to highlight the distinct problems under examination. The first, now entitled "Multiple Discoveries in Science as a Strategic Research Site," covers the theoretical significance of multiples as it lays out eight aspects of the phenomenon which can be investigated to clarify facets of science as a social institution. Of particular interest is the way Merton illuminates some policy implications of such investigations. Building upon his previous analyses, he indicates the damage that can be done to scientific development by centralized decisions to eliminate, as inevitably wasteful, all "duplication in research."

In this paper we also find a clear indication of respects in which the substantive organization of scientific knowledge forms a strategic site for sociological research. In discussing an aspect of the study of multiples—the fourth in his list of eight—Merton suggests that the community of science cannot continue to be treated sociologically as though it were a wholly homogeneous entity. He argues that important sociological differences exist among the various branches of science (owing partly to the nature and organization of their particular bodies of knowledge) and that these differences can be elucidated; for example, as they are reflected in differing rates of multiple discoveries and, presumably, in differing frequency with which priority is contested.

Concluding this section is "The Ambivalence of Scientists." It takes up the curious failure of social scientists to investigate systematically the oft-noted phenomenon of multiple discoveries, taking this as a question worth study in its own right, and uses this as an occasion to look more searchingly into the complexities of the normative structure of science. Merton's focus is on the ways in which the operation of the reward system exacerbates tensions and conflicts among components of the normative structure to produce inner conflicts among scientists. For example, the legitimate quest for recognition of scientific contributions, which serves both to maintain the value placed on originality and to keep the reward system operating, runs head on into the value placed on restrained claims for scientific work (a derivative of the norm of disinterestedness and of organized scepticism as it applies to the individual's own work). This results in a deep ambivalence in which "scientists are contemptuous of the very attitudes which they have acquired from the institutions [of science] to which they subscribe."

These problems of stress and conflict are not so epidemic among scientists as to deprive them of all morale and motivation. For the vast majority of them, it is not a choice between defending one's priority of discovery or losing it, but the challenge of trying to make any sort of contribution at all to a body of knowledge. The grades of contribution are of course a matter of evaluation within the various communities of science. The papers in the final part of this volume move more closely to a sociological interest in the substance of scientific knowledge by examining the evaluation process in science.

N. W. S.

14 Priorities in Scientific Discovery

1957

We can only guess what historians of the future will say about the condition of present-day sociology. But it seems safe to anticipate one of their observations. When the Trevelyans of 2050 come to write that history—as they well might, for this clan of historians promises to go on forever—they will doubtless find it strange that so few sociologists (and historians) of the twentieth century could bring themselves, in their work, to treat science as one of the great social institutions of the time. They will observe that long after the sociology of science became an identifiable field of inquiry,[1] it remained little cultivated in a world where science loomed large enough to present mankind with the choice of destruction or survival. They may even suggest that somewhere in the process by which social scientists take note of the world as it is and as it once was, a sense of values appears to have become badly scrambled.

This spacious area of neglect may therefore have room for a paper which tries to examine science as a social institution, not in the large but in terms of a few of its principal components.

Priority Disputes as Social Conflict

A calendar of disputes over priority

We begin by noting the great frequency with which the history of science

This essay was read as a presidential address at the annual meeting of the American Sociological Society, August 1957. It was first published in *American Sociological Review* 22, no. 6 (December 1957): 635–59, and is reprinted by permission of the American Sociological Association.

1. The rudiments of a sociology of science can be found in an overview of the subject by Bernard Barber, *Science and the Social Order* (Glencoe: The Free Press,

is punctuated by disputes, often by sordid disputes, over priority of discovery. During the last three centuries in which modern science developed, numerous scientists, both great and small, have engaged in such acrimonious controversy. Recall only these few. Keenly aware of the importance of his inventions and discoveries, Galileo became a seasoned campaigner as he vigorously defended his rights to priority first, in his *Defense against the Calumnies and Impostures of Baldassar Capra*, where he showed how his invention of the "geometric and military compass" had been taken from him, and then, in *The Assayer,* where he flayed four other would-be rivals: Father Horatio Grassi, who tried "to diminish whatever praise there may be in this [invention of the telescope for use in astronomy] which belongs to me"; Christopher Scheiner, who claimed to have been first to observe the sunspots (although, unknown to both Scheiner and Galileo, Johannes Fabricius had published such observations before); an unspecified villain (probably the Frenchman Jean Tarde) who "attempted to rob me of that glory which was mine, pretending not to have seen my writings and trying to represent themselves as the original discoverers of these marvels"; and finally, Simon Mayr (Marius), who "had the gall to claim that he had observed the Medicean planets which revolve about Jupiter before I had [and who used] a sly way of attempting to establish his priority."[2]

The peerless Newton fought several battles with Robert Hooke over priority in optics and celestial mechanics and entered into a long and painful controversy with Leibniz over the invention of the calculus. Hooke,[3] who has been described as the "universal claimant" because "there was scarcely a discovery in his time which he did not conceive himself to claim" (and, it might be added, often justly so, for he was one of the most inventive men in his century of genius), Hooke, in turn, con-

1952); and Bernard Barber, "Sociology of Science: A Trend Report and Bibliography," *Current Sociology* 5, no. 2 (Paris: UNESCO, 1957).

2. Galileo, "The Assayer," 1623, translated by Stillman Drake in *Discoveries and Opinions of Galileo* (N.Y.: Doubleday, 1957), pp. 232–33, 245. Galileo thought it crafty of Mayr to date his book as published in 1609 by using the Julian calendar without indicating that, as a Protestant, he had not accepted the Gregorian calendar adopted by "us Catholics" which would have shifted the date of publication to January 1610, when Galileo had reported having made his first observations. Later in this paper, I shall have more to say about the implications of attaching importance to such short intervals separating rival claims to priority.

3. For scholarly reappraisals of Hooke's role in developing the theory of gravitation, see Louise Diehl Patterson, "Hooke's Gravitation Theory and Its Influence on Newton," *Isis* 40 (November 1949): 327–41; *Isis* 41 (March 1950): 32–45; and E. N. da C. Andrade, "Robert Hooke," Wilkins Lecture, *Proceedings of the Royal Society*, series B, Biological Sciences, 137 (24 July 1950). The recent biography by Margaret 'Espinasse is too uncritical and defensive of Hooke to be satisfactory; see *Robert Hooke* (London: Heinemann, 1956).

tested priority not only with Newton but with Huygens over the important invention of the spiral-spring balance for regulating watches to eliminate the effect of gravity.

The calendar of disputes was full also in the eighteenth century. Perhaps the most tedious and sectarian of these was the great "Water Controversy" in which that shy, rich, and noble genius of science, Henry Cavendish, was pushed into a three-way tug-of-war with Watt and Lavoisier over the question of which one had first demonstrated the compound nature of water and thereby removed it from its millennia-long position as one of the elements. Earthy battles raged also over claims to the first discovery of heavenly bodies, as in the case of the most dramatic astronomical discovery of the century in which the Englishman John Couch Adams and the Frenchman Urban Jean LeVerrier inferred the existence and predicted the position of the planet now known as Neptune, which was found where their independent computations showed it would be. Medicine had its share of conflicts over priority; for example, Jenner believed himself first to demonstrate that vaccination afforded security against smallpox, but the advocates of Pearson and Rabaut believed otherwise.

Throughout the nineteenth century and down to the present, disputes over priority continued to be frequent and intense. Lister knew he had first introduced antisepsis, but others insisted that Lemaire had done so before. The sensitive and modest Faraday was wounded by the claims of others to several of his major discoveries in physics: one among these, the discovery of electro-magnetic rotation, was said to have been made before by Wollaston; Faraday's onetime mentor, Sir Humphrey Davy (who had himself been involved in similar disputes) actually opposed Faraday's election to the Royal Society on the ground that his was not the original discovery.[4] Laplace, several of the Bernoullis, Legendre, Gauss, Cauchy were only a few of the giants among mathematicians embroiled in quarrels over priority.

What is true of physics, chemistry, astronomy, medicine, and mathematics is true also of all other scientific disciplines, not excluding the social and psychological sciences. As we know, sociology was officially born only after a long period of abnormally severe labor. Nor was the post-partum any more tranquil. It was disturbed by violent controversies between the followers of St-Simon and Comte as they quarreled over the delicate question of which of the two was the father of sociology and which merely the obstetrician. And to come to the very recent past, Janet is but one among several who have claimed that they had the essentials of psychoanalysis before Freud.

4. Bence Jones, *The Life and Letters of Faraday* (London: Longmans, Green, 1870), 1:336–52.

To extend the list of priority fights would be industrious and, for this occasion, superfluous. For the moment, it is enough to note that these controversies, far from being a rare exception in science, have long been frequent, harsh, and ugly. They have practically become an integral part of the social relations between scientists. Indeed, the pattern is so common that the Germans have characteristically compounded a word for it, *Prioritätsstreit.*

On the face of it, the pattern of conflict over priority can be easily explained. It seems to be merely a consequence of the same discoveries being made simultaneously, or nearly so, a recurrent event in the history of science which has not exactly escaped the notice of sociologists, or of others, at least since the definitive work of William Ogburn and Dorothy Thomas. But on second glance, the matter does not appear quite so simple.

The bunching of similar or identical discoveries in science is only an *occasion*[5] for disputes over priority, not their *cause* or their *grounds.* After all, scientists also know that discoveries are often made independently. (As we shall see, they not only know this but fear it, and this often activates a rush to ensure their priority.) It would therefore seem a simple matter for scientists to acknowledge that their simultaneous discoveries were independent and that the question of priority is consequently beside the point. On occasion, this is just what has happened, as we shall see in that most moving of all cases of noblesse oblige in the history of science, when Darwin and Wallace tried to outdo one another in giving credit to the other for what each had separately worked out. Fifty years after the event, Wallace was still insisting upon the contrast between his own hurried work, written within a week after the great idea came to him, and Darwin's work, based on twenty years of collecting evidence. "*I* was then (as often since) the 'young man in a hurry,' " said the reminiscing Wallace; "*he,* the painstaking and patient student seeking ever the full demonstration of the truth he had discovered, rather than to achieve immediate personal fame."[6]

On other occasions, self-denial has gone even further. For example, the incomparable Euler withheld his long-sought solution to the calculus of variations until the twenty-three-year-old Lagrange, who had developed a new method needed to reach the solution, could put it into print, " 'so as not to deprive you,' Euler informed the young man, 'of any part of the

5. And not always even the occasion. Disputes over priority have occurred when alleged or actual anticipations of an idea have been placed decades or, at times, even centuries or millennia earlier, when they are generally described as "rediscoveries."

6. This remark is taken from Wallace's commentary at the semicentenary of the joint discovery, a classic of self-abnegation that deserves to be rescued from the near-oblivion into which it has fallen. For a transcript, see James Marchant, *Alfred Russel Wallace: Letters and Reminiscences* (New York: Harper, 1916), pp. 91–96.

glory which is your due.' "[7] Apart from these and many other examples of generosity in the annals of science, there have doubtless been many more that never found their way into the pages of history. Nevertheless, the recurrent struggles for priority, with all their intensity of affect, far overshadow these cases of noblesse oblige, and it still remains necessary to account for them.

Alleged sources of conflicts over priority

One explanation of these disputes would regard them as mere expressions of human nature. On this view, egotism is natural to the species; scientists, being human, will have their due share and will sometimes express their egotism through self-aggrandizing claims to priority. But, of course, this interpretation does not stand up. The history of social thought is strewn with the corpses of those who have tried, in their theory, to make the hazardous leap from human nature to particular forms of social conduct, as has been observed from the time of Montesquieu, through Comte and Durkheim, to the present.[8]

A second explanation derives these conflicts not from the original nature shared by all men, but from propensities toward egotism found among some men. It assumes that, like other occupations, the occupation of science attracts some ego-centered people, and assumes further that it might even attract many such people, who, hungry for fame, elect to enter a profession that promises enduring fame to the successful. Unlike the argument from nature, this one, dealing with processes of self-selection and of social selection, is not defective in principle. It is possible that differing kinds of personalities tend to be recruited by various occupations, and, though I happen to doubt it, it is possible that quarrelsome or contentious personalities are especially apt to be attracted to science and recruited into it. The extent to which this is so is a still unanswered question, but developing inquiry into the type of personality characteristic of those entering the various professions may in due course discover how far it is so.[9] In any event, it should not be difficult to find *some* aggressive men of science.

7. E. T. Bell, *Men of Mathematics* (New York: Simon and Schuster, 1937), pp. 155–56. And see the comparable act of generosity on the part of the venerable Legendre toward the mathematical genius, Niels Abel, then in his twenties (ibid., p. 337).

8. Émile Durkheim had traced this basic theme in sociological theory as early as his Latin thesis of 1892, which has fortunately been translated into French for the benefit of some of us later sociologists. See his *Montesquieu et Rousseau: Précurseurs de la Sociologie* (Paris: Marcel Rivière, 1953), esp. chapter 1.

9. Information about this is sparse and unsatisfactory. As a bare beginning, a study of the Thematic Apperception Test protocols of 64 eminent biological, physical, and social scientists found no signs of their being "particularly aggressive." See Anne Roe, *The Making of a Scientist* (New York: Dodd, Mead, 1953), p. 192.

But even should the processes of selection result in the recruitment of contentious men, there are theoretical reasons for believing that this does not adequately account for the great amount of contention over priority that flares up in science. For one thing, these controversies often involve men of ordinarily modest disposition who act in seemingly self-assertive ways only when they come to defend their rights to intellectual property. This has often been remarked, and sometimes with great puzzlement. As Sir Humphrey Davy asked at the time of the great Water Controversy between Cavendish and Watt, how does it happen that this conflict over priority should engage such a man as Cavendish, "unambitious, unassuming, with difficulty . . . persuaded to bring forward his important discoveries . . . and . . . fearful of the voice of fame"?[10] And the biographer of Cavendish, writing about the same episode, describes it as "a perplexing dilemma. Two unusually modest and unambitious men, universally respected for their integrity, famous for their discoveries and inventions, are suddenly found standing in a hostile position towards each other."[11] Evidently, ingrained egotism is not required to engage in a fight for priority.

A second strategic fact shows the inadequacy of explaining these many struggles as owing to egotistic personalities. Very often, the principals themselves, the discoverers or inventors, take no part in arguing their claims to priority (or withdraw from the controversy as they find that it places them in the distasteful role of insisting upon their own merits or of deprecating the merits of their rivals). Instead, it is their friends and followers, or other more detached scientists, who commonly see the assignment of priority as a moral issue that must be fought to a conclusion. For example, it was Wollaston's friends, rather than the distinguished

10. Sir Humphrey Davy, *Collected Works*, vol. 7, p. 128, quoted in George Wilson, *The Life of the Honorable Henry Cavendish* (London, 1851), p. 63.

11. Wilson, *Henry Cavendish*, p. 64. There can be little doubt about the unassuming character of Cavendish, the pathologically shy recluse, whose unpublished notebooks were crowded with discoveries disproving then widely held theories and anticipating discoveries not made again for a long time to come. He stands as the example *a fortiori*, for even such a man as this was drawn into a controversy over priority.

The history of science evidently has its own brand of chain-reactions. It was the reading of Wilson's *Life of Cavendish* with its report of Cavendish's long-forgotten experiment on the sparking of air over alkalis which led Ramsay (just as the same experiment led Rayleigh) to the discovery of the element argon. Both Rayleigh and Ramsay delicately set out their respective claims to the discovery, claims not easily disentangled since the two had been in such close touch. They finally agreed to joint publication as "the only solution" to the problem of assigning appropriate credit. The episode gave rise to a great controversy over priority in which neither of the discoverers would take part; the debate is continued in the biographies of the two: by the old friend and collaborator of Ramsay, Morris W. Travers, in *A Life of Sir William Ramsay* (London: Edward Arnold Ltd., 1956), pp. 100, 121–122, 292, passim; and, by R. J. Strutt, the son of Lord Rayleigh, *John William Strutt: Third Baron Rayleigh* (London: Edward Arnold, 1924), chapter 11.

scientist himself, who insinuated that the young Faraday had usurped credit for the experiments on electro-magnetic rotation.[12] Similarly, it was Priestley, De Luc, and Blagden, "all men eminent in science and of unblemished character," who embroiled the shy Cavendish and the unassertive Watt in the Water Controversy.[13] Finally, it was the quarrelsome, eminent, and justly esteemed scientist François Arago (whom we shall meet again) and a crowd of astronomers, principally in France and England but also in Germany and Russia, rather than Adams, "the shy, gentle and unaffected" co-discoverer of Neptune, who heated the pot of conflict over priority until it boiled over and then simmered down into general acknowledgement that the planet had been independently discovered by Adams and LeVerrier.[14] And so on, in one after another of the historic quarrels over priority in science.

Now these argumentative associates and bystanders stand to gain little or nothing from successfully prosecuting the claims of their candidate, except in the pickwickian sense of having identified themselves with him or with the nation of which they are all a part. Their behavior can scarcely be explained by egotism. They do not suffer from rival claims to precedence. Their personal status is not being threatened. And yet, over and again, they take up the cudgels in the status-battle[15] and, uninhibited by any semblance of indulging in self-praise, express their great moral indignation over the outrage being perpetrated upon their candidate.

This is, I believe, a particularly significant fact. For, as we know from the sociological theory of institutions, the expression of disinterested moral indignation is a signpost announcing the violation of a social norm.[16] Although the indignant bystanders are themselves not injured by what they take to be the misbehavior of the culprit, they respond with hostility. They want to see "fair play," to see that behavior conforms to the rules of the

12. Jones, *Faraday*, pp. 351–52; see also the informative book by T. W. Chalmers, *Historic Researches: Chapters in the History of Physical and Chemical Discovery* (New York: Scribner's, 1952), p. 54.

13. This is the contemporary judgment by Wilson, *Henry Cavendish*, pp. 63–64.

14. Sir Harold Spencer Jones, "John Couch Adams and the Discovery of Neptune," reprinted in James R. Newman, *The World of Mathematics* (New York: Simon and Schuster, 1956), 2:822–39. A list of cases in which associates, rather than principals, took the lead in these conflicts is a very long one. I do not include it here.

15. Sometimes, of course, they act as judges and arbitrators rather than advocates, as was true of Lyell and Hooker in the episode involving Darwin and Wallace. But, as we shall see, the same institutional norms are variously called into play in all these cases.

16. For an acute analysis of the theoretical place of moral obligation and its correlate, moral indignation, in the theory of institutions, particularly as this was developed in the long course of Durkheim's work, see Talcott Parsons, *The Structure of Social Action* (Glencoe: The Free Press, 1949), pp. 368–470; for further formulations and citations of additional literature, see R. K. Merton, *Social Theory and Social Structure*, rev. ed. (Glencoe: The Free Press, 1957), pp. 361 ff.

game. The very fact of their entering the fray goes to show that science is a social institution with a distinctive body of norms exerting moral authority and that these norms are invoked particularly when it is felt that they are being violated. In this sense, fights over priority, with all their typical vehemence and passionate feelings, are not merely expressions of hot tempers, although these may of course raise the temperature of controversy; basically, they constitute responses to what are taken to be violations of the institutional norms of intellectual property.

Institutional Norms of Science

To say that these frequent conflicts over priority are rooted in the egotism of human nature, then, explains next to nothing; to say that they are rooted in the contentious personalities of those recruited by science may explain part, but not enough; to say, however, that these conflicts are largely a consequence of the institutional norms of science itself comes closer, I think, to the truth. For, as I shall suggest, it is these norms that exert pressure upon scientists to assert their claims, and this goes far toward explaining the seeming paradox that even those meek and unaggressive men, ordinarily slow to press their own claims in other spheres of life, will often do so in their scientific work.

The ways in which the norms of science help produce this result seem clear enough. On every side the scientist is reminded that it is his role to advance knowledge and his happiest fulfillment of that role, to advance knowledge greatly. This is only to say, of course, that in the institution of science originality is at a premium. For it is through originality, in greater or smaller increments, that knowledge advances. When the institution of science works efficiently—and like other social institutions, it does not always do so—recognition and esteem accrue to those who have best fulfilled their roles, to those who have made genuinely original contributions to the common stock of knowledge. Then are found those happy circumstances in which self-interest and moral obligation coincide and fuse.

Recognition of what one has accomplished is thus largely a motive derived from institutional emphases. Recognition for originality becomes socially validated testimony that one has successfully lived up to the most exacting requirements of one's role as scientist. The self-image of the individual scientist will also depend greatly on the appraisals by his scientific peers of the extent to which he has lived up to this exacting and critically important aspect of his role. As Darwin once phrased it, "My love of natural science . . . has been much aided by the ambition to be esteemed by my fellow naturalists."

Interest in recognition,[17] therefore, need not be, though it can readily become, simply a desire for self-aggrandizement or an expression of egotism. It is, rather, the motivational counterpart on the psychological plane to the emphasis upon originality on the institutional plane. It is not necessary that individual scientists begin with a lust for fame; it is enough that science, with its abiding and often functional emphasis on originality and its assigning of large rewards for originality, makes recognition of priority uppermost. Recognition and fame then become symbol and reward for having done one's job well.

This means that long before we know anything about the distinctive personality of this or that scientist, we know that he will be under pressure to make his contributions to knowledge known to other scientists and that they, in turn, will be under pressure to acknowledge his rights to his intellectual property. To be sure, some scientists are more vulnerable to these pressures than others—some are self-effacing, others self-assertive; some generous in granting recognition, others stingy. But the great frequency of struggles over priority does not result merely from these traits of individual scientists but from the institution of science, which defines originality as a supreme value and thereby makes recognition of one's originality a major concern.[18]

When this recognition of priority is either not granted or fades from view, the scientist loses his scientific property. Although this kind of property shares with other types general recognition of the "owner's" rights, it contrasts sharply in all other respects. Once he has made his contribution, the scientist no longer has exclusive rights of access to it. It becomes part of the public domain of science. Nor has he the right of regulating its use by others by withholding it unless it is acknowledged as his. In short, property rights[19] in science become whittled down to just this

17. It is not only the institution of science, of course, that instills and reinforces the concern with recognition; in some degree, all institutions do. This is evident since the time W. I. Thomas included "recognition" as one of what he called "the four wishes" of men. The point is, rather, that with its emphasis on originality, the institution of science greatly reinforces this concern and indirectly leads scientists to vigorous self-assertion of their priority. For Thomas's fullest account of the four wishes, see *The Unadjusted Girl* (Boston: Little, Brown, 1925), chapter 1.

18. In developing this view, I do not mean to imply that scientists, any more than other men, are merely obedient puppets doing exactly what social institutions require of them. But I do mean to say that, like men in other institutional spheres, scientists tend to develop the values and to channel their motivations in directions the institution defines for them. For an extended formulation of the general theory of institutionalized motivation, see Talcott Parsons, *Essays in Sociological Theory*, rev. ed. (Glencoe: The Free Press, 1954), esp. chapters 2 and 3.

19. That the notion of property is part and parcel of the institution of science can be seen from the language employed by scientists in speaking of their work. Ramsay, for example, asks Rayleigh's "permission to look into atmospheric nitrogen" on which Rayleigh had been working; the young Clerk Maxwell writes William Thomson, "I

one: the recognition by others of the scientist's distinctive part in having brought the result into being.

It may be that this concentration of the numerous rights ordinarily bound up in other forms of property into the one right of recognition by others helps produce the great concentration of affect that commonly characterizes disputes over priority. Often, the intensity of affect seems disproportionate to the occasion; for example, when a scientist feels he has not been given enough recognition for what is, in truth, a minor contribution to knowledge, he may respond with as much indignation as the truly inventive scientist, or even with more, if he secretly senses that this is the outermost limit of what he can reasonably hope to contribute.[20] This same concentration of property-rights into the one right of recognition may also account for the deep moral indignation expressed by scientists

---

do not know the Game laws and Patent laws of science . . . but I certainly intend to poach among your electrical images"; Norbert Wiener describes "differential space, the space of the Brownian motion" as "wholly mine in its purely mathematical aspects, whereas I was only a junior partner in the theory of Banach spaces." Borrowing, trespassing, poaching, credit, stealing, a concept which "belongs" to us—these are only a few of the many terms in the lexicon of property adopted by scientists as a matter of course.

20. Some of this had occurred to Galileo in his counterattack on Sarsi (pseudonym for Grassi): "Only too clearly does Sarsi show his desire to strip me completely of any praise. Not content with having disproved our reasoning set forth to explain the fact that the tails of comets sometimes appear to be bent in an arc, he adds that nothing new was achieved by me in this, as it had all been published long ago, and then refuted, by Johann Kepler. In the mind of the reader who goes no more deeply than Sarsi's account, the idea will remain that I am not only a thief of other men's ideas, but a petty, mean thief at that, who goes about pilfering even what has been refuted. And who knows; perhaps in Sarsi's eyes the pettiness of the theft does not render me more blameworthy than I would be if I had bravely applied myself to greater thefts. If, instead of filching some trifle, I had more nobly set myself to search out books by some reputable author not as well known in these parts, and had then tried to suppress his name and attribute all his labors to myself, perhaps Sarsi would consider such an enterprise as grand and heroic as the other seems to him cowardly and abject" (Galileo, "The Assayer," pp. 261–62).

This type of reaction to what I describe as the "professional adumbrationist" (in the unpublished part of this paper) was expressed also by Benjamin Franklin after he had suffered from claims by others that they had first worked out the experiment of the lightning kite. As he said in part (the rest of his observations are almost equally in point), "The smaller your invention is, the more mortification you receive in having the credit of it disputed with you by a rival, whom the jealousy and envy of others are ready to support against you, at least so far as to make the point doubtful. It is not in itself of importance enough for a dispute; no one would think your proofs and reasons worth their attention: and yet if you do not dispute the point, and demonstrate your right, you not only lose the credit of being in that instance *ingenious*, but you suffer the disgrace of not being *ingenuous*; not only of being a plagiary but of being a plagiary for trifles. Had the invention been greater it would have disgraced you less; for men have not so contemptible an idea of him that robs for gold on the highway, as of him that can pick pockets for half-pence and farthings." (Quoted in the informed and far-reaching monograph by I. B. Cohen, *Franklin and Newton* [Philadelphia: The American Philosophical Society, 1956], p. 76.)

when one of their number has had his rights to priority denied or challenged. Even though they have no personal stake in the particular episode, they feel strongly about the single property-norm and the expression of their hostility serves the latent function of reaffirming the moral validity of this norm.

National claims to priority

In a world made up of national states, each with its own share of ethnocentrism, the new discovery redounds to the credit of the discoverer not as an individual only, but also as a national. From at least the seventeenth century, Britons, Frenchmen, Germans, Dutchmen, and Italians have urged their country's claims to priority; a little later, Americans and Russians entered the lists to make it clear that they had primacy.

The seventeenth-century English scientist Wallis, for example, writes: "I would very fain that Mr. Hooke and Mr. Newton would set themselves in earnest for promoting the designs about telescopes, that others may not steal from us what our nation invents, only for the neglect to publish them ourselves." So, also, Halley says of his comet that "if it should return according to our prediction about the year 1758 [as of course it did], impartial posterity will not refuse to acknowledge that this was first discovered by an Englishman."[21]

Or to move abruptly to the present, we see the Russians, now that they have taken a powerful place on the world-scene, beginning to insist on the national character of science and on the importance of finding out who first made a discovery. Although the pattern of national claims to priority is old, the formulation of its rationale in a Russian journal deserves quotation if only because it is so vigorously outspoken:

> Marxism-Leninism shatters into bits the cosmopolitan fiction concerning supra-class, non-national, "universal" science, and definitely proves that science, like all culture in modern society, is national in form and class in content. . . . The slightest inattention to questions of priority in science, the slightest neglect of them, must therefore be condemned, for it plays into the hands of our enemies, who cover their ideological aggression with cosmopolitan talk about the supposed non-existence of questions of priority in science, *i.e.*, the questions concerning which peoples [here, be it noted, collectivities displace the individual scientist] made what contribution to the general store of world culture . . . [And summarizing the answers to these questions in compact summary] The Russian people has the richest history. In the course of this history, it has created the richest culture, and all the other countries of the world have drawn upon it and continue to draw upon it to this day.[22]

21. Louis T. More, *Isaac Newton* (New York: Scribner's, 1934), pp. 146–47, and pp. 241, 477–78.

22. An editorial, "Against the Bourgeois Ideology of Cosmopolitanism," *Voprosy filosofii*, 1948, no. 2, as translated in the *Current Digest of the Soviet Press* 1, no. 1 (1 February 1949): 9–10, 12. For an informed account, see David Joravsky, "Soviet

Against this background of affirmation, one can better appreciate the recent statement by Khrushchev that "we Russians had the H-bomb before you" and the comment by the New York Times that "the question of priority in the explosion of the hydrogen bomb is . . . a matter of semantics," to be settled only when we know whether the "prototype-bomb" or the "full-fledged bomb" is in question.[23]

The recent propensity of the Russians to claim priority in all manner of inventions and scientific discoveries thus energetically reduplicates the earlier, and now less forceful though far from vanished, propensity of other nations to claim like priorities. The restraint often shown by individual scientists in making such claims becomes rather inconspicuous when official or self-constituted representatives of nations put in their claims.

The Reward System of Science

Like other institutions, the institution of science has developed an elaborate system for allocating rewards to those who variously live up to its norms. Of course, this was not always so. The evolution of this system has been the work of centuries, and it will of course never be finished. In the early days of modern science, Francis Bacon could explain and complain all in one by saying that "it is enough to check the growth of science, that efforts and labours in this field go unrewarded. . . . And it is nothing strange if a thing not held in honour does not prosper."[24] And a half-century later, much the same could be said by Thomas Sprat, the Bishop of Rochester, in his official history of the newly-established Royal Society:

It is not to be wonder'd, if men have not been very zealous about those studies, which have been so farr remov'd, from present benefit, and from the applause of men. For what should incite them, to bestow their time, and Art, in revealing to mankind, those Mysteries; for which, it may be, they would be onely

Views on the History of Science," *Isis* 46 (March 1955): 3–13, esp. at pp. 9n. and 11, which treat of changing Russian attitudes toward priority and simultaneous invention; see also Merton, *Social Theory and Social Structure*, pp. 556–60 [and chapter 13 in this volume—ED.].

23. *New York Times*, 27 July 1957, p. 3, col. 1. When it comes to proud claims of having first constructed the hydrogen bomb, national pride truly goeth before destruction.

24. Francis Bacon, *Novum Organum*, trans. by Ellis and Spedding (London: Routledge, n.d.), book 1, aphorism 91. The ellipsis in the text above was for brevity's sake; it should be filled out here below because of the pertinence of what Bacon went on to say: "For it does not rest with the same persons to cultivate sciences and to reward them. The growth of them comes from great wits, the prizes and rewards of them are in the hands of the people, or of great persons, who are but in very few cases even moderately learned. Moreover this kind of progress is not only unrewarded with prizes and substantial benefits; it has not even the advantage of popular applause. For it is a greater matter than the generality of men can take in, and is apt to be overwhelmed and extinguished by the gales of popular opinions."

despis'd at last? How few must there needs be, who will be willing, to be impoverish'd for the common good? while they shall see, all the rewards, which might give life to their Industry, passing by them, and bestow'd on the deserts of easier studies?[25]

The echo of these complaints still reverberates in the halls of universities and scientific societies, but chiefly with regard to material rather than honorific rewards. With the growth and professionalization of science, the system of honorific rewards has become diversely elaborated, and apparently at an accelerated rate.

Heading the list of the immensely varied forms of recognition long in use is eponymy,[26] the practice of affixing the name of the scientist to all or part of what he has found, as with the Copernican system, Hooke's law, Planck's constant, or Halley's comet. In this way, scientists leave their signatures indelibly in history; their names enter into all the scientific languages of the world.

At the rugged and thinly populated peak of this system of eponymy are the men who have put their stamp upon the science and thought of their age. Such men are naturally in very short supply, and these few sometimes have an entire epoch named after them, as when we speak of the Newtonian epoch, the Darwinian era, or the Freudian age.

The gradations of eponymy have the character of a Guttman scale in which those men assigned highest rank are also assigned lesser degrees of honorific recognition. Accordingly, these peerless scientists are typically included also in the next highest ranks of eponymy, in which they are credited with having fathered a new science or a new branch of science (at times, according to the heroic theory, through a kind of parthenogenesis for which they apparently needed no collaborators). Of the illustrious Fathers of this or that science (or of this or that specialty), there is an end, but an end not easily reached. Consider only these few, culled from a list many times this length:

Morgagni, the Father of Pathology;
Cuvier, the Father of Palaeontology;
Faraday, the Father of Electrotechnics;
Daniel Bernoulli, the Father of Mathematical Physics;
Bichat, the Father of Histology;
van Leeuwenhoek, the Father of Protozoology and Bacteriology;

25. Thomas Sprat, *The History of the Royal Society* (London, 1667), p. 27.

26. In his dedication to "The Starry Messenger," announcing his discovery of the satellites of Jupiter, Galileo begins with a paean to the practice of eponymy which opens with these words: "Surely a distinguished public service has been rendered by those who have protected from envy the noble achievements of men who have excelled in virtue, and have thus preserved from oblivion and neglect those names which deserve immortality" (Drake, *Discoveries and Opinions of Galileo*, p. 23). He then proceeds to call the satellites "the Medicean Stars" in honor of the Grand Duke of Tuscany, who soon becomes his patron.

Jenner, the Father of Preventive Medicine;
Chladni, the Father of Modern Acoustics;
Herbart, the Father of Scientific Pedagogy;
Wundt, the Father of Experimental Psychology;
Pearson, the Father of Biometry;
and, of course,
Comte, the Father of Sociology.

In a science as farflung and differentiated as chemistry, there is room for several paternities. If Robert Boyle is the undisputed Father of Chemistry (and, as his Irish epitaph has it, also the Uncle of the Earl of Cork), then Priestley is the Father of Pneumatic Chemistry, Lavoisier the Father of Modern Chemistry, and the nonpareil Willard Gibbs, the Father of Physical Chemistry.

On occasion, the presumed father of a science is called upon, in the persons of his immediate disciples or later adherents, to prove his paternity, as with Johannes Müller and Albrecht von Haller, who are severally regarded as the Father of Experimental Physiology.

Once established, this eponymous pattern is stepped up to extremes. Each new specialty has its own parent, whose identity is often known only to those at work within the specialty. Thus, Manuel Garcia emerges as the Father of Laryngoscopy, Adolphe Brongiart as the Father of Modern Palaeobotany, Timothy Bright as the Father of Modern Shorthand, and Father Johann Dzierson (whose important work may have influenced Mendel) as the Father of Modern Rational Beekeeping.

Sometimes, a particular form of a discipline bears eponymous witness to the man who first gave it shape, as with Hippocratic medicine, Aristotelian logic, Euclidean geometry, Boolean algebra, and Keynesian economics. Most rarely, the same individual acquires a double immortality, both for what he achieved and for what he failed to achieve, as in the cases of Euclidean and non-Euclidean geometries, and Aristotelian and non-Aristotelian logics.

In rough hierarchic order, the next echelon is comprised by thousands of eponymous laws, theories, theorems, hypotheses, instruments, constants, and distributions. No short list can hope to be representative of the wide range of these scientific contributions that have immortalized the men who made them. But a few examples in haphazard array might include the Brownian movement, the Zeeman effect, Rydberg's constant, Moseley's atomic number, and the Lorenz curve or, to come closer home, where we refer only to assured contemporary recognition rather than to possibly permanent fame, the Spearman rank-correlation coefficient, the Rorschach ink-blot, the Thurstone scale, the Bogardus social-distance scale, the Bales categories of interaction, the Guttman scalogram, and the Lazarsfeld latent-structure analysis.

Each science, or art based on science, evolves its own distinctive patterns of eponymy to honor those who have made it what it is. In the medical sciences, for example, the attention of posterity is assured to the discoverer or first describer of parts of the body (as with the Eustachian tube, the circle of Willis, Graffian follicles, Wharton's duct, and the canal of Nuck) though, oddly enough, Vesalius, commonly described as the Father of Modern Anatomy, has been accorded no one part of the body as distinctly his own. In medicine, also, eponymy registers the first diagnostician of a disease (as with Addison's, Bright's Hodgkin's, Menière's, and Parkinson's diseases); the inventor of diagnostic tests (as with Romberg's sign, the Wassermann reaction, the Calmette test, and the Babinski reflex); and the inventor of instruments used in research or practice (as with the Kelly pad, the Kelly clamp, and the Kelly rectoscope). Yet, however numerous and diversified this array of eponyms in medicine,[27] they are still reserved, of course, to only a small fraction of the many who have labored in the medical vineyard. Eponymy is a prize that, though large in absolute aggregate, is limited to the relatively few.

Time does not permit, nor does the occasion require, detailed examination of eponymous practices in all the other sciences. Consider, then, only two other patterns: In a special branch of physics, it became the practice to honor great physicists by attaching their names to electrical and magnetic units (as with volt, ohm, ampere, coulomb, farad, joule, watt, henry, maxwell, gauss, gilbert, and oersted). In biology, it is the long-standing practice to append the name of the first describer to the name of a species, a custom which greatly agitated Darwin since, as he saw it, this put "a premium on hasty and careless work" as the "species-mongers" among naturalists try to achieve an easy immortality by "miserably describ[ing] a species in two or three lines."[28] (This, I may say, will not be the last occasion for us to see how the system of rewards in science can be stepped up to such lengths as to get out of hand and defeat its original purposes.)

Eponymy is only the most enduring and perhaps most prestigious kind of recognition institutionalized in science. Were the reward system confined to this, it would not provide for the many other distinguished scientists

27. It has been suggested that, in medicine at least, eponymous titles are given to diseases only so long as they are poorly understood. "Any disease designated by an eponym is a good subject for research" (O. H. Perry, *Medical Etymology* [Philadelphia: W. B. Saunders Co., 1949], pp. 11–12).

28. Exercised by the excesses eponymy in natural history had reached, the usually mild Darwin repeatedly denounced this "miserable and degrading passion of mere species naming." What is most in point for us is the way in which the pathological exaggeration of eponymizing highlights the normal role of eponymy in providing its share of incentives for serious and sustained work in science. Francis Darwin, ed., *The Life and Letters of Charles Darwin* (New York: Appleton, 1925), vol. 1, pp. 332–44.

without whose work the revolutionary discoveries could not have been made. Graded rewards in the coin of the scientific realm—honorific recognition by fellow-scientists—are distributed among the stratified layers of scientific accomplishment. Merely to list some of these other but still considerable forms of recognition will perhaps be enough to remind us of the complex structure of the reward system in science.

In recent generations, the Nobel Prize, with nominations for it made by scientists of distinction throughout the world, is perhaps the preeminent token of recognized achievement in science.[29] There is also an iconography of fame in science, with medals honoring famous scientists and the recipients of the award alike (as with the Rumford medal and the Arago medal). Beyond these, are memberships in honorary academies and sciences (for example, the Royal Society and the French Academy of Sciences), and fellowships in national and local societies. In those nations that still preserve a titled aristocracy, scientists have been ennobled, as in England since the time when Queen Anne added laurels to her crown by knighting Newton, not, as might be supposed, because of his superb administrative work as Master of the Mint, but for his scientific discoveries. These things move slowly; it required almost two centuries before another Queen of England would, in 1892, confer a peerage of the realm upon a man of science for his work in science, and thus transform the preeminent Sir William Thomson into the no less eminent Lord Kelvin.[30] Scientists themselves have distinguished the stars from the supporting cast by issuing directories of "starred men of science," and universities have been known to accord honorary degrees to scientists along with the larger company of philanthropists, industrialists, businessmen, statesmen, and politicians.

Recognition is finally allocated by those guardians of posthumous fame, the historians of science. From the most disciplined scholarly works to the vulgarized and sentimentalized accounts designed for the millions, great attention is paid to priority of discovery, to the iteration and reiteration of "firsts." In this way, many historians of science help maintain the prevailing institutional emphasis on the importance of priority. One of the most eminent among them, the late George Sarton, at once expresses and exemplifies the commemorative function of historiography when he writes that "the first scholar to conceive that subject [the history of science] as an independent discipline and to realize its importance was . . . Auguste

29. On the machinery and results of the Nobel and other prize awards, see Barber, *Science and the Social Order*, pp. 108 ff.; and Leo Moulin, "The Nobel Prizes for the Sciences, 1901–1950," *British Journal of Sociology* 6 (September 1955): 246–63.

30. For caustic comment on the lag in according such recognition to men of science, see excerpts from newspapers of the day in Silvanus P. Thompson, *The Life of William Thomson: Baron Kelvin of Largs* (London: Macmillan, 1910), vol. 2, pp. 906–7.

Comte." He then goes on to propose that great scholar, Paul Tannery, as most deserving to be called "the father of our studies," and finally states the thesis that "as the historian is expected to determine not only the relative truth of scientific ideas at different chronological states, but also their relative novelty, he is irresistibly led to the fixation of *first* events."[31]

Although scientific knowledge is impersonal in the sense that its claim to truth must be assessed entirely apart from its source, the historian of science is called upon to prevent scientific knowledge from sinking (or rising) into anonymity, to preserve the collective memory of its origins. Anonymous givers have no place in this scheme of things. Eponymity, not anonymity, is the standard. And, as we have seen, outstanding scientists, in turn, labor hard to have their names inscribed in the golden book of firsts.[32]

Seen in composite, from the eponyms enduringly recording the names of scientists in the international language of science to the immense array of parochial and ephemeral prizes, the reward system of science reinforces and perpetuates the institutional emphasis upon originality. It is in this specific sense that originality can be said to be a major institutional goal of modern science, at times the paramount one, and recognition for originality a derived, but often as heavily emphasized, goal. In the organized competition to contribute to man's scientific knowledge, the race *is* to the swift, to him who gets there first with his contribution in hand.

31. George Sarton, *The Study of the History of Science* (Cambridge: Harvard University Press, 1936), pp. 3–4, 35–36. Sarton goes on to observe that this practice of identifying first events "never fails to involve him [the historian] in new difficulties, because creations absolutely *de novo* are very rare, if they occur at all; most novelties are only novel combinations of old elements and the degree of novelty is thus a matter of interpretation, which may vary considerably according to the historian's experience, standpoint, or prejudices. . . . It is always risky, yet when every reasonable precaution has been taken one must be willing to run the risk and make the challenge, for this is the only means of being corrected, if correction be needed" (ibid., p. 36). This is a telling sign of the deep-rooted sentiment that recognition for originality in science must be expressed, that it is an obligation—"the historian is expected . . ."—to search out the "first" to contribute an idea or finding, even though a comprehensive view of the cumulative and interlocking character of scientific inquiry suggests that the attribution of "firsts" is often difficult and sometimes arbitrary. For a further statement on this matter of priority, see George Sarton, *The Study of the History of Mathematics* (Cambridge: Harvard University Press, 1936), pp. 33–36.

I cannot undertake here to examine the attitudes commonly manifested by historians of science toward this emphasis on searching out priorities. It can be said that these too are often ambivalent.

32. This was presumably not always so. As is well known, medieval authors often tried to cloak their writing in anonymity. But this is not the place to examine the complex subject of variations in cultural emphases upon originality and recognition. For some observations on this, see George Sarton, *A Guide to the History of Science* (Waltham, Mass: Chronica Botanica Co., 1952), p. 23, who reminds us of ancient and medieval practices in which "modest authors would try to pass off their own compositions under the name of an illustrious author of an earlier time," ghost-writing in reverse. See also R. K. Merton, *Science, Technology and Society in Seventeenth-Century England* (Bruges, Belgium: Osiris, 1938), pp. 360–632, at p. 528.

Institutional norm of humility

If the institution of science placed great value *only* on originality, scientists would perhaps attach even more importance to recognition of priority than they do. But, of course, this value does not stand alone. It is only one of a complex set making up the ethos of science—disinterestedness, universalism, organized scepticism, communism of intellectual property, and humility being some of the others.[33] Among these, the socially enforced value of humility is in most immediate point, serving, as it does, to reduce the misbehavior of scientists below the rate that would occur if importance were assigned only to originality and the establishing of priority.

The value of humility takes diverse expression. One form is the practice of acknowledging the heavy indebtedness to the legacy of knowledge bequeathed by predecessors. This kind of humility is perhaps best expressed in the epigram Newton made his own: "If I have seen farther, it is by standing on the shoulders of giants" (this, incidentally, in a letter to Hooke who was then challenging Newton's priority in the theory of colors).[34] That this tradition has not always been honored in practice can be inferred from the admiration that Darwin, himself lavish in such acknowledgments, expressed to Lyell for "the elaborate honesty with which you quote the words of all living and dead geologists."[35] Exploring the literature of a field of science becomes not only an instrumental practice, designed to learn from the past, but a commemorative practice, designed to pay homage to those who have prepared the way for one's work.

Humility is expected also in the form of the scientist's insisting upon his personal limitations and the limitations of scientific knowledge altogether. Galileo taught himself and his pupils to say, "I do not know." Perhaps another often-quoted image by Newton most fully expresses this kind of humility in the face of what is yet to be known:

> I do not know what I may appear to the world, but to myself I seem to have been only like a boy playing on the seashore, and diverting myself in now and then finding a smoother pebble or a prettier shell than ordinary, whilst the great ocean of truth lay all undiscovered before me.[36]

33. For a review of other values of science, see Barber, *Science and the Social Order*, chapter 4; Merton, *Social Theory and Social Structure*, pp. 552–61; H. A. Shepard, "The Value System of a University Research Group," *American Sociological Review* 19 (August 1954): 456–62.

34. Alexandre Koyré, "An Unpublished Letter of Robert Hooke to Isaac Newton," *Isis* 43 (December 1952): 312–37, at 315.

35. F. Darwin, *Charles Darwin*, 1:263.

36. David Brewster, *Memoirs of the Life, Writings, and Discoveries of Sir Isaac Newton* (Edinburgh and London, 1855), vol. 2, chapter 27. For our purposes, unlike those of the historian, it is a matter of indifference whether Newton actually felt modest or was merely conforming to expectation. In either case, he expresses the

If this contrast between public image ("what I may appear to the world") and self-image ("but to myself I seem") is fitting for the greatest among scientists, it is presumably not entirely out of place for the rest. The same theme continues unabated. Laplace, the Newton of France, in spite of what has been described as "his desire to shine in the constantly changing spotlight of public esteem," reportedly utters an epigrammatic paraphrase of Newton in his last words, "What we know is not much; what we do not know is immense."[37] Lagrange summarizes his lifetime of discovery in the one phrase, "I do not know." And Lord Kelvin, at the Jubilee celebrating his fifty years as a distinguished scientist in the course of which he was honored by scores of scientific societies and academies, characterizes his lifelong effort to develop a grand and comprehensive theory of the properties of matter by the one word, "Failure."[38]

Like all human values, the value of modesty can be vulgarized and run into the ground by excessive and thoughtless repetition. It can become merely conventional, emptied of substance and genuine feeling. There really *can* be too much of a good thing. It is perhaps this excess which led Charles Richet, himself a Nobel laureate, to report the quiet self-appraisal by a celebrated scientist: "I possess every good quality, but the one that distinguishes me above all is modesty."[39] Other scientists—for example, the great Harvard mathematician, George Birkhoff—would have no truck with modesty, whether false, prim, or genuine. Having been told by a Mexican physicist of his hope that the United States would continue "to send us savants of your stature," Birkhoff sturdily replied, "Professor Erro, in the States I *am* the only one of my stature." And as Norbert Wiener is reported to have said in his obituary address for Birkhoff, "He was the first among us and he accepted the fact. He was not modest."[40] Nevertheless, such forthright acknowledgement of one's eminence is not quite the norm among scientists.

---

norm of personal humility, which is widely held to be appropriate. I. B. Cohen (*Franklin and Newton*, pp. 47–58, passim) repeatedly and incisively makes the point that both admirers and critics of Newton have failed to make the indispensable distinction between what he said and what he did.

37. Bell, *Men of Mathematics*, p. 172. Bell refers to "a common and engaging trait of the truly eminent scientist in his frequent confession of how little he knows." What he describes as a trait of the scientist can also be seen as an expectation on the part of the community of scientists. It is not that many scientists *happen* to be humble men; they are *expected* to be humble. See E. T. Bell, "Mathematics and Speculation," *The Scientific Monthly* 32 (March 1931): pp. 193–209, at p. 204.

38. G. F. Fitzgerald, *Lord Kelvin, 1846–99*, Jubilee commemoration volume, with an essay on his works, 1899; S. P. Thompson, *Life of William Thomson*, vol. 2, chapter 24.

39. See the gallery of trenchant pen-portraits of scientists in Charles Richet, *The Natural History of a Savant*, trans. by Sir Oliver Lodge (New York: Doran, 1927), p. 86.

40. Carlos Graef Fernandez (as transcribed by Samuel Kaplan), "My Tilt with Albert Einstein," *American Scientist* 44 (April 1956): pp. 204–11, at p. 204.

It would appear, then, that the institution of science, like other institutions, incorporates potentially incompatible values: among them, the value of originality, which leads scientists to want their priority to be recognized, and the value of humility, which leads them to insist on how little they have been able to accomplish. These values are not real contradictories, of course—" 'tis a poor thing, but mine own"—but they do call for opposed kinds of behavior. To blend these potential incompatibles[41] into a single orientation, to reconcile them in practice, is no easy matter. Rather, as we shall now see, the tension between these kindred values—kindred as Cain and Abel were kin—creates an inner conflict among men of science who have internalized both of them and generates a distinct ambivalence toward the claiming of priorities.

Ambivalence toward Priority

The components of this ambivalence are fairly clear. After all, to insist on one's originality by claiming priority is not exactly humble and to dismiss one's priority by ignoring it is not exactly to affirm the value of originality.[42] As a result of this conflict, scientists come to despise themselves for wanting that which the institutional values of science have led them to want.

With the rare candor that distinguishes him, Darwin so clearly exhibits this agitated ambivalence in its every detail that this one case can be taken as paradigmatic for many others (which are matters of less-detailed and less candid record). In his *Autobiography*, he writes that, even before his historic voyage on the Beagle in 1831, he was "ambitious to take a fair place among scientific men—whether more ambitious or less so than most of my fellow-workers, I can form no opinion."[43] A quarter of a century

41. For further examination of the problem of blending incompatible norms into stable patterns of behavior, in this case among physicians, see R. K. Merton, "Some Preliminaries to a Sociology of Medical Education," in R. K. Merton, G. G. Reader, and P. L. Kendall, eds., *The Student-Physician* (Cambridge: Harvard University Press, 1957), p. 72 ff. As is well known, R. S. Lynd has set forth the general notion that institutional norms are organized as near-incompatibles; see his *Knowledge for What?* (Princeton: Princeton University Press, 1939), chapter 3.

42. Strictly speaking, originality and priority are of course not the same thing. Belated independent rediscoveries of what was long since known may represent great originality on the part of the rediscoverer, as is perhaps best shown in the remarkable case of the self-taught twentieth-century Indian mathematician, Srinivasa Ramanujan, who, all unknowing that it had been done before, re-created much of early nineteenth-century mathematics, and more besides. Cf. G. H. Hardy *Ramanujan: Twelve Lectures Suggested by His Life and Work* (Cambridge: Harvard University Press, 1940). Edwin G. Boring, who has long been interested in the subject of priority in science, has, among many other perceptive observations, noted the lack of identity between originality and priority. See, for example, his early paper, "The Problem of Originality in Science," *American Journal of Psychology* 39 (December 1927): pp. 70–90, esp. at p. 78.

43. F. Darwin, *Charles Darwin*, p. 54.

after this voyage, he is still wrestling with his ambition, exclaiming in a letter that "I wish I could set less value on the bauble fame, either present or posthumous, than I do, but not, I think, to any extreme degree."[44]

Two years before the traumatizing news from Wallace, reporting his formulation of the theory of evolution, Darwin writes his now-famous letter to Lyell, explaining that he is not quite ready to publish his views, as Lyell had suggested he do in order not to be forestalled, and again expressing his uncontrollable ambivalence in these words: "I rather hate the idea of writing for priority, yet I certainly should be vexed if any one were to publish my doctrines before me."[45]

And then, in June 1858, the blow falls. What Lyell warned would happen and what Darwin could not bring himself to believe could happen, as all the world knows, did happen. Here is Darwin writing Lyell of the crushing event:

> [Wallace] has today sent me the enclosed, and asked me to forward it to you. It seems to me well worth reading. Your words have come true with a vengeance—that I should be forestalled. . . . I never saw a more striking coincidence; if Wallace had my MS. sketch written out in 1842, he could not have made a better short abstract! Even his terms now stand as heads of my chapters. . . . So all my originality, whatever it may amount to, will be smashed.[46]

Humility and disinterestedness urge Darwin to give up his claim to priority; the wish for originality and recognition urges him that all need not be lost. At first, with typical magnanimity, but without pretense of equanimity, he makes the desperate decision to step aside altogether. A week later, he is writing Lyell again; perhaps he might publish a short version of his long-standing text, "a dozen pages or so." And yet, he says in his anguished letter, "I cannot persuade myself that I can do so honourably." Torn by his mixed feelings, he concludes his letter, "My good dear friend, forgive me. This is a trumpery letter, influenced by trumpery feelings." And in an effort finally to purge himself of his feelings, he appends a postscript, "I will never trouble you or Hooker on the subject again."[47]

The next day he writes Lyell once more, this time to repudiate the postscript. Again, he registers his ambivalence: "It seems hard on me that I should lose my priority of many years' standing, but I cannot feel at all sure that this alters the justice of the case. First impressions are generally right, and I at first thought it would be dishonourable in me now to publish."[48]

44. Ibid., p. 452.
45. Ibid., pp. 426–27.
46. Ibid., p. 473.
47. Ibid., pp. 474–75.
48. Ibid., p. 475.

As fate would have it, Darwin is just then prostrated by the death of his infant daughter. He manages to respond to the request of his friend Hooker and sends the Wallace manuscript and his own original sketch of 1844, "solely," he writes, "that you may see by your own handwriting that you did read it. . . . Do not waste much time. It is miserable in me to care at all about priority."[49]

Other members of the scientific community do what the tormented Darwin will not do for himself. Lyell and Hooker take matters in hand and arrange for that momentous session in which both papers are read at the Linnean Society. And as they put it in their letter prefacing the publication of the joint paper of "Messrs. C. Darwin and A. Wallace," "in adopting our present course . . . we have explained to him [Darwin] that we are not solely considering the relative claims to priority of himself and his friend, but the interests of science generally."[50] Despite this disclaimer of interest in priority, be it noted that scientific *knowledge* is not the richer or the poorer for having credit given where credit is due; it is the social *institution* of science and individual men of science that would suffer from repeated failures to allocate credit justly.

This historic and not merely historical episode so plainly exhibits the ambivalence occasioned by the double concern with priority and modesty that it need not be examined further. Had the institutionalized emphasis on originality been alone in point, the claim to priority would have invited neither self-blame nor self-contempt; publication of the long antecedent work would have proclaimed its own originality. But the value of originality was joined with the value of humility and modesty. To insist on priority would be to trumpet one's own excellence, but scientific peers and friends of the discoverers, acting as a third party in accord with the institutional norms, could with full propriety announce the joint claims to originality that the discoverers could not bring themselves to do. Underneath it all lies a deep and agitated ambivalence toward priority.

I have not yet counted the recorded cases of debates about priority in science and the manner of their outcome. Such a count, moreover, will not tell the full story, for it will not include the doubtless numerous instances in which independent ideas and discoveries were never announced by those who found their ideas anticipated in print. Nevertheless, I have the strong impression that disputes, even bitter disputes, over priority outnumber the cases of despondent but unreserved admission that the other fellow had made the discovery first.

49. Ibid., p. 476.

50. "On the Tendency of Species to Form Varieties and on the Perpetuation of Varieties and Species by Natural Means of Selection," by C. Darwin and A. R. Wallace. Communicated by Sir C. Lyell and J. D. Hooker, *Journal of the Linnean Society* 3 (1859): 45. Read 1 July 1858.

The institutional values of modesty and humility are apparently not always enough to counteract both the institutional emphasis upon originality and the actual workings of the system of allocating rewards. Originality, as exemplified by the new idea or the new finding, is more readily observable by others in science and is more fully rewarded than the often unobservable kind of humility that keeps an independent discoverer from reporting that he too had had the same idea or the same finding. Moreover, after publication by another, it is often difficult, if not impossible, to demonstrate that one had independently arrived at the same result. For these and other reasons, it is generally an unequal contest between the values of recognized originality and of modesty. Great modesty may elicit respect, but great originality promises everlasting fame.

In short, the social organization of science allocates honor in a way that tends to vitiate the institutional emphasis upon modesty. It is this, I believe, which goes far toward explaining why so many scientists, even those who are ordinarily men of the most scrupulous integrity, will go to great lengths to press their claims to priority of discovery. As I have often suggested, perhaps too often, any *extreme* institutional

> emphasis upon achievement—whether this be scientific productivity, accumulation of wealth or, by a small stretch of the imagination, the conquests of a Don Juan—will attenuate conformity to the institutional norms governing behavior designed to achieve the particular form of 'success,' especially among those who are socially disadvantaged in the competitive race.[51]

Or more specifically and more completely, great concern with the goal of recognition for originality can generate a tendency toward sharp practices just inside the rules of the game or sharper practices far outside. That this has been the case with the behavior of scientists who were all-out to have their originality recognized, the rest of this paper will try to show.

51. Merton, *Social Theory and Social Structure*, p. 166. Scientists do not all occupy similar positions in the social structure; there are, consequently, differentials in access to *opportunity* for scientific achievement (and, of course, differences of individual capacity for achievement). The theory of the relations of social structure to anomie requires us to explore differential pressures upon those scientists variously located in the social structure. Contrast only the disputatious Robert Hooke, a socially mobile man whose rise in status resulted wholly from his scientific achievements, and the singularly undisputatious Henry Cavendish, high-born and very rich (far richer, and, by the canons of Burke's peerage, more elevated even than that other great aristocrat of science, Robert Boyle) who, in the words of Biot, was *"le plus riche de tous les savans; et probablement aussi, le plus savant de tous les riches."* Or consider what Norbert Wiener has said of himself, "I was competitive beyond the run of younger mathematicians, and I knew equally that this was not a very pretty attitude. However, it was not an attitude which I was free to assume or to reject. I was quite aware that I was an out among ins and I would get no shred of recognition that I did not force " (*I Am a Mathematician* [New York: Doubleday, 1956], p. 87). But these are only straws in the wind; once again, limitations of space allow me only to identify a problem, not to examine it.

Types of Response to Cultural Emphasis on Originality

Fraud in science

The extreme form of deviant behavior in science would of course be the use of fraud to obtain credit for an original discovery. For reasons to be examined, the annals of science include very few instances of downright fraud, although, in the nature of the case, an accurate estimate of frequency is impossible. Darwin, for example, said that he knew of only "three intentionally falsified statements" in science.[52] Yet, some time before, his contemporary, Charles Babbage, the mathematician and inventor of calculating machines (one of which prophetically made use of perforated cards), had angrily taken a classified inventory of fraud in science.[53]

At the extreme are hoaxes and forgery: the concocting of false data in science and learning—or, more accurately, in pseudo-science and anti-scholarship. Literary documents have been forged in abundance, at times by men of previously unblemished reputation, in order to gain money or fame. Though no one can say with confidence it appears that love of money was at the root of the forgery of fifty or so rare nineteenth-century pamphlets by that prince of bibliographers, that court of last appeal for the authentication of rare books and manuscripts, Thomas J. Wise. Of quite another stripe was John Payne Collier, the Shakespearean scholar who, unrivaled for his genuine finds in Elizabethan drama and "encouraged by the steadily growing plaudits of his colleagues," could not rest content with this measure of fame and proceeded to forge, with great and knowledgeable skill, a yet-uncounted array of literary papers.[54] But these rogues seem idle alongside the fecund and audacious Vrain-Lucas who, in the space of eight years, created more than 27,000 pieces of manuscript, all duly sold to Michel Chasles, perhaps the outstanding French geometer of the mid-nineteenth century, whose credulity stretches our own, inasmuch as this vast collection included letters by Pontius Pilate, Mary Magdalene, the resurrected Lazarus, Ovid, Luther, Dante, Shakespeare, Galileo, Pascal, and Newton, all written on paper and in modern French. Most provocative among these documents was the correspondence between Pascal and the then eleven-year-old Newton (all in French, of course, although even at the advanced age of thirty-one Newton could struggle through French

52. F. Darwin, *Charles Darwin*, p. 84.

53. Charles Babbage, *The Decline of Science in England* (London, 1830), pp. 174–83. George Lundberg has independently noted that "a scientist's greed for applause [sometimes] becomes greater than his devotion to truth" (*Social Research* [New York: Longmans, Green, 1929], p. 34; and in less detail, in the second edition, 1946, p. 52).

54. I have drawn these examples of frauds in anti-scholarship from the zestful and careful account by Richard D. Altick, *The Scholar Adventurers* (New York: Macmillan, 1951), chapters 2 and 6.

only with the aid of a dictionary), for these letters made it plain that Pascal, not Newton, had, to the greater glory of France, first discovered the law of gravitation, a momentous correction of history, which for several years excited the interest of the *Académie des Sciences* and usurped many pages of the *Comptes Rendus* until, in 1869, Vrain-Lucas was finally brought to book and sentenced to two years in prison. For our purposes, it is altogether fitting that Vrain-Lucas should have had Pascal address this maxim to the boy Newton: *"Tout homme qui n'aspire pas à se faire un nom n'exécutera jamais rien de grand."*[55]

Such lavish forgery is unknown to science proper, but the pressure to demonstrate the truth of a theory or to produce a sensational discovery has occasionally led to the faking of scientific evidence. The biologist Paul Kammerer produced specimens of spotted salamanders designed to prove the Lamarckian thesis experimentally; was thereupon offered a chair at the University of Moscow where in 1925 the Lamarckian views of Michurin held reign; and upon proof that the specimens were fakes, attributed the fraud to a research assistant.[56] Most recently, the Piltdown man—that is, the skull and jaw from which his existence was inferred—has been shown, after forty years of uneasy acceptance, to be a carefully contrived hoax.[57]

Excessive concern with "success" in scientific work has on occasion led to the types of fraud Babbage picturesquely described as "trimming" and "cooking." The trimmer clips off "little bits here and there from observations which differ most in excess from the mean, and [sticks] . . . them on to those which are too small . . . [for the unallowable purpose of] 'equitable adjustment.' " The cook makes "multitudes of observations" and selects only those which agree with an hypothesis, and, as Babbage says, "the cook must be very unlucky if he cannot pick out fifteen or twenty

55. The definitive reports on the Vrain-Lucas affair by M. P. Faugère and by Henri Bordier and Mabille are not available to me at this telling; substantial details, including extracts from the court proceedings, are given by the paleographer, Étienne Charavay, *Affair Vrain-Lucas: Etude Critique* (Paris, 1870); a more accessible summary that does not, however, do full justice to the prodigious inventiveness of Vrain-Lucas is provided by J. A. Farrer, *Literary Forgeries* (London: Longmans Green, 1907), chapter 12. The biographer of Newton, Sir David Brewster, at the age of 87, did his share to safeguard the integrity of historical scholarship, but this did not prevent Chasles from prizing the three thousand letters of Galileo which he had acquired from his friend, although they happened to be in French, rather than in the Latin or Italian in which Galileo wrote.

56. (Kammerer's suicide may not have resulted from this traumatic episode. See the recent effort to vindicate Kammerer by Arthur Koestler, *The Case of the Midwife Toad* [New York: Random House, 1972].) Martin Gardner, *In the Name of Science* (New York: G. P. Putnam's Sons, 1952), p. 143; W. S. Beck, *Modern Science and the Nature of Life* (New York: Harcourt, Brace, 1957), pp. 201–2; Conway Zirkle, "The Citation of Fraudulent Data," *Science* 120 (30 July 1954): 189–90.

57. William L. Straus, Jr., "The Great Piltdown Hoax," *Science* 119 (26 February 1954): 265–69.

which will do for serving up." This eagerness to demonstrate a thesis can, on occasion, lead even truth to be fed with cooked data, as it did for the neurotic scientist, described by Lawrence Kubie, "who had proved his case, but was so driven by his anxieties that he had to bolster an already proven theorem by falsifying some quite unnecessary additional statistical data."[58]

The great cultural emphasis upon recognition for original discovery can lead by gradations from these rare practices of outright fraud to more frequent practices just beyond the edge of acceptability, sometimes without the scientist's being aware that he has exceeded allowable limits. Scientists may find themselves reporting only "successful experiments or results, so-called, and neglecting to report 'failures.' " Alan Gregg, that informed observer of the world of medical research, practice, and education, reports the case of

> the medical scientist of the greatest distinction who told me that during his graduate fellowship at one of the great English universities he encountered for the first time the idea that in scientific work one should be really honest in reporting the results of his experiments. Before that time he had always been told and had quite naturally assumed that the point was to get his observations and theories accepted by others, and published.[59]

Yet, these deviant practices should be seen in perspective. What evidence there is suggests that they are extremely infrequent, and this temporary focus upon them will surely not be distorted into regarding the exceptional case as the typical. Apart from the moral integrity of scientists themselves —and this is, of course, the major basis for honesty in science—there is much in the social organization of science that provides a further compelling basis for honest work. Scientific research is typically, if not always, under the exacting scrutiny of fellow experts, involving, as it usually though not always does, the verifiability of results by others. Scientific inquiry is in effect subject to rigorous policing, to a degree perhaps unparalleled in any other field of human activity. Personal honesty is supported by the public and testable character of science. As Babbage remarked, "the cook would [at best] procure a temporary reputation . . . at the expense of his permanent fame."

Competition in the realm of science, intensified by the great emphasis on original and significant discoveries, may occasionally generate incentives for eclipsing rivals by illicit or dubious means. But this seldom occurs in the form of preparing fraudulent data; instead, it appears in quite other forms of deviant behavior involving spurious claims to discovery. More

58. Lawrence S. Kubie, "Some Unsolved Problems of the Scientific Career," *American Scientist* 41 (1953): 596–613; ibid. 42 (1954): 104–12.

59. Alan Gregg, *Challenges to Contemporary Medicine* (New York: Columbia University Press, 1956), p. 115.

concretely, it is an occasional theft rather than forgery, and more often, libel and slander rather than theft that are found on the small seamy side of science.

Plagiary: fact and slander

Deviant behavior most often takes the form of occasional plagiaries and many slanderous charges or insinuations of plagiary. The historical record shows relatively few cases (and of course the record may be defective) in which one scientist actually pilfered another. We are assured that in the *Mécanique céleste* (until then, outranked only by Newton's *Principia*) "theorems and formulae are appropriated wholesale without acknowledgement" by Laplace.[60] Or, to take a marginal case, Sir Everard Home, the distinguished English surgeon who was appointed custodian of the unpublished papers of his even more distinguished brother-in-law, John Hunter, published 116 papers of uncertain origin in the *Philosophical Transactions* after Hunter's death, and burned Hunter's manuscripts, an action greatly criticized by knowledgeable and suspicious contemporaries.[61] It is true also that Robert Boyle, not impressed by the thought that theft of his ideas might be a high tribute to his talent, was in 1688 driven to the desperate expedient of printing "An advertisement about the loss of many of his writings," later describing the theft of his work and reporting that he would from then on write only on loose sheets, in the hope that these would tempt thieves less than "bulky packets" and, going on to say that he was resolved to send his writings to press without extensive revision in order to avoid prolonged delays.[62] But even with such cases of larceny on the

60. As stated by the historian of astronomy, Agnes Mae Clerke, in her article on Laplace in the eleventh edition of the *Encyclopaedia Britannica.* Some of Clerke's further observations are much in point: "In the delicate task of apportioning his own large share of merit, he certainly does not err on the side of modesty; but it would perhaps be as difficult to produce an instance of injustice, as of generosity in his estimate of others. Far more serious blame attaches to his all but total suppression in the body of the work—and the fault pervades the whole of his writings—of the names of his predecessors and contemporaries . . . a production which may be described as the organized result of a century of patient toil presents itself to the world as the offspring of a single brain." And yet, since these matters are seldom all of a piece, "Biot relates that, when he himself was beginning his career, Laplace introduced him at the Institute for the purpose of explaining his supposed discovery of equations of mixed differences, and afterwards showed him, under a strict pledge of secrecy, the papers, then yellow with age, in which he had long before obtained the same results" (vol. 16, pp. 201–2). As we shall see, Gauss, who was meticulous in acknowledging predecessors, treated the young Bolyai as did Laplace the young Biot.

61. Ralph H. Major, *A History of Medicine* (Oxford: Blackwell Scientific Publications, 1954), 2:703.

62. The account by A. M. Clerke in the article on Boyle in the *Dictionary of National Biography* is somewhat mistaken in attributing charges of plagiary to the

grand scale, the aggregate of demonstrable theft in modern science is not large.

What does loom large is the repeated practice of charging others with pilfering scientific ideas. Falsely accused of plagiarizing Harvey in physiology, Snell in optics, and Harriot and Fermat in geometry, Descartes in turn accuses Hobbes and the teen-age Pascal of plagiarizing him.[63] To maintain his property, Descartes implores his friend Mersenne, "I also beg you to tell him [Hobbes] as little as possible about what you know of my unpublished opinions, for if I'm not greatly mistaken, he is a man who is seeking to acquire a reputation at my expense and through shady practices."[64] All unknowing that the serene and unambitious Gauss had long since discovered the method of least squares, Legendre, himself "a man of the highest character and scrupulously fair," practically accuses Gauss of having filched the idea from him and complains that Gauss, already so well-stocked with momentous discoveries, might at least have had the decency not to adopt his brainchild.[65]

At times, the rivalrous concern with priority can go so far as to set, not the Egyptians against the Egyptians, but brother against brother, as in the case of the great eighteenth-century mathematicians, the brothers Jacob and Johannes Bernoulli, who repeatedly and bitterly attacked one another's claims to priority. (Johannes improved on this by throwing his own son out of the house for having won a prize from the French Academy on which he himself had had his eye.)[66]

Or to turn to our own province, Comte, tormented by the suggestion that his law of three stages had really been originated by St. Simon, denounces his one-time master and describes him as a "superficial and de-

---

published advertisement. This speaks only of losses of manuscript through "unwelcome accidents" (e.g., the upsetting of corrosive liquors over a file of manuscripts) and at most hints at less impersonal sources of loss. But a later unpublished paper by Boyle, dug up by his biographer Birch, is levelled against the numerous plagiarists of his works. This document, running to three folio pages of print, is a compendium of the ingenious devices for thievery developed by the grand larcenists of seventeenth-century science. See *The Works of the Honourable Robert Boyle*, 6 vols. With the Life of the author, by J. Birch (London, 1772), 1: cxxv–cxxviii, ccxxii–ccxxiv.

63. For the case of Harvey, see A. R. Hall, *The Scientific Revolution, 1500–1800* (London: Longmans, Green, 1954), p. 148; for Hobbes, see Descartes, *Oeuvres*, ed. Charles Adam and Paul Tannery, vol. 3, *Correspondance* (Paris, 1899), pp. 283 ff.; for Pascal, see ibid., vol. 5 (1903), p. 366.

64. Descartes, *Oeuvres*, 3:320.

65. Bell, *Men of Mathematics*, pp. 259–60. Legendre seems to have been particularly sensitive to these matters, perhaps because he was often victimized; note Clerke's remark that between Laplace and Legendre "there was a feeling of 'more than coldness,' owing to his appropriation, with scant acknowledgement, of the other's labors." *Encyclopaedia Britannica*, vol. 16, p. 202.

66. Bell, *Men of Mathematics*, p. 134.

praved charlatan."[67] Again, to take Freud's own paraphrase, Janet claims that "everything good in psychoanalysis repeats, with slight modifications, the views of Janet—everything else in psychoanalysis being bad."[68] Freud refuses to lock horns with Janet in what he describes as "gladiator fights in front of the noble mob," but some years later, his disciple, Ernest Jones, reports that at a London congress he has "put an end to" Janet's pretensions, and Freud applauds him in a letter that urges him to "strike while the iron is hot," in the interests of "fair play."[69]

So the almost changeless pattern repeats itself. Two or more scientists quietly announce a discovery. Since it is often the case that these are truly independent discoveries, with each scientist having separately exhibited originality of mind, the process is sometimes stabilized at that point, with due credit to both, as in the instance of Darwin and Wallace. But since the situation is often ambiguous with the role of each not easy to demonstrate, and since each *knows* that he had himself arrived at the discovery, and since the institutionalized stakes of reputation are high and the joy of discovery immense, this is often not a stable solution. One or another of the discoverers—or frequently, his colleagues or fellow-nationals—suggests that he rather than his rival was really first, and that the independence of the rival is at least unproved. Then begins the familiar deterioration of standards governing conflictful interaction: the other side, grouping its forces, counters with the opinion that plagiary had indeed occurred, that let him whom the shoe fits wear it and furthermore, to make matters quite clear, the shoe is on the other foot. Reinforced by group loyalties and often by chauvinism, the controversy gains force, mutual recriminations of plagiary abound, and there develops an atmosphere of thoroughgoing hostility and mutual distrust.

On some occasions, this can lead to outright deceit in order to buttress valid claims, as with Newton in his controversy with Leibniz over the invention of the calculus. When the Royal Society finally established a committee to adjudicate the rival claims, Newton, who was then president

67. Frank E. Manuel, *The New World of Henri Saint-Simon* (Cambridge: Harvard University Press, 1956), pp. 340–42; also Richard L. Hawkins, *Auguste Comte and the United States* (Cambridge: Harvard University Press, 1936), pp. 81–82, as cited by Manuel.

68. Sigmund Freud, *History of the Psychoanalytic Movement* (London: Hogarth Press, 1949); also, Freud, *An Autobiographical Study* (London: Hogarth Press, 1948), pp. 54–55, where he seeks "to put an end to the glib repetition of the view that whatever is of value in psycho-analysis is merely borrowed from the ideas of Janet. . . . Historically psycho-analysis is completely independent of Janet's discoveries, just as in its content it diverges from them and goes far beyond them." For Janet's not always delicate insinuations, see his *Psychological Healing* (New York: Macmillan, 1925), 1:601–40.

69. Ernest Jones, *Sigmund Freud: Life and Work* (London: Hogarth Press, 1955), 2:112.

of the Royal Society, packed the committee, helped direct its activities, anonymously wrote the preface for the second published report—the draft is in his handwriting—and included in that preface a disarming reference to the old legal maxim that "no one is a proper witness for himself [and that] he would be an iniquitous Judge, and would crush underfoot the laws of all the people, who would admit anyone as a lawful witness in his own cause."[70] We can gauge the immense pressures for self-vindication that must have operated for such a man as Newton to have adopted these means for defense of his valid claims. It was not because Newton was so weak but because the institutionalized values were so strong that he was driven to such lengths.

This interplay of offensive and defensive maneuvers—no doubt students of the theory of games can recast it more rigorously—thus gives further emphasis to priority. Scientists try to exonerate themselves in advance from possible charges of filching by going to great lengths to establish their priority of discovery. Often this kind of anticipatory defense produces the very result it was designed to avoid by inviting others to show that prior announcement or publication need not mean there was no plagiary.

The effort to safeguard priority and to have proof of one's integrity has led to a variety of institutional arrangements designed to cope with this strain on the system of rewards. In the seventeenth century, for example, and even as late as the nineteenth, discoveries were sometimes reported in the form of anagrams—as with Galileo's "triple star" of Saturn and Hooke's law of tension—for the double purpose of establishing priority of conception and of yet not putting rivals on to one's original ideas, until they had been further worked out.[71] Then, as now, complex ideas were quickly published in abstracts, as when Halley urged Newton to do so in

70. There is a sizeable library discussing the Newton-Leibniz controversy. I have drawn chiefly upon More, *Isaac Newton*, who devotes the whole of chapter 15 to this subject; Augustus de Morgan, *Essays on the Life and Works of Newton* (Chicago: Open Court Pub. Co., 1914), esp. appendix 2; and Brewster, *Memoirs of Newton*, chapter 22; cf. Cohen, *Franklin and Newton*, who is properly critical of the biography by More at various points (e.g., pp. 84–85). On the basis of his examination of the Portsmouth Papers, More concludes that "the principals, and practically all those associated with them wantonly made statements which were false; and not one of them came through with a clean record" (p. 567). E. N. da C. Andrade has aptly summed up Newton's ambivalence in this judgment: "Evidence can be cited for the view that Newton was modest or most overweening; the truth is that he was a very complex character . . . when not worried or irritated he was modest about his achievements." See Andrade's *Sir Isaac Newton* (London: Collins, 1954), esp. pp. 131–32.

71. The earlier widespread use of anagrams is well known. As late as the nineteenth century, the physicists Balfour Stewart and P. G. Tait reintroduced this practice and "to secure priority . . . [took] the unusual step of publishing [their idea] as an anagram in *Nature* some months before the publication of the book" (Sir J. J. Thomson, *Recollections and Reflections* [London: G. Bell, 1936], p. 22).

order to secure "his invention to himself till such time as he would be at leisure to publish it."[72] There is also the long-standing practice of depositing sealed and dated manuscripts with scientific academies in order to protect both priority and idea.[73] Scientific journals often print the date on which the manuscript of a published article was received, thus serving, even apart from such intent, to register the time it first came to notice. Numerous personal expedients have been developed: for example, letters detailing one's own ideas are sent off to a potential rival, thus disarming him; preliminary and confidential reports are circulated among a chosen few; personal records of research are meticulously dated (as by Kelvin). Finally, it has often been suggested that the functional equivalent of a patent-office be established in science to adjudicate rival claims to priority.[74]

In prolonged and yet overly quick summary, these are some of the forms of deviance invited by the institutional emphasis on priority and some of the institutional expedients devised to reduce the frequency of these deviations. But as we would expect from the theory of alternative responses to excessively emphasized goals, other forms of behavior, verging toward deviance though still well within the tacit rules and not as subject to moral disapproval as the foregoing, have also made their appearance.

Alternative responses to emphasis on originality

The large majority of scientists, like the large majority of artists, writers, doctors, bankers and bookkeepers, have little prospect of great and decisive originality. For most of us artisans of research, getting things into print becomes a symbolic equivalent to making a significant discovery. Nor could science advance without the great unending flow of papers reporting careful investigations, even if these are routine rather than distinctly original. The indispensable reporting of research can, however, become converted into an itch to publish that, in turn, becomes aggravated by the tendency, in many academic institutions, to transform the sheer number of publications into a ritualized measure of scientific or scholarly accomplishment.[75]

72. Thomas Birch, *The History of the Royal Society of London* (London: 1756–57), 4:437.

73. For a recent instance, see the episode described by Wiener in which the race between Bouligand and Wiener to contribute new concepts "in potential theory" ended in a "dead heat," since Bouligand had submitted his "results to the [French] Academy in a sealed envelope, after a custom sanctioned by centuries of academy tradition" (Wiener, *I Am a Mathematician*, p. 92).

74. J. Hettinger, "Problems of Scientific Property and Its Solution," *Science Progress* 26 (January 1932), pp. 449–61; also the paper by Dr. A. I. Sotesi, of the New York Academy of Medicine, cited by Bernhard J. Stern in *Social Factors in Medical Progress* (New York: Columbia University Press, 1927), p. 108.

75. There is not space here to examine the institutional conditions which lead the piling up of publications to become a virtually ritualistic activity.

The urge to publish is given a further push by the moral imperative of science to make one's work known to others; it is the obverse to the culturally repudiated practice of jealously hoarding scientific knowledge for oneself. As Priestley liked to say, "whenever he discovered a new fact in science, he instantly proclaimed it to the world, in order that other minds might be employed upon it besides his own."[76] Indeed, John Aubrey, that seventeenth-century master of the thumbnail biography and member of the Royal Society, could extend the moral imperative for communication of knowledge to justify even plagiary if the original author will not put his ideas into print. In his view it was better to have scientific goods stolen and circulated than to have them lost entirely.[77]

To this point (and I provide comfort by reporting that the end of the paper is in sight), we have examined types of deviant responses to the institutional emphasis on priority that are *active* responses: the fabrication of "data," aggressive self-assertion, the denouncing of rivals, plagiary, and charges of plagiary. Other scientists have responded to the same pressures *passively* or at least by internalizing their aggressions and directing them against themselves.[78] Since these passive responses, unlike the active ones, are private and often not publicly observable, they seldom enter the historical record. This need not mean, of course, that passive withdrawal from the competition for originality in science is infrequent; it might simply mean that the men responding in this fashion do not come to public notice, unless they do so after their accomplishments have qualified them for the pages of history.

Chief among these passive deviant responses is what I have described, on occasion, as *retreatism*, the abandoning of the once-esteemed cultural goal of originality and of practices directed toward reaching that goal. In

76. Priestley's remark as paraphrased by his longtime friend, T. L. Hawkes, and reported by George Wilson, *Henry Cavendish*, p. 111. The seventeenth-century Dutch genius of microscopy, Anton van Leeuwenhoek, also adopted a policy, as he described it, that "whenever I found out anything remarkable, I have thought it my duty to put down my discovery on paper, so that all ingenious people might be informed thereof" (Quoted by Major, *History of Medicine*, 1:531). The same sentiment was expressed by Saint-Simon, among many others. Cf. Manuel, *Saint-Simon*, pp. 63–64.

77. Aubrey could say, irresponsibly and probably without malice, that the mathematician John Wallis "may stand with much glory upon his owne basis, and need not be beholding to any man, for Fame, yet he is so greedy of glorie, that he steales feathers from others to adorn his own cap; e.g. he lies at watch, at Sir Christopher Wren's discours, Mr. Robert Hooke's, etc.; putts down their notions in his note booke, and then prints it, without owneing the authors. This frequently, of which they complaine. But though he does an Injury to the Inventors, he does good to Learning, in publishing such curious notions, which the author (especially Sir Christopher Wren) might never have the leisure to write himselfe" (John Aubrey, *Brief Lives*, ed. Andrew Clark [Oxford, 1898], 2:281–82).

78. The distinction between active and passive forms of deviant behavior is drawn from Talcott Parsons, *The Social System* (Glencoe: The Free Press, 1951), pp. 256–67.

such instances, the scientist withdraws from the field of inquiry, either by giving up science altogether or by confining himself to some alternative role in it, such as teaching or administration. (This does not say, of course, that teaching and administration do not have their own attractions, or that they are less significant than inquiry; I refer here only to the scientists who reluctantly abandon their research because it does not measure up to their own standards of excellence.)

A few historical instances of such retreatism must stand in place of more. The nineteenth-century physicist Waterston, his classic paper on molecular velocity having been rejected by the Royal Society as "nothing but nonsense," becomes hopelessly discouraged and leaves science altogether.[79] Deeply disappointed by the lack of response to his historic papers on heredity, Mendel refuses to publish the now permanently lost results of his further research and, after becoming abbot of his monastery, gives up his research on heredity.[80] Robert Mayer, tormented by refusals to grant him priority for the principle of conservation of energy, tries a suicide leap from a third-story window and succeeds only in breaking his legs and being straitjacketed, for a time, in an insane asylum.[81]

Perhaps the most telling instance of retreatism in mathematics is that of Janos Bolyai, inventor of one of the non-Euclidean geometries. The young Bolyai tries to obey his mathematician-father who, out of the bitter fruits of his own experience, warns his son to give up any effort to prove the postulate on parallels—or, as his father more picturesquely put it, to "detest it just as much as lewd intercourse; it can deprive you of all your leisure, your health, your rest, and the whole happiness of your life." He dutifully becomes an army officer instead, but his demon does not permit the twenty-one-year-old Bolyai to leave the postulate alone. After years of work, he develops his geometry, sends the manuscript to his father who in turn transmits it to Gauss, the prince of mathematicians, for a magisterial opinion. Gauss sees in the work proof of authentic genius, writes

79. R. H. Murray, *Science and Scientists in the Nineteenth Century* (London: Sheldon, 1925), pp. 346–48; and David L. Watson, *Scientists are Human,* (London: Watts and Co., 1938), pp. 58, 80; Strutt, *John William Strutt*, pp. 169–71. Evidently, Sidney Lee, the editor of the *Dictionary of National Biography* by the time it reached the volume in which Waterston should have had an honored place, could not penetrate the obscurity into which the great discoverer was plunged by the unfounded rejection of his work; there is no biography of Waterston in the *DNB*.

80. Hugo Iltis, *Life of Mendel* (New York: W. W. Norton, 1932), pp. 111–12; and see Mendel's prophetic remark, "My time will come" (Ibid., p. 282).

81. Mayer's having been rejected by his liberal friends who took part in the revolution of 1848, which he as a conservative opposed, may have contributed to his disturbance. For some recent evidence on how Mayer's priority was safeguarded by the lay-sociologist Josef Popper, see Otto Blüh, "The Value of Inspiration: A Study on Julius Robert Mayer and Josef Popper-Lynkeus," *Isis* 43 (September 1952): 211–20. Blüh's opinion that claims of priority in science are no longer taken seriously seems exaggerated.

the elder Bolyai so, and adds, in all truth, that he cannot express his enthusiasm as fully as he would like, for "to praise it, would be to praise myself. Indeed, the whole contents of the work, the path taken by your son, the results to which he is led, coincide almost entirely with my meditations, which have occupied my mind partly for the last thirty or thirty-five years . . . I am very glad that it is just the son of my old friend, who takes the precedence of me in such a remarkable manner." Delighted by this accolade, the elder Bolyai sends the letter to his son, innocently saying that it is "very satisfactory and redounds to the honor of our country and our nation." Young Bolyai reads the letter, but has no eye for the statements which say that his ideas are sound, that in the judgment of the incomparable Gauss he is blessed with genius. He sees only that Gauss has anticipated him. For a time, he believes that his father must have previously confided his ideas to Gauss who had thereupon made them his own.[82] His priority lost, and, with the further blow, years later, of coming upon Lobachevsky's non-Euclidean geometry, he never again publishes any work in mathematics.[83]

Apart from historical cases of notable scientists retreating from the field after denial of the recognition owing them, there are many contemporary cases that come to the notice of psychiatrists rather than historians. Since Lawrence Kubie is almost alone among psychiatrists to have described

82. The principal source on the Bolyais, including the germane correspondence, is Paul Stäckel, *Wolfgang und Johann Bolyai, Geometrische Untersuchungen*, 2 vols. (Leipzig: 1913), which was not available to me at this writing. An excellent short account is provided by Roberto Bonola, *Non-Euclidean Geometry*, trans. H. S. Carslaw, 2d rev. ed. (La Salle, Illinois: Open Court Publishing Company, 1938), pp. 96–113; see also Dirk J. Struik, *A Concise History of Mathematics* (New York: Dover Publications, 1948), 2:251–54; Franz Schmidt, "Lebensgeschichte des Ungarischen Mathematikers Johann Bolyai de Bolya," *Abhandlungen zur Geschichte der Mathematik* 8 (1898): 135–46.

83. Two letters provide context for Bolyai's great fall from the high peak of exhilaration into the slough of despond. In 1823 he writes his father: ". . . the goal is not yet reached, but I have made such wonderful discoveries that I have been almost overwhelmed by them, and it would be the cause of continual regret if they were lost. When you will see them, you too will recognize it. In the meantime I can say only this: *I have created a new universe from nothing.* All that I have sent you till now is but a house of cards compared to the tower. I am as fully persuaded that it will bring me honor, as if I had already completed the discovery." And just as, a generation later, Lyell was prophetically to warn Darwin of being forestalled, so does the elder Bolyai warn the younger: "If you have really succeeded in the question, it is right that no time be lost in making it public, for two reasons: first, because ideas pass easily from one to another, who can anticipate its publication; and secondly, there is some truth in this, that many things have an epoch, in which they are found at the same time in several places, just as the violets appear on every side in spring. Also every scientific struggle is just a serious war, in which I cannot say when peace will arrive. Thus we ought to conquer when we are able, since the advantage is always to the first comer." (Quoted by Bonola, *Non-Euclidean Geometry*, pp. 98, 99.) Small wonder that though young Bolyai continued to work sporadically in mathematics, he never again published the results of his work.

these in print, I shall draw upon his pertinent account of the maladaptations of scientists suffering from an unquenched thirst for original discovery and ensuing praise.

When the scientist's aspirations become too lofty to be realized, the result sometimes is apathy, imbued with fantasy. In Kubie's words:

> The young scientist may dwell for years in secret contemplation of his own unspoken hope of making great scientific discoveries. As time goes on, his silence begins to frighten him; and in the effort to master his fear, he may build up a secret feeling that his very silence is august, and that once he is ready to reveal his theories, they will shake the world. Thus a secret megalomania can hide among the ambitions of the young research worker.[84]

Perhaps most stressful of all is the situation in which the recognition accorded the scientist is not proportioned to his industry or even to the merit of his work. He may find himself serving primarily to remove obstacles to fundamental discoveries by others. His "negative experiments clear the road for the steady advance of science, but at the same time they clear the road for the more glamorous successes of other scientists, who may have used no greater intelligence, skill or devotion; perhaps even less."[85] Like other men, scientists become disturbed by the pan-human problem of evil, in which "the fortunes of men seem to bear practically no relation to their merits and efforts."[86]

Kubie hazards some further observations that read almost as if they were describing the behavior of delinquents in response to a condition of relative anomie. "Success or failure, whether in specific investigations or in an entire career may be almost accidental, with chance a major factor in determining not what is discovered, but when and by whom. . . . Yet young students are not warned that their future success may be determined by forces which are outside their own creative capacity or their willingness to work hard."[87] As a result of all this, Kubie suspects the emergence of

84. Kubie, "Some Unsolved Problems of the Scientific Career," p. 110.
85. Ibid.
86. Gilbert Murray, quoted in a similar theoretical context by Merton, *Social Theory and Social Structure*, p. 147.
87. Ibid., pp. 111–12. This reading of the case is not inconsistent with the facts of multiple independent discoveries and inventions. As the long history of multiple discoveries makes clear, and as W. F. Ogburn and D. S. Thomas among the sociologists have shown, certain discoveries become almost "inevitable" when the cultural base cumulates to a certain level. But this still leaves some indeterminacy in the matter of *who* will *first* make the discovery. Kubie mentions some "near-misses" of discoveries that suggest undoubted merit is not all when it comes to the *first* formulation of a discovery, and this list can be greatly extended. In the nature of the case, moreover, we often do not know of those scientists who have abandoned a line of inquiry that was moving toward a particular discovery when they found it had been made and announced by another. These "personal tragedies" of near-discovery—tragedy in terms of the prevailing cutural belief that all credit is due him who is "first,"—are the silent tragedies that leave no mark in the historiography of science.

what he calls a "new psychosocial ailment among scientists which may not be wholly unrelated to the gangster tradition of dead-end kids. Are we witnessing the development of a generation of hardened, cynical, amoral, embittered, disillusioned young scientists?"

Lacking the evidence, this had best be left as a rhetorical question. But the import of the question needs comment. In "Social Structure and Anomie," I have set out diagnoses of the ways in which a culture giving emphasis to aspirations for all, aspirations which cannot be realized by many, exerts a pressure for deviant behavior and for cynicism, for rejection of the reigning moralities and the rules of the game. We see here the possibility that the same pressures may in some degree be at work in the institution of science. But even though the pressures are severe, they need not produce much deviant behavior. There are great differences between the social structure of science and other social structures in which deviance is frequent. Among other things, the institution of science continues to have an abiding emphasis on other values that curb the culturally induced tendency toward deviation, an emphasis on the value of truth by whomsoever it is found, and a commitment to the disinterested pursuit of truth. Simply because we have focused on the deviant behavior of scientists, we should not forget how relatively rare this is. Only a few try to gain reputation by means that will lose them repute.

Functions and Dysfunctions of Emphasis on Priority

It has sometimes been said that the emphasis upon recognition of priority has the function of motivating scientists to make discoveries. For example, Sir Frederick Banting, the major figure in the discovery of insulin-therapy for diabetes, was long disturbed by the conviction that the chief of his department had been given too much credit for what he had contributed to the discovery. Time and again, Banting returned to the importance of allocating due credit for a discovery: ". . . it makes Research men," he said. "It stimulates the individuality and develops personality. Our religion, our moral fabric, our very basis of life are centered round the idea of reward. It is not abnormal therefore that the Research man should desire the kudos of his own work and his own idea. If this is taken away from him, the greatest stimulant for work is withdrawn."[88]

88. Quoted in Lloyd Stevenson, *Sir Frederick Banting* (London: Heinemann Medical Books, 1947), p. 301. Two hundred years before, John Morgan, the celebrated founder of the first American medical school, had expressed the same conception, but in sociologically more acceptable terms. To his mind, personal motivation for fame was linked with the social benefit of the advancement of science. Men of science, he said, "have the highest motives that can animate the pursuits of a generous

From this, it would seem that the institutional emphasis is maintained with an eye to its functional utility. But as I have tried to show, the emphasis upon priority is often not confined within functional limits. Once it becomes established, forces of rivalrous interaction lead it to get out of hand. Recognition of priority, operating to reward those who advanced science materially by being the first to make a significant discovery, becomes a sentiment in its own right. Rationalized as a means of providing incentives for original work and as expressing esteem for those who have done much to advance science, it becomes transformed into an end-in-itself. It becomes stepped up to a dysfunctional extreme far beyond the limits of utility.[89] It can even reach the revealing extreme where, for example, the permanent secretary of the French Academy of Sciences, François Arago, could exclaim (apropos of the controversy involving Cavendish and Watt) that to describe discoveries as having been made " 'about the same time' proves nothing; questions as to priority may depend on weeks, on days, on hours, on minutes."[90]

When the criteria of priority become as finely discriminated as this—and Arago only put in words what many others have expressed in behavior—then priority has lost all functional significance. For when two scientists independently make the same discovery months or weeks apart, to say nothing of days or hours, it can scarcely be thought that one has exhibited greater originality than the other or that the short interim that separates them can be used to speed up the rate of scientific development.

Conclusion

The interpretation I have tried to develop here is not, I am happy to say, a new one. Nor do I consider it fully established and beyond debate. After

---

mind. They consider themselves as under the notice of the public, to which every ingenious person labours to approve himself. A love of fame and a laudable ambition allure him with the most powerful charms. These passions have, in all ages, fired the souls of heroes, of patriots, of lovers of science, have made them renowned in war, eminent in government and peace, justly celebrated for the improvement of polite and useful knowledge." In effect, "other-directedness" can be functional to the society, providing that the criteria of judgment by others are sound. See John Morgan, *A Discourse Upon the Institution of Medical Schools in America* (1765; photo-offset reprint of first edition, Baltimore: The Johns Hopkins Press, 1937), pp. 59–60.

89. For suggestive observations on the process of "stepping up patterns to unanticipated extremities," a process which he called "perseveration," see W. I. Thomas, *Primitive Behavior* (New York: McGraw-Hill, 1937), p. 9 and passim; see also, Merton, *Social Theory and Social Structure*, pp. 199 ff. As I have tried to show in this paper, science has experienced this stepping-up of functional norms to an extreme at which they become dysfunctional to the workings of the institution.

90. M. [F.] Arago, *Historical Eloge of James Watt*, trans. J. P. Muirhead (London, 1839), p. 106. The whole of this document and Arago's role in the Adams-LeVerrier controversy clearly exemplify the forces producing conflicts over priority.

all, neither under the laws of logic nor under the laws of any other realm, must one become permanently wed to an hypothesis simply because one has tentatively embraced it. But the interpretation does seem to account for some of the otherwise puzzling aspects of conflicts over priority in science, and it is closely bound to a body of sociological theory.

In short review, the interpretation is this: Like other social institutions, the institution of science has its characteristic values, norms, and organization. Among these, the emphasis on the value of originality has a self-evident rationale, for it is originality that does much to advance science. Like other institutions also, science has its system of allocating rewards for performance of roles. These rewards are largely honorific, since even today, when science is largely professionalized, the pursuit of science is culturally defined as being primarily a disinterested search for truth and only secondarily a means of earning a livelihood. In line with the value-emphasis, rewards are to be meted out in accord with the measure of accomplishment. When the institution operates effectively, the augmenting of knowledge and the augmenting of personal fame go hand in hand; the institutional goal and the personal reward are tied together. But these institutional values have the defects of their qualities. The institution can get partly out of control, as the emphasis upon originality and its recognition is stepped up. The more thoroughly scientists ascribe an unlimited value to originality, the more they are in this sense dedicated to the advancement of knowledge, the greater is their involvement in the successful outcome of inquiry and their emotional vulnerability to failure.

Against this cultural and social background, one can begin to glimpse the sources, other than idiosyncratic ones, of the misbehavior of individual scientists. The culture of science is, in this measure, pathogenic. It can lead scientists to develop an extreme concern with recognition which is in turn the validation by peers of the worth of their work. Contentiousness, self-assertive claims, secretiveness lest one be forestalled, reporting only the data that support an hypothesis, false charges of plagiarism, even the occasional theft of ideas and, in rare cases, the fabrication of data,—all these have appeared in the history of science and can be thought of as deviant behavior in response to a discrepancy between the enormous emphasis in the culture of science upon original discovery and the actual difficulty many scientists experience in making an original discovery. In this situation of stress, all manner of adaptive behaviors are called into play, some of these being far beyond the mores of science.

All this can be put more generally. We have heard much in recent years about the dangers brought about by emphasis on the relativity of values, about the precarious condition of a society in which men do not believe in values deeply enough and do not feel strongly enough about what they do believe. If there is a lesson to be learned from this review of some consequences of a belief in the absolute importance of originality, perhaps it is the old lesson that unrestricted belief in absolutes has its dangers too. It can produce the kind of fanatic zeal in which anything goes. In its way, the absolutizing of values can be just as damaging as the decay of values to the life of men in society.[91]

91. It is of some interest that just when this paper was in galley proof, all the world came to experience the social, political, and scientific repercussions of a spectacular "first" in science-based technology, when Russian scientists put a man-made sphere into space.

15 Behavior Patterns of Scientists

1968

The history of science indelibly records 1953 as the year in which the structure of the DNA molecule was discovered. But it is 1968 that will probably emerge as the year of the double helix in the history that treats the behavior of scientists, for James Watson's deeply personal account of that discovery, now in its ninth printing, has evidently seized the public imagination. Widely and diversely reviewed in journals of science and para-science, it has been discussed in scores of monthlies, weeklies, and daily newspapers, from the London *Times* to the Erie, Pennsylvania, *Times*, from the *Village Voice* to the *Wall Street Journal* (which, aptly enough, manages to give a faintly financial slant to the book, concluding that "Watson, in the long run, may have done science a favor. In these days when the public is asked to allocate billions for scientific research, it's of some comfort to know that the spenders are human.").

To judge from the popular reviews, that indeed was taken to be the essential message of the book: scientists are human, after all. This phrasing, it turns out, does not mean that scientists can be assigned at long last to the species Homo sapiens. Many Americans and some Englishmen were apparently prepared to entertain that serviceable hypothesis even before the appearance of *The Double Helix*. Evidently, what is meant by the Watson-induced thought that scientists too are human is that scientists are all too human; that, in the succinct, jaundiced words of the St. Louis

Presented as the annual Phi Beta Kappa–Sigma Xi address before the American Association for the Advancement of Science in December 1968. Copublished in *American Scientist* 58 (Spring 1969): 1–23, and *The American Scholar* 38 (Spring 1969): 197–225. This paper also contains passages from Robert K. Merton and Richard Lewis, "The Competitive Pressures: (1) The Race for Priority," in *Impact of Science on Society* 21, no. 2 (1971): 151–60, © 1971 by UNESCO, reprinted by permission of the United Nations Educational, Scientific and Cultural Organization.

Post-Dispatch, "they can be boastful, jealous, garrulous, violent, [and even] stupid."

What, then, are the stories Watson tells about the social and intellectual interactions that entered into the discovery, stories eliciting the popular response that scientists are all too human? Above all else, he tells of the race for priority; a close awareness of the champion rival who must be defeated in this contest of minds; a driving insistence on getting needed data from sometimes reluctant, sometimes inadvertent collaborators; a competition for specific discoveries over the years between the Cavendish and Caltech; an allegedly English sense of private domains for scientific investigation that bear no-poaching signs; an express ambition for that ultimate symbol of accomplishment, the Nobel. He tells, too, about alternating periods of intense thought and almost calculated idleness (while the gestation of ideas pursues its course); about false starts and errors of inference; about quickly getting up needed scientific knowledge despite an impressive inventory of initial ignorance; about the complementarity of talents, skills, and character-structure of the symbiotic collaborators; about an unfailing sense for the key problem, and an intuitive and stubbornly maintained imagery of the nature of its solution, together with the implications as these were expressed in that masterstroke of calculated understatement wrought by Francis Crick: "It has not escaped our notice that the pairing we have postulated immediately suggests a possible copying mechanism for the genetic material."

The stories detailed in *The Double Helix* have evidently gone far to dispel a popular mythology about the complex behavior of scientists. That this response should have occurred among the public at large is not surprising. Embodying as they do some of the prime values of world civilization, scientists have long been placed on pedestals where they may have no wish to be perched—not, at least, the more thoughtful among them.

In part, too, the imagery of scientists moving coolly, methodically, and unerringly to the results they report may stem from the etiquette that governs the writing of scientific papers. This etiquette, as we know, requires them to be works of vast expurgation, stripping the complex events and behaviors that culminated in the report of everything except their cognitive substance. Compare only the lean, taut, almost laconic, nine-hundred-word article that appeared in *Nature* that momentous April in 1953 with the tangled web of events reported in Watson's forty-thousand-word account of the same discovery.

The sense of popular revelation upon learning that scientists are actually human testifies, then, to the prevalence of an earlier belief to the contrary. Ironically enough, that older mythology now threatens to be displaced by a somewhat new variant, expressed in responses to the Watson memoir by scientists and humanists alike. (I use the term mythology in its decidedly untechnical sense to denote a set of ill-founded beliefs held uncritically by an interested group.) The new variant has several interrelated components. The patterns of motives and behavior set out in Watson's irreverent, naturalistic narrative are held to be distinctive of the newest era of science, staffed by "a new kind of scientist and one that could hardly have been thought of before science became a mass occupation." Only in our highly competitive age, allegedly, are appreciable numbers of scientists concerned to "scoop" others at work in the field and so to gain recognition for their accomplishments. As another scientist-reviewer, Jerome Lettvin, sees it, part of what Watson reports expresses "no more than the general opportunism that is the hallmark of modern competitive science"—a statement in which the governing phrase is "*modern* competitive science." And in still another version—this one the response of a humanist to *The Double Helix*—it is suggested, with unconcealed reluctance, that "a keenness of early recognition may even be, these days, as essential to discovery as intelligence. Science, like all other activities *now*"—again, I accent the temporal qualifier—"is crowded and accelerated. There is no sitting alone *anymore* and letting apples fall down."

All apart from Watson's provocative book, some scientists have contrasted what they remember as the comparatively quiet times of science before the Second World War or thereabouts with the intensely competitive situation today. Hans Gaffron, in one nostalgic and admirably succinct example, writes:

> The student now, in 1970, finds it difficult to believe that, at least with many of us in the 1920s, there was never the thought of having to hurry, or of having to publish results prematurely and more than once lest they be overlooked or taken over in their entirety by somebody else. Even important discoveries were left for a year or two in the hands of a man with whom they originated so that he could develop them according to his means and abilities. We used to say: "An apple already bitten into is not very attractive." The man who had the first bite was expected to keep and eat his apple. But then more and more people appeared on the scene who felt no compunction to bite into every apple within reach and then often drop it just as quickly. It was considered very bad manners, but they were the men of the future. . . .
>
> The phenomenal increase in the number of people whose work brings them in contact with scientific investigations has changed not only the image of the average scientist but also his motives and relationship with his colleagues. The latter are not fellows working in neighboring fields—their fields—but all too often are direct competitors engaged in simultaneous, absolutely identical, experiments. Not only has the ruthlessness of accomplished business techniques invaded the areas where industrial exploitation overlaps research, but this kind of behavior is no longer considered alien to science.

There is a certain plausibility to this view that the mores of science and the behavior of scientists must surely have changed in the recent past. For plainly, all the basic demographic, social, economic, political, and organizational parameters of science have acquired dramatically new values. The size of the population of working scientists has increased exponentially from the scattered hundreds three centuries ago to the hundred or more myriads today. The time of the amateur is long since past; scientists are now professionals all, their work providing them with a livelihood and, for some, a not altogether impoverished one. The social organization of scientific inquiry has greatly changed, with collaboration and research teams the order of the day. As just another pale reflection of this changed organization of scientific inquiry, each decade registers more and more multi-authored articles in decided contrast to the almost unchanging character of single-authored papers in the humanities. The monumental budgets assigned to science—although never large enough, as all of us know—are orders of magnitude greater than the straitened budgets of only a few generations ago, to say nothing of the immense contrast with those of the more remote past. The vast increase in numbers of scientists and in funds for science practically dictates the exponential increase in the quantity of published research. As science has become more institutionalized, it has also become more intimately interrelated with the other institutions of society. Science-based technologies and the partial diffusion of a scientific outlook have become great social forces that move our history and greatly affect the relations obtaining between the nations of the world. Scientists do not, of course, make the major political decisions, but they now affect them significantly. The Szilard-Einstein letter to President Roosevelt, for example, would be described by some as one of the most consequential communications in recorded history (although evidence now indicates that work on the atomic bomb would have been pushed forward in any case).

But there is no need to continue with this truncated list of particulars in which science today so conspicuously departs from the science of an earlier time. With all these profound changes, as any sociologist is apt to tell you if you give him half a chance, there must also be a new ethos of science abroad, a new set of values and institutionally patterned motives. And, as I have noted, practicing scientists in biology and physics and chemistry have indeed suggested that we now have a new breed of scientists, actuated by new motives, oriented to the main chance, and gravely agitated by failures to achieve. Like other men, scientists become disturbed by the panhuman problem of evil, in which, to assume the language of Gilbert Murray, "the fortunes of men seem to bear practically no relation to their merits and efforts."

Without at all adopting the new mythology of science, the psychiatrist Lawrence Kubie asks: "Are we witnessing the development of a generation of hardened, cynical, amoral, embittered, disillusioned young scientists?"

The question is not unrelated to the new mythology which maintains that behavior of the kind candidly described by Watson is something new to our time, and so, we must suppose, altogether alien to the earlier, heroic age of science, say, the seventeenth century. It is an intriguing and, as I have said, not altogether implausible thought, one that the rest of this paper is designed to examine.

And here I must interrupt these introductory observations with a personal confession. It was just thirty years ago that I suggested, in a footnote tucked away in a monograph on science in seventeenth-century England, that the race for priority might constitute a strategic subject for study and might provide clues to ways in which the institution of science shapes the motives, passions, and social relations of scientists. So far as I can tell, the youthful author of that footnote proved to be its only reader. At any rate, no one, not even he, harkened to the muted clarion call. Some ten years ago, when addressing a captive audience of a thousand sociologists, I tried to make amends for this lapse of two decades and examined the import of priority races for an understanding of both the institution of science and the behavior of scientists. In more recent years, my colleagues at Columbia and I have examined these implications in a series of investigations. In what follows, I shall draw mercilessly upon these inquiries.

Now to return to the belief system that regards the rough-and-tumble of contest and competition in science as peculiar to our own deteriorating times, that treats such contest as inevitably self-aggrandizing and takes the drive to be first in reaching a discovery as necessarily displacing that "relish of knowledge" (of which John Locke spoke), as doing away with intrinsic joy in discovery or pleasure in the beauty of a powerful simplifying idea.

As with most mythologies, this one is not altogether out of touch with the world of everyday experience. Although it may have surprised the outsider, Watson's unabashed report on the race for priority scarcely came as news to his fellow scientists. They know from hard-won experience that multiple independent discoveries at about the same time constitute one of their occupational hazards. They not only know it, but often act on that premise. That the consequent rush to achieve priority is common in our time hardly needs documentation. The evidence is there on every side. A few years before Watson reached his much wider audience, Arthur Schawlow casually noted in the public prints that Charles Townes and he had been "in a hurry, of course. We feared that it might be only a matter of time before others would come up with the same idea. So we decided to publish before building a working model. . . . Subsequently, Theodore Maiman won the frantic race between many experimenters to build the first laser. Our theory was verified." Townes had ample biographical reason to be in a hurry. After all, in the early 1950s, he had been involved in that fivefold independent discovery of the maser, along with Willis Lamb, Joseph Weber, Nikolai Basov, and Aleksandr Prokhorov.

The contemporary annals of science are peppered with cases of scientists spurred on to more intense effort by the knowledge that others were on much the same track. Harriet Zuckerman's interviews with Nobel laureates find many of them testifying, in the words of one of them, that "it was bound to happen soon. Had I not done it, . . . it was there, waiting for somebody . . . [probably] at the Rockefeller Institute." Or to turn from the moving frontiers of science to its interior regions, Warren Hagstrom found that two-thirds of a sample of some fourteen hundred scientists had been anticipated by others in their own contributions, a good number of these on more than one occasion. And, if there were need of any further sign that contemporary scientists are often engaged in the race for priority, we need only turn to the periodic editorials by Samuel Goudsmit in the *Physical Review Letters*, where he notes the drive for quick publication to ensure priority, sometimes at the expense of physicists "working along the same lines who want to do a more complete job before publishing their findings." Some of his editorials are touched with anguish as he reviews expedients adopted by physicists seeking publication in the *Letters* in order to " 'scoop' a competitor who has already submitted a full article" or by some who use the newspapers for the first announcement of their findings or ideas.

Changes in the social structure of science appear to have counteracting effects on this form of competition, some serving to intensify it, others to dampen it. The exponential increase in the number of scientists has been accompanied by more and more specialization in research, quite along the lines of both Spencerian and Durkheimian theories of role differentiation. Although the process of differentiation has reduced the numbers of those engaged in direct competition for discovery in a given narrow field, it has probably intensified rivalry by increasing awareness of the work going forward elsewhere on the same problem. (The young Watson's candid account of his abiding sense that the great Pauling might get there first exemplifies this to the full.)

Various differences in the intellectual and social structure of scientific specialties probably affect the extent and intensity of competition for discovery within them. The various fields can be thought of as differing in their "population density" of scientists. This refers not to the obvious differences in the absolute numbers of scientists at work in this or that discipline or specialty. Population density refers rather to the numbers at work in relation to the significant problematics of the field, so that some fields are more "crowded" than others in the sense that many workers are focusing on the same problems.

In such specialties, competition tends to be particularly intense and the tensions generated by the race for priority greater. And, as always with intense direct competition, this may elicit competitive behaviors that evade —if they do not actually violate—the norms of science. Thus, we find editorials in the *Physical Review Letters* like one asserting that "the Letters on experimental high-energy physics are getting out of hand. The competition in this branch of physics is so fierce that speedy publication is requested even for unimportant contributions and unfinished work."

In the workshop vernacular of many scientists, these highly competitive fields are "hot fields" and deal with "hot subjects." The decisive characteristic of a hot field seems to be its high rate of significant discoveries (and, possibly, a lower ratio of routine to highly consequential ideas and findings). Thus, as an interested observer, Alvin Weinberg could write of one such hot field that "hardly a month goes by without a stunning success in molecular biology being reported in the *Proceedings* of the National Academy of Sciences."

Hot fields are not only more active than "cold" ones, but their results are taken to have implications that reach well beyond the borders of the specialty. They tend, at least for a time, to attract larger proportions of talented scientists who have an eye for the jugular, concerned to work on highly consequential problems rather than ones of less import. Hot fields also have a high rate of immigration and a low rate of emigration, again until they show signs of cooling off.

Levels of interaction between workers in the hot field, particularly at the leading edge, are unusually high. Access to informal channels of communication may intensify rivalry but reduce anxiety, since workers in the field need not speculate about how rivals are faring or what new developments are under way.

Kinds and degrees of competition differ not only among specialties but probably, also, among the various prestige strata of scientists within all fields.

Races for priority in finding solutions to deep and consequential problems amongst the frontiersmen of science tend to differ from the kind of competition occurring in the much larger middle and journeyman ranks. Working at the leading edge of a field, even one which happens not to be advancing at a rapid rate, usually means having readier access to what is going on in it. It thus enables leading scientists to know their chief competitors as rivals.

Among the middle and lower ranks, this shades off toward more impersonal competition. These scientists often do not know who else is engaged in similar work, and this lack of information generates its own brand of pressures and anxieties. Competition is less often experienced as a sportive, though often intense, personal rivalry; it tends to become a diffuse pressure to publish quickly in order not to be preempted by unknown others.

This is not to suggest that the tensions of competition are uniformly greater at the frontiers of a science than in its interior regions. The concern with being preempted is relative and not necessarily correlated to the intellectual significance of the work in hand. So, too, anxiety is relative to levels of aspiration and attainment, and can be just as intense, or even more so, for seemingly small stakes as large ones.

The rapid growth of scientific research on the large scale has affected the mode of competition in yet another way. Zuckerman notes that "as the social organization of scientific work becomes more complex—more often collaborative and sometimes intricately organized—the visibility of *individual* role-performance is reduced." This creates its own variety of tensions in science.

The institution of science has long worked in such a way as to reward scientists by having knowledgeable peers grant them recognition for their distinctive contributions. Scientists have correspondingly developed a passion for eponymity rather than anonymity.

But a side-effect of large-scale research, with its often sizable teams of workers, is a kind of anonymity for its members. Thus, the editor of *Physical Review Letters* reports "the difficulty of assigning credit to individual scientists for their contributions" and refers, as a limiting case in point, to "a Letter with, as byline, the names of three institutes; the participating physicists are not mentioned, not even in a footnote. . . . In this same issue . . . we publish another Letter on the identical subject but this Letter gives the names of seventeen authors from two institutions. . . . From these and from previous multiple-author papers it becomes clear that in such cases the role of the individual researcher is almost impossible to evaluate."

The growth of team work not only makes problematic the recognition of individual contributions *by others*; it also makes problematic the evaluation of contributions *by themselves*. To this extent, the changing organization of research may make for estrangement of scientific workers from the scientific inquiry in which they have taken part, after the fashion observed for complex division of labor by Adam Smith, Hegel, and Marx.

The problem of establishing a public identity in science becomes further complicated by exponential growth in the volume of publication. For a published work to become a genuine contribution to science, it must, of course, be visible enough to be utilized by others. Contributions of the first class may be just as visible today as in times of less voluminous publication, since the system of having papers submitted to journals judged and screened by referees, with all its imperfections, generally operates to identify them. But less consequential though useful contributions may now be more often lost to view, and who is to assess the effects of this on the advancement of science?

Under these circumstances, the concern with establishing priority of conception may become deepened for large numbers of research scientists. The threatened absence of a public identity in science may also heighten the competition to publish in journals of high prestige where visibility is greater than in the less-regarded and less widely read journals.

The public response to the Watson personal memoir seldom considered that scientists vary greatly in their attitudes toward competition. Some revel in it; others shun it. For some, it is stimulus; for others, annoyance or threat. Some enjoy the tension of the race; others prefer the tranquility of virtually no competition at all. Freud, for example, wrote nostalgically about the early days of psychoanalysis when, thoroughly neglected, he could enjoy a "splendid isolation" in which "there was nothing to hustle me . . . My publications, which I was able to place with a little trouble, could always lag far behind my knowledge and could be postponed as long as I pleased, since there was no doubtful 'priority' to be defended."

In the same vein, Jacques Hadamard reported that he was primarily attracted by problems in mathematics that had been largely overlooked. And he noted that "after having started a certain set of questions and seeing that several other authors had begun to follow that same line, I . . . [would] drop it and investigate something else."

So, too, Norbert Wiener, although he described himself as competitive, nevertheless maintained that he "did not like to watch the literature day by day in order to be sure that neither Banach nor one of his Polish followers had published some important result before me."

And, as a final instance, Max Planck described the initial lack of interest among his colleagues in his work as an "outright boon," observing that "as the significance of the concept of entropy had not yet come to be fully appreciated, nobody paid any attention to the method adopted by me, and I could work out my calculations completely at leisure, with absolute thoroughness, without fear of interference or competition."

It may be these occasional domains of temporarily unpopular and unpopulated fields of inquiry that have given rise to the nostalgic and overly generalized impression that competition in science is altogether peculiar to our time.

In any case, there is evidence on every side that some unknown proportions of contemporary scientists are actively engaged in trying to get there first. The fact is a commonplace. But does the fact warrant the inference, drawn in the emerging mythology, that intense competition for discovery is in a significant sense distinctive of the new era of science, with its enlarged population of scientists, its grants, prizes, and professional rewards?

I think not. This component of the mythology is the result of parochial perception. It emerges from the simple expedient of not looking at what there is to see throughout the centuries of modern science. It is a mythology achieved by emasculating the history of that science.

For the plain fact is, of course, that the race for priority has been frequent throughout the entire era of modern science. Moving back only a generation or so, we observe the good-natured race between Hahn and Boltwood, for example, to discover the "parent of radium" which Boltwood was able to find first, just as, when Hahn discovered mesothorium, Boltwood acknowledged his having been outdistanced, saying only, "I was almost there myself." There is Ramsay telegraphing Berthelot in Paris "at once" about his isolation of helium; writing Rayleigh to the same effect; and sending a note to the Royal Society to establish priority, just as he and Travers were to announce having nosed out Dewar in the discovery of neon. There is the forthright account by Norbert Wiener of the race between Bouligand and himself in potential theory, making Wiener "aware that he must hurry," but having it end in a "dead heat," since Bouligand had submitted his "results to the [French] Academy in a sealed envelope," just a day before Wiener had gotten off a short note for publication in the *Comptes Rendus*.

At that time, with the epochal voyage to the moon just concluded, we can scarcely forget the race run by the technologist Robert Goddard, the American father of the rocket, to achieve "primacy in outer space," when, after 1923, he was spurred on by the "journalistic claims to priority then made by the German partisans of Hermann Oberth . . . to redouble his efforts" and to launch his first liquid-propellant rocket only three years later.

In this respect the behavior of scientists does not much vary, transcending differences of time and national culture. Peter Kapitza puts it all in the of-course mood as he describes the behavior of Lomonosov, the father of Russian science, saying: "No less importance was attached to priority in scientific work at that time than now." Of this, Lomonosov and his colleagues provide dramatic evidence. When the physicist Richman was killed by lightning in 1763, the Russian Academy of Sciences canceled its general meeting, only to have Lomonosov ask that he nevertheless be given the opportunity to present his paper on electricity, "lest," in his words, "it lose novelty." The president of the Academy saw the point and arranged for a special meeting in order, as he explained, "that gospodin Lomonosov should not be late with his own new productions among scientific people in Europa, and his paper thereby be lost in electrical experiments made meanwhile."

The fact is that almost all of those firmly placed in the pantheon of science—Newton, Descartes, Leibniz, Pascal or Huyghens, Lister, Faraday, Laplace or Davy—were caught up in passionate efforts to achieve priority and to have it publicly registered. Consider only a highly condensed account of how things stood with Newton. Now, I do not undertake to compare Newton and Watson in terms of their nature-given talents or their society-nurtured accomplishments. Such comparison would be not merely odious but downright foolish. But when we are told that the aggressive, prize-seeking, competitive and pathbreaking behavior of Watson is something new unleashed in the mid-twentieth-century world of science, there is some point in examining the apposite behavior of the seventeenth-century giant of science. One incidental similarity of bare chronology is trivial enough to require no more than passing mention. They were both in their golden years, decidedly young men. Just as Jim Watson took up the problem he made his own in his twenty-third year, so we will remember, from Newton's own account, the *annus mirabilis* when at twenty-three or twenty-four he discovered the binomial theorem, started work toward invention of the calculus, took his first steps toward establishing the law of universal gravitation, and began his experiments on optics.

Long after he had made these incomparable contributions to mathematics and physical science, Newton was still busily engaged in ensuring the luster and fame owing him. He was not merely concerned with establishing his priority but was periodically obsessed by it. He developed a corps of young mathematicians and astronomers, such as Roger Cotes, David Gregory, William Whiston, John Keill and, above all, Edmond Halley, "for the energetic building of his fame" (as the historian Frank Manuel has put it in his recent *Portrait of Isaac Newton*). Newton's voluminous manuscripts contain at least twelve versions of a defense of his priority, as against Leibniz, in the invention of the calculus. Toward the end, Newton, then president of the Royal Society, appointed a committee to adjudicate the rival claims of Leibniz and himself, packed the committee with his adherents, directed its every activity, anonymously wrote the preface for the second published report on the controversy—the draft is in his handwriting—and included in that preface a disarming reference to the legal adage that "no one is a proper witness for himself and [that] he would be an iniquitous Judge, and would crush underfoot the laws of all the people, who would admit anyone as a witness in his own cause." We can gauge the pressures for establishing his unique priority that must have operated for Newton to adopt such means for defense of his claims. As I shall presently suggest, this was not so much because Newton was weak as because the newly institutionalized value set upon originality in science was so great that he found himself driven to these lengths.

By comparison, Watson's passing account of a priority skirmish within the Cavendish itself can only be described as tame and evenhanded, almost magnanimous. That conflict largely testified to the ambiguous origins of ideas generated in the course of interaction between colleagues, touched, perhaps, with a bit of cryptomnesia.

For those who believe that the Watson memoir expresses a new and extreme drive for getting there first in science, the antidote will be found in reading the *Philosophical Transactions* of the Royal Society for January and February 1715, devoted almost entirely to the angry quest for priority of Newton over Leibniz. And those who consider that Watson's account converts science into an arena for spectator sport, new or peculiar to our time, have something to learn from the observation by Frank Manuel that

> two of the greatest geniuses of the European world, not only of their own time, but of its whole long history, had been privately belaboring each other with injurious epithets and encouraging their partisans to publish scurrilous innuendoes in learned journals. In the age of reason they behaved like gladiators in a Roman circus. Here were two old bachelors, Leibniz not far from death, Newton with a decade more of life, each fighting for exclusive possession of his brainchild, the right to call the invention of the calculus his own and no one else's.

It is also symbolically fitting that the man who arranged for the recognition in perpetuity of major scientific accomplishments, which are often cases of acknowledged multiple discovery or of barely established priority, should himself have been engaged in a struggle over priority of technological invention. For it happens that the Maecenas who established these prizes—Alfred Nobel himself—was deeply involved in a battle with Frederick Abel and James Dewar over the invention of smokeless nitroglycerin gunpowder. The documentation of this particular scrap "fills several yards of shelves in the Nobel Foundation's archives." And it was small comfort to the agitated Nobel to have the Lord Justice, compelled on technical grounds not to find for Nobel, borrowing and adapting the aphorism made famous by Newton when he declared that "it is obvious that a dwarf who has been allowed to climb up on the back of a giant [the giant being Nobel, of course] can see farther than the giant himself." Nobel's frustrations and resentment over having been deprived of his intellectual property were only slightly siphoned off in his play, *The Patent Bacillus*, which lampooned the British court system.

This sampling of historical evidence is perhaps enough to put into question the belief that science today is competitive to a degree unknown before. If there has been a change in this aspect of the ethos of science, it seems to be of quite another kind. Scientists have apparently become more fully aware that, with growing numbers at work in each special field, any discovery is apt to be made by others as well as themselves, and so are less often apt than before to assume that parallel discoveries must be borrowed ones. Elinor Barber and I have found a secular decline during the last three centuries in the frequency with which multiples are an occasion for intense priority conflicts. Perhaps the culture of science today is not so pathogenic as it once was.

The absence of historical perspective marks another component of the new mythology of science. This one holds that quick, if not premature, publication to ensure priority is peculiar to our new breed of scientists, as witness the manuscript that went off to the editors of *Nature* on that fateful April 2nd of 1953. Again, it will do no harm to examine this opinion from a sociological and historical perspective. Today, as yesterday, scientists are caught up in one of the many ambivalent precepts contained in the institution of science. This one requires that the scientist must be ready to make his newfound knowledge available to his peers as soon as possible, *but* he must avoid an undue tendency to rush into print. (Compare Faraday's motto: "Work, Finish, Publish" with Ehrlich's *"Viel arbeiten, wenig publizieren!"*) To see this in fitting historical context, we must remember that the first scientific journals confronted not an excess but a deficiency of manuscripts meriting publication. The problem did not arise merely from the small number of men at work in science. There was the further restraint that the value set upon the open disclosure of one's scientific work was far from universally accepted.

From its very beginning,[1] the journal of science introduced the institutional device of quick publication to motivate men of science to replace the value set upon secrecy with the value placed upon the open disclosure of the knowledge they had created (a value that, in our own time, has often acquired, through the displacement of goals, a spurious emphasis on publication for its own sake, almost irrespective of the merit of what is published). The concern with getting into print fast is scarcely confined to contemporary science.

Watson fluttered the dovecotes of academia, to say nothing of the wider reading public, by telling us of having joined with Crick in an enthusiastic toast "to the Pauling failure . . . Though the odds still appeared against us, Linus had not yet won his Nobel." Once again, it seems, Watson had violated the mores that govern contest behavior in science and the public disclosure of that behavior. Yet, as we have seen in the preceding chapter, how mild and restrained is this episode by comparison with judgments on contemporaries set out in public by great scientists of the heroic past. Although historical facts to the contrary are abundantly available, there emerges a new mythology that treats competitive behavior of scientists as peculiar to our own competitive age.

This introduces an instructive paradox. These, indeed, are changing times in the ethos of science. But Watson's brash memoir does not testify

1. On this, see chapter 21 of this volume.

to a breakdown of once-prevailing norms that call for discreet and soft-spoken comment on scientific contemporaries. A memoir such as his would have been regarded as a benign model of disciplined restraint by the turbulent scientific community of the seventeenth century. That it should have created the stir it did testifies that, with the institutionalization of science, the austere mores governing the public demeanor of scientists and the public evaluation of contemporaries have become more exacting rather than less. As a result, Watson's little book, so restrained in substance and so mild in tone by comparison with the caustic and sometimes venomous language of, say, Galileo or Newton, violates the sentiments of the many oriented to these more exacting mores.

All of this brings us finally to the question touched off by the responses of many scientists and laymen to the Watson memoir. We are perhaps ready to see now that those responses relate to the long-standing denial that through the centuries scientists, and often the greatest among them, have been concerned with achieving and safeguarding their priority. The question is, of course: what leads to this uneasiness about acknowledging the drive for priority in science? Why the curious notion that a thirst for significant originality and for having that originality accredited by competent colleagues is depraved—somewhat like a thirst for, say, bourbon and 7-Up? Or, in Freud's self-deprecatory words, that it is an "unworthy and puerile" motive for doing science?

In one aspect, the embarrassed attitude of a Darwin or Freud toward his own interest in priority is based upon the implicit assumption that behavior is actuated by a single motive, which can then be appraised as good or bad, as noble or ignoble. It is assumed that the truly dedicated scientist must be moved only by the concern with advancing knowledge. As a result, deep interest in having his priority recognized is seen as marring his nobility of purpose as a man of science (although it might be remembered that "noble" once meant the widely-known).

There is, moreover, a germ of psychological truth in the suspicion enveloping the drive for recognition in science. Any extrinsic reward—fame, money, position—is morally ambiguous and potentially subversive of culturally esteemed values. For as rewards are meted out, they can displace the original motive: concern with recognition can displace concern with advancing knowledge. An excess of incentives can produce distracting conflict.

In another aspect, the ambivalence toward priority means that scientists reflect in themselves the ambivalence built into the social institution of science itself. That ambivalence also derives from the mistaken belief that concern with priority must express naked self-interest, that it is altogether self-serving. On the surface, the hunger for recognition appears as mere personal vanity, generated from within and craving satisfaction from without. But when we reach deeper into the institutional complex that gives added edge to that hunger, it turns out to be anything but personal, repeated as it is with slight variation by one scientist after another. Vanity, so-called, is then seen as the outer face of the inner need for assurance that one's work really matters, that one has measured up to the hard standards maintained by at least some members of the community of scientists. Sometimes, of course, the desire for recognition is stepped up until it gets out of hand. It becomes a driving lust for acclaim; megalomania replaces the comfort of reassurance. But the extreme case need not be taken for the modal one. In providing apt recognition for accomplishment, the institution of science serves several functions, both for scientists and for maintenance of the institution itself.

The community of science thus provides for the social validation of scientific work. In this respect, it amplifies that famous opening line of Aristotle's *Metaphysics*: "All men by nature desire to know." Perhaps, but men of science by culture desire to know that what they know is really so. The organization of science operates as a system of institutionalized vigilance, involving competitive cooperation. It affords both commitment and reward for finding where others have erred or have stopped before tracking down the implications of their results or have passed over in their work what is there to be seen by the fresh eye of another. In such a system, scientists are at the ready to pick apart and appraise each new claim to knowledge. This unending exchange of critical judgment, of praise and punishment, is developed in science to a degree that makes the monitoring of children's behavior by their parents seem little more than child's play. Only after the originality and consequence of his work have been attested by significant others can the scientist feel reasonably confident about it. Deeply felt praise for work well done, moreover, exalts donor and recipient alike; it joins them both in symbolizing the common enterprise. That, in part, expresses the character of competitive cooperation in science.

The function of reassurance by recognition has a dependable basis in the social aspects of knowledge. Few scientists have great certainty about the worth of their work. Even that psychological stalwart, T. H. Huxley, seemingly the acme of self-confidence, tells in his diary what it meant to him to be elected to the Royal Society at the age of twenty-six, by far the youngest in his cohort. It provided him, above all, with much needed reassurance that he was on the right track; in his own language, "acknowledgement of the value of what" he had done. And since, like the rest of us, Huxley was occasionally inclined to doubt his own capacities and to think himself a fool, he concluded that "the only use of honours is as an antidote to such fits of 'the blue devils.' " When he later learned that he was within an ace of receiving the Royal Medal of the Society—he did get it the next year—he went on to say:

What I care for is the justification which the being marked in this position gives to the course I have taken. Obstinate and self-willed as I am, . . . there are times when grave doubts overshadow my mind, and then such testimony as this restores my self-confidence.

The drive for priority is in part an effort to reassure oneself of a capacity for original thought. Thus, rather than being mutually exclusive, as the new mythology of science would have it, joy in discovery and the quest for recognition by scientific peers are stamped out of the same psychological coin. In their conjoint ways, they both express a basic commitment to the value of advancing knowledge.

But authentic reassurance can be provided only by the scientists whose judgment one in turn respects. As we sociologists like to put it, we each have our reference groups and individuals, whose opinions of our performance matter. Our peers and superiors in the hierarchy of accomplishment become the significant judges for us. Darwin writing Huxley about the *Origin of Species* "with awful misgivings" thought that "perhaps I had deluded myself like so many have done, and I then fixed in my mind three judges, on whose decision I determined mentally to abide. The judges were Lyell, Hooker, and yourself." In this, Darwin was replicating the behavior of many another scientist, both before and after him. The astronomer John Flamsteed, before his vendetta with Newton, wrote that "I study not for present applause. Mr. Newton's approbation is more to me than the cry of all the ignorant in the world." In almost the same language, Schrödinger writes Einstein that "your approval and Planck's mean more to me than that of half the world." And a Leo Szilard or a Max Delbrück, widely known as exceedingly tough-minded and demanding judges who, all uncompromising, will not relax their standards of judgment even to provide momentary comfort to their associates, are reference figures whose plaudits for work accomplished have a multiplier effect, influencing in turn the judgments of many another scientist.

Other strategic facts show the inadequacy of treating an interest in recognition of scientific work as merely an expression of egotism. Very often, the discoverers themselves take no part in arguing their claims to the priority or significance of their contributions. Instead, their friends or other more detached scientists see the assignment of priority as a moral issue not to be scanted. For them the assigning of all credit due is a functional requirement for the institution of science itself. After all, to protect the priority of another is only to act in accord with the norm, which has been gathering force since the time of Francis Bacon, that requires scientists to acknowledge their indebtedness to the antecedent work of others. As Kapitza says of his master, "If anybody in publishing his work forgot to mention that the given idea was not his own, Rutherford immediately objected. He saw to it in every possible way that . . . true priority be maintained." Or, to take perhaps the most momentous instance in our day, there is Niels Bohr, agitated by the thought that Meitner and Frisch, and for that matter, Hahn and Strassmann too, might have their priority in the splitting of the atom lost to view in the avalanche of publicity given the Columbia University experiments, going to immense pains to set the record straight (just as he was later to devote himself to the task of getting governments, and physicists too, to consider the human consequences of nuclear weapons).

Chargaff is correct, I believe, in suggesting that the Watson memoir "may contribute to the much-needed demythologizing of modern science." But as I have tried to suggest, to put the accent on "*modern* science" is only to displace the old myth with a new variant. In noting this, I am scarcely alone. Some practicing scientists, both before and after *The Double Helix*, have put aside the myth that competition for originality in science is alien to joy in discovery and that the drive for recognition should occasion self-contempt. Hans Selye asks his peers: "Why is everybody so anxious to deny that he works for recognition? . . . All the scientists I know sufficiently well to judge (and I include myself in this group) are extremely anxious to have their work recognized and approved by others. Is it not below the dignity of an objective scientific mind to permit such a distortion of his true motives? Besides, what is there to be ashamed of?" And, as though he were responding to this rhetorical question, P. B. Medawar goes on to argue: "In my opinion the idea that a scientist ought to be indifferent to matters of priority is simply humbug. Scientists are entitled to be proud of their accomplishments, and what accomplishments can they call 'theirs' except the things they have done or thought of first? People who criticize scientists for wanting to enjoy the satisfaction of intellectual ownership are confusing possessiveness with pride of possession." Himself an inveterate observer of human behavior rather than only of economic numbers, Paul Samuelson also distinguishes cleanly the gold of scientific fame from the brass of popular celebrity. This is how he concludes his presidential address to fellow economists:

> Not for us is the limelight and the applause. But that doesn't mean the game is not worth the candle or that we do not in the end win the game. In the long run, the economic scholar works for the only coin worth having—our own applause.

At a time when scientists and humanists search for common understandings, it is only fitting that the practicing scientist and the practicing poet should both have perceived the deeper implications of the thrust for significant and acknowledged originality in living science. With the poet's inward eye, Robert Frost puts it so:

Would he mind had I
Had him beaten to it?
Could he tell me why
Be original?
Why was it so very,
Very necessary
To be first of all?
How about the lie
Someone else was first?
He saw I was daffing.
He took this from me.
Still it was no laughing
Matter I could see.
He made no reply.

Of all crimes the worst
Is the theft of glory,
Even more accursed
Than to rob the grave[2]

The history of science declares what the poet sings: a care for truth signifies a care for the truth-seeker.

2. From one version of the poem "Kitty Hawk" by Robert Frost, which first appeared in the *Atlantic*. Copyright © 1957 by Estate of Robert Frost. Reprinted by permission of Holt, Rinehart and Winston, Inc.

16 Singletons and Multiples in Science

1961

Bacon's Problematics of Scientific Discovery

Having taken all knowledge to be his province, Francis Bacon made room even for what was to become sociology. His luminous writings include a charter for the social sciences, a proposed division between their several types, a precept guiding the inclusion of problems that might otherwise be lost to the view of social scientists and, finally, an early, incomplete yet instructive formulation of the hypothesis to which the greater part of this paper is devoted.

With the seeming artlessness of the true artist, Bacon set down in his *Novum Organum* what amounts to a charter for the human sciences:

> It may also be asked in the way of doubt rather than objection, whether I speak of natural philosophy only, or whether I mean that the other sciences, logic, ethics, and politics, should also be carried on by this method. Now I certainly mean what I have said to be understood of them all; and as the common logic, which governs by the syllogism, extends not only to natural but to all sciences; so does mine also, which proceeds by induction, embrace everything. For I form a history and tables of discovery for anger, fear, shame, and the like; for matters political; and again for the mental operations of memory, composition and division [this is probably Aristotle's "affirmation and negation," as Fowler makes plain], judgment and the rest; not less than for heat and cold, or light, or vegetation and the like.[1]

Read 23 January 1961 in the Conference on the Influence of Science upon Modern Culture, Commemorating the 400th Anniversary of the Birth of Francis Bacon, sponsored jointly by the American Philosophical Society and the University of Pennsylvania. Originally published as "Singletons and Multiples in Scientific Discovery," in *Proceedings* of the American Philosophical Society 105, no. 5 (October 1961): 470–86; reprinted here by permission of the American Philosophical Society.

1. *Novum Organum*, book 1, aphorism 127.

Not only is Bacon prepared to encompass the human sciences in his plan, but he is careful to distinguish among them. Almost as though he were among us today exploring the differences and connections between the psychological and social sciences, he describes what he calls "Human Philosophy or Humanity" as having "two parts: the one considereth man segregate or distributively; the other congregate, or in society."[2]

So much for Bacon's effort to legitimate social science at a time when its first glimmerings were evident to only a few and before its sporadic development during that century of genius. No economist, Bacon could in 1615 originate the term, if not the concept, of "balance of trade,"[3] in the same year in which Antoyne de Montchrétien christened "political economy" (in his *Traicté de l'Economie Politique*). No psychologist, he could by anticipation appreciate the efforts, in mid-century, of Hobbes, Descartes, and Spinoza to contemplate the human passions introspectively, attending to problems of perception, sensation, imagination, and the like. No great admirer of mathematics but cognizant of the value of quantification, he could write as he did generations before the extraordinary London haberdasher John Graunt, Sir William Petty, and Gregory King could among them fashion the new political arithmetic and so initiate the serious study of demography, urban sociology, and epidemiology.

Bacon did not, of course, foresee all this. Little in his time would allow him to describe social science, in the fashion Galileo described mechanics, as "the very new science dealing with a very ancient subject." But his announced philosophy of investigation allowed for such a conception. In taking note of this, we need not try to fix a particular date on which the birth of the social sciences was authoritatively registered. After all, Bacon had referred, with approving comment, to beginnings of social science before his time, reminding his contemporaries, for example, that "we are much beholden to Machiavel and others, that write what men do, and not what they ought to do," then adding, in that stately and incomparable

2. *Advancement of Learning*, in *The Works of Francis Bacon*, [hereafter cited as *Works*] collected and edited by James Spedding, Robert L. Ellis, and Douglas D. Heath (Boston, 1863), 6:236–37. The pressures of time on this occasion being what they unavoidably are, I resist the temptation to remind ourselves of what Bacon goes on to say about psychosomatics (if the anachronism is allowed), when he follows precedent in writing of "the knowledge concerning the sympathies and concordances between the mind and body, which being mixed cannot be properly assigned to the sciences of either" (Ibid., pp. 154 ff).

3. In his *Letter of Advice to Sir George Villiers*, in *Works*. Bacon's usage was independent of earlier use of the phrase in Italy; see W. H. Price, "The Origin of the Phrase 'Balance of Trade'," *Quarterly Journal of Economics* 20 (November 1905): 157, and the typically informed long footnote on the concept-and-term in Joseph A. Schumpeter, *History of Economic Analysis*, edited from manuscript by Elizabeth Boody Schumpeter (New York: Oxford University Press, 1954), pp. 345–46.

Elizabethan prose from which peak we have achieved a steady decline, "For it is not possible to join serpentine wisdom with the columbine innocency, except when men know exactly all the conditions of the serpent; his baseness and going upon his belly, his volubility and lubricity, his envy and stinge, and the rest; that is, all forms and natures of evil: for without this, virtue lieth open and unfenced."[4] And so, in what follows, but without reference to serpentine evil or columbine good, I shall try to obey the precept of Bacon, and before him of Machiavelli, by examining some of "what men [of science] do, not what they ought to do."

Having legitimatized the social sciences, having divided them into distinct though connected disciplines, and having directed us to examine the actual and not to mistake it for the ideal, Bacon gives us counsel about the scope of inquiry, urging us to give up the "childish fastidiousness" that would have us examine only those things in nature and society that we find good or pleasant or otherwise attractive. You will recall this bit of advice, destined to be echoed or independently reaffirmed in the centuries since his day by many great men of science—by a Claude Bernard or a Pasteur, among the many:

> And for things that are mean or even filthy,—things which (as Pliny says) must be introduced with an apology—such things, no less than the most splendid and costly, must be admitted into natural history. Nor is natural history polluted thereby; for the sun enters the sewer no less than the palace, yet takes no pollution. And for myself, I am not raising a capitol or pyramid to the pride of man, but laying a foundation for a holy temple after the model of the world. That model therefore I follow. For whatever deserves to exist deserves also to be known, for knowledge is the image of existence; and things mean and splendid exist alike. Moreover as from certain putrid substances—musk, for instance, and civet—the sweetest odours are sometimes generated, so too from mean and sordid instances there sometimes emanates excellent light and information. But enough and more than enough of this; such fastidiousness being merely childish and effeminate.[5]

When we consider the particular sense in which scientific discoveries can be said to come about without being dependent upon the undoubted genius of the *particular* scientists who are properly credited with these discoveries, or when we consider, here in passing, what I have considered elsewhere at some length, the sociological import of the frequent clashes over priority of discovery that have marked the history of science—when I examine

4. *Advancement of Learning, Works,* 6: 327.

5. *Novum Organum,* bk. 1, aphorism 120. This same theme was later taken up and amplified by the thoroughgoing Baconian, Robert Boyle, in the first essay of part 1 of *Some Considerations Touching the Usefulness of Experimental Naturall Philosophy, Propos'd in Familiar Discourses to a Friend, by Way of Invitation to the Study of It* (Oxford, 1663).

these and related matters, far from belittling the scientists of genius who have done so much to shape the development of science, I shall only be trying to fathom their distinctive and complex role in that development. Perhaps the precept of Bacon will help us find in these matters seemingly incidental to the work of scientists, "excellent light and information."

After having provided us with an attitude proper to a commemorative occasion such as this one by urging us to take up and develop the force of what the memorialized man has said rather than merely to repeat his words; after having given us a charter for the human sciences in general and having set out a useful though in the end temporary division between the primarily psychological sciences that center on "man segregate" and the primarily social sciences that center on "man congregate"; after having urged us to examine what men do and not merely what they ought to do; and after having warned us, at our peril, not to exclude the apparently mean or trivial from the scope of investigation—after he has done all this, as though it were still not enough, Francis Bacon makes my lot here an easy as well as a pleasant one by practically providing a composite text dealing with the particular subject I wish to examine: the import of a methodical investigation of singleton and multiple discoveries in science for our understanding of how science develops.

Instructed by the ideas that have been developed after Bacon's time, we can piece together from his fragmentary but instructive observations, the prime ingredients of a theory of the social processes making for discovery and invention. I say "piece together" because these ingredients are not to be found in any one place in Bacon's writings, neatly and coherently tied up in a single bundle. In part, my reconstruction is deliberate anachronism. But in part also, it is not so much reading into Bacon as reading him entire to gain a sense of how he conceived scientific discoveries to come about.

To begin with, Bacon wholly rejected the notion that in the new science, discoveries would typically appear at random, dropping down from heaven through the agency of star-touched genius. Instead, he declares that once the right path is followed, discoveries in limitless number will arise from the growing stock of knowledge: it is a process of once fitful and now steady increments in knowledge. This notion of what we should today describe as the accumulative cultural base on which science builds became one of the many Baconian ideas taken up in abundance by his sometimes overly enthusiastic disciples at midcentury. Consider only one of the more devoted of these, John Webster, who in 1654 could pleasurably refer to "our learned Country-man the Lord Bacon" as having made it clear that "every age and generation, proceeding in the same way, and upon the same principles, may dayly go on with the work, to the building up of a well-grounded and lasting Fabrick, which indeed is the only true way for the instauration and advancement of learning and knowledge."[6]

Second, Bacon holds that the individual man of science pursuing his daily labors entirely alone would at best produce small change. As he announces in the *Novum Organum*, "the path of science is not, like that of philosophy, such that only one man can tread it at a time." Consider, he says, "what may be expected from men abounding in leisure"—it would be too much to ask Bacon to foresee the excessively busy life of so many present-day scientists—"and working in association with one another, generation after generation. . . . Men will begin to understand their own strength only when, instead of many of them doing the same things, one shall take charge of one thing and one of another."[7] This theme, too, was repeatedly picked up in the seventeenth century, not least by the first historian of the Royal Society, "fat Tom Sprat," who, happily echoing Bacon, could proclaim that "single labours" in science are not enough to advance science significantly; rather, that it requires the "joynt labours of many," even to the extreme of "*joyning* them into *Committees* (if we may use that word in a Philosophical sence, and so in some measure purge it from the ill sound, which it formerly had)."[8] And still in the Baconian vein, Bishop Sprat notes that social interaction among men of science facilitates originality of conception; or as he puts it less austerely, "In Assemblies, the *Wits* of most men are *sharper*, their *Apprehensions readier*, their *Thoughts fuller*, than in their Closets."[9]

Having formulated two prerequisites for the advancement of science—the accumulating cultural base and the concerted efforts of men of science sharpening their ideas through social interaction—Bacon returns, time and again, to a third component in the social process of discovery. He tells how his proposed methods of scientific inquiry reduce the significance of the undeniably different capacities of men. You will recall the ringing passage in the *Novum Organum* to this effect:

> . . . the course I propose for discovery of sciences is such as leaves but little to the acuteness and strength of wits, but places all wits and understandings

6. John Webster, *Academiarum Examen; or, the Examination of Academies . . . Offered to the Judgment of All Those that Love the Proficiencie of Arts and Science, and the Advancement of Learning* (London, 1654), p. 105. Appropriately enough, the book is dedicated to Bacon.

7. *Novum Organum*, bk. 1, aphorism 113. I take here the instructed translation by Benjamin Farrington, rather than that by Spedding, Ellis, and Heath, or even that by Fowler. See Farrington, *Francis Bacon* (New York: Henry Schuman, 1949), p. 112.

8. Thomas Sprat, *The History of the Royal-Society of London, for the Improving of Natural Knowledge*, [hereafter cited as *History*] (London, 1667), p. 85. The same point of science advancing through the "joynt force of many men" or the "united *Labors* of many" recurs throughout the *History*; e.g., pp. 39, 91, 102, 341.

9. Ibid., p. 98.

nearly on a level. For as in the drawing of a straight line or a perfect circle, much depends on the steadiness and practice of the hand, if it be done by aim of hand only, but if with the aid or rule or compass, little or nothing; so it is exactly with my plan.[10]

Read out of its immediate context, and out of the numerous other contexts in which Bacon expresses the same thought, this can be easily if not perversely misconstrued. It can be taken to claim that all men of science are on the same plane of capacity. It has often been so mistaken. [As was the case, for example in one of my own observations back in 1938; see chapter 11 of this volume, p. 235.] What is more, it has often been held to affirm that all scientists are being *reduced* to the same level by the methods of science rather than being *raised* to a lofty level of competence.

But as we know from the rest of Bacon's writings, both before and after the *Novum Organum*, he meant nothing of the kind. Repeatedly, he recognizes that men have various capacities, and, in his scheme of things scientific, he provides a distinctive place for each kind. That dreamt-of research institute, Solomon's House, allows for all grades of ability and varieties of skills in a complex division of scientific labor. The institute includes "Merchants of Light," who keep up with the work going on in foreign countries (in the language of today, reporters of scientific intelligence); "Mystery-men" who gather up the earlier experiments in science and the mechanical arts (in today's terms, the men who arrange for retrieval of scientific information); "Pioneers" or "Miners" who "try new experiments, such as themselves think good" (the skilled and creative experimentalists); "Compilers," or the lesser theorists, who examine the accumulated materials to draw inferences from them; "Dowry-men" or "Benefactors" who seek to apply this knowledge (men engaged in what we now call "research and development"); the "Lamps," who "after divers meetings and consults of" the whole number, undertake to "direct new experiments, of a higher light, more penetrating into nature than the former" (the experimentalist directing a series of cumulative experiments); "Inoculators," the technicians who "execute the experiments so directed, and report them"; and finally, his "Interpreters of Nature," who "raise the former discoveries by experiments into greater observations, axioms, and aphorisms"—the pure

10. *Novum Organum*, bk. 1, aphorism 61; also 122. The strong-minded Macaulay made this the butt of attack in his—to some famous, to others notorious—essay on Bacon; and the even-tempered Baconian scholar, Fowler, was moved to say, "Bacon's promise never has been and never can be fulfilled." As, of course, it cannot, if it is read out of the context of the rest of Bacon's writings, so that he can be charged with gross exaggeration. But need we forget this context, better known to Fowler than to any of the rest of us? Farrington, above all others known to me, has recognized that only a misplaced and narrowly focused literalism can lead one to assume that Bacon left no place for the great variability in the talents of men engaged in scientific inquiry. See Farrington, *Francis Bacon*, pp. 116–18.

theorist. Solomon's House makes room also for the advanced students, the "novices and apprentices" in order that the "succession of the former employed men not fail."[11]

Evidently, then, Bacon does not put all men of science on a single plane, nor does he foolishly regard them as altogether interchangeable. Rather, he emphasizes his belief that methodical procedures make for greater reliability in the work of science. Once a scientific problem has been defined, profound individual differences among scientists will affect the likelihood of reaching a solution, but the scale of differences in outcome is reduced by the established procedures of scientific work. Only in this sense and to this degree, does the new science, in the Baconian image, place "all wits and understanding nearly on a level."

To the three components of his implicit social theory of discovery—the incremental accumulation of knowledge, the sustained social interaction between men of science and the methodical use of procedures of inquiry—Bacon adds a fourth and even more famous one. All innovations, social or scientific, "are the births of time."[12] "Time is the greatest innovator." He employs the same instructive metaphor to describe both his own work and that of others, as when he accounts his own part in advancing knowledge "a birth of time rather than of wit."[13] Once the needed antecedent conditions obtain, discoveries are offshoots of their time, rather than turning up altogether at random.

To say that discoveries occur when their time has come is to say that they occur only under identifiable requisite conditions. But, of course, these conditions do not always obtain. In the past, says Bacon, inventions and discoveries have made their appearance sporadically, almost accidentally. This is so because there did not then exist the conditions of cumulative knowledge, the association of men of science and the methodical, composite use of empirical and reasoned inquiry. With the new science, all this will change. There are secrets of nature

> . . . lying entirely out of the beat of the imagination, which have not yet been found out. They too no doubt will some time or other, in the course and revolution of many ages, come to light of themselves just as the others did; only by the method of which we are now treating, they can be speedily and suddenly and simultaneously presented and anticipated.[14]

With the increment in this passage, Bacon almost but not quite achieves a sociological conception of the development of science.

To round this out, he need only add the further component that if

11. See *Solomon's House,* in *Works.*
12. *Essays,* in *Works,* 12: 160.
13. *Novum Organum,* bk. 1, aphorism 122.
14. *Novum Organum,* bk. 1, aphorism 109.

discoveries are "a birth of time," they will be effected by more than one discoverer. Never saying this in so many words, Bacon nevertheless intimates it—and more than once. By paraphrasing his language, I anachronize his idea, yet without doing violence to it. What he all but says is that multiple independent discoveries do occur but not nearly so often as people suppose. The erroneous supposal is made both by those who mistakenly identify their own ideas as ancient ones and by others who claim to find in the actually new what is ostensibly old. This is how Bacon puts it:

> That of those that have entered into search, some having fallen upon some conceits [i.e., notions] which they after consider to be the same which they have found in former authors, *have suddenly taken a persuasion that a man shall but with much labour incur and light upon the same inventions which he might with ease receive from others*; and that it is but a vanity and self-pleasing of the wit to go about again, as one that would rather have a flower of his own gathering, than much better gathered to his hand. That the *same humour of sloth and diffidence suggesteth that a man shall but revive some ancient opinion, which was long ago propounded, examined, and rejected.* And that *it is easy to err in conceit* [the view] *that a man's observation or notion is the same with a former opinion,* both because new conceits [notions] must of necessity be uttered in old words, and because upon true and erroneous ground men may meet in consequence or conclusion, as several lines or circles that cut in some one point.[15]

The vice of what we may call "adumbrationism"—the denigrating of new ideas by pretending to find them old—must not be permitted to blind us to the fact that rediscovery does sometimes occur. It does not follow, however, that all newly emerging knowledge is nothing but rediscovery. Plato was mistaken in saying "that all knowledge is but remembrance."[16] In part the error comes from the recurrent practice, particularly in "intellectual matters," of first finding the new idea strange, and then finding it exceedingly familiar.[17] In another part the error comes from the selective perceptions of the reader. "For almost all scholars have this—when anything is presented to them, they will find in it that which they know, not learn from it that which they know not."[18] Yet apart from this common error of mistaking the new for the old in science, the fact remains that "men may meet in consequence or conclusion" despite their initial diver-

15. *Valerius Terminus of the Interpretation of Nature*, in *Works*, 6: 72–73. The emphases are mine.

16. *Essays; or Counsels Civil and Moral*, essay 58, "Of Vicissitude of Things," in *Works*, 12: 273; cf. *Advancement of Learning*, *Works*, 6: 88.

17. *Advancement of Learning*, in *Works*, 6: 130: "In intellectual matters, it is much more common; as may be seen in most of the propositions of Euclid, which till they be demonstrate, they seem strange to our assent; but being demonstrate, our mind accepteth of them by a kind of relation (as the lawyers speak) as if we had known them before."

18. *De Augmentis*, in *Works*, 9: 170.

gence of ideas. In effect, both adumbrationism and the full denial of rediscovery are faulty doctrines; the truth is, in this reconstructed judgment of Bacon, that rediscovery occurs but not as often as the adumbrationists suppose.

Now I am not saying, of course, that Bacon formulated a coherent sociological theory of the composite elements making for discovery in science. That would be adumbrationism with a vengeance. I recognize that I have pieced together his intimations of such a theory from observations scattered through the works he wrote over a span of two decades. But with the advantage of historical hindsight, and of the ideas that were formulated later, we can identify the *ingredients* of such a theory in Bacon. He himself did not see the connections between them. Or, if he saw them, he never recorded them in a form that has come down to us. What is of interest, rather, is that these ingredients should have appeared more than three centuries ago and that many men over a long period of time should have come upon them anew and that they should have begun to compose them into the beginnings of a sociological theory of scientific discovery.[19]

19. Bacon had much else to say that qualifies him as a harbinger of the sociology of science; I cannot deal with these matters here. But at least two sets of observations can be segregated here below to intimate the broad scope of his understanding. First, he notes the problem of the relations between the social structure and the character of knowledge: "Of the impediments which have been in the nature of society and the policies of state. That there is no composition of estate or society, nor order or quality of persons, which have not some point of contrariety towards true knowledge. That monarchies incline wits to profit and pleasure, and commonwealths to glory and vanity. That universities incline wits to sophistry and affectation, cloisters to fables and unprofitable subtilty, study at large to variety; and that it is hard to say, whether mixture of contemplations with an active life, or retiring wholly to contemplations, do disable and hinder the mind more" (*Valerius Terminus*, in *Works*, 6: 76). Thus we must acknowledge that he sees *the problem* of the relations between types of social structure and types of intellectual work, whatever we might think of his hypotheses. And second, he identifies all manner of social considerations that affect the ways in which men of science and learning ordinarily record what they have learned (with the intimation, perhaps, that this sorry variation will have to be sufficiently standardized if the institution of science is to advance knowledge, rather than to congeal it): ". . . as knowledges have hitherto been delivered, there is a kind of contract of error between the deliverer and the receiver; for he who delivers knowledge desires to deliver it in such form as may be best believed, and not as may be most conveniently examined; and he who receives knowledge desires present satisfaction, without waiting for due inquiry; and so rather not to doubt, than not to err; glory making the deliverer careful not to lay open his weakness, and sloth making the receiver unwilling to try his strength. But knowledge that is delivered to others as a thread to be spun on ought to be insinuated (if it were possible) in the same method wherein it was originally invented. And this indeed is possible in knowledge gained by induction; but in this same anticipated and premature knowledge (which is in use) a man cannot easily say how he came to the knowledge which he has obtained. Yet certainly it is possible for a man in a greater or less degree to revisit his own knowledge, and trace over again the footsteps both of his cognition and his consent; and by that means to transplant it into another mind just as it grew in his own" (*De Augmentis*, in *Works*, 9: 122–23; see also pp. 16–18; *Valerius Terminus*, in *Works*, 6: 70–71).

In all this, Bacon had taken hold of a salient truth: the course of scientific development cannot be understood as the work of man segregate. But he exaggerated when he went on to the claim, which remains extravagant even when construed as he evidently intended it, that the new method of science would "level men's wits and leave but little to individual excellence." In this gratuitous overstatement he is not alone. For in the centuries since Bacon, scores of observers have repeatedly stated the matter in much the same disjunctive terms: shall we regard the course of science and technology as a continuing process of cumulative growth, with discoveries tending to come in their due time, *or* as the work of men of genius who, with their ancillaries, bring about basic advances in science? In the ordinary way, these are put as alternatives: *either* the social theory of discovery *or* the "heroic" theory. What Bacon sensed, others glimpsed a little more fully, without questioning the assumed opposition of these theories of discovery. And so for more than three centuries, there has been an intermittent mock battle between the advocates of the heroic theory and the theory of the social determination of discovery in science. In this conflict, truth has often been the major casualty. For want of an alternative theory, we have been condemned to repeat the false disjunction between the heroic theory centered on men of genius and the sociological theory centered on the social determination of scientific discovery.

The Self-Exemplifying Hypothesis of Multiples

At the root of a sociological theory of the development of science is the strategic fact of the multiple and independent appearance of the same scientific discovery—what I shall, for convenience, hereafter describe as a multiple. Ever since 1922 American sociologists have properly associated the theory with William F. Ogburn and Dorothy S. Thomas, who did so much to establish it in sociological thought.[20] On the basis of their compilation of some 150 cases of independent discovery and invention, they concluded that the innovations became virtually inevitable as certain kinds of knowledge accumulated in the cultural heritage and as social developments directed the attention of investigators to particular problems.

Appropriately enough, this is an hypothesis confirmed by its own history. (Almost, as we shall see, it is a Shakespearean play within a play.) For this idea of the sociological significance of multiple independent discoveries and inventions has been periodically rediscovered over a span of centuries. Today I shall not reach back of the nineteenth century for

20. W. F. Ogburn and D. S. Thomas, "Are Inventions Inevitable?" *Political Science Quarterly* 37 (March 1922): 83–98; W. F. Ogburn, *Social Change* (New York: Heubsch, 1922), pp. 90–122. On the same point, see chapters 10 and 17 of this volume.

cases. Let us begin, then, with 1828, when Macaulay, in his essay on Dryden, observes that the independent invention of the calculus by Newton and Leibniz belongs to a larger class of instances in which the same invention or discovery had been made by scientists working apart from one another. For example, Macaulay tells us that

> the doctrine of rent, now universally received by political economists, was propounded, at almost the same moment, by two writers unconnected with each other. Preceding speculators had long been blundering round about it; and it could not possibly have been missed much longer by the most heedless inquirer.

And then he concludes, in truly Macaulayan prose and with the unmistakable Macaulayan flair:

> We are inclined to think that, with respect to every great addition which has been made to the stock of human knowledge, the case has been similar: that without Copernicus we should have been Copernicans—that without Columbus America would have been discovered—that without Locke we should have possessed a just theory of the origin of human ideas.[21]

This is not the time to examine in detail the many occasions on which the fact of multiples with its implications for a theory of scientific development has been noted; on the evidence, often independently noted and set down in print. Working scientists, historians and sociologists of science, biographers, inventors, lawyers, engineers, anthropologists, Marxists and anti-Marxists, Comteans and anti-Comteans have time and again, though with varying degrees of perceptiveness, called attention both to the fact of multiples and to some of its implications. But perhaps a partial listing will bring out the diversity of occasions on which the fact and associated hypothesis of independent multiples in science and technology were themselves independently set forth:

In 1828—as I have said, there was Macaulay, notably in his essay on Dryden;
1835—Auguste Comte, in his *Positive Philosophy*;
1846, 1847, and 1848—the mathematician and logician, Augustus de Morgan;
1855—Sir David Brewster, the physicist, editor of the *Edinburgh Encyclopedia*, and warmly appreciative though not always discriminating biographer of Newton, who was himself involved in several multiples in dioptrics with Malus and Fresnel;
1862–1864—when there was printed an entire cluster of observations upon multiples, growing out of the then-current controversy in England over the patent system, such that the *London Times* ran

21. *Miscellaneous Works of Lord Macaulay*, ed. Lady Trevelyan (New York: Harper, 1880), 1: 110–11.

repeated leaders on the subject, remarking the common notoriety of the fact "that the progress of mechanical discovery is constantly marked by the simultaneous revelation to many minds of the same method of overcoming some practical difficulty" (13th September 1865);

1864—Samuel Smiles, that immensely popular Victorian biographer and apostle of self-help, repeatedly touched upon the fact of multiples;

1869—François Arago, the astronomer, physicist, biographer, and permanent secretary of the Academy of Sciences, made much of multiples;

1869—Francis Galton who, in his *Hereditary Genius*, considered "it notorious that the same discovery is frequently made simultaneously and quite independently, by different persons" as attested by famous cases in point during the few years preceding, and who returned to the same subject in 1874, in his *English Men of Science*;

1885—by the now little-known American anthropologists, Babcock and Pierce;

1894—Friedrich Engels, in his letter to Heinz Starkenburg, wrote of his partner in ideas that "while Marx discovered the materialist conception of history, Thierry, Mignet, Guizot, and all the English historians up to 1850 are the proof that it was being striven for, and the discovery of the same conception by Morgan proves that the time was ripe for it and that indeed it *had* to be discovered";

1904—François Mentré, the French social philosopher and historian, whose basic paper, "La simultanéité des decouvertes," *Revue scientifique*, supplies a list of some 50 cases;

1905—Albert Venn Dicey, English jurist and political scientist in his magisterial *Lectures on the Relation between Law and Public Opinion*;

1906–1913—Pierre Duhem, the physico-chemist and one of the fathers of the modern history of science, who examines the fact and implications of multiples in every one of his major works;

1906—the distinguished German physiologist, Emil Du Bois-Reymond;

1913—the man who was to become the dean of American historians of science, George Sarton;

1917—the dean of American anthropologists, A. L. Kroeber;

1921—by Einstein; and then, as we near the formulation best known in the United States, in

1922—the fact and associated hypothesis of multiples as stated by the historian of science Abel Rey in France; by the then leading exponent of Marxist theory in Russia, Nicolai Bukharin; by the authoritative political scientist and essayist, Viscount Morley in England; and, of course, by Ogburn and Thomas in the United States.

The limits of time have required me to confine this partial list to the nineteenth century and the early twentieth. But this self-set rule must be breached at least once. For on this occasion, we can scarcely exclude the observations on the subject made by the chief founder of both the American Philosophical Society and the University of Pennsylvania. Of Franklin's several versions of the matter, I select one that bears his unmistakable imprint. Writing to the Abbé de la Roche, he remarks:

> I have often noted, in reading the works of M. Helvétius, that, though we were born and brought up in two countries so remote from each other, we have often hit upon the same thoughts; and it is a reflection very flattering to me that we have loved the same studies and, so far as we have known them, the same friends, and the same woman.[22]

Here, as elsewhere, Franklin takes the occurrence of multiples as a matter of course.

Just so do most of the others in the truncated list of multiple discoveries of the theory of multiple discoveries. That many, indeed most, of them came upon the idea independently is at least suggested by the form in which they present it, as something they have found worthy of note. Its independence is suggested also by the interest which each succeeding formulation of the idea excited among those readers who happened to comment on it in print, either in book reviews or articles. The fact is that the theory was most unevenly diffused among scholars and scientists. By the middle of the nineteenth century, it had become, for some, a commonplace and often deplored truth; for others, it represented an entirely new conception of how science advances through the uneven accumulation of knowledge and through immanently or socially induced foci of attention to particular problems by many scientists at about the same time.

Further evidence that the idea—which, in a sense, has been "in the air" for about three centuries—was being independently rediscovered is also inadvertently supplied by those critics who attacked it as thoroughly unsound or at least as ideologically suspicious. Down to the present day, some authors can bring themselves to describe the hypothesis as essentially Marxist and so, we are invited to suppose, as necessarily false. That Marx was a precocious boy of ten when Macaulay first set down his ideas on the subject and a high-spirited youth of eighteen or so when Comte asserted the same ideas—the same Comte destined to be the butt of Marx's ire—all this would appear unknown to those critics who describe the theory of multiples as entirely Marxist. What the early Victorian writers of leaders for the *London Times* would have said of this description of the

22. Albert Henry Smyth, *The Writings of Benjamin Franklin* (New York: Macmillan, 1905–07), 7: 434–35.

hypothesis they put in print can unfortunately only be conjectured. In short, despite the many distinct occasions on which the theory of multiples was published, it has periodically emerged as an idea new to many observers who worked it out for themselves.

Even so, the fact of multiple discoveries in science continues to be regarded by some, including minds of a high order, as something surpassing strange and almost unexplainable. Here is the great pathologist and historian of medicine, William Henry Welch, on the subject:

> The circumstances that a long-awaited discovery or invention has been made by more than one investigator, independently and almost simultaneously, and with varying approach to completeness, is a curious and not always explicable phenomenon familiar in the history of discovery.[23]

Other scholars tacitly assume that the pattern of multiples is both curious and distinctive of their own field of inquiry, if not entirely confined to it. As one example, consider the observation by the notable historian of geometry, Julian Lowell Coolidge:

> It is a curious fact in the history of mathematics that discoveries of the greatest importance were made simultaneously by different men of genius.[24]

And recently, the sociologist Talcott Parsons is recorded as having described the threefold, or possibly fivefold, discovery of "the internalization of values and culture as part of the personality" as "a very remarkable phenomenon because all of these people were independent of each other and their discovery is . . . fundamental."[25]

In part, of course, observations of this kind are merely casual remarks, not to be taken literally. But I should like now to develop the hypothesis that, far from being odd or curious or remarkable, the pattern of independent multiple discoveries in science is in principle the dominant pattern rather than a subsidiary one. It is the singletons—discoveries made only once in the history of science—that are the residual cases, requiring special explanation. Put even more sharply, the hypothesis states that all scientific discoveries are in principle multiples, including those that on the surface appear to be singletons.

Evidence on the Hypothesis of Multiples

Stated in this extreme form, the hypothesis must at first sound extravagant,

23. William Henry Welch, *Papers and Addresses* (Baltimore: Johns Hopkins Press, 1920), 3: 229.

24. Julian Lowell Coolidge, *A History of Geometrical Methods* (Oxford: Clarendon Press, 1940), p. 122.

25. Talcott Parsons, in *Alpha Kappa Deltan: A Sociological Journal* 29 (Winter 1959): 3–12, at 9–10.

not to say incorrigible, removed from any possible test of competent evidence. For if even historically established singletons are declared to be multiples-in-principle—potential multiples that happened to emerge as singletons—it would seem that this is a self-sealing hypothesis, immune to investigation. And yet, it may be that things are not really as bad as all that.

An incorrigible hypothesis is, of course, not an hypothesis at all, but only a dogma or perhaps an incantation. I suggest, however, that, far from being incorrigible and therefore outrageous, this hypothesis of multiples is actually held much of the time by working scientists. The evidence for this is ready to hand and once its pertinence is seen, it can be gathered in abundance. Here, then, are ten kinds of related evidence that bears upon the hypothesis that discoveries in science are in principle multiples, with the singletons being the exceptional type requiring special explanation.

First is the class of discoveries long regarded as singletons that turn out to be rediscoveries of previously unpublished work. Cases of this kind abound. But here, I allude only to two notable instances: Cavendish and Gauss. Much of Cavendish's vast store of unpublished experiments and theories became progressively known only after his death in 1810, as Harcourt published some of his work in chemistry in 1839; Clerk Maxwell, his work in electricity in 1879; and Thorpe, his complete chemical and dynamical researches in 1921.[26] But in the meanwhile, many of Cavendish's unpublished discoveries were made independently by contemporary and later investigators, among them, Black, Priestley, John Robison, Charles, Dalton, Gay-Lussac, Faraday, Boscovich, Larmor, Pickering, to cite only a few. And in most cases, the rediscoveries were regarded as singletons until Cavendish's records were belatedly published. The case of Gauss, as we know, is much the same. Loath to rush into print, Gauss crowded his notebooks with mathematical inventions and other discoveries that turned up independently in work by Abel, Jacobi, Laplace, Galois, Dedekind, Franz Neumann, Grassmann, Hamilton, and others.[27] Again, presumed singletons turned out to be multiples, as once unpublished work became

26. The detailed cases of rediscovery can be garnered from G. Wilson, *The Life of the Hon. Henry Cavendish*, 2 vols. (London, 1851); Henry Cavendish, *Scientific Papers*, ed. from the published papers and the Cavendish manuscripts (Cambridge: At the University Press, 1921), vol. 1, *The Electrical Researches*, ed. J. Clerk Maxwell, rev. Sir Joseph Larmor, vol. 2, *Chemical and Dynamical*, ed. Sir Edward Thorpe and others; A. J. Berry, *Henry Cavendish: His Life and Scientific Work* (London, Hutchinson, 1960).

27. A preliminary list of such rediscoveries of Gauss' unpublished work has been compiled from the details in his voluminous letters—e.g., *Briefwechsel zwischen Gauss und Bessel* (Leipzig: Wilhelm Engelmann, 1889); *Briefwechsel zwischen Carl Friedrich Gauss und Wolfgang Bolyai* (Leipzig: Teubner, 1899)—and in Waldo G. Dunnington, *Carl Friedrich Gauss* (New York: Exposition Press, 1955). I shall return to the further implications of such repeated involvement of the same scientists in multiple discoveries later in this paper, when I propose a sociological concept of scientific genius.

known. Far from being exceptions, Cavendish and Gauss are instances of a larger class.

What holds for unpublished work often holds also for work which, though published, proved relatively neglected or inaccessible, owing either to its being at odds with prevailing conceptions, or its difficulty of apprehension, or its having been printed in little-known journals, and so on. Here, again, singletons become redefined as multiples when the earlier work is belatedly identified. In this class of cases, to choose among the most familiar, we need only recall Mendel and Gibbs. The case of Mendel[28] is too well known to need review; that of Gibbs almost as familiar, since Ostwald, in his preface to the German edition of the *Studies in Thermodynamics*, remarked, in effect, that "it is easier to re-discover Gibbs than to read him."[29]

These are all cases of seeming singletons which then turn out to have been multiples or rediscoveries. Other, more compelling, classes of evidence bear upon the apparently incorrigible hypothesis that singletons, rather than multiples, are the exception requiring distinctive explanation and that discoveries in science are, in principle, potential multiples. These next classes of evidence are all types of forestalled multiples, discoveries that are historically identified as singletons only because the public report of the discovery forestalled others from making it independently. These are the cases of which it can be said: There, but for the grace of swift diffusion, goes a multiple.[30]

Second, then, and in every one of the sciences, including the social sciences, there are reports in print stating that a scientist has discontinued an inquiry, well along toward completion, because a new publication has anticipated both his hypothesis and the design of inquiry into the hypothesis. The frequency of such instances cannot be firmly estimated, of course, but I can report having located many.

Third, and closely akin to the foregoing type, are the cases in which the scientist, though he is forestalled, goes ahead to report his original, albeit anticipated, work. We can all call to mind those countless footnotes in the literature of science that announce with chagrin: "Since completing this experiment, I find that Woodworth (or Bell or Minot, as the case may be) had arrived at this conclusion last year and that Jones did so fully sixty years ago." No doubt many of us here today have experienced one or more

28. See Hugo Iltis, *Life of Mendel* (New York: W. W. Norton, 1932); Conway Zirkle, "Gregor Mendel and His Precursors," *Isis* 42 (June 1951): 97–104.

29. This is the entirely apt paraphrase by Muriel Rukeyser in *Willard Gibbs* (New York: Doubleday Doran, 1942), 4: 314.

30. It is only appropriate that the original saying—"There, but for the grace of God, . . ."—should itself be, with minor variations, a repeatedly reinvented expression.

of these episodes in which we find that our best and, strictly speaking, our most original inquiries have been anticipated. On this assumption, I single out only one case in point:

The experience of Lord Kelvin as an undergraduate of 18, when he was still the untitled William Thomson, who sent his first paper on mathematics to the Cambridge *Journal* only to find that he "had been anticipated by M. Chasles, the eminent French geometrician in two points . . . [and] when the paper appeared some months later, prefixed a reference to M. Chasles' memoirs, and to another similar memoir by M. Sturm. Still later, Thomson discovered that the same theorems had been also stated and proved by Gauss; and, after all this, he found that these theorems had been discovered and fully published more than ten years previously by Green, whose scarce work he never saw till 1845."[31]

Far from being rare, these voyages of subsequent and repeated discovery of an entire array of multiples are frequent enough to be routine.

Fourth, these publicly recorded instances of forestalled multiples do not, of course, begin to exhaust the presumably great, perhaps vast, number of unrecorded instances. Many scientists cannot bring themselves to report in print that they were forestalled. These cases are ordinarily known only to a limited circle, closely familiar with the work of the forestalled scientists. Interview studies of communication among scientists have begun to identify the frequency of such ordinarily unknown forestalling of multiples. Systematic field studies of this kind have turned up large proportions of what is often described as "unnecessary duplication" in research resulting from imperfections in the channels of communication between contemporary scientists. One such study[32] of American and Canadian mathematicians, for example, found 31 percent of the more productive mathematicians reporting that delayed publication of the work of others had resulted in such "needless duplication," that is, in multiples.

Fifth, we find seeming singletons repeatedly turning out to be multiples, as friends, enemies, co-workers, teachers, students, and casual scientific acquaintances have reluctantly or avidly performed the service of a candid friend by acquainting an elated scientist with the fact that his original finding or idea is not the singleton he had every reason to suppose it to

31. Silvanus P. Thompson, *The Life of William Thomson, Baron Kelvin of Largs* (London: Macmillan, 1910), 1: 44–45.

32. See Herbert Menzel, *Review of Studies in the Flow of Information Among Scientists*, Columbia University Bureau of Applied Social Research, a report prepared for the National Science Foundation (January 1960), 1: 21, 2: 48. Much other apposite information summarized in the Menzel monograph cannot be crowded into this paper. It should be added, however, that these data were uncovered in studies that were not focused on the matter of multiple and singleton discoveries; judging from the personal reports of previously undisclosed multiples that spontaneously came my way after I had published another paper on this general subject, I should judge that these occur on a scale so large that it has scarcely begun to be appreciated.

be, but rather a doubleton or larger multiple, with the result that this latest independent version of the discovery never found its way into print. So, the young W. R. Hamilton hits upon and develops an idea in optics and as he plaintively describes the episode:

> A fortnight ago I believed that no writer had ever treated of Optics on a similar plan. But within that period, my tutor, the Reverend Mr. Boyton, has shown me in the College Library a beautiful memoir of Malus on the subject. . . . With respect to those results which are common to both, it is proper to state that I have arrived at them in my own researches before I was aware of his.[33]

What his tutor did for Hamilton, others have done for innumerable scientists through the years. The diaries, letters, and memoirs of scientists are crowded with cases of this pattern (and with accounts of how they variously responded to these carriers of bad news).

Sixth, the pattern of forestalled multiples emerges as part of the oral tradition rather than the written one in still another form: as part of lectures. Here again, one instance must stand for many. Consider only the famous lectures of Kelvin at the Johns Hopkins where, it is recorded, he enjoyed "the surprise of finding [from members of his audience] that some of the things he was newly discovering for himself had already been discovered and published by others."[34]

A *seventh* type of pattern, tending to convert potential multiples into singletons, so far as the formal historical record goes, occurs when scientists have been diverted from a clearly developed program of investigation which, from all indications, was pointed in the direction successfully taken up by others. It is, of course, conjecture that the discoveries actually made by others would in fact have been made by the first but diverted investigator. But consider how such a scientist as Sir Ronald Ross, persuaded that his discoveries of the malarial parasite and the host mosquito were only the beginning, reports his conviction that, but for the interference with his plan by the authorities who employed him, he would have gone on to the discoveries made by others:

> The great treasure-house had been opened, but I was dragged away before I could handle the treasures. Scores of beautiful researches now lay open to me. I should have followed the "vermicule" in the mosquito's stomach—that was left to Robert Koch. I intended to mix the "germinal threads" with birds'

33. Robert Perceval Graves, *Life of Sir William Rowan Hamilton*, 3 vols. (Dublin: Hodges, Figgis, 1882), 1: 177.

34. Thompson, *William Thomson*, 2: 815–16. Kelvin tells of one such episode, thus: "I was thinking about this three days ago, and said to myself, 'There must be bright lines of reflexion from bodies in which we have those molecules that can produce intense absorption.' Speaking about this to Lord Rayleigh at breakfast, he informed me of this paper of Stokes's, and I looked and saw that what I had thought of was there. It was perfectly well known, but the molecule first discovered it to me."

blood—that was left to Schaudinn. I wished to complete the cycle of the human parasites—that was left to the Italians and others.[35]

Conjectural, to be sure, but with some indications that extraneous circumstances terminated a program of research that would have resulted in some of these discoveries becoming multiples rather than remaining adventitious singletons.

These several patterns of forestalled multiples, however, provide us with only sketchy evidence bearing on the apparently incorrigible hypothesis that multiples, both potential and actual, are the rule in scientific discovery and singletons the exception requiring special explanation. I turn now to evidence of quite another sort, the behavior of scientists themselves and the assumptions underlying that behavior. And here I suggest that, far from being outrageous, the hypothesis is in fact commonly adopted as a working assumption by scientists themselves. I suggest that in actual practice, scientists, and perhaps especially the greatest among them, themselves assume that singleton discoveries are imminent multiples. Granted that it is a difficult and unsure task to infer beliefs from behavior; almost as difficult and unsure as to infer behavior from beliefs. But in this case, we shall see that the behavior of scientists clearly testifies to their underlying belief that discoveries in science are potential multiples.

After all, scientists have cause to know that many discoveries are made independently. They not only know it, but act on it.[36] Since the culture of science puts a premium not only on originality but on chronological firsts in discovery, this awareness of multiples understandably activates a rush to ensure priority. Numerous expedients have been developed to ensure not being forestalled: for example, letters detailing one's new ideas or findings are dispatched to a potential rival, thus disarming him; preliminary reports are circulated; personal records of research are meticulously dated (as by Abel or Kelvin).

The race to be first in reporting a discovery testifies to the assumption that if the one scientist does not soon make the discovery, another will. This, then, provides an *eighth* kind of evidence bearing on our hypothesis [evidence set out in chapter 14 of this volume]. The many instances detailed there are quite typical; Norbert Wiener is no more circumstantial and outspoken about his experience than were Wallis, Wren, Huygens, Newton, the Bernoullis, and an indefinitely large number of other scientists through the centuries whose diaries, autobiographies, letters, and notes testify to the same effect.

35. Ronald Ross, *Memoirs, with a Full Account of the Great Malaria Problem and Its Solution* (London: John Murray, 1923), p. 313.

36. The following paragraphs are based on "Priorities in Scientific Discovery," chapter 14 of this volume.

In all this, I exclude those cases in which scientists move to establish their priority only to ensure that their discoveries not be diffused in the community of scientists before their own creative role in them is made eminently visible or to ensure that they not be later accused of having derived their own ideas from fellow-scientists who have borrowed them or cases in which, like that of Priestley, scientists publish quickly in order to advance science rapidly by making their work available to others at once. In this class of cases pertinent to the hypothesis, I refer only to those in which the rush to establish priority is avowedly motivated by the concern not to be forestalled, for this alone is competent evidence that scientists in fact assume that their initial singletons are destined not to remain singletons for long; that, in short, a multiple is definitely in the making.

But *ninth,* not all scientists who see themselves involved in a potential multiple are prepared to be outspoken about the matter. In many cases of this sort, their scientific colleagues, or kin, are. We have only to remember the elder Bolyai, himself a mathematician of some consequence, prophetically warning his son that "no time be lost in making it [his non-Euclidean geometry] public, for two reasons:

> first, because ideas pass easily from one to another, who can anticipate its publication, and secondly, there is some truth in this, that many things have an epoch, in which they are found at the same time in several places, just as the violets appear on every side in spring. . . . Thus we ought to conquer when we are able, for the advantage is always to the first comer."[37]

Almost we hear in these words the echoed warning by other faithful colleagues of the imminent danger of being forestalled: his friend Robinson urging Oughtred to make his work on logarithms public;[38] Wallis and Halley warning Newton;[39] Halley warning Flamsteed;[40] Bache warning Joseph Henry that "no time be lost in publishing his remarks before the American Philosophical Society" now that word has come of Faraday's work on self-induction;[41] Lyell warning Darwin (Edward Blyth notwith-

37. The letter is quoted in Roberto Bonola, *Non-Euclidean Geometry* 2d rev. ed. (La Salle, Ill.: Open Court Publishing Co., 1938), pp. 98–99. See chapter 14 of this volume.

38. Stephen Peter Rigaud, ed., *Correspondence of Scientific Men of the 17th Century,* 2 vols. (Oxford: Oxford Univ. Press, 1841), 1: 7, 2: 27.

39. Charles R. Weld, *A History of the Royal Society* (London: Parker, 1848), 1: 408–9.

40. Francis Baily, *An Account of the Revd. John Flamsteed, the First Astronomer-Royal, Compiled from His Own Manuscripts* (London, 1835), p. 161. This case has particular point since Halley and Flamsteed were of course devoted enemies, but Halley thought it important that no English scientist be forestalled by a foreign scientist.

41. Thomas Coulson, *Joseph Henry: His Life and Work* (Princeton, N.J.: Princeton Univ. Press, 1950), pp. 109–10, 47–48.

standing) that he must publish lest he be forestalled;[42] Bessel and Schumacher warning Gauss that he will be anticipated (as he was) on every side;[43] the elderly Legendre warning the young Karl Jacobi that the younger Niels Abel would overtake him in the race for discoveries in the theory of elliptic functions unless "you take possession of that which belongs to you by letting your book appear at the earliest possible date."[44]

Between them, Gauss and Bessel supply a beautifully ironic instance of how apt it is for scientists to assume that their original discoveries will be duplicated by others if they do not put them into print soon. For years on end the faithful Bessel has been haranguing Gauss to publish his new discoveries on pain of being forestalled. At last, Gauss behaves as Bessel would have him behave. He publishes a treatise on dioptrics and sends a copy to Bessel who, after heroically congratulating him on the work, ruefully reports that it thoroughly anticipates Bessel's own current but still unpublished investigations.[45]

Gauss supplies us with another striking instance of the scientist's or mathematician's firm belief that a discovery or invention is not reserved to himself alone. In 1795, at the ripe age of eighteen, he works out the method of least squares. To him the method seems to flow so directly from antecedent work that he is persuaded others must already have hit upon it; he is willing to bet, for example, that Tobias Mayer must have known it.[46] In this he was, of course, mistaken, as he learned later; his invention of least squares had not been anticipated. Neverthless, he was abundantly right in principle: the invention was bound to be a multiple. As things turned out, it proved to be a quadruplet, with Legendre inventing it independently in 1805 before Gauss had got around to publishing it, and with Daniel Huber in Basel and Robert Adrain in the United States coming up with it a little later.[47]

There is a final and perhaps most decisive kind of evidence that the community of scientists does in fact assume that discoveries are potential multiples. This evidence is provided by the institutional expedients de-

42. Francis Darwin, ed., *The Life and Letters of Charles Darwin* (New York: Appleton, 1925), 1: 426–27, 473.

43. Dunnington, *Gauss*, p. 216; C. A. F. Peters, ed., *Briefwechsel zwischen C. F. Gauss und H. C. Schumacher* (Altona: Gustav Esch, 1860), 2: 82–83, 299–300, 3: 69, 75, 6: 10–11, 55.

44. Ore, *Niels Henrik Abel*, p. 203.

45. *Briefwechsel zwischen Gauss und Bessel*, Herausgegeben auf Veranlassung der Königlichen Preussischen Akädemie der Wissenschaften (Leipzig: Wilhelm Engelmann, 1880), pp. 531–32.

46. *Briefwechsel zwischen Gauss und Schumacher*, 3:387.

47. Dunnington, *Gauss*, p. 19. Adrain, the outstanding American mathematician of his day, was involved in several multiples. See J. L. Coolidge, "Robert Adrain and the Beginning of American Mathematics," *American Mathematical Monthly* 33 (Feb. 1926): 61–76.

signed to protect the scientist's priority of conception. Since the seventeenth century, scientific academies and societies have established the practice of having sealed and dated manuscripts deposited with them in order to protect both priority and idea. As this was described in the early minutes of the Royal Society:

> When any fellow should have a philosophical notion or invention, not yet made out, and desire that the same sealed up in a box might be deposited with one of the secretaries, till it could be perfected, and so brought to light, this might be allowed for the better securing inventions to their authors.[48]

From at least the sixteenth century and as late as the nineteenth, it will also be remembered, discoveries were often reported in the form of anagrams—as with Galileo's "triple star" of Saturn and Hooke's law of tension—for the double purpose of establishing priority of conception and yet of not putting rivals on to one's original ideas, until they had been worked out further.[49] From the time of Newton, scientists have printed short abstracts for the same purpose.[50] These and comparable expedients all testify that scientists, even those who manifestly subscribe to the contrary opinion, in practice assume that discoveries are potential multiples and will remain singletons only if prompt action forestalls the later independent discovery. It would appear, then, that what might first have seemed to be an incorrigible, perhaps outrageous, hypothesis about multiples in science is in fact widely assumed by scientists themselves.

A great variety of evidence—I have here set out only ten related kinds—testifies, then, to the hypothesis that, once science has become institutionalized, and significant numbers are at work on scientific investigation, the same discoveries will be made independently more than once and that singletons can be conceived of as forestalled multiples.

Patterns of Multiple Discoveries

Before turning to the last part of this paper—the part dealing with a sociological conception of the role of genius in the advancement of science—I think it useful to report some findings from a methodical study of multiple discoveries. Of the multitude of multiples, Dr. Elinor Barber and I have undertaken to examine 264 intensively. The greatest part of these—179 of them—are doublets; 51, triplets; 17, quadruplets; 6, quintuplets;

48. Thomas Birch, *The History of the Royal Society of London* (London: A. Millar, 1756), 2:30. The French Academy of Sciences made extensive use of this arrangement; among the many documents deposited under seal was Lavoisier's on combustion; see Lavoisier *Oeuvres de Lavoisier. Correspondance*, ed., René Fric (Paris: Michel, 1957), fasc. 2, pp. 388–89.

49. See chapter 14 of this volume.

50. See Birch, *History of the Royal Society* 4:437.

8, sextuplets. This aggregate of multiples also includes one septuplet and two nonaries, in which most of the nine independent co-discoverers were presumably ready to entertain the hypothesis that if any one of them had not arrived at the discovery, it would probably have been made in any case.

Each of these 264 multiples has been variously classified, after a search of the monographic evidence dealing with it. It has been classified in the particular discipline in which it occurred; the historical period of the multiple; the interval of time elapsing between the repeated discoveries; the number of co-discoverers; whether or not it gave rise to a contest over priority; the nationality of the co-discoverers, distinguishing those who were fellow nationals from the rest; the ages of the co-discoverers; and so on. The information about each multiple obtained through historical inquiry has been coded and transferred to punchcards, in this way permitting detailed statistical analysis.

This is not the occasion to report the findings in hand; my purpose here is only to suggest that the intensive study of particular cases of multiple discovery can be instructively supplemented by methodical analysis of large numbers of cases. It may be of interest, for example, that 20 percent of the multiples under review occurred within an interval of one year; some of them on the same day or within the same week. Another 18 percent occurred within a two-year span and, to turn to the other end of the scale, 34 percent of them involved an interval of ten years or more. The shorter the interval between the several appearances of a multiple, the *less* often does it lead to a debate over independence or other aspects of priority: of those made within a year of each other, just about half were subject to a contest over priority; of those more than 20 years apart, four in every five were contested. Ethnocentrism notwithstanding, if the independent co-discoverers are from different nations, there is slightly *less*, rather than more, probability of a conflict over priority. And to allude to just one other preliminary finding—this one, on the whole, rather encouraging—there seems to be a secular decline in the frequency with which multiples are an occasion for priority conflicts between scientists. Of the 36 multiples before 1700 which we have examined, 92 percent were strenuously contested; this figure drops to 72 percent in the eighteenth century; remains at about the same level (74 percent) in the first half of the nineteenth century and declines notably to 59 percent in the latter half; and reaches the low of 33 percent in the first half of this century. It may be that scientists are becoming more fully aware that, with growing numbers of investigators at work in each special field, any particular discovery is apt to be made by others as well as by themselves.

In any case, this inquiry has been enough to persuade us that the statistical analysis of historical data bearing on discovery is a feasible and instructive next step in the sociology of science.

Sociological Theory of Genius In Science

After this short interlude, I return to the last part of the sociological theory of scientific development, dealing with the role of genius in that development. As I have intimated, the hypothesis of multiples has long been tied to the companion hypothesis that the great men of science, the undeniable geniuses, are altogether dispensable, for had they not lived, things would have turned out pretty much as they actually did. For generations the debate has waxed hot and heavy on this point. Scientists, philosophers, men of letters, historians, sociologists, and psychologists have all at one time or another taken a polemical position in the debate. Emerson and Carlyle, Spencer and William James, Ostwald and de Candolle, Galton and Cooley—these are only a few among the many who have placed the social theory in opposition to the theory that provides ample space for the individual of scientific genius. That so many acute minds should have for so long regarded this as an authentic debate must not keep us from noticing how the issues have been falsely drawn; and that once the two theories are clearly stated, there is no necessary opposition between them. Instead, it is proposed that once scientific genius is conceived of sociologically, rather than, as the practice has commonly been, psychologically, the two ideas of the environmental determination of discovery can be consolidated into a single theory. Far from being incompatible, the two complement one another.

In this enlarged sociological conception, scientists of genius are precisely those whose work in the end would be eventually rediscovered. These rediscoveries would be made not by a single scientist but by an entire corps of scientists. On this view, the individual of scientific genius is the functional equivalent of a considerable array of other scientists of varying degrees of talent. On this hypothesis, the undeniably large stature of great scientists remains acknowledged. It is not cut down to size in order to fit a Procrustean theory of the environmental determination of scientific discovery. At the same time, this enlarged conception does not abandon the sociological theory of discovery in order to provide for the indisputable, great differences between scientists of large talent and of small; it does *not*, in the phrase of Bacon, "place all wits and understandings nearly on a level."

This enlarged sociological conception holds that great scientists will have been repeatedly involved in multiples. First, because the genius will have made many scientific discoveries altogether; and since each of these is, on the first part of the theory, a potential multiple, some will have become actual multiples. Second, this means that each scientist of genius will have contributed the functional equivalent to the advancement of science of what a considerable number of other scientists will have contributed in the aggregate, some of these having been caught up in the repeated multiples in which the genius was actually involved.

In a word, the greatest men of science have been involved in a multiplicity of multiples. This is true for Galileo and Newton, for Faraday and Clerk Maxwell, for Hooke, Cavendish, and Stensen, for Gauss and Laplace, for Lavoisier, Priestley, and Scheele—in short, for all those whose place in the pantheon of science is beyond dispute, however much they may differ in the measure of their genius.

Once again, I can only allude to the pertinent evidence rather than report it in full. But consider the case of Kelvin, by way of illustration. After examining some 400 of his 661 scientific communications and addresses—the rest have still to be studied—Dr. Elinor Barber and I find him testifying to at least 32 multiple discoveries in which he eventually found that his independent discoveries had also been made by others. These 32 multiples involved an aggregate of 30 other scientists, some, like Stokes, Green, Helmholtz, Cavendish, Clausius, Poincaré, Rayleigh, themselves men of undeniable genius, others, like Hankel, Pfaff, Homer Lane, Varley and Lamé being men of talent, no doubt, but still not of the highest order. The great majority of these multiples of Kelvin were doublets, but some were triplets and a few, quadruplets. For the hypothesis that each of these discoveries was destined to find expression, even if the genius of Kelvin had not obtained, there is the best of traditional proof: each was in fact made by others. Yet Kelvin's stature as a scientist remains undiminished. For it required a considerable number of others to duplicate these thirty-two discoveries which Kelvin himself made.

Following out the logic of this kind of fact, we can set up a matrix of multiple discoveries, with the entries in the matrix indicating the particular scientists involved in each of the multiples. Some of these others are themselves men of genius, in turn often involved in still other multiples. Others in the matrix are the men of somewhat less talent who, on the average, are involved in fewer multiples. And toward the lower end of the scale of demonstrated scientific talent are the far more numerous men of science, who in the aggregate are indispensable to the advancement of science and whose one moment of prime achievement came when they found for themselves one of the many discoveries that the man of genius had made independently of them.

To continue for a moment with the specimen case of Kelvin, these 32 multiples are of course only a portion of the multiples in which he was eventually involved. For, as I have said, they are only the ones which Kelvin himself found to have been made by others. Beyond these are the discoveries by Kelvin which were only later made independently by others. Of these we do not yet have a firm estimate. And beyond these still are what I have described as the forestalled multiples: the discoveries of

Kelvin which were not, so far as the record shows, made independently by others but which, on our hypothesis, would have been made had it not been for the widespread circulation of Kelvin's prior findings. Yet, even on this incomplete showing, it would seem that this one man of scientific genius was, in a reasonably exact sense, functionally equivalent to a sizable number of other scientists. And still, by the same token, his individual accomplishments in science remain undiminished when we note that he was not individually indispensable for these discoveries (since they were in fact made by others). This is the sense in which an enlarged sociological theory can take account both of the environmental determination of discovery while still providing for great variablity in the intellectual stature of individual scientists.

Just a few words about another like instance, in quite another field of science. Whatever else may be said about Sigmund Freud, he is undeniably the prime creator of psychoanalysis. And still, only a first examination of about a hundred of his publications finds him reporting that he was involved in an aggregate of more than thirty multiples, discoveries which he made all unknowing that they had been made by others. Once again, the pattern is much like that we found for Kelvin. Some of Freud's subsequently discovered anticipators were themselves minds of acknowledged highest order: Schiller, von Hartmann, Schopenhauer, Fechner. But many of the rest of his independent co-discoverers or anticipators are scarcely apt to be known to most of us as distinguished for the highest quality of scientific achievement; men such as Watkiss Lloyd, Kutschin, E. Hacker, Grasset, Neufeld, and so on and on. It required a Freud to achieve individually what a large number of others achieved severally; it required a Freud to focus the attention of many on ideas which might otherwise not have come to their notice; in these and kindred aspects lay his genius. But that he was not individually indispensable to the intellectual developments for which he, more than any other, was historically responsible is indicated by the many multiples in which he was in fact engaged and the many others which, presumably, he forestalled by his individually incomparable genius.

What has been found to hold for Kelvin and Freud is being found to hold for other scientists of the first rank who are now being examined in the light of the theory. They are all scientists of multiple multiples; their undeniable stature rests in doing individually what must otherwise be done and, as we have reason to infer, at a much slower pace, by a substantial number of other scientists, themselves of varying degrees of demonstrated talent. The sociological theory of scientific discovery has no need, therefore, to retain the false disjunction between the cumulative development of science and the distinctive role of the scientific genius.

There is perhaps time for a few needed and self-imposed caveats. For I cannot escape the uneasy sense that this short though, you will grant me, not entirely succinct, summary of masses of data on scientific discovery must lend itself to misunderstanding. This is so, if only because so much has unavoidably been left unsaid. As a preventive to such misunderstanding, may I conclude by listing some *seeming* implications which are anything but implicit in what I have managed to report?

First, in presenting this modified version of a three-century-old conception of the course of scientific discovery, I do not imply that all discoveries are inevitable in the sense that, come what may, they will be made, at the time and the place, if not by the individual(s) who in fact made them. Quite the contrary: there are, of course, cases of scientific discoveries which could have been made generations, even centuries, before they were actually made, in the sense that the principal ingredients of these discoveries were long present in the culture. This recurrent fact of long-delayed discovery raises distinctive problems for the theory advanced here, but these are not unsolvable problems.

Second, and perhaps contrary to the impression I have given, the theory rejects the pointless practice of what I have called "adumbrationism," that is, the practice of claiming to find dim anticipations of current scientific discoveries in older, and preferably ancient, work by the expedient of excessively liberal interpretations of what is being said now and of what was said then. The theory is not a twentieth-century version of the seventeenth- and eighteenth-century quarrel between the ancients and the moderns.

Third, the theory is not another version of Ecclesiastes, holding that "there is no new thing under the sun." The theory provides for the growth, differentiation, and development of science just as it allows for the fact that new increments in science are in principle or in fact repeated increments. It allows also for occasional mutations in scientific theory which are significantly new even though they are introduced by more than one scientist.

Fourth, the theory does not hold that to be truly independent, multiples must be chronologically simultaneous. This is only the limiting case. Even discoveries far removed from one another in calendrical time may be instructively construed as "simultaneous" or nearly so in social and cultural time, depending upon the accumulated state of knowledge in the several cultures and the structures of the several societies in which they appear.

Fifth, the theory allows for differences in the probability of actual, rather than potential multiples according to the character of the particular discovery. Discoveries in science are of course not all of a piece. Some flow directly from antecedent knowledge in the sense that they are widely visible implications of what has gone just before. Other discoveries involve more of a leap from antecedent knowledge, and these are perhaps less apt to be actual multiples. But it is suggested that, in the end, these too manifest the same processes of scientific development as the others.

Sixth, and above all, the theory rejects the false disjunction between the social determination of scientific discovery and the role of the genius or "great man" in science. By conceiving scientific genius sociologically, as one who in his own person represents the functional equivalent of a number and variety of often lesser talents, the theory maintains that the genius plays a distinctive role in advancing science, often accelerating its rate of development and sometimes, by the excess of authority attributed to him, slowing further development.

Seventh and finally, the diverse implications of the theory are subject to methodical investigation. The basic materials for such study can be drawn from both historical evidence and from field inquiry into the experience of contemporary scientists. What Bacon obliquely noticed and many others recurrently examined can become a major focus in the contemporary sociology of science.

17 Multiple Discoveries as Strategic Research Site

1963

The pages of the history of science record thousands of instances of similar discoveries having been made by scientists working independently of one another. Sometimes the discoveries are simultaneous or almost so; sometimes a scientist will make anew a discovery which, unknown to him, somebody else had made years before. Such occurrences suggest that discoveries become virtually inevitable when prerequisite kinds of knowledge and tools accumulate in man's cultural store and when the attention of an appreciable number of investigators becomes focused on a problem, by emerging social needs, by developments internal to the science, or by both.

As we have seen in the preceding chapter, this is an hypothesis confirmed by its own history.[1] The idea of the sociological import of independent multiple discoveries—for brevity's sake, I shall continue to refer to them as "multiples"—has itself been periodically rediscovered over a span of centuries. Moreover, this repeated rediscovery of the same facts and associated hypothesis has remained all these years in a static condition, as though it were permanently condemned to repetition without extension. After all, fifty years have elapsed since Ogburn and Thomas compiled their list of independent discoveries.[2] It has been at least a century and a half since observers began taking formal note of the fact of multiples—

First published as part of "Resistance to the Systematic Study of Multiple Discoveries in Science," *European Journal of Sociology* 4 (1963): 237–49; reprinted with permission.

1. Its subject matter, problematics, and identity as a discipline require the sociology of science to be abundantly stocked with such self-exemplifying ideas.

2. W. F. Ogburn and D. S. Thomas, "Are Inventions Inevitable?" *Political Science Quarterly* 37 (1922): 83–100; W. F. Ogburn, *Social Change* (New York: B. W. Huebsch, 1922), pp. 80–102.

even to the extent of compiling short lists of cases in point—and began to draw out the implications of the fact. And, as we have just seen, it has been at least 350 years since Francis Bacon set down some of the principal ingredients of the hypothesis in a set of luminous aphorisms. Why, then, has the idea remained static all this while?

It may be, of course, that this is so because the last word has been said about the implications of multiples for a theory of how science develops. Or again, the idea may have remained undeveloped because the familiar fact of multiples is quite incidental and lacks significant import; that it is seemingly as trivial and insignificant, say, as the equally familiar fact that people occasionally make slips of the tongue or pen. All this is possible. But I want now to examine the position that although it is possible, it is not so. Instead, I suggest, first, that the facts of multiples and priorities in scientific discovery provide a research site that is more strategic for advancing the sociology and psychology of science than appears to be generally recognized and, second, that the failure to build on this research site results largely from nonrational resistance to the systematic scrutiny of these facts. This paper, then, deals with the intellectual uses of the methodical investigation of multiple discoveries; the next paper, with the hypothesis that the neglect stems from identifiable forms of resistance. In short, I try first to answer the question: why bother with *systematic* study of the subject? and then try to answer the next question: in view of its theoretical significance, why don't social scientists bother with it?

Something is known about the social, cultural, and economic sources of "resistance" (in the colloquial sense of "opposition") to new ideas and findings in science, both in the large community of laymen[3] and in the smaller community of scientists themselves.[4] In examining the resistance of scientists to the detailed study of multiples, I shall be considering "resistance" in the more technical, psychosocial sense of motivated neglect

3. Notably, in a series of papers and monographs by Bernhard J. Stern: e.g., *Social Factors in Medical Progress* (New York: Columbia University Press, 1927); *Should We Be Vaccinated?* A survey of the controversy in its historical and scientific aspects (New York: Harper and Brothers, 1927); "Resistance to the Adoption of Technological Innovations," *Technological Trends and National Policy* (Washington: Government Printing Office, 1937), pp. 39–66. See also R. K. Merton, "Science and the Social Order," chapter 12 of this volume; "The Machine, the Worker, and the Engineer," *Science* 105 (1947): pp. 79–84 (for the resistance of laymen to social research).

4. Again, Bernhard J. Stern, *Society and Medical Progress* (Princeton: Princeton University Press, 1941), esp. chap. 9 ("Resistances to medical change"); Bernard Barber, "Resistance by Scientists to Scientific Discovery," *Science* 134 (1 Sept. 1961): 596–602; on the reluctance to develop the sociology of science, see "The Neglect of the Sociology of Science," chapter 10 of this volume. See also Philipp Frank, "The Variety of Reasons for the Acceptance of Scientific Theories," *Scientific Monthly* 79 (1954): 139–45; Alexandre Koyré, "Influence of Philosophic Trends on the Formulation of Scientific Theories," *Scientific Monthly* 80 (1955): 107–11.

or denial of an accessible but painful reality,[5] in this case the reality of multiples and of the frequent conflicts over priority of scientific discovery.

Multiples as a Strategic Research Site

In describing multiple discoveries as affording a strategic research site, I mean only that the data they provide can be investigated to good advantage in order to clarify the workings of social and cultural processes in the advancement of science.[6] We can identify at least eight connected respects in which this is so.

First, the methodical study of multiples supplements the current emphasis of research in the psychology and sociology of science on "creativity" which is largely focused on (a) the psychological traits that appear to be distinctive of creative talents in science; (b) the psychological processes of scientific thought adapting, in one form or another, Poincaré's and then Graham Wallas's four-step process of preparation, incubation, illumination and verification; and (c) the social statuses of creative scientists.[7] Now, this array of inquiries into the endopsychic and social attributes of individual scientists of course has its place. But we also know that it is only one, and not necessarily an exclusively apt, type of inquiry. Indeed, much of the recent work on "creativity in science" is a little reminiscent of the early work on "leadership" which, for all its suggestive leads, resulted in palpably few definitive findings about the traits and qualities of "leaders" in human affairs.

Second, the study of multiples supplements, in ways that will soon become evident, the current research focus on the interpersonal relations in which scientists are engaged while they are at work; a focus on the "milieu"[8] of the scientist. This emphasis has been reinforced by the

5. Sigmund Freud, "The Resistances to Psycho-analysis," *Imago* 11 (1925): 222–33, reprinted in Freud, *The Standard Edition of the Complete Psychological Works of Sigmund Freud*, ed. James Strachey (London: Hogarth Press, 1961), vol. 19 [1923–25], pp. 213–22.

6. For the notion of "strategic research site," see R. K. Merton, "Problem-finding in Sociology," *Sociology Today,* ed. R. K. Merton, L. Broom, and L. S. Cottrell, Jr. (New York: Basic Books, 1959), pp. xxvi–xxix.

7. For an extensive review of these inquiries, see Morris I. Stein and Shirley J. Heinze, *Creativity and the Individual: Summaries of Selected Literature in Psychology and Psychiatry* (Glencoe: The Free Press, 1960).

8. For the notion of the "milieu" as the network of personal relations which intervenes between the individual and the larger social structure, see H. H. Gerth and C. W. Mills, *Character and Social Structure* (New York: Harcourt, Brace, 1953); and on the tendency of some social scientists to focus on the milieu, as contrasted with the larger structure, in dealing with social environments, see R. K. Merton, "The Social-cultural Environment and *Anomie,*" in H. L. Witmer and Kotinsky, eds., *New Perspectives for Research on Juvenile Deliquency* (Washington, D.C.: Government Printing Office, 1955), pp. 25–26, 42. As I note there, current over-emphasis on

tradition of small-group research, with its established theory and research instruments which are, understandably enough, being applied to study of groups of research-scientists.

Third, the study of multiples can supplement the pattern of fitting new research on the behavior of scientists into another established tradition of social science investigation, this time the study of the formal organization of research establishments and the bearing of this organization on the productive work of scientists.

I do not propose to question the uses of these three major types of research on the behavior and productivity of scientists: their traits and psychological processes of creative work, the effects upon them of local interpersonal relations and of the formal organization of their workplaces. But to recognize these uses need not obscure the fact that they all deal either with the endopsychic and social traits and processes of individual scientists or with the immediate social environments in which scientists find themselves. Yet we know, as a decisive fact, that scientists live and work in larger social and cultural environments than those comprised by their local *milieux*. And this seems to be true particularly of the most creative among them. Outstanding scientists tend to be "cosmopolitans," oriented to the wider national and trans-national environments, rather than "locals," oriented primarily to their immediate band of associates.[9]

K. E. Clark's study of America's psychologists, for example, found that especially productive psychologists were more apt than a control group to report that their significant reference groups and reference individuals—the people "whose opinions of their work they care about"—were composed by other outstanding psychologists in the United States and in other countries, rather than by their local colleagues.[10]

---

the milieu, in contrast to larger social structures, is a "little like the prevailing resistance among physical scientists in the 17th century to the notion of action at a distance." The milieu is not the same as what has been called "informal groups," since it includes formal personal relations as well. It overlaps, but is not identical, with what has been called the "ambiance": the collection of *all* people, and not only those in the immediate social environment, with whom a person interacts. See Theodore Caplow, "The Definition and Measurement of Ambiances," *Social Forces* 34 (1955): 28–33.

9. R. K. Merton, *Social Theory and Social Structure,* chap. 10 ("Patterns of Influence: Local and Cosmopolitan Influentials"). For use of these concepts in studying the behavior of scientists, academicians, and other professionals, see A. W. Gouldner, "Cosmopolitans and Locals: Toward an Analysis of Latent Social Roles," *Administrative Science Quarterly* 2 (1957): 281–306, 2 (1958): 444–80; W. G. Bennis et al., ibid., 2 (1958): 481–500; Herbert A. Shepard, "Nine Dilemmas in Industrial Research," ibid., 1 (1956): 295–309; Armond Fields, "Eine Untersuchung über administrative Rollen," *Kölner Zeitschrift für Soziologie und Sozialpsychologie* 8 (1956): 113–23. For the relation of these types to "effective scope" see P. F. Lazarsfeld and W. Thielens, Jr., *The Academic Mind* (Glencoe: The Free Press, 1958), pp. 263–65.

10. K. E. Clark, *America's Psychologists: A Survey of a Growing Profession* (Washington, D.C.: American Psychological Association, 1957), pp. 85–86.

The history of science attests that this has typically been the case for outstanding investigators in every science through the last three centuries. The theoretic import of this should not be overlooked. It would be an egregious blunder to allow the otherwise useful emphasis on trait-analysis or on small-group research to deflect attention from the presumably great part played, in scientific work, of social interaction with others who are *not* in the local milieu. To do so would be to impose convenient existing tools of investigation upon a problem for which they may not be the most appropriate and, surely, not the exclusively appropriate ones. The data, instruments, and theory dealing with larger aggregates of interacting scientists and of spatially distant reference groups and individuals would seem particularly in point for studying the behavior of scientists for whom patterns of social interaction at a distance seem empirically central. This is only a special case of a general hypothesis about "effective scope."[11] People in various social statuses differ in the radius of their significant social environments: some, the locals, being primarily oriented toward their local milieux, others, the cosmopolitans, being primarily oriented toward the larger society and responsive to it. The systematic study of multiples and priorities in scientific discovery—which of course typically engage scientists with others outside their local environment—thus provides one basis for investigating extended social relations between scientists and the effects of these upon their work.

It may be useful to put much the same point in a slightly different context. Historians of science and other scholars have long used the phrase, "the community of scientists." For the most part, this has remained an apt metaphor rather than becoming a productive concept. Yet it need not remain a literary figure of speech: apt and chaste, untarnished by actual use. For we find that *the* community of scientists is a dispersed rather than a geographically compact collectivity. The structure of this community cannot, therefore, be adequately understood by focusing only on the small local groups of which scientists are a part. The sheer fact that multiple discoveries are made by scientists working independently of one another testifies to the further crucial fact that, though remote in space, they are responding to much the same social and intellectual forces that impinge upon them all. In a word, the Robinson Crusoe of science is just as much a figment as the Robinson Crusoe of old-fashioned economics. He is an illusion, created by a scheme of thought that requires us to look only inward at thought processes and so to abstract entirely from the wider social and cultural contexts of that thought. Occasional scientists may suppose that they really work alone, meaning by this not the evident fact that only individual men and women, not "the group," think and

11. Lazarsfeld and Thielens, *Academic Mind*, pp. 262–65.

develop imaginative ideas but that they do so, all apart from environing structures of values, social relations, and socially as well as intellectually induced foci of attention. But, as multiple discoveries testify, this image of the man of science is just as much a case of the fallacy of misplaced concreteness as is the equivalent image of the man of business "who ascribes his achievements to his own unaided efforts, in bland unconsciousness of a social order without whose continuous support and vigilant protection he would be as a lamb bleating in the desert."[12] For scientists, even the most lonely of lone wolves among them, are all "members of one another." The study of multiples shows how scientists are bound to the past by building upon a deposit of accumulated knowledge, how they are bound to the present by interacting with others in the course of their work and having their attention drawn to particular problems and ideas by socially and intellectually accentuated interests, and how they are bound to the future by the obligation inherent in their social role to pass on an augmented knowledge and a more fully specified ignorance. The community of scientists extends both in time and in space.

These three respects in which the study of multiples provides a strategic research site are simply different facets of the same guiding conception: they supplement current emphases in research on the behavior of scientists by conceiving that behavior as a resultant not only of the idiosyncratic characteristics and the local ambiance of scientists, but also of their place within the wider social structure and culture. Beyond these are quite other uses of the study of multiples.

A *fourth* use is to help us identify certain significant similarities and differences between the various branches of science. To the extent that the rate of multiples and the types of rediscoveries are much the same in the social and psychological sciences as in the physical and life sciences, we are led to similarities between them, just as differences in such rates and types alert us to differences between them. In short, the study of multiples can supplement the traditional notion of the unity of all science, a notion usually formulated in terms of the logic of method. It can lead us to reexamine this unity from the standpoint of the actual behavior of scientists in each of the major divisions of science and so to identify their distinctive relations to their respective social and cultural environments. This type of behavioral inquiry does not, of course, replace inquiries into the philosophy of science or the logical foundations of scientific method. It supplements them, by attending to what men in the various sciences actually do, rather than by limiting us to what textbooks of scientific method tell us they should do, as they go about their work.

12. R. H. Tawney, *Religion and the Rise of Capitalism* (New York: Harcourt, Brace, 1926).

This brings us to a *fifth* use of studying multiples. As we have noted, scientists typically experience multiples as one of their occupational hazards. They are occasions for acute stress. Few scientists indeed react with equanimity when they learn that one of their own best contributions to science—what they know to have been the result of long hard work—is "only" (as the telling phrase has it) a rediscovery of what was found some time before, or "just" another discovery of what others have found at about the same time. No one who systematically examines the disputes over priority can ever again accept as veridical the picture of the scientist as one who is exempt, by his social role and his socially patterned personality, from affective involvement with *his* ideas and *his* discoveries of once unknown fact. The value of examining the behavior of men under stress in order to understand them better in all manner of other situations need not be recapitulated here.[13] By observing the behavior of scientists under what they experience as the stress of being forestalled in a discovery, we gain clues to ways in which the social institution of science shapes the motives, social relations and affect of men of science. I have tried to show elsewhere,[14] for example, how the values and reward system of science, with their sometimes pathogenic emphasis upon originality, help account for certain deviant behaviors of scientists: secretiveness during the early stages of inquiry, lest they be forestalled; violent conflicts over priority; an unending flow of premature publications designed to establish grounds for later claims to having been first. These, I suggest, are normal responses to a badly integrated institution of science, such that we can better understand the fact that a sample of American "starred men of science" reports that, next to what they describe as "personal curiosity," "rivalry" is most often the spur to their work.[15]

To the *sixth* use of the study of multiples, I should like to devote some little time, for it has implications both for a sociological theory of scientific discovery and for social policy governing the support of scientific work. With the vast increases in public and private funds for the support of scientific research, there has emerged a great concern to avoid what is called "wasteful duplication"[16] in allocating these funds. This is a wide-

13. See, for example, Hans Selye, *The Stress of Life* (New York: McGraw-Hill, 1956).

14. See "Priorities in Scientific Discovery," chapter 14 of this volume.

15. S. S. Visher, *Scientists Starred 1903–1943 in "American Men of Science,"* (Baltimore: Johns Hopkins University Press, 1947), pp. 531–32.

16. See, for example, the extensive *Proceedings* of the International Conference on Scientific Information, Washington, D.C., 16–21, Nov. 1958 (Washington, D.C.: National Academy of Sciences–National Research Council, 1959). Not, of course, that this problem is now being recognized for the first time. So-called "universal catalogues" of scientific papers and books have a long history. Even by 1828 the followers of Saint-Simon were complaining: "In the absence of any official inventory

spread concern, recently expressed in the planned society of the U.S.S.R. as it has been expressed for some time in unplanned societies of the West. It has given rise to new organizations for improving communication among scientists; in the United States, for example, the Bio-Sciences Information Exchange. One of the explicit functions of this Exchange is to protect the individual scientist from the "distress" that comes from being "just about to ship off a manuscript only to discover that someone else had done his work for him."[17] This function requires no comment here: it is designed to improve the system of scientific communication and so to prevent the unintended repetition of already *completed* scientific investigations. But the Exchange is also thought of as having the function of guiding those who allocate funds for research so that they may reduce (or, ideally, eliminate) what is usually described as "the wasteful duplication of scientific effort."

Often enough this notion of duplication conceals a premise that should be further examined in the light of research on multiples in science before it is adopted at face value as a guide to policy. For it is not at all apparent that it is "wasteful" for several individual scientists, or teams of scientists, to work toward and to arrive at solutions of the same problem. Consider only four items of relevance which must be tucked out of sight and out of mind in order to arrive at the deceptively cogent conclusion that multiple discoveries necessarily signify "waste" of duplicative (or unknowingly replicative) scientific effort.

Item: True, the theory of multiples in science leads us to conclude that these repeated discoveries were "inevitable," since if one scientist involved in the multiple had not made the discovery, another would have (as we know from the fact that he did). But this "inevitability" holds only under certain, still poorly identified, conditions. In reviewing the facts of multiples, we ordinarily know only of the several scientists who actually *did* make the same discovery; we usually do not know how many others were at work on the same problem without having solved it. In short, we really do not know how many scientists of what degrees of competence are

---

of ascertained discoveries, the isolated men of science daily run the risk that they may be repeating experiments already made by others. If they were acquainted with other experiments, they would be spared efforts as laborious as they are useless, and it would be easier for them to obtain means for forging ahead." Nor is this all. The complaint about wasteful duplication is coupled with an observation on the quest for priority in science: "Let us add here," say the early Saint-Simonians, "that the security of men of science is not complete. They are haunted by the work of a competitor. Possibly someone else is gleaning the same field and may, as the saying goes, 'get there first'. The man of science has to hide himself and conduct in haste and isolation work requiring deliberation and demanding aid from association with others" (*The Doctrine of Saint-Simon: An Exposition, First Year, 1828–1829,* trans. Georg G. Iggers [Boston: Beacon Press, 1958] p. 9).

17. "Who? What? Where?: An Editorial," *Science* 128 (8 August 1958): 277.

required to focus on a particular kind of problem in order to ensure a high probability that it will be solved in a given span of time. If that number is progressively reduced, through what may at first seem to be a rational policy of allocating only one grant or very few grants for research on the problem, the discovery may become anything but inevitable, at least during a given interval.

Item: That duplication of scientific effort is wasteful may be true when the problems in hand are fairly routine, and bound to yield to a solution, once a scientist elects to work seriously on them. But these, of course, are the small change of science. When it comes to basic problems which are far from routine and, once solved, will have far-reaching implications for further inquiry, duplication, triplication or a higher multiplication of effort may be anything but wasteful.

Item: It would be ironic if current planned efforts to achieve efficiency in creative scientific work were to prove self-defeating. In the past, when the support of science was slight and thinly dispersed, the efficiency-of-the-seemingly-inefficient pattern resulted in many multiples, partly because many scientists, often unknowing that this was so, elected to work on the same problems. A superficial notion of "wasteful duplication" might result in substituting a policy of the inefficiency-of-the-seemingly-efficient, by so allocating funds for research as drastically to restrict the range of scientists at work on the same problem, thus reducing the probability not only of multiple independent solutions but of any solution altogether at the time. The theory of multiples provides one basis for reexamining policies governing the allocation of funds in support of science.

Item: The fallacy of wasteful duplication is much like the fallacy that has long afflicted the interpretation of multiples in science. This fallacy made use of an old-fashioned concept of redundancy—strictly old-fashioned, for it has been going the rounds of philosophers, historians and sociologists for a couple of hundred years. The argument went as follows: The occurrence of a multiple discovery is proof in itself that all but one of the actual discoverers were redundant (i.e. superfluous). For if all the other co-discoverers had not made the discovery, it would have been made in any case. Ergo, the fact attests their superfluity.

I shall not bother with those knotty points of method which might protect the unwary against drawing such false conclusions from sound evidence. I note only that this old notion of redundancy ordinarily merged two distinct meanings. For one thing, it meant abundant, copious, plentiful, more than is abstractly needed to achieve a purpose. For another, it meant superfluous, that which can be *safely* done away with. The merger of these two meanings smuggled in a fallacy. This was the absolutistic fallacy of assuming that something was either redundant or not, once and for all, and irrespective of the situations in which it is found. The newer, more differentiated concept of redundancy is relative and statistical. It recognizes that efficiency increases the prospect of error; that redundancy (or reduced efficiency) makes for safety from error. It leads us to think of and then, in certain cases, to measure a functionally optimum amount of redundancy under specified conditions: that amount which will approximate a maximum probability of achieving the wanted outcome but not so great an amount that the last increment will fail appreciably to enlarge that probability. Multiples in science comprise a particular kind of redundancy that can be thought of in terms of the newer, fruitful concept which opens our eyes to what was presumably there all along, but went unnoticed. There is safety as well as truth in numbers of similar independent discoveries.

Once we use this concept, we see the fallacy of the apparently cogent thesis that in multiples all discoveries but one are superfluous. This is seen to be logically airtight and sociologically false. For it assumes what remains to be demonstrated. It assumes that a discovery has only to be made in order for it to enter the public domain of science. But the history of science is checkered with cases that show this is not so. Often a new idea or a new empirical finding has been achieved and published, only to go unnoticed by others, until it is later uncovered or independently rediscovered and only then incorporated into the science. After all, this is what we mean by rediscovery: the signals provided by a discovery are lost in the noise of the great information system that constitutes science, and so must be issued anew. Multiples—that is, redundant discoveries—have a greater chance of being heard by others in the social system of science and so, then and there, to affect its further development. From this standpoint, multiples are redundant, but not necessarily superfluous (or wasteful). When the all-but-one versions of the same discovery are described as superfluous, this refers only to the discoverer's psychological experience: he has indeed made the discovery. But this account neglects the sociological components of the process of discovery which deal with the probability of the discovery being made in the first place and, once made, of its being assimilated as a functional part of the science.

Multiple discoveries can thus be seen to have several and varied social functions for the system of science. They heighten the likelihood that the discovery will be promptly incorporated in current scientific knowledge and will so facilitate the further advancement of knowledge. They confirm the truth of the discovery (although on occasion errors *have* been independently arrived at). They help us detect a problem which I have barely and far from rigorously formulated, to say nothing of having solved it: how to calculate the functionally optimum amount of redundancy in independent efforts to solve scientific problems of designated kinds, such that the probability of the solution is approximately maximized without entailing so much replication of effort that the last increment will not appreciably increase that probability. They help us distinguish between the psychological experience of individual scientists who originate a new and fruitful idea or make a new and fruitful observation from the independent social process through which this discovery succeeds or fails to become incorporated in the then-current body of scientific knowledge. So much, then, for this sixth set of uses that make up the rationale for the systematic study of multiples in science.

A *seventh* use I have examined at some length elsewhere,[18] and will therefore only summarize here. The methodical investigation of multiples enables us to develop a sociological theory of the role of scientific genius in the development of science. This new theory does away with the false disjunction between a heroic theory of science, that ascribes all basic advances to genius, and an environmental theory, that holds these geniuses to have been altogether dispensable, since if they had not lived, things would have turned out pretty much as they did. These traditionally opposed theories are not inherently opposed; they become so only when, as has been the case, they are pushed to indefensible extremes. In an enlarged sociological conception, men of scientific genius are precisely those whose discoveries, had they remained contemporaneously unknown, would eventually be rediscovered. *But* these rediscoveries would be made not by a single scientist but by an aggregate of scientists. On this view, the individual man of scientific genius is the functional equivalent of a considerable array of other scientists of varying degrees of talent. The evidence for this conception is in part provided by the multiplicity of multiples in which men of undeniable scientific genius have been involved.

An *eighth* and, for present purposes, final use has to do with what might be described as the therapeutic function which the study of multiples serves for the community of scientists. But I shall postpone further examination of this use until the close of the next chapter, when we shall have covered some of the evidence indicating that there is ample need for this therapeutic function among scientists of our own day just as there was among scientists of the past.

Perhaps enough has been said about the rationale for the systematic study of multiples. If there is any merit to the opinion that the subject has at least an eight-fold promise for enlarging our understanding of how science develops, there naturally arises the question: why, then, has such systematic[19] study been largely absent? Like many other questions in the

18. See chapter 16 of this volume. See also an abbreviated version of this paper, "The Role of Genius in Scientific Advance," *New Scientist,* no. 259 (2 Nov. 1961): 306–8.

19. To avoid misunderstanding, it should be reiterated that I refer only to the *systematic* investigation of multiples and frequent conflicts over priority. The ubiquity

sociology of science, this one is self-exemplifying: it calls for application of ideas in the sociology of science to the cognitive and social behavior of sociologists of science themselves.

---

of the events themselves has required historians of science and biographers of scientists to record a good deal of evidence on the subject. But the methodical study of the sources of multiples and priority-conflicts, of their structure and consequences for the advancement of science, has remained in much the same undeveloped state for a long time.

18 The Ambivalence of Scientists

1963

Many of the endlessly recurrent facts about multiples and priorities are readily accessible—in the diaries and letters, the note-books, scientific papers, and biographies of scientists. This only compounds the mystery of why so little systematic attention has been accorded the subject. The facts have been noted, for they are too conspicuous to remain unobserved, but then they have been quickly put aside, swept under the rug, and forgotten. We seem to have here something like motivated neglect of this aspect of the behavior of scientists and that is precisely the hypothesis I want to examine now.

This resistance to the study of multiples and priorities can be conceived as a resultant of intense forces pressing for public recognition of scientific accomplishments that are held in check by countervailing forces, inherent in the social role of scientists, which press for the modest acknowledgment of limitations, if not for downright humility. Such resistance is a sign of malintegration of the social institution of science which incorporates potentially incompatible values: among them, the value set upon originality, which leads scientists to want their priority to be recognized, and the value set upon due humility, which leads them to insist on how little they have in fact been able to accomplish. To blend these potential incompatibles into a single orientation and to reconcile them in practice is no easy matter. Rather, as we shall now see, the tension between these kindred values creates an inner conflict among men of science who have internalized both of them. Among other things, the tension generates a

First published as a part of "Resistance to the Systematic Study of Multiple Discoveries in Science," *European Journal of Sociology* 4 (1963): 250–82; reprinted with permission. A condensed version of part of this paper appears under this title in the *Bulletin* of the Johns Hopkins Hospital, 112 (February 1963): 77–97.

distinct resistance to the systematic study of multiples and often associated conflicts over priority.[1]

Various kinds of overt behavior can be interpreted as expressions of such resistance. For one thing, it is expressed in the recurrent pattern of trying to trivialize or to incidentalize the facts of multiples and priority in science. When these matters are discussed in print, they are typically treated as though they were either rare and aberrant (although they are extraordinarily frequent and typical) or as though they were inconsequential both for the lives of scientists and for the advancement of science (although they are demonstrably significant for both).

Understandably enough, many scientists themselves regard these matters as unfortunate interruptions to their getting on with the main job. Kelvin, for example, remarks that "questions of priority, however interesting they may be to the persons concerned, sink into insignificance" as one turns to the proper concern of advancing knowledge.[2] As indeed they do: but sentiments such as these also pervade the historical and sociological study of the behavior of scientists so that systematic inquiry into these matters also goes by default. Or again, it is felt that "the question of priority plays only an insignificant role in the scientific literature of our time"[3] so that, once again, this becomes regarded as a subject which can no longer provide a basis for clarifying the complex motivations and behavior of scientists (if indeed it ever was so regarded).

Now the practice of seeking to trivialize what can be shown to be significant is a well-known manifestation of resistance. Statements of this sort read almost as though they were a paraphrase of the old maxim that the law does not concern itself with exceedingly small matters; *de minimis non curat scientia* [*lex*]. Not that there has been a conspiracy of silence about these intensely human conflicts in the world of the intellect and especially in science. These have been far too conspicuous to be denied altogether. Rather, the repeated conflict behavior of great and small men of science has been incidentalized as not reflecting any conceivably significant aspects of their role as scientists.

Resistance is expressed also in various kinds of distortions: in motivated misperceptions or in an hiatus in recall and reporting. It often leads to those wish-fulfilling beliefs and false memories that we describe as illusions. And of such behavior the annals that treat of multiples and priorities are uncommonly full. So much so that I have arrived at a rule of thumb that

1. This paragraph draws upon a fuller account of the workings of these values in the social institution of science in "Priorities in Scientific Discovery," chapter 14 of this volume.

2. Silvanus P. Thompson, *The Life of William Thomson, Baron Kelvin of Largs* (London: Macmillan, 1910), 2:602.

3. Otto Blüh, "The Value of Inspiration: A Study of Julius Robert Mayer and Josef Popper-Lynkeus," *Isis* 43 (1952): 211–20, at 211.

seems to work out fairly well. The rule is this: whenever the biography or autobiography of a scientist announces that he had little or no concern with priority of discovery, there is a reasonably good chance that, not many pages later in the book, we shall find him deeply embroiled in one or another battle over priority. A few cases must stand here for many:

Of the great surgeon, W. S. Halsted (who together with Osler, Kelly, and Welch founded the Johns Hopkins Medical School), Harvey Cushing writes: he was "overmodest about his work, indifferent to matters of priority."[4] Our rule of thumb leads us to expect what we find: some twenty pages later in the book in which this is cited, we find a letter by Halsted about his work on cocaine as an anesthesia: "I anticipated all of Schleich's work by about six years (or five). . . . [In Vienna,] I showed Wölfler how to use cocaine. He had declared that it was useless in surgery. But before I left Vienna he published an enthusiastic article in one of the daily papers on the subject. It did not, however, occur to him to mention my name."[5]

Or again, the authoritative biography of that great psychiatrist of the Salpêtrière, Charcot, approvingly quotes the eulogy which says, among other things, that despite his many discoveries, Charcot "never thought for a moment to claim priority or reward." Alerted by our rule of thumb, we find some thirty pages later an account of Charcot insisting on his having been the first to recognize exophthalmic goiter and, a little later, emphatically affirming that he "would like to claim priority" for the idea of isolating patients who are suffering from hysteria.[6]

But perhaps the most apt case of such denial of an accessible reality is that of Ernest Jones, writing in his comprehensive biography that "although Freud was never interested in questions of priority, which he found merely boring"—surely this is a classic case of trivialization at work—"he was fond of exploring the source of what appeared to be original ideas, particularly his own."[7] This is an extraordinarily illuminating statement. For, of course, no one could have "known" better than Jones—"known" in the narrowly cognitive sense—how very often Freud turned to matters of priority: in his own work, in the work of his colleagues (both friends and enemies), and in the history of psychology altogether.

4. In his magisterial biography, *Harvey Cushing* (Springfield: Charles C. Thomas, 1946), pp. 119–20, John F. Fulton describes Cushing's biographical sketch of Halsted, from which this excerpt is quoted, as "an excellent description."

5. Ibid., p. 142.

6. Georges Gullain, *J.-M. Charcot: His Life, His Work*, ed. and trans. Pearce Bailey (New York: Paul B. Hoeber, 1959), pp. 61, 95–96, 142–43.

7. Ernest Jones, *Sigmund Freud: Life and Work*, 3 vols. (London: Hogarth Press, 1957), 3:105. Contrast David Riesman, who takes ample note of Freud's interest in priority, in *Individualism Reconsidered* (Glencoe: The Free Press, 1954), pp. 314–15, 378.

In point of fact, Dr. Elinor Barber and I have identified more than one hundred fifty occasions on which Freud exhibited an interest in priority. Freud himself reports, with characteristic self-awareness, that he even dreamt about priority and the due allocation of credit for accomplishment in science.[8] He oscillates between the poles of his ambivalence toward

8. Sigmund Freud, *The Interpretation of Dreams*, trans. A. A. Brill, 3rd ed. (London: Allen & Unwin, 1932), p. 175. "Now [my dream] means: 'I am indeed the man who has written that valuable and successful treatise (on cocaine).'" This near-miss in being recognized as the discoverer of cocaine as a local anesthetic is of periodic interest to Freud throughout the greater part of his life. Freud simply cannot put it to rest. At the time he is moving toward the idea, in 1884, he writes his fiancée, Martha, about his "toying with a project . . . ; perhaps nothing will come of this, either. It is a therapeutic experiment involving the use of cocaine. . . . There may be any number of other people experimenting on it already; perhaps it won't work. But I am certainly going to try it and, as you know, if one tries something often enough and goes on wanting it, one day it may succeed" (*Letters of Sigmund Freud*, ed. Ernst L. Freud [New York: Basic Books, 1960], pp. 107–8). Seven months later, he writes his future sister-in-law that "'Cocaine has brought me a great deal of credit, but the lion's share has gone elsewhere'" (quoted by Ernest Jones in his detailed chapter on "The Cocaine Episode," *Freud: Life and Work*, 1:98). Two years later he is writing Martha about an episode in the Salpêtrière when the distinguished American ophthalmologist, Hermann Knapp, "who has written a lot about cocaine" says to another of Freud, "it was he who started it all" (Ibid., 209). Evidently the episode stung, for not to cite the other intervening allusions to it, Freud is writing Fritz Wittels about "the cocaine story," some thirty-eight years later, on the occasion of an English translation of Wittels' objectionable biography of Freud: "I guessed its usefulness for the eye, but for private reasons (in order to travel) had to drop the experiment and personally charged my friend Königstein to test the drug on the eye. . . . Königstein (it was *he*, not I, who so deeply regretted having missed winning these laurels) then claimed to be considered the codiscoverer [with Koller] and . . . both Königstein and Koller chose Julius Wagner *and myself* as the arbitrators. I think it did us both honor that each of us took the side of the opposing client. Wagner, as Koller's delegate, voted in favor of recognizing Königstein's claim, whereas I was wholeheartedly in favor of awarding the credit to Koller alone. I can no longer remember [reports Freud] what compromise we decided on" (*Letters of Sigmund Freud*, p. 351). About the same time, Freud puts all this in print (in *An Autobiographical Study* [1925; London: Hogarth Press, 1948], pp. 24–25), explaining that, "While I was in the middle of this work, an opportunity arose for making a journey to visit my fiancée, from whom I had been parted for two years. I hastily wound up my investigation of cocaine and contented myself in my book on the subject with prophesying that further uses for it would soon be found. I suggested, however, to my friend Königstein, the ophthalmologist, that he should investigate the question of how far the anaesthetizing properties of cocaine were applicable in diseases of the eye. When I returned from my holiday, I found that not he, but another of my friends, Carl Koller (now in New York), whom I had also spoken to about cocaine, had made the decisive experiments. . . . Koller is therefore rightly regarded as the discoverer of local anaesthesia by cocaine, which has become so important in minor surgery; [but, adds Freud in so many words] I bore my fiancée no grudge for the interruption of my work." All apart from the cocaine story, Freud, with the resolute self-scrutiny that left little place for self-deception, analyzes another of his dreams as having at its root "an arrogant phantasy of ambition, but that in its stead only its suppression and abasement has reached the dream-content." *Interpretation of Dreams*, p. 440.

priority: occasionally seeing multiples as more or less inevitable, as when he reports a fantasy in which "science would ignore me entirely during my lifetime; some decades later, someone else would infallibly come upon the same things—for which the time was not now ripe—, would achieve recognition for them and bring me honour as a forerunner whose failure had been inevitable."[9] On other occasions he sometimes reluctantly, sometimes calmly and insistently, acknowledges anticipations of his own ideas or reports his own anticipations of others;[10] he "implores" his disciple Lou Andreas-Salomé to finish an essay in order "not to give me precedence in time";[11] he admonishes Adler for what he describes as his "uncontrolled craving for priority,"[12] just as he admonishes Georg Groddeck for being unable to conquer "that banal ambition which hankers after originality and priority";[13] he assesses and repeatedly reassesses the distinctive roles of Breuer and himself in establishing psychoanalysis;[14] he returns time

9. Sigmund Freud, "On the history of the psycho-analytic movement," in *The Standard Edition of the Complete Works of Sigmund Freud,* ed. James Strachey, 20 vols. to date (London: Hogarth Press, 1953–), 14:22.

10. The dozens of such instances need not be cited here, but see only the remarkable paper in which Freud reports that "careful psychological investigation . . . reveals hidden and long-forgotten sources which gave the stimulus to the apparently original ideas, and it replaces the ostensible new creation by a revival of something forgotten applied to fresh material. There is nothing to regret in this; we had no right to expect that what was 'original' could be untraceable and undetermined.

"In my case, too, the originality of many of the new ideas employed by me in the interpretation of dreams and in psycho-analysis has evaporated in this way. I am ignorant of the source of only one of these ideas. It was no less than the key to my view of dreams and helped me to solve their riddles. . . . I started out from the strange, confused and senseless character of so many dreams, and hit upon the notion that dreams were bound to become like that because something was struggling for expression in them which was opposed by a resistance from other mental forces. . . .

"Precisely this essential part of my theory of dreams was, however, discovered by Popper-Lynkeus independently. . . . [His story, *Träumen wie Wachen,*] was certainly written in ignorance of the theory of dreams which I published in 1900, just as I was then in ignorance of Lynkeus's *Phantasien* " (Sigmund Freud, "Josef Popper-Lynkeus and the theory of dreams," in *Standard Edition . . . of Freud,* 19:261–63).

11. In his letter of 25 May 1916, *Letters of Sigmund Freud,* p. 313.

12. Freud, "History of the psycho-analytic movement," *Standard Edition . . . of Freud,* 14:51.

13. *Letters of Sigmund Freud,* p. 317. I shall have occasion to return to the rest of this letter later on in this paper, when we examine the basic *uncertainty* of genuinely independent originality in science

14. It would take a paper in itself to trace out in detail and to interpret Freud's repeated and developing efforts, over a span of more than thirty years, to disentangle Breuer's and his own contributions to the emergence of psychoanalysis. As he became the object of social pressure to identify the contributions of the two, and as the differences gradually became clear to him, he worked toward more discriminating distinctions between their respective intellectual roles in that development. Consider only these few cases in point:

[1896] "I owe my conclusions to the use of the new psycho-analytic method, the probing procedure of J. Breuer" ("Heredity and the aetiology of the neuroses," in

Freud, *Collected Papers*, 4 vols. [London: Hogarth Press, 1949], 1:148). This, as the editor indicates, is the first use of the term, "psycho-analytic," and since the thirty-year-old Freud cannot yet know what will eventually turn out to be encompassed by this method, he simply identifies it with the "probing procedure" of Breuer.

[1896] In his paper "The Aetiology of Hysteria," published in the same year, Freud of course continues to refer to "Breuer's method" and starts with "the momentous discovery of J. Breuer: that the symptoms of hysteria (apart from stigmata) are determined by certain experiences of the patient's which operate traumatically and are reproduced in his psychic life as memory-symbols of these experiences." This is the paper in which he reports, without reservations, that "at the bottom of every case of hysteria will be found one or more experiences of premature sexual experience, belonging to the first years of childhood, which may be reproduced by analytic work though whole decades have intervened"—a judgment which he was of course to find mistaken and one which he was to retract and, courageously and imaginatively, to convert into the problem of why these traumatic experiences were so often a matter of phantasy. In it he refers to "Breuer's method" on a half-dozen or so occasions, but we see how he begins to differentiate some of his own ideas from those of Breuer (*Collected Papers*, 1:183–219).

[1904] By this time, Freud becomes clear and makes it clear to others how he has moved beyond Breuer: e.g. "The particular method of psychotherapy which Freud practises and terms psycho-analysis is an outgrowth [*n.b.*] of the so-called cathartic treatment discussed by him in collaboration with J. Breuer. . . . At the personal suggestion of Breuer, Freud revived this method and tried it with a large number of patients. . . . The changes which Freud introduced in Breuer's cathartic method of treatment were at first changes in technique; these, however, brought about new results and have finally necessitated a different though not contradictory conception of the therapeutic task" ("Freud's psycho-analytic method," *Collected Papers*, 1:264–65, this being Freud's contribution to Löwenfeld's *Psychische Zwangerscheinungen*).

[1905] There is something of a regression here, from the newly perceived differentiation, when Freud refers to "that cathartic or psycho-analytic investigation, discovered by J. Breuer and me" ("Three Contributions to the Theory of Sex," in *The Basic Writings of Sigmund Freud*, trans. and ed. by A. A. Brill [New York: Modern Library, 1938], p. 573).

[1905] But in the same year, Freud definitely dissociates himself from one of Breuer's ideas, saying that: "If, where a piece of joint work is in question, it is legitimate to make a subsequent division of property, I should like to take this opportunity of stating that the hypothesis of 'hypnoid states'—which many reviewers were inclined to regard as the central portion of our work—sprang entirely from the initiative of Breuer. I regard the use of such a term as superfluous and misleading" ("Fragment of an analysis of a case of hysteria," *Collected Papers*, 3:35n).

[1909] Attaching great importance to the international recognition accorded psychoanalysis by the invitation to speak at the celebration of the twentieth anniversary of Clark University, Freud was carried away, temporarily abandoning the distinctive roles he had gradually assigned Breuer and himself, and said unequivocally: "Granted that it is a merit to have created psycho-analysis, it is not my merit. I was a student busy with the passing of my last examinations, when another physician of Vienna, Dr. Joseph Breuer, made the first application of this method to a case of an hysterical girl (1880–1882)." Sigmund Freud, "Origin and development of psycho-analysis," *American Journal of Psychology* 21 (1910): 181–218, at 181. The paper, with this statement, appeared simultaneously in English and German and was soon translated into Dutch, Hungarian, Polish, Russian, and Italian.

[1914] Five years later, Freud expressed second thoughts on the matter: "In 1909, in the lecture-room of an American university, I had my first opportunity of speaking in public about psycho-analysis. The occasion was a momentous one for my work, and moved by this thought I then declared that it was not I who had brought psycho-analysis into existence: the credit for this was due to someone else;

to Joseph Breuer. . . . Since I gave those lectures, however, well-disposed friends have suggested to me a doubt whether my gratitude was not expressed too extravagantly on that occasion. In their view, I ought to have done as I had previously been accustomed to do: treated Breuer's 'cathartic procedure' as a preliminary stage of psycho-analysis. . . . It is of no great importance in any case [*n.b.*, in the light of Freud's repeated worrying of the matter over a period of twenty years] whether the history of psycho-analysis is reckoned as beginning with the cathartic method or with my modification of it; I refer to this uninteresting point [*n.b.*] merely because certain opponents of psycho-analysis have a habit of occasionally recollecting that, after all, the art of psycho-analysis was not invented by me, but by Breuer. This only happens, of course, if their views allow them to find something in it deserving attention; if they set no such limits to their rejection of it, psycho-analysis is always without question my work alone. I have never heard that Breuer's great share in psycho-analysis has earned him a proportionate measure of criticism and abuse. As I have long recognized that to stir up contradiction and arouse bitterness is the inevitable fate of psycho-analysis, I have come to the conclusion that I must be the true originator of all that is particularly characteristic in it. I am happy to be able to add that none of the efforts to minimize my part in creating this much-abused analysis have ever come from Breuer himself or could claim any support from him.

"Breuer's discoveries [include a 'fragment of theory' holding that symptoms of hysteria] represented an abnormal employment of amounts of excitation which had not been disposed of (conversion). Whenever Breuer, in his theoretical contribution to the *Studies on Hysteria* (1895), referred to this process of conversion, he always added my name in brackets after it, as though the priority for this first attempt at theoretical evaluation belonged to me. [The editor notes: "There seems to be some mistake here. In the course of Breuer's contribution he uses the term 'conversion' (or its derivatives) at least fifteen times. But only once (the first time he uses it, *Standard Ed.*, II, p. 206) does he add Freud's name in brackets. It seems possible that Freud saw some preliminary version of Breuer's manuscript and dissuaded him from adding his name more than once in the printed book." Whether this last conjecture is true or not, the fact attests once again Freud's abiding interest with matters of priority and its corollary, the meticulous effort to have 'credit' for originality properly allocated.] I believe that actually this distinction relates only to the name, and that the conception came to us simultaneously and together" ("On the history of the psycho-analytic movement," *Standard Edition of . . . Freud*, 14:7–9).

[1924] Ten years later, Freud reverts to all this in a settled and consistent fashion, writing: "Soon after the publication of the studies in hysteria the collaboration of Breuer and Freud came to an end. Breuer, who was really a general practitioner, gave up the treatment of nervous diseases, while Freud took pains to further perfect the instrument left to him by his older colleague. The technical innovations which he initiated and the new discoveries which he made transformed the cathartic method into psycho-analysis" ("Psycho-analysis: exploring the hidden recesses of the mind," *These Eventful Years* [London and New York: 1924], 2:513).

[1925] Freud's obituary of Breuer will be taken as a final source in point: "I have repeatedly attempted . . . to define my share in the *Studies* which we published jointly. My merit lay chiefly in reviving in Breuer an interest which seemed to have become extinct, and in then urging him on to publication. . . . I found reason later to suppose that a purely emotional factor, too, had given him an aversion to further work on the elucidation of the neuroses. He had come up against something that is never absent—his patient's transference on to her physician, and he had not grasped the impersonal nature of the process. . . . Besides the case history of his first patient Breuer contributed a theoretical paper to the *Studies*. It is very far from being out of date; on the contrary, it conceals thoughts and suggestions which have even now not been turned to sufficient account. Anyone immersing himself in this speculative essay will form a true impression of the mental build of this man, *whose scientific interests were, alas, turned in the direction of our psychopathology*

and again to his priority conflict with Janet,[15] reporting that he had brought the recalcitrant Breuer to agree to an early publication of their joint monograph because "in the meantime, Janet's work had anticipated some of his [Breuer's] results";[16] he writes nostalgically about the days of "my splendid isolation" when "there was nothing to hustle me. . . . My publications, which I was able to place with a little trouble, could always lag far behind my knowledge and could be postponed as long as I pleased, since there was no doubtful 'priority' to be defended";[17] he repeatedly allocates priorities among others (Le Bon, Ferenczi, Bleuler, Stekel, being only a few among the many);[18] he even credits Adler with priority for an error;[19]

---

during only one short episode of his long life" ("Josef Breuer," *Standard Edition . . . of Freud,* 19:279–80).

This short synopsis of Freud's recurring attempts over a span of some forty years to distinguish his contributions from those of Breuer's suggests the possibility that, partly owing to the social pressures upon him to establish the nature of his own originality, he was not altogether uninterested in what he described as "of no great importance" and as an "uninteresting point"; not, at least, if matters of "interest" are those which engage the attention.

15. Of the many occasions on which Freud returned to this matter of Pierre Janet's claim to priority, I cite only "On the history of the psycho-analytic movement," *Standard Edition . . . of Freud,* 14:32–33; and *An Autobiographical Study,* pp. 21, 33, 54–55, where he seeks "to put an end to the glib repetition of the view that whatever is of value in psycho-analysis is merely borrowed from the ideas of Janet. . . . Historically, psycho-analysis is completely independent of Janet's discoveries, just as in its content it diverges from them and goes far beyond them." For some of Janet's not always delicate insinuations, see his *Psychological Healing* (New York: Macmillan, 1925), 1:601–40.

16. *An Autobiographical Study,* pp. 36–37; "Josef Breuer," *Standard Edition . . . of Freud,* 19:279–80: "At the date of the publication of our *Studies,* we were able to appeal to Charcot's writings and to Pierre Janet's investigations, which had by that time deprived Breuer's discoveries of some of their priority. But when Breuer was treating his first case (in 1881–2) none of this was as yet available. Janet's *Automatisme psychologique* appeared in 1889 and his second work, *L'état mental des hystériques,* not until 1892. It seems that Breuer's researches were wholly original, and were directed only by the hints offered to him by the material of his case."

17. "On the history of the psycho-analytic movement," *Standard Edition . . . of Freud,* 14:22. With regard to the pattern of biographers and disciples imposing their illusory convictions upon the actual experience of men of science, consider that the translation of this passage by A. A. Brill completely omits, presumably as inconsequential, the phrase: "There was no doubtful 'priority' to be defended." See *The Basic Writings of Sigmund Freud,* p. 943.

18. References to these will be found scattered through Freud's publications and letters: e.g., *Group Psychology and Analysis of the Ego* (London: Hogarth Press, 1921), pp. 23–24, alludes to Le Bon having been anticipated by Sighele in his most important idea of "the collective inhibition of intellectual functioning and the heightening of affectivity in groups." On this case, see R. K. Merton, introduction to Gustave Le Bon, *The Crowd* (New York: Viking Press, 1960), pp. vii–xviii. To Ferenczi, he writes: "Your priority in all this is evident." Jones, *Sigmund Freud,* 3:353–54.

19. Grimly, Freud writes in the midst of his counter-attack on the secessionist: "Adler must also be credited with priority in confusing dreams with latent dream-

and, to prolong the types of occasions no further, he repeatedly intervenes in priority battles among his disciples and current or former colleagues (for example, between Abraham and Jung),[20] saying that he could not "stifle the disputes about priority for which there were so many opportunities under these conditions of work in common."[21]

In view of even this small sampling of cases in point, it may not be audacious to interpret as a sign of resistance Jones's remarkable statement that "Freud was never interested in questions of priority, which he found merely boring." That Freud was ambivalent toward priority, true; that he was pained by conflicts over priority, indisputable; that he was concerned to establish the priority of others as of himself, beyond doubt; but to describe him as "never interested" in the question of priority and as "bored" by it requires the extraordinary feat of denying, as though they had never occurred, scores of occasions on which Freud exhibited profound interest in the question, many of these being occasions which Jones himself has detailed with the loving care of a genuine scholar. True, Freud appears to have been no more concerned with these matters than were Newton or Galileo, Laplace or Darwin, or any of the other giants of science about whom biographers and others have announced their entire lack of interest in priority just before, as honest scholars, they inundate us with a flood of evidence to the contrary. This denial of the realities they report and segregate seems to be an instance of that keeping of intellect and perception in abeyance which so typically reflects deep-seated resistance.

To propose that such resistance helps account for the studied neglect of systematic study of multiples and priority is still, of course, to leave open the question of what brings the resistance about. It would seem to have obvious parallels with other occasions in the history of thought, not least with psychoanalysis itself, when amply available facts, having far-reaching theoretical implications, were experienced as unedifying or unsavory, ignoble or trivial, and so were conscientiously ignored. It is a little like psychologists having once largely ignored sexuality because it was not a subject fit for polite society or having regarded dreams or incomplete actions as manifestly trivial and so undeserving of thorough inquiry.

What complicates the problem in the case of multiples and priority is that the study calls for detached examination of the behavior of some scientists by other scientists. Even to assemble the facts of the case is to be charged with blemishing the record of undeniably great men of science; as

thoughts" ("On the history of the psycho-analytic movement," *Standard Edition . . . of Freud*, 14:57).

20. Jones, *Sigmund Freud*, 2:52–56.

21. "On the history of the psycho-analytic movement," *Standard Edition . . . of Freud*, 14:25.

though one were a raker of muck that a gentleman would pass by in silence. Even more, to investigate the subject systematically is to be regarded not merely as a muckraker, but as a muckmaker.[22]

The behavior of fellow scientists involved in multiples and priority contests tends to be condemned or applauded rather than analyzed. It is morally evaluated, not systematically investigated. Disputes over priority are simply described as "unfortunate," and the moral judgment is substituted for the effort to understand what this implies for the psychology of scientists and the sociology of science as an institution. We find Goethe referring to "all those foolish quarrels about earlier and later discovery, plagiary, and quasi-purloinings."[23] We are free, of course, to find this behavior unfortunate or foolish or comic or sad. But these affective responses to the behavior of our ancestors- or brothers-in-science seem to have usurped the place that might be given over to analysis of this behavior and its implications for the ways in which science develops. It is a little as though the physician were to respond only evaluatively to illness, describe it as unfortunate or painful, and consider his task done; or as though the psychiatrist were to describe the behavior of schizophrenics as absurd and let it go at that; or as though the criminologist were to substitute his sentiment that certain crimes are appalling and despicable for the

22. Historians of scientific and other ideas are nevertheless rebelling against bowdlerized versions of the life and work of scientists. George Sarton, for example, urges attention "to the long travail and maybe the suffering which led to each [discovery], the mistakes which were made, the false tracks which were followed, the misunderstandings, the quarrels, the victories and the failures; . . . the gradual unveiling of all the contingencies and hazards which constitute the warp and woof of living science" (*A Guide to the History of Science* [Waltham, Mass.: Chronica Botanica Co., 1952], p. 41). A. C. Crombie observes that "we must completely misunderstand Newton the man, and we run the risk of missing the essential processes of a mind so profoundly original and individual as his, if we exclude all those influences and interests that may be distasteful to us, or seem to us odd in a scientist. On closer examination it may turn out in fact that it was those very things that were his chief interest and that most profoundly affected his scientific imagination" ("Newton's Conception of Scientific Method," *Bulletin of the Institute of Physics*, Nov. 1957, pp. 350–62, at 361). And Jacques Barzun finds merely tiresome the homilies that pass as descriptions of scientists at work, reminding us that "science is made by man, in the light of interests, errors and hopes, just like poetry, philosophy and human history itself. To say this is not to degrade science, as naive persons might think; it is on the contrary to enhance its achievements by showing that they sprang not from patience on a monument but from genius toiling in the mud" (*Teacher in America* [Garden City, N.J.: Doubleday, Anchor Books, 1954], p. 90). As far back as the 1840s Augustus de Morgan had complained about the "curious tendency of biographers [particularly of scientists] to exalt those of whom they write into monsters of perfection." No one could ever accuse de Morgan of this practice, particularly when he was writing about Newton. See his *Essays on the Life and Work of Newton* (Chicago: Open Court Publishing Co., 1914), pp. 62–63.

23. Goethe's *Briefe* in *Werke* (Weimar: Hermann Boehlaus, 1903), 27:219–23. I am indebted to Aaron Noland, of the *Journal of the History of Ideas*, for calling my attention to this passage.

effort to discover what brings these crimes about. The history of the sciences shows that the provisional emancipation from sentiment in order to investigate phenomena methodically has been a most difficult task and has occurred at different times in the various sciences and at different times for selected problems within each of the sciences. Emancipation from sentiment came fairly early in the history of much of medicine; it came very late in the history of the treatment of the mentally ill and the analysis of criminal behavior. I suggest that only now are we beginning to emancipate the study of the concrete behavior of scientists from the altogether human tendency to respond to it in terms of the sentiments and values which we have made our own rather than to examine some of that behavior in reasonably detached fashion.

In regard to the study of multiples and priorities, apparently, we must remember again—what we all know in the abstract but are sometimes inclined to forget when we get down to new cases—that, as Clerk Maxwell noted, "It was a great step in science when men became convinced that, in order to understand the nature of things, they must begin by asking, not whether a thing is good or bad, noxious or beneficial, but of what kind it is? and how much is there of it? Quality and quantity were then first recognized as the primary features to be observed in scientific inquiry."[24]

Contributing to the substitution of sentiment for analysis and so to the resistance against systematic study of multiples and their often connected disputes over priority is the often painful contrast between the actual behavior of scientists and the behavior ideally prescribed for them.

For all of us who harbor the ideal image of the scientist, it may be disconcerting to have Edmond Halley forthrightly described by the first Astronomer Royal, John Flamsteed, as being just as "lazy and slothful as he is corrupt." And then, bringing an even greater name into the drama, he goes on to write:

> With my lunar observations he [Halley] gives her true places and latitudes, which are copied from the three large synopses that I imparted to Sir Isaac Newton, under this condition that he should not impart them to anybody, without my leave. Yet so true to his word, and so candid is the Knight, that he immediately imparted it to Halley; who has printed them as far as they reach. . . . The lazy and malicious thief would scarce be at the pains to gather them himself.[25]

24. James Clerk Maxwell, "Relation of mathematics and physics," an address to the British Association for the Advancement of Science, 1870.

25. Francis Baily, *An Account of the Rev. John Flamsteed, the First Astronomer-Royal; Compiled from his own Manuscripts, and other Authentic Documents, never before published* (London: Printed by Order of the Lords Commissioners of the Admiralty, 1835), pp. 323–24. Much of this volume is devoted to the "notorious" and angry disputes over priority and intellectual property engaging Flamsteed and Newton and Halley, among others.

As Flamsteed put it most plainly, he found Newton, "always insidious, ambitious, and excessively covetous of praise."[26]

Nor do matters fare differently in the emerging social sciences. We find the eighteenth-century Adam Ferguson replying to the charge of having plagiarized the lectures of his friend, Adam Smith, by admitting that "he had derived many notions from a French author, and that Smith had been there before him."[27] (Incidentally, this polemic pattern of "you too" or "you, rather than I" was by then well established; Newton, for example, had adopted this kind of counterattack in reply to Robert Hooke.) Another friend, Adam Robertson, is held to have made unacknowledged use of ideas which Smith had set forth in lectures and conversations, and "in order to establish priority," Smith is provoked into a public lecture which "gave a fairly long list of his new ideas."[28]

In the same vein, we find Saint-Simon ironically extending his "sincere thanks" to the historian Guizot for having "popularized the observations I published in the *Organisateur*" and asking Guizot to read the letter of thanks "with great care [since] it is highly desirable, both for the public and for me, that he appropriate its content as fully as he did my first ideas."[29]

Or again, we are at the turn of our century, listening to the long and bitter dispute between Gustave Le Bon and Scipio Sighele, in which the Italian complains that Le Bon "uses my observations on the psychology of crowds without citing me," and adds, that "without any trace of irony, I believe that no higher or less suspect praise can be given than by this adoption of my ideas without citing me."[30]

26. Ibid., pp. 73–74.

27. William Robert Scott, *Adam Smith as Student and Professor* (Glasgow: Jackson, Son and Co., 1937), p. 119. In this circle of friends, Alexander Carlyle cites this equivocal "defense" of Ferguson; cf. Carlyle's *Autobiography* (Boston: Ticknor and Fields, 1860), p. 285. On Newton's use of swift counterattack, see his letter of 20 June 1686 to Halley in which he writes: "I am told by one who had it from another lately present at one of your meetings, how that Mr. Hooke should there make a great stir, pretending that I had it all from him, and desiring they would see that he had justice done him. This carriage toward me is very strange and undeserved; so that I cannot forbear in stating the point of justice, to tell you further, that he has published Borell[i]'s hypothesis in his own name; and the asserting of this to himself, and completing it as his own, seems to me the ground of all the stir he makes." The letter is reproduced in David Brewster, *Memoirs of the Life, Writings, and Discoveries of Sir Isaac Newton* (Edinburgh: Thomas Constable, 1855), 1:442.

28. Scott, *Adam Smith*, pp. 55, 101, 119.

29. *The Doctrine of Saint-Simon: An Exposition. First Year, 1828–1829*, trans. Georg G. Iggers (Boston: Beacon Press, 1958), p. 23n.

30. Scipio Sighele, *La foule criminelle: essai de psychologie collective*, 2d ed. (Paris: Alcan, 1901), pt. 2, chap. 11, under the title "physiologie du succès," which is introduced by a note stating that the chapter first appeared in *Revue des Revues*, 1 Oct. 1894, the date being cited to safeguard his priority from Le Bon.

In another time and place, Lester Ward writes to E. A. Ross about Albion Small's *General Sociology* that "I suppose I ought to be amused instead of provoked. But a big volume filled with nothing but the things that you and I have been saying for years, only said over again in a verbose language which strains to avoid the particular words used by others and to palm off some other words for new ideas, is certainly exasperating."[31]

And so it goes, on and on. Ignoring for the moment the great volume of such angry complaints in the physical and life sciences, we hear Comte denouncing Saint-Simon, and the Saint-Simonians, Comte; Spencer in turn upbraiding the Comtists for holding him to be a mere imitator of Comte;[32] Marx berating Hyndman as an out-and-out robber of his ideas;[33] the usually equable Gaetano Mosca fuming at "the Marquess Pareto" over his double crime of first having appropriated Mosca's theory of "the political class" and then rechristening the idea by the far more popular term, "elite";[34] Jungians accusing Freud,[35] Freud accusing Adler (with the further and by now familiar charge that the borrowed ideas become "labelled as his own by a change in nomenclature"),[36] and Adlerians accusing Freud and a variety of others.[37]

31. In a letter of 18 March 1906, reprinted in "The Ward-Ross Correspondence, IV, 1906–1912," ed. B. J. Stern, in *American Sociological Review* 14 (1949): 88–119, at 90. In his reply, Ross writes that "I agree with you about Small's book. Having no quarrel with the matter of the book I resolutely shut my eyes to the form. But there is no denying that the cloudiness and prolixity will hurt the book with the public and may give sociology something of a black eye. Already I notice a feeling of 'If this be sociology, Good Lord deliver us.' However sociology has endured many things like it and my faith in its ultimate triumph never wavers" (Ibid., p. 93).

32. For example, in David Duncan, *The Life and Letters of Herbert Spencer* (London: Methuen and Co., 1908), appendix B, pp. 565–72.

33. On this, see Isaiah Berlin, *Karl Marx*, 2nd ed. (London: Oxford University Press, 1948), p. 267. Berlin goes on to note: "Marx held violent opinions on plagiarism," as we know from his unrestrained attacks on Malthus and Bastiat, among others.

34. This prolonged conflict over priority rankled enough for Mosca to return to it over a span of more than thirty-five years. A detailed account will be found in chapter 8 of James H. Meisel, *The Myth of the Ruling Class: Gaetano Mosca and the 'Elite'* (Ann Arbor: University of Michigan Press, 1958). Mosca did not hesitate to note that other contemporaries, among them J. Novicov and Otto Ammon, had independently reached much the same conclusions but that for these and other ideas, "the only case in which I was not able to convince myself of that same spontaneity is that of Professor Pareto." Mosca then goes on to explain: "Plagiarism in the social sciences cannot be as easily established as in literary productions, because what matters most in the former is the concept, not the form, and it is always possible to repeat and to reproduce a concept by changing words around. . . . An educated and shrewd man may always introduce modifications and even add a little something of his own" (Quoted by Meisel, p. 173).

35. For one detailed account of this polemic, see Edward Glover, *Freud or Jung* (New York: W. W. Norton, 1950).

36. See Freud's all-out attack on Adler in which he says, among much else, "At the Vienna Psycho-analytical Society we once actually heard him claim priority for

As we approach our own day in the social sciences, we hear echoes of these angry and agitated words reverberating through the corridors of the peaceful temple of science. Since these episodes involve our contemporaries and often our associates, they become, we must suppose, even more painful to observe and more difficult to analyze with detachment than episodes of the distant past. Only a few present-day conflicts over priority in psychology and sociology, with their intimations or outright assertions of unacknowledged borrowing, need be reviewed to reinstitute the embarrassment and wriggling discomfort experienced by social scientists who are onlookers.

J. L. Moreno, for example, and S. R. Slavson are deep in conflict over the question of who originated group psychotherapy, with Moreno describing Slavson as "liking my concepts and terms group therapy and group psychotherapy and a few years later [beginning] to use them without quotation."[38] Slavson, in his turn, retorts that Moreno was not really the inventor of psychodrama and that priority actually belongs to Karl Joergensen of Sweden, thus following, perhaps unwittingly, the established practice of countering with the claim of a still earlier priority.[39] Or again, Moreno maintains that his ideas and some of Kurt Lewin's are not really cases of independent multiple discovery and that by the time Lewin had published his work on group or action dynamics, he, Moreno, was "the acknowledged leader . . . in the new developments of action and group theory."[40] Some of Lewin's students, Moreno goes on to say, "attended his workshops" and adopted his ideas and techniques under camouflaging labels employed in group dynamics. As Moreno puts it, in colorful lan-

---

the conception of the 'unity of the neuroses' and for the 'dynamic view' of them. This came as a great surprise to me, for I had always believed that these two principles were stated by me before I ever made Adler's acquaintance " ("On the history of the psycho-analytic movement," *Standard Edition . . . of Freud*, 14: 51–58).

37. See, for example, Heinz L. and Rowena R. Ansbacher, *The Individual Psychology of Alfred Adler* (New York: Basic Books, 1956), and the counter-attack by David Rapaport on it and on a review by R. W. White, who "in certain respects" gave the palm to Adler, in *Contemporary Psychology* 2 (November 1957): 303–4. See also Jones, *Sigmund Freud*, 3:296.

38. J. L. Moreno, *Who Shall Survive? Foundations of Sociometry, Group Psychotherapy and Sociodrama*, 2nd ed. (Beacon, New York: Beacon House, Inc., 1953), p. lxi. Moreno goes on to observe: "He who claims priority (which is a form of superiority), however justified, becomes unpopular with the majority" (p. lxii). Moreno's position is defended at length in Didier Anzieu, *Le psychodrame analytique chez l'enfant* (Paris: Presses Universitaires de France, 1946), pp. 28–29.

39. S. R. Slavson, "A preliminary note on the relation of psychodrama and group psychotherapy," *International Journal of Group Psychotherapy* 5 (1955): 361–66. And see the reply by Joseph Meirs, "Scandinavian myth about psychodrama: a counter-statement to S. R. Slavson's 'preliminary note'," *Group Psychotherapy* 10 (1957): 349–52.

40. Moreno, *Who Shall Survive?*, p. ci.

guage reminiscent of that we have seen employed by angry scientists of the past:

It was a shrewd device to plant, at least in the mind of *some* people, the idea that by sheer coincidence of circumstances the same ideas developed independently. By using a technique of quoting only each other, that is, those who belong to their clique, and not quoting any of my close associates or myself, their double game became the laughing stock of the connoisseurs.[41]

In much the same fashion but covering a broader scope, Pitirim Sorokin attacks what he describes as "amnesia and the discoverer's complex" in modern sociology and psychosocial science. He flails social scientists whom he sees as having borrowed from past observers without acknowledgement and aims his heaviest guns at those he regards as having filched ideas from their contemporaries which are then put forward as their own. Once again, the language is that of an angry Galileo, Flamsteed, or Hooke. Thus, the concept of the "basic personality structure" is described as "a vague variation of a very old concept 'pilfered' from sociologists."[42] Leopold von Wiese is approvingly quoted as having written that certain social theorists have "a strange lack of references to their predecessors" and, despite the "essential similarity" of sociological framework, they have a "complete lack of references to theories of mine [Sorokin's] published many years before."[43] Most often, Sorokin says, claims to priority are probably due to "the ignorance of our pseudo-discoverers, many of whom are newcomers from other fields," such as statistics, and have failed to live up to the obligation to find out what has gone before.[44] Beyond this merely ignorant group, he writes, the "wouldbe Columbuses" of social science today include "an insignificant fraction of deliberate plagiarists." Some of these

pseudo-discoverers are the victims of ambitions far exceeding their creative potential and of our society's competitive mores and its cult of success. Driven by their Narcissistic complex and by the ever-operating forces of rivalry, they are eager to overestimate their achievements, to advertise them as 'discoveries made for the first time,' and with a semirational naïveté they are apt sincerely to fool themselves and others with their claims.[45]

And finally, almost as an echo of Ward writing to Ross about Small, or of Mosca writing about Pareto, or Freud about Adler, Sorokin refers to the "technique of using new terms for old concepts to give them a look of

41. Ibid., p. cii.
42. Pitirim A. Sorokin, *Fads and Foibles in Modern Sociology and Related Sciences* (Chicago: Henry Regnery Co. 1956), p. 13.
43. Ibid., p. 14.
44. Ibid., p. 17.
45. Ibid., p. 19.

originality. These and similar devices help to sell, especially to a credulous public, the old intellectual merchandise as the new."[46]

In the climate created by such denunciations, even the social scientists who are not directly involved, at least for the moment, feel acutely uncomfortable. Uneasy and distressed, they can hardly bring themselves to study this behavior. For when sociological analysis is stripped bare of sentiment, it often leaves the sociologist shivering in the cold. Since his own sentiments and allegiances are involved, it becomes all the more difficult to examine the hot conflicts of associates with required detachment. The sociological and psychological study of multiples and priorities, with all it could tell us about the cultural and social behavior of scientists, accordingly tends to remain undeveloped.

All this creates a degree of ambivalence toward the subject and its implications. Thus, Freud also recognizes his own ambivalence when he writes of his work on the Moses of Michelangelo that, having come upon a little book (of 46 pages) published in 1863 by an Englishman, Watkiss Lloyd, he read it

> with mixed feelings. I once more had occasion to experience in myself what unworthy and puerile motives enter into our thoughts and acts even in a serious cause. My first feeling was of regret that the author should have anticipated so much of my thought, which seemed precious to me because it was the result of my own efforts; and it was only in the second instance that I was able to get pleasure from its unexpected confirmation of my opinion. Our views, however, diverge on one very important point.[47]

Ambivalence is otherwise expressed when Moreno concludes his assault on those who do not acknowledge his originality of conception. He remarks that the

> motives for exposing interpersonal conflicts with former associates have little to do with 'priority' or 'recognition.' My cravings for ego-satisfaction, for 'being loved and admired,' has been comfortably reciprocated. If a father of ideas gets fifty percent-returns he can consider himself lucky, and I got more than this.[48]

It is almost as though we were once again reading Descartes as he manages to write both that he "does not boast of being the first discoverer" and then proceeds on other occasions to insist on his priority over Pascal.[49]

The ambivalence toward claims of priority means that scientists are contemptuous of the very attitudes they have acquired from the institution to which they subscribe. The sentiments they have acquired from the insti-

46. Ibid., p. 19.
47. "The Moses of Michelangelo," in Freud, *Collected Papers*, 4:284–85.
48. Moreno, *Who Shall Survive?*, p. cvi.
49. Cf. Chapter 14 in this volume.

tution of science, with its great premium on originality, makes it difficult to give up a claim to a new idea or new finding. Yet the same institution emphasizes the selfless dedication to the advancement of knowledge for its own sake. Concern with priority and ambivalence toward that concern register in the individual what is generated by the value system of science.[50]

The self-contempt often expressed by scientists as they observe with dismay their own concern with having their originality of discovery recognized is evidently based upon the widespread though uncritical assumption that behavior is actuated by a single motive, which can then be appraised as good or bad, as noble or ignoble. They assume that the truly dedicated scientist must be concerned only with advancing knowledge. As a result, their deep interest in having their priority of discovery recognized by peers is seen as marring their nobility of purpose as men of science (although it might be remembered that "noble" means the widely-known). This assumption has a germ of psychological truth: any reward—money, fame, position—is morally ambiguous and potentially subversive of culturally esteemed motives. For as rewards are meted out—fame, for example—the motive of seeking the reward can displace the original motive, concern with recognition can displace concern with advancing knowledge.[51] But this is only a possibility, not an inevitability. When the institution of science works effectively—and like other social institutions it does not always do so—recognition and esteem accrue to those scientists who have best fulfilled their roles, to those who have made important contributions to the common stock of knowledge. Then are found those happy circumstances in which moral obligation and self-interest coincide and fuse. The observed ambivalence of scientists toward their own interest in having their priority recognized—an ambivalence we have seen registered even by that most astute of psychologists, Freud—shows them to assume that such an ancillary motive somehow tarnishes the "purity" of their interest in scientific inquiry. Yet it need not be that scientists seek only to win the applause of their peers but, rather, that they are comforted and gratified by it when it does ring out.

Occasionally a scientist senses all this and vigorously challenges the assumption underlying the shame over interest in recognition; for example, a Hans Selye, who asks his peers:

> Why is everybody so anxious to deny that he works for recognition? In my walk of life, I have met a great many scientists, among them some of the most

50. Lionel Trilling has observed that the "scientist also loves fame, but illicitly: it is not in accord with his professional legend that he should do so, and he is ashamed if his guilty passion is discovered" (*A Gathering of Fugitives* [Boston: Beacon Press, 1956], pp. 143–44).

51. On the displacement of goals, see Merton, *Social Theory and Social Structure*, pp. 199–200.

prominent scholars of our century; but I doubt if any one of them would have thought that public recognition of his achievements—by a title, a medal, a prize, or an honorary degree—played a decisive role in motivating his enthusiasm for research. When a prize brings both honor and cash, many scientists would even be more inclined to admit being pleased about the money ('one must live') than about the public recognition ('I am not sensitive to flattery'). Why do even the greatest minds stoop to such falsehoods? For, without being conscious lies, these ratiocinations are undoubtedly false. Many of the really talented scientists are not at all money-minded; nor do they condone greed for wealth either in themselves or in others. On the other hand, all the scientists I know sufficiently well to judge (and I include myself in this group) are extremely anxious to have their work recognized and approved by others. Is it not below the dignity of an objective scientific mind to permit such a distortion of his true motives? Besides, what is there to be ashamed of?[52]

Dr. Selye's final question need not remain a rhetorical one. Shame is experienced when one's identity and self-image are suddenly violated by one's actual behavior—as in the case of the shame we have seen expressed by Darwin when his own behavior forced him to realize that recognition of his priority meant more to him than he had ever been willing to suppose. To admit to a deep-seated wish for recognition may seem to prefer recognition to the joy of discovery as an end in itself, activating the further awareness that the pleasure of recognition for accomplishment could, and perhaps momentarily did, replace the pleasure of scientific work for its own sake.

On the surface, this hunger for recognition appears as mere personal vanity, generated from within and craving satisfaction from without. But this is truly a superficial diagnosis, compounded of a moralizing deprecation of self or others and representing a classic instance of the fallacy of misplaced concreteness in which relevant sociological details are suppressed by exclusive attention to the feeling-states of the particular individual scientist. When we reach deeper and wider, into the institutional complex that gives point to this hunger for recognition, it turns out to be anything but personal and individual, repeated as it is with slight variation by one scientist after another. Vanity, so-called, is then seen as the outer face of the inner need for assurance that one's work really matters, that one has measured up to the hard standards maintained by a community of scientists. It then becomes clear that the institution of science reinforces, when it does not create, this deep-rooted need for validation of work accomplished. Sometimes, of course, the need is stepped up until it gets out of hand: the desire for recognition becomes a driving lust for acclaim (even when unwarranted), megalomania replaces the comfort of reassurance. But the extreme case need not be mistaken for the modal one.

52. Hans Selye, *The Stress of Life* (New York: McGraw-Hill, 1956), p. 288. On the same point, see chapter 15 of this volume.

In general, the need to have accomplishment recognized, which for the scientist means that his knowing peers judge his work worth the while, is the result of deep devotion to the advancement of knowledge as an ultimate value. Rather than necessarily being at odds with dedication to science, the concern with recognition is ordinarily a direct expression of it. This becomes evident only if one does not stop analysis by characterizing this concern as a matter of vanity or self-aggrandizement but goes on to consider that, sociologically, recognition of accomplishment by informed others represents a mechanism of social validation of that accomplishment. Science in particular is a social world, not an aggregate of solipsistic worlds. Continued appraisal of work and recognition for work well done constitute one of the mechanisms that unite the world of science.

The Eureka Syndrome

All this can be seen in somewhat different context: the deep concern with establishing priority or at least independence of discovery is only the other side of the coin of the socially reinforced elation that comes with having arrived at a new and true scientific idea or result. And the deeper the commitment to a discovery, the greater, presumably, the reaction to the threat of having its novelty denied. Concern with priority is often only the counterpart to elation in discovery—the eureka syndrome. We have only to remember what is perhaps the most ecstatic expression of joy in discovery in the annals of science: here is Kepler on his discovery of the third planetary law:

> What I prophesied 22 years ago as soon as I found the heavenly orbits were of the same number as the five (regular) solids, what I fully believed long before I had seen Ptolemy's Harmonics, what I promised my friends in the name of this book, which I christened before I was 16 years old, what I urged as an end to be sought, that for which I joined Tycho Brahe, for which I settled in Prague, for which I have spent most of my life at astronomical calculations—at last I have brought to light, and seen to be true beyond my fondest hopes. It is not 18 months since I saw the first ray of light, three months since the unclouded sun-glorious sight burst upon me! Let nothing confine me; I will indulge my sacred ecstasy. I will triumph over mankind by the honest confession that I have stolen the golden vases of the Egyptians to raise a tabernacle for my God far away from the lands of Egypt. If you forgive me, I rejoice; if you are angry, I cannot help it. The book is written; the die is cast. Let it be read now or by posterity, I care not which. It may well wait a century for a reader, as God has waited 6000 years for an observer.[53]

We can only surmise how deep would have been Kepler's anguish had another claimed that he had long before come upon the third law. So, too,

53. As translated in William S. Knickerbocker, ed., *Classics of Modern Science* (New York: Knopf, 1927), p. 30.

with a Gay-Lussac, seizing upon the person nearest him for a victory waltz so that he could "express his ecstasy on the occasion of a new discovery by the poetry of motion."[54] Or, to come closer home, William James "all aflame" with his idea of pragmatism and hardly able to contain his exhilaration over it.[55] Or, in more restrained exuberance, Joseph Henry, once he had hit upon a new way of constructing electromagnets, reporting that "when this conception came into my brain, I was so pleased with it that I could not help rising to my feet and giving it my hearty approbation."[56] Or finally, the young Freud writing his "darling girl," Martha, of his "joy" in a "discovery which may not be insignificant": a new technique of staining nervous tissue with a solution of gold chloride,[57] or, years later, reminding Karl Abraham that "we have the incomparable pleasure of gaining the first insights."[58]

In short, when a scientist has made a genuine discovery, he is as happy as a scientist can be. But the peak of exhilaration may only deepen the plunge into despair should the discovery be taken from him.[59] If the loss is occasioned only by finding that it was, in truth, not a first but a later independent discovery, the blow may be severe enough, though mitigated by the sad consolation that at least the idea has been confirmed by another. But this is as nothing, of course, when compared with the traumatizing charge that not only was the discovery later than another of like kind but that it really was borrowed or even stolen. Rather than being mutually exclusive, joy in discovery and eagerness for recognition by scientific peers are stamped out of the same psychological coin. They can both express a basic commitment to the value of advancing knowledge.

Cryptomnesia ("Unconscious Plagiary")

Further complicating the already complex emotions that attend multiple discoveries is the phenomenon of so-called "unconscious plagiary." Inter-

54. Edward Thorpe, *Essays in Historical Chemistry* (London: Macmillan, 1931).

55. See James's letter to Flournoy, in Ralph Barton Perry, *The Thought and Character of William James* (Boston: Little, Brown, 1936), 2:452.

56. Thomas Coulson, *Joseph Henry: His Life and Work* (Princeton: Princeton University Press, 1950), 49–50. The self-effacing Henry, it will be remembered, was periodically involved in multiples and, on occasion, in disputes over priority as evidenced not least in his candid report of great disappointment over Faraday's having been regarded as the prior discoverer of electromagnetic induction.

57. *Letters of Sigmund Freud*, p. 72.

58. Ibid., p. 286.

59. For an example, witness the account sent to Gauss by Schumacher, of Niels Abel's dismay upon learning that he had been anticipated by Jacobi, with Abel needing some brandy to sustain himself, and Schumacher's concluding remark: "Wenn Sie einmal Ihre Untersuchungen bekannt machen, wird es ihm wahrscheinlich noch mehr an Schnapps kosten" (*Briefwechsel zwischen C. F. Gauss und H. C. Schumacher*, ed. C. A. F. Peters [Altona: Gustav Esch, 1860], 2:179).

estingly enough, the potpourri term itself testifies to the admixture of moralizing and analysis that commonly enters into discussions of the subject. It is compounded of a loosely conceived psychological component ("unconscious") and a legal-moralistic component ("plagiary", with all its connotations of violating a code and attendant guilt). As a concept, "unconscious plagiary" is just as misplaced or obsolete in psychosocial studies as that of insanity, which was rightly relegated to the sphere of law, where it continues to lead a harrowing existence. The neutral and analytical term, cryptomnesia, serves us better, referring as it does to seemingly creative thought in which ideas based upon unrecalled past experience are taken to be new.

The fact that cryptomnesia can occur at all subjects the scientist (or other creative minds) to the ever-present possibility that his most cherished original idea may actually be the forgotten residue of what he had once read or heard elsewhere. This fear may give rise to either of two conflicting patterns of behavior. In some cases, it may lie behind the emphatic insistence of an imaginative mind that he is beholden to no one else for his newfound ideas.[60] This pattern of a possibly cryptomnesic scientist who protests his originality too much, not knowing whether he is right or not, differs of course from the pattern of the-lady-who-doth-protest-too-much, knowing as she does that her act will belie her words. In other cases the scientist who knows that cryptomnesia can occur may assume that he has unwittingly assimilated an idea which he once believed to have been original with him. This may hold for big ideas or small ones. I know that the statistician, W. Allen Wallis, will not mind my citing such a minor episode from his experience. In the well-known textbook of statistics which Harry Roberts and he published in 1956, they introduced the convenient practice of numbering tables and charts, not seriatim as is ordinarily done, but by the number of the page on which they appear. This has the advantage that later cross-references to the tables or charts at once indicate the page on which they are found. This useful little idea also turns out to be a multiple. The book is published, and Wallis soon receives a friendly letter from an economist notifying him that this system of numbering had been employed by Dunlap and Kurtz in a handbook of statistics published back in 1932. Wallis's reply exemplifies the uncertainty that comes from realizing that cryptomnesia is an ever-present possibility:

> I was much interested in your letter. . . . The numbering method seemed to me so good that it obviously must have been thought of before. . . . The Dunlap and Kurtz book is, I am virtually certain, in my office in Chicago. When I arrived at Chicago ten years ago as Ted Yntema's successor, he very kindly

60. But, to take one of the most familiar cases, it is by no means clear that Montesquieu intended by his motto—*Prolem sine matre creatam*—that the *Spirit of the Laws* was only a source and indebted to none before him.

left a considerable portion of his statistical library for me. I have noticed this volume, though I do not recall ever looking into it. Nevertheless, there does seem to me to be a real possibility that on some occasion I did look at it, note the numbering system, forget it, but then did think it up 'fresh' when faced with a numbering problem.[61]

What holds for this little instance holds also for discoveries of consequence to science: the possibility of cryptomnesia leads some to doubt their own powers of recall and to assume that what they once thought to be their original idea may be, after all, the trace of a forgotten exposure to the idea as set forth by another.

Among the many cases in point, consider only these few. Having had the experience at age nineteen of learning that his discovery in optics was 'only' a rediscovery, William Rowan Hamilton, the mathematical genius who discovered quaternions (in part, independently invented by Grassmann), developed a lifelong preoccupation with the twin fear of being plagiarized and of unwittingly plagiarizing others. As he put it on one of the many occasions on which he turned to this subject in his correspondence with de Morgan: "As to myself, I am *sure* that I *must* have often reproduced things which I had read long before, without being able to identify them as belonging to other persons."[62] Or again: "But about the 'sighing'—am I to quarrel with Dickens, or figure in one of his publications of a later date? Where is the priority business to end? I am sick of it as you can be; but still, in anything important as regards science, I should take it as a favour to be *warned,* if I were inadvertently exposing myself to the charge of plagiarising."[63]

Turning from mathematics to psychology, we find Freud characteristically examining his own experience, remembering that he had been given Börne's works when he was fourteen and still had the book fifty years later, so that although "he could not remember the essay in question," which dealt with free association as a procedure for creative writing, "it does not appear impossible to us that this hint may perhaps have uncovered that piece of cryptomnesia which, in so many cases, may be suspected behind an apparent originality."[64] In reviewing the multiple

61. W. Allen Wallis, personal communication.

62. R. P. Graves, *Life of Sir William Rowan Hamilton* (Dublin: Hodges, Figgis, 1882), 3:297. This extensive biography includes scores of letters by Hamilton which report his pervasive concern with matters of priority, rediscovery, the giving of credit for originality in science, plagiary, scientists' desire for immortal fame, anticipations, fear of being forestalled, etc. As de Morgan observed, Hamilton was obsessed by possible cryptomnesia: "He had a morbid fear of being a plagiarist; and the letters which he wrote to those who had treated like subjects with himself sometimes contained curious and far-fetched misgivings about his own priority" (Ibid., 3:217).

63. Ibid., 3:368.

64. Quoted from Freud's anonymous paper, "A note on the pre-history of the technique of analysis," by Lewis W. Brandt in his instructive paper dealing with

discovery of part of the "theory of dreams" by Popper-Lynkeus and himself, Freud has this to say:

The subjective side of originality also deserves consideration. A scientific worker may sometimes ask himself what was the source of the ideas peculiar to himself which he has applied to his material. As regards some of them he will discover without much reflection the hints from which they were derived, the statements made by other people which he has picked out and modified and whose implications he has elaborated. But as regards others of his ideas he can make no such acknowledgements; he can only suppose that these thoughts and lines of approach were generated—he cannot tell how—in his own mental activity, and it is on them that he bases his claim to originality.

Careful psychological investigation, however, diminishes this claim still further. It reveals hidden and long-forgotten sources which gave the stimulus to the apparently original ideas, and it replaces the ostensible new creation by a revival of something forgotten applied to fresh material. There is nothing to regret in this; we had no right to expect that what was 'original' could be untraceable and undetermined.

In my case, too, the originality of many of the new ideas employed by me in the interpretation of dreams and in psycho-analysis has evaporated in this way. I am ignorant of the source of only one of these ideas ["dream-censorship"].[65]

Most incisively, Freud exemplifies the basic uncertainty inherent in the fact that cryptomnesia *can* occur, when he writes in "Analysis Terminable and Interminable":

My delight was proportionally great when I recently discovered that that theory [of the "death instinct"] was held by one of the great thinkers of ancient Greece. For the sake of this confirmation I am happy to sacrifice the prestige of originality, especially as I read so widely in earlier years that I can never be quite certain that what I thought was a creation of my own mind may not really have been an outcome of cryptomnesia.[66]

It was this sort of thing, no doubt, that prompted the irrepressible Mark Twain to declare: "What a good thing Adam had—when he said a thing he knew nobody had said it before."

Celebrated cases of seeming cryptomnesia abound in all fields of creative work. To take only one dramatic example, Helen Keller writes in despair of having published a story that was "so much alike in thought and language [to another] that it was evident" the earlier one must have been read to her and that "mine was a—" [even in print she pauses, draws a deep breath, and only then can bring herself to say] that "mine was a—plagiary." "No one drank deeper of the cup of bitterness than I did," she

---

Schiller as a possibly cryptomnesic source for Freud: "Freud and Schiller," *Psychoanalysis and the Psychoanalytic Review* 46 (Winter 1960): 97–101.

65. Freud, "Joseph Popper-Lynkeus and the Theory of Dreams," *Standard Edition . . . of Freud*, 19:261. This same passage is translated from the German in the paper by Brandt.

66. I am indebted to Lewis W. Brandt for calling my attention to this passage.

concludes, and concludes this although it proved impossible to find anyone who had actually read her the story.[67]

Contributing further to the uncertainty about the extent of one's originality is the recurrence of episodes in which a scientist has unwittingly borrowed ideas from himself. Many scientists and scholars have found, to their combined chagrin and disbelief, that an idea which seemed to have come to them out of the blue had actually been formulated by them years before, and then forgotten. An old notebook, a resurrected paper, a colleague cursed with total recall, a former student—any of these can make it plain that what was thought to be a new departure was actually a repetition (or at most, an extended and improved version) of what they had worked out for themselves in the past. Of many such cases, consider only a few, some of a century or more ago, others of contemporary vintage:

Joseph Priestley records with chagrin that

> I have so completely forgotten what I have myself published, that in reading my own writings, what I find in them often appears perfectly new to me, and I have more than once made experiments, the results of which had been published by me.[68]

The ingenious and jovial mathematician, Augustus de Morgan, has his own lively version of the experience:

> I have read a Paper (but not on mathematics) before now, have said to myself, I perfectly agree with this man, he is a very sensible fellow, and have found out at last that it was an old Paper of my own I was reading, and very much flattered I was with my own unbiased testimony to my own merits.[69]

And it is told of the distinguished mathematician, James Joseph Sylvester, that he "had difficulty in remembering his own inventions and once even disputed that a certain theorem of his own could possibly be true."[70]

Or consider a brace of cryptomnesic borrowings from self in our own day:

The Nobel laureate, Otto Loewi, reports his waking in the middle of the night, jotting down some notes on what he sensed to be a momentous discovery, going back to sleep, awaking to find that he could not possibly decipher his scrawl, spending the day in a miserable and unavailing effort to remember what he had had in mind, being again aroused

67. Helen Keller, *The Story of My Life* (New York: Doubleday, Page and Co., 1908), pp. 63–72, at 65.

68. *Life of Priestley*, Centenary Edition, p. 74.

69. Letter of Augustus de Morgan to W. R. Hamilton, in Graves, *Life of Hamilton*, 3:494.

70. E. T. Bell, *Men of Mathematics* (New York: Simon and Schuster, 1937), p. 386.

from his slumber at three the next morning, racing to the laboratory, making an experiment and two hours later conclusively proving the chemical transmission of nervous impulse. So far, so good: another case, evidently, of the pattern of subconscious creativity unforgettably described by Poincaré. But some years later, when Loewi, upon request, reported all this to the International Physiological Congress, he was reminded by a former student that, eighteen years before that nocturnal discovery, he had fully reported his basic idea. "This," says Loewi, "I had entirely forgotten."[71]

The psychologist Edwin Boring writes me of a colleague who came to him in an excited Eureka frame of mind, announcing that he had just worked out a new technique for scales of sensory measurement, and that he is now hunting for a name for it. And then, before "the shine of the new idea had rubbed off, he discovers that he had discussed this in print some six years before and had even given it a tentative name."

And to advert to Freud, as I have so often done if only because his intellectual experience is uncommonly documented, Jones reports several instances of his "obtaining a clear insight which he subsequently forgot, and then later suddenly coming across it again as a new revelation."[72] As Freud noted in another connection, "it is familiar ground that a sense of conviction of the accuracy of one's memory has no objective value."[73]

If cryptomnesia is possible in relation to one's own earlier work, then it is surely possible in relation to the work of others. And this can undermine the calm assurance that one has, in truth, worked out a new idea for oneself when confronted with another version of the same idea worked out by someone else.

Various contexts may affect the probability of cryptomnesia in relation to one's own work. It may be the more probable the more the scientist has worked in a variety of problem-areas rather than narrowly restricting his research focus to problems having marked continuity. Looking at this hypothesis, not in terms of the individual scientist but in terms of the

71. Otto Loewi, *From the Workshop of Discoveries* (Lawrence, Kansas: University of Kansas Press, 1953), pp. 33–34.

72. Jones, *Sigmund Freud*, 3:271. One such case, for example, is Freud's conception of paranoid jealousy as an instance of repressed homosexuality.

73. This observation appears in his paper of 1913, on "Fausse Reconnaissance ('Déjà raconté') in psycho-analytic treatment," *Collected Papers*, 2:334–41. This same paper, devoted to paramnesia, has Freud reporting a multiple discovery and assuring the reader (and himself) that it is just that, and not a case of cryptomnesia: "In 1907, in the second edition of my *Psychopathologie des Alltagslebens*, I proposed an exactly similar explanation for this form of apparent paramnesia without mentioning Grasset's paper [of 1904] or knowing of its existence. By way of excuse I may remark that I arrived at my conclusion as the result of a psychoanalytic investigation which I was able to make of an example of *déjà vu* . . . [that] had occurred twenty-eight years earlier" (Ibid., p. 337).

relative frequency of self-cryptomnesia in different sciences, we should expect it to be more frequent in the newer sciences, with their comparatively uncodified and therefore more nearly empirical knowledge than the better-established sciences. To the extent that these patterned differences in degree of theoretical integration occur, we should expect more cryptomnesia in relation to one's own work in the social sciences.

The frequency of such cryptomnesia should also be affected by the social organization of scientific work, which seems to affect every aspect of multiple discoveries in science. When research is organized in teams, it would be less likely, we must suppose, that earlier ideas and findings would be altogether forgotten. For if some members of the team forget them, others will not. Moreover, repeated interaction between collaborators will tend to fix these ideas and findings in memory.

The conspicuous changes in the social organization of scientific research should have a marked effect, not only on this matter of self-cryptomnesia, but on every aspect of multiples and priorities in science. The trend toward collaborative investigation in research organization is reflected in patterns of publication, with more and more research papers being by several authors rather than by only one. The extent of this change differs among the various disciplines. The sciences which have developed cogent theory, complex and often costly instrumentation and rigorous experiments or sets of observations seem to have experienced this change earlier and at a more rapid rate than the sciences which are less well developed in these respects. By way of illustration, consider the pattern of publication in just three cases: one drawn from the measurement of constants in physics; a second, from psychology; and the third, from sociology. I have tabulated the number of authors of each of 414 papers on the measurement of physical constants cited in an authoritative monograph on the subject.[74] The results, in brief, are these: of the papers published before 1920, 93 percent were by single authors; for those between 1920 and 1940, this declines to 65 percent; and for those since 1940, to 26 percent. Taking only the most recent period, we find that 28 percent were by two authors; 19 percent by three; 14 percent by four; and 13 percent by five or more authors (with some 2 percent of these being by ten or more coauthors).

Much the same trend, but far less marked, is found for the papers published in the *Journal of Abnormal and Social Psychology* since 1936: taken by consecutive five-year periods, single-author papers decline from 80 percent of all in 1936 to 75 percent to 69 percent to 54 percent and finally to 49 percent during the last five years. And the *American Socio-*

74. E. Richard Cohen, Kenneth M. Crowe, and Jesse W. M. Dumond, *The Fundamental Concepts of Physics* (New York: Interscience Publishers, Inc., 1957), pp. 92–102.

logical Review for the same period witnesses a similar but even more restrained trend, with single-authored papers declining from 92 percent in 1936, to 90 percent to 87 percent to 76 percent, and in the last five years, to 65 percent.[75]

Although the facts are far from conclusive, this continuing change in the social structure of research, as registered by publications, seems to make for a greater concern among scientists with the question of "how will my contribution be identified" in collaborative work than with the historically dominant pattern of wanting to ensure their priority over others in the field. Not that the latter has been wholly displaced, as we have seen. But it may be that institutionally induced concern with priority is becoming overshadowed by the structurally induced concern with the allocation of credit among collaborators. One study of a team of thirty economists and behavioral scientists, for example, found that "the behavioral scientists were apt to be less concerned about 'piracy' and 'credit' than economists. This difference may be due to the greater emphasis on joint authorship in the behavioral sciences than in economics."[76]

For our purposes, the import of these changes in collaboration is, first, that the degree of concern with priority in science is probably not an historical constant; second, that it varies with the changing organization of scientific work; and third, that these changes may eventually and indirectly help make for the dispassionate and methodical study of multiples and priority in science, as resistance to that study is undercut by widespread recognition of the ubiquity of multiples in science.

Nevertheless, although scientists *know* that genuinely independent discoveries in science occur, many of them do not manage, as we have seen, to draw the implications of this for their own work. For reasons I have tried to intimate, they find it difficult, and sometimes impossible, to accept the fact that they have been anticipated, or that a contemporary has come to the same result just at the time they did, or that the others were truly independent of them. As we have also seen, the values in the social institution of science and the penumbra of uncertainty that surrounds the independence of thought combine to prevent the ready acceptance of events that undercut one's assurance of unique originality, an assurance

75. The extent of these differences between patterns of collaboration in the major scientific and humanistic disciplines has been investigated by Harriet Zuckerman; see her forthcoming *Scientific Elite* (Chicago: University of Chicago Press, in press). Bernard Berelson has found that for the year, 1957–58, among a sample of those who had received their doctorate ten years before, the relative numbers of publications with single authors ranged from 17 percent in chemistry and 30 percent in biology to 96 percent in history and 97 percent in English. See Berelson, *Graduate Education in the United States* (New York: McGraw-Hill, 1960), p. 55.

76. Warren G. Bennis, "Some Barriers to Teamwork in Social Research," *Social Problems* 3 (1956): 223–35, at 228–29.

born of the hard labor required to produce the new idea or new result. Consequently, multiple discoveries are experienced at best as an unpleasant reality and at worst as proof that deliberate or cryptomnesic borrowing has occurred. The reasonably detached study of multiples and priorities may possibly counter these tendencies to dismay or suspicion.

Such studies will probably not create the Olympian mood of a Goethe vigorously reaffirming Ecclesiastes: "No one can take from us the joy of the first becoming aware of something, the so-called discovery. But if we also demand the honor, it can be utterly spoiled for us for we are usually not the first. What does discovery mean, and who can say that he has discovered this or that? After all, it's pure idiocy to brag about priority; for it's simply unconscious conceit, not to admit frankly that one is a plagiarist."[77] But multiple discoveries can be recognized as having their uses, not only, as we noted before, for enlarging the likelihood that the discovery will be promptly caught up in the advancement of science but also for the individual discoverers. For, as we have seen Freud affirming in an effort to rouse himself from his ambivalence toward having been anticipated by Watkiss Lloyd, independent multiples do seem to lend confirmation to an idea or finding. Furthermore, even W. R. Hamilton, tormented his life long by the fear that he was being plagiarized or by the anxiety that he himself might be an "innocent plagiarist," managed on at least one occasion to note, as did Freud, the secondary benefits of a multiple, when, in an effort to dissolve his ambivalence, he wrote Herschel:

> I persuade myself that, if those results had been anticipated, the learning [of] it would have given me no pain; for it was, so far as I could analyze my sensations, without any feeling of vexation that I learned that the result respecting the relation of the lines of curvature to the circular sections was known before. The field of pure, not to say of mixed, mathematics is far too large and rich to leave one excusable for sitting down to complain, when he finds that this or that spot which he was beginning to cultivate as his own has been already appropriated. [And now comes his hard-won and, sad to tell, temporary, insight:] There is even a stronger feeling inspired of the presence of that Truth to which we all profess to minister, when we find our own discoveries, such as they are, coincide independently with the discoveries of other men. The voice which is heard by two at once appears to be more real and external—one is more *sure* that it is no personal and private fancy, no

77. Quoted in the epigraph of Lancelot Law Whyte's *The Unconscious Before Freud* (New York: Basic Books, 1960). We need not mark the irony that the maxim, there is nothing new under the sun, has itself variously recurred: remember only Terence, beset by charges of wholesale theft, saying: "nihil est dictum quod non sit dictum prius." Or five centuries later, Donatus exclaiming: "Pereant qui ante nos nostra dixerunt." Or Shakespeare, in Sonnet 59:

> If there be nothing new, but that which is
> Hath been before, how are our brains beguil'd,
> Which, labouring for invention, bear amiss
> The second burthen of a former child!

idiosyncratic peculiarity, no ringing in sick ears, no flashes seen by rubbing our own eyes.[78]

And then, unable to contain himself, Hamilton goes on to announce in the same letter that he had anticipated the work on ellipsoids by Joachimstal in "a long extinct periodical of whose *existence* he [Joachimstal] probably never heard, with a date which happened to be a *precise decennium* earlier."[79]

If the fluctuating ailment of that genius Hamilton proves that the knowledge of multiples is no panacea for ambivalence toward priority, his moment of insight suggests that it may be some small help. The mathematician, R. L. Wilder, is, to my knowledge, the only one who has seen this clearly and has, to my mingled pleasure and discomfiture, anticipated me in suggesting that the study of multiples may have a therapeutic function for the community of scientists. Since he has anticipated my observation, let me then borrow his words:

> I wish to inquire, above the individual level, into the manner in which mathematical concepts originate, and to study those factors that encourage their formation and influence their growth. I think that much benefit might be derived from such an inquiry. For example, if the individual working mathematician understands that when a concept is about to make its appearance, it is most likely to do so through the medium of more than one creative mathematician; and if, furthermore, he knows the reasons for this phenomenon, then we can expect less indulgence in bad feelings and suspicion of plagiarism to ensue than we find in notable past instances. Mathematical history contains numerous cases of arguments over priority, with nothing settled after the smoke of battle has cleared away except that when you come right down to it practically the same thing was thought of by someone else several years previously, only he didn't quite realize the full significance of what he had, or did not have the good luck to possess the tools wherewith to exploit it. . . . [Yet] it is exactly what one should expect if he is acquainted with the manner in which concepts evolve.[80]

All this does not deny, of course, the possibility that in particular cases the unwitting or deliberate use of ideas and findings without acknowledgement may occur. I have tried elsewhere[81] to show how the institution of science, with its premium upon originality, indirectly motivates just that kind of deviant behavior among some scientists. But for our understanding of how scientific knowledge develops, we have long since needed, among other things, to overcome the resistance toward the dispassionate and methodical study of multiples and attendant priority conflicts, rather than

78. Graves, *Life of Hamilton,* 2:533, in a letter to Herschel, 23 Nov. 1846.
79. Ibid., 2:534.
80. R. L. Wilder, "The origin and growth of mathematical concepts," *Bulletin of the American Mathematical Society* 69 (1953): 423–48, at 425.
81. See chapter 14.

to neglect this study altogether or to come to it only when we plunge, as emotionally involved participants, into conflicts over rights to intellectual property. After all, one of the roles assigned the sociologist is to investigate the behavior of all manner of men, including men of science, without giving way to the entirely human tendency to substitute for that investigation a clucking of tongues and a condemning of that which is and ought not to be.[82]

82. This investigation has been aided by a fellowship from the John Simon Guggenheim Memorial Foundation. I am especially indebted to Dr. Elinor Barber, who has contributed greatly to my studies in the sociology of science. Harriet Zuckerman, Jerald T. Hage, and Cynthia Epstein have provided able assistance at one or another part of the investigation.

The Processes of Evaluation in Science

Part 5

Prefatory Note

The four papers making up this final section represent not only the extension of an interest at the core of the basic paradigm but a reaching out to examine the ways in which additional variables may interact with factors identified in the paradigm. Matters of role performance (potential and achieved) and of recognition (instrumental and honorific) are central to all the papers, with attention focusing sequentially on the consequences of accumulated honorific recognition, the processes through which such recognition is acquired or denied, and the relevance of age (biological and social) to the preceding topics.

The first paper, prepared originally for a conference on "Recognition of Excellence" and published in 1960, considers research questions important for the general problem of how excellence is recognized and rewarded or neglected and even penalized in society and how the effectiveness of these processes might be increased. The viewpoint covers role-performance in every sector of society, but much of the reasoning that went into the paper—and the illustrative materials—come from Merton's earlier work on the topic within the context of the scientific institution. It provides a useful background for the more concentrated attention on science that characterizes the following papers.

The connection between the first and second papers is obvious. While a concern for ensuring that excellence be recognized within society animates the former, it is the consequences of the accumulation of recognition specifically within science that forms the focus of "The Matthew Effect." This paper, published in 1968, is the fruit of a longstanding interest of Merton's, for there are references to the phenomenon in other papers at least as early as 1961. Any reward system operating in society will produce an uneven distribution of the coinage with which it deals, be this money, power, or esteem, and it is always an appropriate question to inquire into the consequences of this fact for people and for the institution in question. Although Merton does not draw parallels between the reward system of science and those of the economic and political institutions, it is useful to recall that they can be drawn—the rich tend to get richer in all three systems—even though at times they have been overdrawn.

Although professional recognition leads to influence and can lead to power, it appears less directly linked to basic life-chances than are money and political power, and it is much more difficult to acquire through illegitimate means. Yet recognition is the coin of the realm, and within the confines of science it is of supreme importance to scientists. The Matthew effect points out that such recognition has a tendency to be self-reinforcing (and that its lack can be negatively reinforcing) so that career-lines, in terms of scientific success or failure, may come to resemble logistic curves rather than straight lines. While indicating—and documenting—the dysfunctions of this pattern for individual scientists, Merton makes a clear distinction between its individual and social consequences.

It is a difficult unsolved problem to compare the degree of equity in the reward systems of science and other institutional areas. In science, according to the ideal norm of universalism, it is the quality of work, judged by peers against the current state of the art, that should entirely determine the kind of recognition received. One major institutional device for the competent appraisal of the quality of scientific work is the referee system, the subject of the third paper in this part.

In "Patterns of Evaluation in Science," Zuckerman and Merton first contribute to our understanding of the process of institutionalization in science by tracing the evolution of the referee system from its origins. They then analyze the archives of the *Physical Review*, rather than only the distribution of papers actually published in that journal, to determine whether referees' acceptance or rejection of manuscripts is affected by normatively extraneous characteristics of authors. The paper breaks new ground by developing a sociometric analysis of refereeing: a statistical matrix relating the status (age, reputation, institutional affiliation) of scientist-authors to that of scientists refereeing their papers. In the case of this outstanding journal of physics, they find that "the relative status of referee and author had no perceptible influence on patterns of evaluation."

A next obvious step is to compare the operation of the referee system in various disciplines. The paper does not take that step, but after comparing the rates of rejection of manuscripts submitted to journals in different scientific and humanistic disciplines, it suggests that these differ according to gross differences in the organization of knowledge. The general point, however, is not pursued here; instead the various functions of the referee system as well as the problems inherent in its operation are explored at length.

The authors might have noted an additional consequence of the referee system. It is this editor's own experience, anyway, that a referee may often acquire a sense of avuncular pride in a paper that he has approved for publication and is thus more likely to draw upon it after it appears in print, to cite it, and so on, than if he had not had the prior exposure to it. If, further, referees tend to have a higher rank than authors (as is the case with nearly two-thirds of the referee-author pairs studied by Zuckerman and Merton in physics), this fact should have a small but positive influence on the visibility of those papers.

The last paper, "Age, Aging, and Age Structure in Science," also written with Harriet Zuckerman, brings the demography of science into conjunction with its social structure. It was contributed to the project on aging and society conducted by Matilda White Riley, Anne Foner, and Marilyn Johnson under the auspices of the Russell Sage Foundation, and appeared in 1972. In it is clearly demonstrated one of the advantages provided by a tested paradigm: not only does it facilitate the "specification and organization of ignorance"—the systematic survey of questions posed by the addition of a new variable to the matrix of what is already known—it also sensitizes the scientist to a wide range of materials which are relevant to the questions newly framed.

The authors describe this paper as more of a research prospectus than a set of firm conclusions. But because it is based on a paradigm, it does provide an indicative preview of what empirical research can be expected to find. A concentrated attack on a problem that employs a set of interdependent hypotheses is of course more productive than one which touches upon the same phenomena in an entirely ad hoc manner.

The factor of age is obviously relevant to behavior within the social structure of science in several ways: affecting the stage of a scientist's career, locating his peers in terms of career and of the life cycle, and relating him to other scientists of different ages. Zuckerman and Merton are not content, however, to explore only these questions. More systematically than in the preceding paper, they bring in attributes of scientific knowledge—specifically, the relative degree to which different bodies of knowledge are codified—as a critical intervening variable to help explain the differential role of age in various fields of science.

The last two papers point to a possible new phase in the sociology of science, one in which an evolving paradigm for investigating the social structure of science will link up with an evolving paradigm for investigating the cognitive structure of scientific disciplines. With the fast-growing interest in the sociology of science, this development may occur more rapidly than could have been expected only a few years ago. If it does occur, perhaps the safest prediction to be made here is that major contributions to that development will come from Robert K. Merton.

N. W. S.

19 "Recognition" and "Excellence": Instructive Ambiguities

1960

Each of the words *recognition* and *excellence* carries a pair of primary denotations. What is more, the terms in each pair of meanings are symmetrically related. To examine these meanings will turn out to be something more than a semantic exercise. If we are clear about them and about what they imply, we are the better prepared, I suggest, to understand the particular tasks involved in advancing the recognition of excellence.

Recognition in Its Instrumental Sense

Take, first, the instructive ambiguity of the word *recognition*. In one sense, *recognition* refers to the fact of apprehending or identifying something, of placing it in a particular category, of seeing it as having a certain character. (In the present case, the still undefined character is that of "excellence.") Here we deal with detecting qualities of excellence that might ordinarily go unnoticed. This sense of the word carries with it the further implication that if these qualities of excellence do not come to the attention of appropriate other people, they will often remain undeveloped and unrealized. The concern is to find ways of detecting potential excellence and of doing so early enough to help potentiality become actuality.

In short, this is the instrumental sense of recognition. It directs our attention to the possibility that, to some unknown extent, much human capacity for socially valued accomplishment remains latent and undeveloped. It assumes that much talent fails to find expression because it is subjected to adverse conditions. The policies attached to the instrumental

Reprinted, with permission, from Adam Yarmolinsky, ed., *Recognition of Excellence: Working Papers* (New York: The Free Press, 1960), pp. 297–328.

sense of recognition have to do with the early detection of qualities of excellence, with the removal of normally existing obstacles to their development, and with the provision of opportunities for their optimum development.

The instrumental sense of recognition thus implies at least three connected and difficult types of research designed to afford an intelligent basis for these policies.

First, it calls for research on means of identifying talent that ordinarily goes unnoticed or is noticed too late for it to become an appropriately trained capacity. This is "detective research."

Second, it calls for research designed to identify the principal current obstacles to the effective development of various kinds of socially valued talent. This is "remedial research."

Third, it calls for research designed to discover the kinds of human and organizational environments that help bring out kinds of creativity that are socially valued (technological invention or business acumen) and that "should be" more highly valued (scientific discovery or poetry). It would presumably not include investigations designed to discover environments that facilitate socially disvalued forms of "creativity" (the organizational acumen of the gang-leader or the capacity for envisaging new kinds of genocide). This is, in the broadest sense, "educational" or "evocative" research.

Much research work is being expended on recognition of excellence in its instrumental sense: the identifying of talent, the removal of road-blocks to its expression, and the provision of literally educative environments. Much more systematic investigation along these lines, of course, needs to be done. Nevertheless, these instrumental aspects of the recognition of excellence cannot be said to suffer from inattention. It is only that comparatively small inroads have yet been made on a complex of difficult problems.

It is evidently otherwise with the problems caught up in the second major sense of "the recognition of excellence," what I shall call its *honorific* meaning.

Recognition in Its Honorific Sense

Provisionally distinct from the instrumental sense of recognition, although ultimately connected with it, the honorific sense refers to the high evaluation of positive accomplishments chiefly through the public and private institutions of a society. In this aspect, recognition looks to the rewarding of achievement. Correlatively, lack of recognition does not refer to talent or genius being thwarted by uncontrolled circumstances, as it does in the instrumental sense. Rather, it assumes that the accomplishments are there but go without the public notice due them in their own time. The accomplishments, and the talent or genius giving rise to them, have not been "recognized"; they have not been accorded the esteem owing them.

The failure of recognition, construed in this honorific sense, represents a special case of the general problem of appropriate public acknowledgment of great and middling achievements. To pin down this special and possibly widespread type of instance, we might affix a special tag to it: the *elegiac* sense of the lack of recognition. The gem *is* of purest ray serene, but it is condemned to remain in the dark unfathom'd caves of ocean; the flower *has* its sweetness, although this is wasted on the desert air. Too late, we mourn our neglect of demonstrated greatness. The loss is chiefly ours. Only secondarily is it the loss of genius expressed but unnoticed. Beyond the loss, the guilt is also principally ours. We have not lived up to our tacit obligation, as the recipients and carriers of culture, to signalize and to honor genuine accomplishment. For the public recognition of great achievement constitutes one way in which we give expression to our cultural values. It follows, then, that the failure to recognize such achievement is an indictment of us. Preoccupied with our private concerns and with lesser values, we fail to respond adequately to occasional greatness among the men and women around us.

The honorific sense of the recognition of excellence and the elegaic sense of nonrecognition shade over into the instrumental sense, through an interesting and informative assumption. Many of us assume that if talent and genius of certain kinds go long unrecognized and unhonored, this will produce a world of values in which the potential of talent and genius will increasingly go unrealized. A world unfriendly to poetic genius is a world in which we find a mute inglorious Milton. Or as, in the early days of modern science, Francis Bacon could complain and explain all in one:

> . . . it is enough to check the growth of science that efforts and labours in this field go unrewarded . . . , it is nothing strange if a thing not held in honour does not prosper.

As I shall suggest in the next part of these notes, we know next to nothing, in a methodical way, about the effects of various systems for the honorific recognition of excellence upon persons of talent or genius. This is one place where thoroughgoing inquiry is badly needed. After all, three centuries have gone by since Bishop Sprat implied the hypothesis that certain kinds of reward systems will, more than others, give life to the industry of men engaged in currently unpopular but culturally consequential work. Perhaps it would not be rashly rushing into research to take up this hypothesis and see where it leads us.

The echo of the complaints by Bacon and Sprat still reverberates in the halls of universities and scientific societies, and especially so in those often shabby quarters given over to the humanities and the social sciences. With the recent catapulting of physical science into enforced prominence through the interworld-shaking drama of the Space Age, popular applause (and popular attack) has become so noisy as to compare with the unmuffled roar in modern amphitheaters of baseball and harness-racing, of football and boxing. Of course, the applause directed to men of science and technology is often misdirected. For, as Bacon noted of his own time, and unlike the public activities of sport, we, the people, are ordinarily not able to make sound independent judgments of merit in scientific work. We mistake the readily observable feats of technology, the hardware of the Space Years, for their less conspicuous but basic underpinnings in the sciences that are thoroughly unintelligible to most of us. All this only highlights one of the problems inherent in developing new systems of rewards for accomplishment in science, in the arts, humanities, and social sciences, in all activities making for the public good. Who shall judge? And what criteria shall they use for their judgments?

Enough has perhaps been said to suggest how the two meanings of "recognition"—the instrumental and the honorific—link up with distinct though connected problems of research and action in providing for the enlarged and appropriate recognition of excellence. By and large, the instrumental meaning directs our attention to the detecting of potentials for excellence of achievement and to the providing of opportunities such that these potentials more often become realities than they otherwise would. The honorific meaning directs our attention to the signalizing and rewarding of demonstrated excellence. Honorific recognition presumably honors givers as well as receivers. It expresses and demonstrates the donor's soundness of values; in this definite sense, it is more blessed to give honor than to receive it. In its assumed effects upon recipients who have demonstrated excellence and upon novices who have yet to show the stuff of which they are made, recognition in its honorific sense merges with recognition in its instrumental sense. This is on the not unreasonable but still untested assumption that arrangements for the public recognition of excellence will result in more young men and women of ability turning to the pursuit of excellence. For, to transpose the Baconian aphorism into a more hopeful key, it is nothing strange if a thing held in honor, prospers.

Excellence in the Sense of Quality

The double meaning of *excellence* has only been intimated in the preceding pages. Many of us are persuaded that we *know* what we mean by excellence and would prefer not to be asked to explain. We act as though we believe that close inspection of the idea of excellence will cause it to dissolve into nothing, much as the idea of simultaneity of two events some distance apart came to nothing under the penetrating eye of a wise physicist. Yet we need not be and, perhaps fortunately, cannot be as penetrating as all that. For immediate purposes, it is enough to note that two of the meanings implied by the word *excellence* correspond loosely to the two meanings of *recognition* that have been briefly examined.

Corresponding to the instrumental sense of recognition is the sense of excellence as a personal quality. This refers to qualities of men and women that are (or, in the judgment of the judge, should be) highly prized. The quality must have found at least some minimal expression, for how else could we infer that Jones or Smith "have" it. Indeed, it is the office of the entire vast array of tests and measurements of aptitudes, personality, character, and the like to try to work out reliable outward indications of human qualities that are not visible to the naked eye. The task is immense, and work on it has scarcely begun. On this there is general consent, not least among skilled and thoughtful makers of these tests and measurements themselves.

Perhaps there are better ways of identifying these qualities of excellence before they become so manifest as to be inescapable. Perhaps the pedestrian labors of armies of psychologists and statisticians will, in the end, get nowhere. It is, of course, only a collective bet by a society which underwrites the large costs of these labors that this is one of the means—perhaps the most promising means—of identifying talent early enough to provide needed opportunities for its fullest possible flowering. Some of us refuse to be party to this bet, particularly those of us who are appalled by the unholy alliance between IBM and psychologists, and between them and the consumers of talent in schools and in business, government, and almost every other sphere of social life. But we should know that we can't have it both ways. If we find all this pointless, and claim that in the multitude of tests and measurements there is no truth, then we might make it our business to work for some better means for the early identification of excellence. If, contrariwise, we adopt the second line of attack and believe that danger to man and society resides in the validity of these tests and measurements, then we might consider how else special opportunity for the development of these qualities can be provided. And finally, if we reject the entire conception of special opportunities for talent as "meritocratic" or "undemocratic" and altogether obnoxious, then we might forthrightly acknowledge this and consider the further implications of closing off special facilities from the talent that requires them.

Whatever our position on current methods of trying to detect qualities of excellence as early as possible, we are driven back to the reasons for wanting to detect *qualities* of excellence at all. Why not wait until quality has become performance; capacity, actuality? As I have intimated, the various reasons that have been advanced all seem to boil down to one; we want to identify qualities of human excellence in order to provide the fullest opportunity for their effective expression and development. In a word, the psychological research designed to identify human aptitudes of individuals has value for society only if it is coupled with sociological and psychological research on the nature of social environments that evoke or curb the expression and development of these aptitudes.

It is a matter of demonstrable fact that far more work by far more investigators has been devoted to the problem of identifying individual differences in capacity than to the correlative problem of identifying differences in social environments that evoke or suppress the effective development of identifiable aptitudes. This is not, I believe, the soured report of a sociological chauvinist. After all, sociologists are plagued with more subjects and problems requiring study than they are presently prepared to cope with. They are not in avid search of new "fields" in which to labor. But it is true that the great emphasis on studies of individual aptitudes is not matched by emphasis on studies of social environments in which people of the same or differing aptitudes live, learn, and work. Yet the psychological investigations on the first set of problems have little significance for enlarging the fulfillment of individual promise unless it is connected with sociological investigations on the second set of problems. It is only so that we shall come to understand the frequent gap between individual promise and individual achievement, between excellence as quality and excellence as performance.

Excellence in the Sense of Performance

As you can see, I have not found it possible to consider the first sense of *excellence* without touching upon the second. Excellence in the sense of performance refers, of course, to the preeminent possession of some quality that has been demonstrated by achievement. If the first sense of excellence as quality embodies a doctrine of justification by faith in the individual who has yet to prove himself, the second sense of excellence as performance manifestly embodies a doctrine of justification by works. Enduring issues of the relations of man and the universe have a way of cropping up in the most unexpected places: the concept of excellence as quality is reminiscent of Luther, just as the concept of excellence as performance reminds us of Calvin.

The essential point that we all know but sometimes forget is that capacity and performance can and not infrequently do diverge. Each of us has his own favorite examples of plainly talented people who somehow failed to realize their promise and of apparently mediocre people who somehow outdid themselves. Psychologists have recently tagged these as the "underachievers" and the "over-achievers": those who fail to do as well as they ought, so far as one can judge from their potential, and those who do far better than might have been supposed from what one knew of them. Sociological and psychological research on the conditions making for one or another of these observed patterns of performance is only in its beginnings. Much more research on this is needed that we might the better understand how excellence of quality, a personal attribute, can be more generally expressed in excellence of performance, with its social and cultural consequences.

Recognition and Excellence: The Structure and Functions of their Multiple Connections

One way of trying to clarify our thinking about the recognition *of* excellence is to examine methodically the connections between recognition *and* excellence. The bare outlines of these connections can be drawn by use of a familiar kind of logical syntax in which each meaning of the one word is diversely combined with each meaning of the other, as this is done in the following table:

1. Recognition as instrumental, excellence as quality
2. Recognition as honorific, excellence as quality
3. Recognition as instrumental, excellence as performance
4. Recognition as honorific, excellence as performance

By exploring the relations between recognition and excellence, we can identify four major types of problems that are evidently involved in the enterprise under review. In considering each of these types briefly, I do not suppose, of course, that I have earmarked all of the problems that require consideration.

1. *Recognition as instrumental and excellence as quality*

This refers to the search for identifying excellence of capacities before it has become abundantly manifest. The first problem, obviously, is to have a sound appraisal of current procedures designed to identify talent early in the game. Much effort is being put into this kind of inquiry. For the most part, this presently emphasizes the assessment of the efficacy of various tests and measurements for identifying the individual capacities of youth primarily and of adults secondarily. This kind of effort is subjected to censure by seemingly influential critics, such as William Whyte, who argue both (a) that their efficacy is slight and so they are misleading and (b) that their efficacy is great and so they presage the police-state in which each of us is put once and for all in an occupational niche by the experts, all this at the expense of our individuality and our autonomy. I do not undertake to comment on this criticism; garbled as it typically is, parts of it seem well founded.

It does seem to me, however, that comparatively little work is now devoted to developing other means for the early identification of people having a capacity for excellence. One has the impression that some people in almost every community have an eye for seeing talent of various kinds before this has become apparent to everyone else. Perhaps an effort should be made to search out these human counterparts to the truffle-dog, these men and women who somehow see below the surface of appearance to underlying quality, who somehow know excellence when they first encounter it and before others are cognizant of it. Who are they? Some, no doubt, are teachers who in the ordinary course of working at their job can't help being exposed to youngsters and some of whom have not yet had their sensibilities trained out of them by courses on techniques of teaching or dulled by an excess of paper work that keeps them from seeing what is there to see. But if we may judge from some early sociological work on identifying "influentials" in a community, it could turn out that these informal spotters of talent are found in many other walks of life as well. They may be ministers or artists, newspaper editors or physicians. The range of their occupations is probably wide, but they have in common this eye for genuine ability of various kinds. They are the unappointed talent scouts of America.

How many there are and who they are is anybody's guess. Perhaps they exist in such small numbers that they could in the best of circumstances, do little. But again, perhaps not. The fact is that I don't know and (therefore?) I strongly suspect that no one really knows. In any event, we ought to find out. What we need here, I believe, is some careful research on "the talent to spot talent" and to spot it early. Explorations in two or three communities might be enough to see whether this is worth the doing and, what is not altogether irrelevant, whether it can be done.

If these unofficial talent scouts do exist in sizable numbers and if means of identifying them can be developed, they could become an important part of the common enterprise to identify excellence of capacity. For this to happen, it is necessary that they be given some official standing. As isolated individuals, remote from the institutional machinery for the selection and rewarding of ability, they will not be listened to. Perhaps we have something to learn from the first homespun and now increasingly professionalized experience of talent scouts in baseball and other sports. From all indications, they do a fairly good job of locating comparatively undeveloped talent.

The local talent scouts might, for example, occasionally set us straight on youngsters who do poorly in formal tests of aptitude and who nevertheless have a capacity for doing some things superbly well. In other words, the talent scout might redress the occasional errors that are bound to turn up in the mass measurement of human capacity. Above all, they might counteract one of the biases, of unknown magnitude, that I believe to be inherent in our system of locating excellence as quality and rewarding it. This bias, if it in fact exists, condemns an unknown but appreciable number of able youngsters to oblivion, which is unjust, and loses to society the fruits of the excellence of capacity, which is wasteful. Since considerations of morality and of expediency rarely coincide, we should seize upon this seeming case of such coincidence for special attention.

The case of the "late bloomer" and its implications for recognition of excellence. That wise and good man, Alan Gregg, was to my knowledge the only one to glimpse the bias in favor of precocity that is built into our institutions for detecting and rewarding talent. He locates the problem by first identifying four types of "emergent ability" which are severally related to the "recognition of ability." (In "For Future Doctors" he is speaking of medical men, but this, I believe, can be generalized to hold for all occupations.)

1. *The rampart type of ability*: this shows a rapid rise to an early maximum, and then a gradual slope downward after the summit is reached. Or, more invidiously, up with a blaze of glory and down with a dull thud.
2. *The plateau type*: also declares itself promptly but maintains itself with steady continuity. Dependable, consistent performers, serene and solid, beautifully free from bad luck or bad management.
3. *The slow crescendo type*: shows steady, slow improvement throughout his life; a slow starter, he creates neither great expectations nor later great disappointment. He is apt to become the workhorse of his community.
4. *The late-blooming type*: he manages to mix surprise with success, since his ability shows itself so unexpectedly and so late as to excite but little jealousy and few dependents to abuse his time.

There is something germane to our subject in each of these four types, but I want to center primarily on the "late bloomer," for this type raises some critical questions about the operation of our institutions for identifying and rewarding ability. Since the late bloomer is slow in getting started, he is of course the type that is apt to be overlooked. This inherent oversight is all the more probable owing to the character of our social institutions, which put a premium on early manifestations of ability, in a word, on precocity. I cannot improve on Gregg's statement of the case, so here it is:

> By being generous with time, yes, lavish with it, Nature allows man an extraordinary chance to learn. What gain can there be, then, in throwing away this natural advantage by rewarding precocity, as we certainly do when we gear the grades in school to chronological age by starting the first grade at the age of six and so college entrance for the vast majority at seventeen and a half to nineteen? For, *once you have most of your students the same age, the academic rewards*—from scholarships to internships and residencies—go to those who are uncommonly bright *for their age*. In other words, you have rewarded precocity, which may or may not be the precursor of later ability. So, in effect, you have unwittingly belittled man's cardinal educational capital—time to mature.

Gregg argues further that "precocity may succeed in the immediate competitive struggle but, in the long run, at the expense of mutants having a slower rate of development but greater potentialities." By claiming that a slow development is significantly associated with greater potentialities, Gregg may be assuming what he then concludes. But his argument cuts deeply, nevertheless. For we know only of the "late bloomers" who have eventually come to bloom at all; we don't know the potential late bloomers who, cut off from support and response in their youth, never manage to come into their own at all. Judged ordinary by comparison with their precocious age peers, they are treated as youth of small capacity. They slip through the net of our institutional sieves for the location of ability, since this is a net that makes chronological age the basis for assessing relative ability. Treated by the institutional system as mediocrities with little promise of improvement, many of these potential late bloomers presumably come to believe it of themselves and act accordingly. At least what little we know of the social psychology of the formation of self-images suggests that this is so. For most of us most of the time, and not only the so-called "other-directed men" among us, tend to form our self-image—our image of potentiality and of achievement—as a reflection of the images others make plain they have of us. And it is the images that institutional authorities have of us that in particular tend to become self-fulfilling images: if the teachers, inspecting our Iowa scores and our aptitude-test figures and comparing our record with that of our age-peers, conclude that we're run-of-the-mine and *treat us accordingly*, then they lead us to become what they think we are.

What is more, I think it likely that the bias toward precocity in our educational and other institutions that Gregg has detected has notably different consequences for people in different social classes. The potential late bloomers in the lower economic strata are more apt to lose out altogether than their counterparts in the middle and upper strata. If poor children are not precocious, if they don't exhibit great ability early in their lives and so are not rewarded by scholarships and other sustaining grants, they drop out of school and in many instances never get to realize their potentialities. The potential late bloomers among the well-to-do have a better prospect of belated recognition. Even if they do poorly in their school work at first, they are apt to go on to college in any case. The values of their social class dictate this as the thing to do, and their families can see them through. By remaining in the system, they can eventually come to view. But many of their more numerous counterparts in the lower strata are probably lost for good. The bias toward precocity in our institutions thus works profound damage on the late bloomers with few economic or social advantages. And it is this group that our informal talent scouts can perhaps help salvage.

I've dwelt upon these aspects of the instrumental recognition of excellence as capacity because they seem most neglected. These are problems that psychological tests and measurements do not solve (nor are they so intended). That is why I believe that we need new kinds of sociological research centered on the social processes of current identification and selection of talent. At any rate, this is worth considering.

2. *Recognition as honorific and excellence as quality*

Inevitably, I've touched upon the honorific recognition—the giving of symbolic and material rewards—of high capacity. I shall add little to this in these notes, if only because I want to get on to other matters. Just a few words, then, on some problems deserving notice here.

One problem is this: how far down the scale of ability shall honorific recognition be given? Does our system provide too much for winners in a competition and give too little recognition to those who only place and show? And what about the contingent below these?

I do not suggest that meaningless consolation prizes be distributed to all, irrespective of ability. This would serve small purpose and might, in fact, damage the significance of the public recognition of great promise. Nevertheless, one gains the impression that the cut-off points for the recognition of promise have been artifacts of other decisions rather than decided upon after careful review.

Another problem has been considered in the staff-paper: how can local climates of values be modified to accord heightened significance to intellectual and artistic interests and abilities? It is often said that this *should* be done, but little has been said about how it *can* be done. I should think it useful to discuss the kind of research and the kind of action that seem indicated to achieve the purpose.

3. *Recognition as instrumental and excellence as performance*

This refers to the problems of identifying not merely the capacity for excellence but excellence of actual performance. At first glance, this would seem to be something that takes care of itself. Unlike the not easily observable capacity for excellence, demonstrated excellence must presumably come to notice. But it is precisely this assumption that should be put as a question rather than assumed as a fact.

What kinds of excellent work go largely unrecognized, except, perhaps, by the few who are the direct beneficiaries of that work?

As I have said, efforts to identify and reward persons who foster excellence in others have long interested me. I am therefore apt to make much of this hobby-horse and he may throw me in the process. But even with as much restraint as I can muster, I still think it a subject of peculiar interest to any enterprise for the recognition of excellence.

One kind of excellence that may be going relatively unnoticed is the talent some people have for evoking superior or maximum performance in others. I put it this broadly—evoking maximum performance in others—rather than more narrowly—evoking excellence in others—because this seems to be the nature of the type. It is not only that he may be one of our informal talent scouts who finds excellence and then does much to help it thrive. Often, this type of person works even greater though less dramatic miracles: they bring out the best in the people around them, including those of distinctly small talent.

This hypothetical person might be described as a catalyst. (I suppose that the term catalyst cannot be strictly applied, for I am persuaded that he too undergoes some change in the process. But for want of a better tag, let this one stand.) We all know of the catalysts of human worth, the agents who accelerate positive growth in the people about them. A few of us are lucky enough to have encountered one or another of them in the flesh. History, early and late, occasionally records their existence and their undeniable role in the facilitation of others' accomplishments. Perhaps they need to remain unsung heroes in order to do their self-chosen job. If so, their honorific recognition might be self-defeating.

I refer, to take some cases quite casually—that is, as they happen to have come to my attention—to people like Edward Marsh, or Isaac Beeckman or Father Mersenne or Cardinal de Bérulle. Marsh, to judge from the just-published biography of him by Christopher Hassall, was in no sense a genius himself but, as the reviewer Harold Hobson noted in *The Christian Science Monitor*, "he was the cause, or the liberator of genius in others." (One of these "others," whom one would have thought needed little assistance in achieving his destiny, was Winston Churchill.) Another evoker of excellence, in quite a different time and place, was the seventeenth-century Isaac Beeckman, a mathematician of some ability in his own right but notable chiefly for having seen and having called forth in the young Descartes the mathematical genius not yet apparent to others. Let's stay with Descartes for a moment, for he was the beneficiary of a series of catalytic agents. After Beeckman, there was Father Mersenne, that great clearing-house for seventeenth-century science. It was Mersenne who supplied Descartes (and others) with an unceasing flow of scientific and mathematical problems that proved effective challenges to him. Later, it is said, Descartes was brought to publish some of his treatises only because of the insistence by Cardinal de Bérulle that he do so as *une obligation de conscience* and that Descartes would be held responsible on the day of accounting *du tort qu'il ferait au genre humain en le privant du fruit de ses méditations*. Such allegedly verbatim conversations in an earlier day are necessarily suspect, if only because the tape recorder wasn't there to supply the evidence. But the main import seems confirmed by Descartes' own letters: before Bérulle, Descartes contemptuously rejected the notion of becoming a *faiseur de livres*; after Bérulle, he wrote as never before.

Among these few examples, Father Mersenne must be considered the very archetype of the catalyst. He recognized the merit of Campanella, Bacon, Galileo, Herbert of Cherbury and sought fitting recognition of them. He was the friend of great men who were at odds with one another and communicated only through him: Beaugrand and Desargues, Descartes and Gassendi, Roberval and Hobbes. He stimulated others to achieve what he himself could not achieve. Pascal's eulogy of Father Mersenne is a eulogy for a disinterested evoker of excellence in others:

> Il avoit un talent particulier pour former de belles questions, en quoy il n'avoit peut-être pas de semblable; mais encore qu'il n'eust pas un pareil bonheur a les resoudre et que ce soit proprement en cecy que consiste tout l'honneur, il est vray neantmoins qu'on luy a obligation et qu'il a donne l'occasion de plusieurs belles decouvertes qui peut-être n'auroient jamais este faites s'il n'y eust excite les scavans.

There is no great difficulty in accumulating lists of these evokers of excellence as they appear in the biographies of great men and women. But except for occasional anecdotes, we know little of the catalysts who have brought out the best in contemporary talents. An investigation of this would be worth undertaking, and it is distinctly feasible. It could follow the pattern of investigation worked out in studies of "influentials" in local communities: that is, of people who are habitually sought out for advice by people around them. Adapting the procedure to present purposes, one would begin by compiling lists of outstanding people in the sciences, arts and humanities, perhaps, in government, administration, education, and so forth. A sample of these men and women could be interviewed to find out whether they had, in the course of their career, encountered catalysts—people who had brought out the best in them. This derived list would form the basis for further investigation, with special attention to those men and women who turn up in the canvass on more than one occasion. These, then, would comprise a collection (a not necessarily representative sample) of evokers of excellence in our time. It would remain to be seen what manner of men and women these are, which of them had received public recognition fitting their roles as evokers of excellence, etc. Perhaps we could find some of these socially invisible men and women who do much to bring about the excellence of performance by others that is publicly visible.

All this does not assume, of course, that these evokers of excellence are known only within their immediate social circle. Some may be, and these local eductors of the best there is in those about them, would be of particular interest. But the rest, and these may turn out to be the large majority, are cosmopolitans rather than locals, already recognized outside their immediate locale. After all, Lytton Strachey, described by Clive Bell as the sort of listener who leads one to say things worth listening to, was scarcely without honor and repute in the world outside the tight little Bloomsbury group. The case of Strachey reminds us too that the evoker of excellence may also be the author of excellence in his own right. These need not be mutually exclusive roles.

Consider only one other case of this double role, this time in one of the sciences. The many contributions of John Newport Langley to experimental biology which he made almost to the time of his death in 1925 were enough to ensure him a lasting place in the annals of that science; this at least is the judgment of C. S. Sherrington, a man who should know. But beyond his own scientific discoveries, Langley was the source of discoveries by others. His school of physiology at Cambridge was "remarkably productive of physiologists of distinction," largely owing, it appears, to his own direct influence as a teacher in the research rooms. (As a lecturer, he was evidently ineffectual.) And beyond this, in turn, he helped raise the calibre of physiological work and of the reporting of physiological experiments throughout Britain.

These people whose excellence of accomplishment is widely recognized are perhaps apt to receive independent recognition of their capacity for evoking excellence in others. But it may be quite otherwise with many of those who create little that is publicly visible but whose hidden genius consists in their being the cause, or at least the occasion, of bringing out the best in the people around them. These, I assume, are the overlooked men and women, though many colleges and universities provide for recognition of their "great teachers" who may not have themselves contributed to learning and science. And these, in particular, might be identified by the dragnet of inquiry directed to the men and women who have achieved acknowledged excellence.

Before leaving this subject, I should say again that it is an open question whether these evokers of excellence *should* be given public recognition. We should consider the possible consequences of putting them in the public spotlight. Perhaps they want comparative obscurity and need it to do their work. Perhaps a formal system of acclaim for these assumedly unsung heroes would bring about imitations of the real thing. This would not be the first time that public rewards, either in the market of money or the market of prestige, have resulted in a multitude of imitations, in which the outer appearance is substituted for the genuine article. This danger, as I shall suggest in the next section, is inherent in every reward system. But, apparently, it can be damped, if not entirely eliminated, by devising systems that take this danger into account and counter it in advance.

4. *Recognition as honorific, excellence as performance*

I shall raise a few questions about the assumptions and problems historically associated with formal systems for the public recognition of excellence of achievement.

To begin with, there is the question of "unit" of achievement that is to form the basis for judgment of quality. Is it to be *a* discovery, *a* paper, *a* book, *a* painting, building, sculpture, or symphony, *a* notable achievement in keeping the peace or in ending the war? This principle of the single achievement seems to govern the award of a Nobel prize in the sciences. But in literature it is more commonly a person's life-work that is being recognized. The question of *what* is to receive recognition must be settled in every reward system, and yet the comparative merits of the various criteria commonly employed are not altogether clear. This is one question that deserves review.

Second is the question, What qualities of a seeming achievement are to be judged? Is it popularity, orthodoxy, heterodoxy? The appropriate answers to this question do not seem self-evident to me. Take the issue of "orthodoxy" and "heterodoxy." Each of us can cite cases where later history has reversed the contemporaneous judgment of the worth of particular people and their works of heterodoxy. Much (not all) innovation is in the nature of the case heterodox, and we all know that the lot of the heterodox innovator is often a most unhappy one. It is said that when Hamilton developed his radical idea of quaternions, after having won a recognized place for his work in optics and dynamics, he managed to find just one mathematician who would take the idea seriously enough to work on it. And Sydenham, probably second only to Harvey among seventeenth-century English physicians, was never made a Fellow of the Royal College of Physicians. And so on through an easily extended list of instances in which true excellence went comparatively unnoticed or stood condemned because it involved a great departure from reigning ideas.

But there is no joy or merit in escaping the error of taking heterodoxy to be inevitably false or ugly or sinister only to be caught up in the equal and opposite error of thinking the heterodox to be inevitably true or beautiful or altogether excellent. Put in so many words, this is a commonplace wrapped in banality. Yet, not infrequently, people alienated from the world about them do regard heterodoxy as a good in itself, whatever its character. And others, perhaps in recoil from being tagged as philistines or in reaction to the familiar cases of true merit, in each age, being neglected because they were unorthodox, are quick to see worth in heterodoxy, counter-cyclicalism, or "originality," all apart from its substance. In every time, apparently, shrewd men have known that an appropriate kind of seeming heterodoxy is amply rewarded. It is often a ticket to fame (and not necessarily notoriety). The Digbys in seventeenth-century England—among them, Sir Kenelm Digby in particular—were among the most adept in becoming heroes through nonconformity of a kind that received the applause of the crowd. And today every lecturer to women's clubs, literary circles, and businessmen's associations knows that there is no better way to win their hearts than by attacking part of what they stand for while letting it be known that they are not beyond redemption.

All this is some distance from the question of the criteria of excellence, but it is not, I believe, entirely unconnected with it. There is a difficult problem, in many spheres of human activity, of discriminating the authentic innovation that merits recognition from the mere novelty that doesn't.

Coordinate with the questions of what is to be judged and the criteria by which it is to be judged is a third question: who shall judge? An essential point here is that judgments of the worth of performance go on incessantly in society. Moreover, each society, partly unwittingly and partly by plan, has its distinctive sets of status-judges. At times, these judges have little official standing but much influence. Burckhardt tells us, for example, that in Renaissance Italy, the poet-scholar had "the fullest consciousness that he was the giver of fame and immortality, or, if he chose, of oblivion." Subject to little control and willing to exploit this role of publicity-agent for contemporaries eager to win a secure place in posterity, an Aretino could be for hire to blast the reputation of one's enemies or to establish one's own reputation.

When status-judges are formally designated, on grounds of assumed competence and responsibility, there still remains the problem of ensuring that the criteria for judging excellence will be effectively employed. Beyond this there is the problem of deciding whether there should be a fixed or flexible number of people singled out for recognition. The decision of the French Academy, for example, that only a cohort of forty could qualify for immortality at any one time carried with it the exclusion, through the centuries, of many writers who won their own immortality. When the fixed number is coupled with a growing tendency toward conservatism, it results in the academicians of the forty-first chair—the "also rans"—exhibiting a level of excellence that would be hard to match among the officially designated academicians. The familiar list of incumbents of the 41st chair would include Descartes, Pascal, Molière, La Rochefoucauld,

Bayle, Rousseau, Saint-Simon, Diderot, Stendhal, Flaubert, Zola, Proust. It would be no great task to compile similar lists of men and women of genius or great distinction who were not accorded official recognition by institutions presumably designed to recognize excellence of the very kind they so eminently exhibited. In part this results from errors of judgment. There is probably some irreducible minimum of such mistakes. But in part these errors result from the limitations inherent in having a fixed number of places for recognition. Should it happen that a particular generation is rich in achievements of the highest order in one or another department of life, it follows that some people whose accomplishments are of high rank on an absolute scale will be excluded from official recognition. Their accomplishments might in some sense far outrank those which, in another time of comparatively less productivity, proved enough to qualify talents for the highest recognition. But whether the number of designations is fixed or flexible, there still remains the difficulty of deciding where and how to establish a cut-off point. This will be arbitrary and, I suspect, inevitably so. Even if we could "measure" degrees of excellence with fine precision, there would still remain the problem of where to draw the line between those to be accorded a particular form of recognition and those to be granted a less exalted form or none at all.

This question can be put in another form, which I only cite rather than discuss, since I have touched upon it earlier. How should formal reward systems provide for recognition of grades of achievement rather than being confined to those of topmost rank?

Functions and Dysfunctions of Reward Systems

In part one of this memorandum, I commented briefly on the rationale that might underlie any system of recognition of excellence, either as capacity or as achievement. It might be useful now to consider some of the specific questions that come to mind as one reviews the major purposes of arrangements for the recognition of excellence of achievement. For the moment it may be enough to examine two major purposes and the questions that arise in connection with each. The first of these purposes is the presumed benefit which recognition will have for the recipient of recognition; the second is the presumed beneficial effect it will have on others, thus creating a climate conducive to the cultivation of excellence.

An obvious question here is the decision on the *timing* of the award. At which stage in the career might it prove most helpful to the recipient? I cannot even try to examine this question in the detail it requires, but here are a few remarks.

An observation by Faraday is in point. Just about a century ago, he was asked by the parliamentary committee of the British Association for the Advancement of Science "whether any measures could be adopted by the Government or the Legislature, to improve the position of science, or of the cultivators of science, in this country?" Faraday replied out of his own experience—he was then 63—saying that he himself had long since received all the help and recognition he needed; that he could not say he had not "valued such distinctions"; indeed, he esteemed them very highly, although he thought he had never worked for them or sought them out; but that even if new distinctions were "now created here, the time is past when these would possess any attraction for me."

To the extent that Faraday is typical, and whether he is is precisely the question, it would seem that new awards mean little to accomplished people who, nearing the close of life, have already received substantial recognition.

Often, those who have accomplished a great deal will respond bitterly to the offer, late in their career, of a symbolic kind of recognition that "ought" to have been given them long before. As is befitting a man who "wrote about himself in exactly the same tone as he wrote about the Universe," Herbert Spencer supplies the most elaborate account of the injustice of belated honors in particular and of systems for honoring achievement in general. In his letter to the *Académie des Sciences Morales et Politiques* explaining why he must decline the honor of election, he says that such honors would at best be useful to "men of much promise" by sometimes saving them "from sinking in their struggles with adverse circumstances in the midst of a society prepossessed in favour of known men"; but that this aid ordinarily comes when it is no longer needed, when the obstacles have been surmounted. Presumably for this reason, Spencer declined almost all the honors offered him in his later life.

This kind of response is perhaps more common than we suppose, if we can judge from the instances I happen to have come across. When Veblen was belatedly asked to serve as President of the American Economic Association, he replied: "They didn't offer it to me when I needed it." The naturalist and scholar, D'Arcy Wentworth Thompson, author of the classic *Growth and Form*, was embittered at his lack of formal recognition and particularly by his not having been selected a Fellow of the Royal Society until he was in his mid-fifties. And for all his usual equanimity, Mr. Justice Holmes said that he could feel no "real triumph" over the LL.D. bestowed on him by Harvard (and more particularly, by Eliot) because the "honor came too late."

Are there systems of honors that avoid either the "too little" because "too late" as well as the possible counter-mistakes of "too much" because "too soon" (in which the recipient, having received the highest accolade, rests complacently on his oars)?

If Spencer suffered acutely from official inattention until it was too late, his associate, Huxley, received high honors very early in his life as a scientist. His diary tells what it meant to him to be elected an F.R.S. at the age of 26, by far the youngest of his competitors. It evidently provided, above all, much needed reassurance that he was on the right track in his work. It was "acknowledgement of the value of what" he had done. And since, like many of the rest of us, Huxley was occasionally inclined to think himself a fool and to doubt his own capacities, he concluded that "the only use of honours is as an antidote to such fits of the 'blue devils.' " When shortly afterward he learned that he was within an ace of receiving the Royal Medal of the Society—he did get it only a year later—he went on to say: "Except for its practical value as a means of getting a position I care little enough for the medal. What I do care for is the justification which the being marked in this position gives to the course I have taken. Obstinate and self-willed as I am, . . . there are times when grave doubts overshadow my mind, and then such testimony as this restores my self-confidence." The function of curbing acute self-doubts may be one of the more significant results of effectively administered systems of public recognition.[1]

This matter of the appropriate timing of high honors has long been of interest to scientists concerned to provide incentives for scientific work. Take the world of science in France, just after the turn of the nineteenth century. There were not many who would dare contradict the magisterial Laplace on any views he happened to express, but Lagrange was one of this courageous band. The two men—both, of course, Academicians of eminence—are debating the merits of François Arago as a prospective member of the Academy. Laplace saw importance and utility in Arago's work but regarded it as nothing more than a sign of promise; Lagrange takes it up from there:

> "Even you, M. de Laplace, when you entered the Academy, had done nothing brilliant; you only gave promise. Your grand discoveries did not come till afterwards."
>
> [Laplace turns from his own case to the principle he believes in issue:] "I maintain that it is useful to young savants to hold out the position of member of the Institute as a future recompense, to excite their zeal."
>
> [The retort] "You resemble the driver of the hackney coach, who, to excite his horses to a gallop, tied a bundle of hay at the end of his carriage pole; the poor horses redoubled their efforts, and the bundle of hay always flew on before them. Eventually, his plan made them fall off, and soon after brought on their death."

The transcript of this dialogue may be less than verbatim, but it again puts the question of the most effective timing of major awards if they are to serve as incentives and as means of reassurance to people of excellence who have yet to come into their own.

1. [On the same point, see chapter 15 of this volume.—Ed.]

These scattered observations center on the functions of awards for their recipients. As a transition to a passing comment on their functions for the climate of values and for novices just beginning their work, consider the remarks by Czeslaw Milosz about the public recognition of "the intellectual."

Honorific awards should not be regarded only psychologically, Milosz says in effect, as providing incentives for excellent work. They also serve a social function, by testifying to the merit of kinds of excellence that might otherwise be regarded as having small significance in society. Recognition may counter tendencies of the intellectual to feel himself alienated from his society.

20 The Matthew Effect in Science

1968

This paper develops a conception of ways in which certain psychosocial processes affect the allocation of rewards to scientists for their contributions—an allocation which in turn affects the flow of ideas and findings through the communication networks of science. The conception is based upon an analysis of the composite of experience reported in Harriet Zuckerman's interviews with Nobel laureates in the United States[1] and upon data drawn from the diaries, letters, notebooks, scientific papers, and biographies of other scientists.

The Reward System and "Occupants of the Forty-First Chair"

We might best begin with some general observations on the reward system in science, basing these on earlier theoretical formulations and empirical investigations. Some time ago[2] it was noted that graded rewards in the realm of science are distributed principally in the coin of recognition accorded research by fellow scientists. This recognition is stratified for varying grades of scientific accomplishment, as judged by the scientist's

Based on a paper read before the American Sociological Association in San Francisco, August 1967. Reprinted, with permission, from *Science* 159, no. 3810 (5 January 1968): 56–63, copyright 1968 by the American Association for the Advancement of Science.

1. The methods of obtaining these tape-recorded interviews and the character of their substance are described in H. A. Zuckerman, *Scientific Elite: Nobel Laureates in the United States* (Chicago: University of Chicago Press, in press), and in "Interviewing an Ultra-Elite," *Public Opinion Quarterly* 36 (1972): 159–75. [It is now (1973) belatedly evident to me that I drew upon the interview and other materials of the Zuckerman study to such an extent that, clearly, the paper should have appeared under joint authorship.]
2. See chapter 14 of this volume.

peers. Both the self-image and the public image of scientists are largely shaped by the communally validating testimony of significant others that they have variously lived up to the exacting institutional requirements of their roles.

A number of workers, in empirical studies, have investigated various aspects of the reward system of science as thus conceived. Glaser[3] has found, for example, that some degree of recognition is required to stabilize the careers of scientists. In a case study Crane[4] used the quantity of publication (apart from quality) as a measure of scientific productivity and found that highly productive scientists at a major university gained recognition more often than equally productive scientists at a lesser university. Hagstrom[5] has developed and partly tested the hypothesis that material rewards in science function primarily to reinforce the operation of a reward system in which the primary reward of recognition for scientific contributions is exchanged for access to scientific information. Storer[6] has analyzed the ambivalence of the scientist's response to recognition "as a case in which the norm of disinterestedness operates to make scientists deny the value to them of influence and authority in science." Zuckerman[7] and the Coles[8] have found that scientists who receive recognition for research done early in their careers are more productive later on than those who do not. And the Coles have also found that, at least in the case of contemporary American physics, the reward system operates largely in accord with institutional values of the science, inasmuch as quality of research is more often and more substantially rewarded than mere quantity.

In science as in other institutional realms, a special problem in the workings of the reward system turns up when individuals or organizations take on the job of gauging and suitably rewarding lofty performance on behalf of a large community. Thus, that ultimate accolade in twentieth-century science, the Nobel prize, is often assumed to mark off its recipients from all the other scientists of the time. Yet this assumption is at odds with the well-known fact that a good number of scientists who have not received the prize and will not receive it have contributed as much to the advancement of science as some of the recipients, or more. This can be described as the phenomenon of "the 41st chair." The derivation of this

3. B. G. Glaser, *Organizational Scientists: Their Professional Careers* (Indianapolis: Bobbs-Merrill, 1964).

4. D. Crane, *Amer. Sociol. Rev.* 30 (1965): 699.

5. W. O. Hagstrom, *The Scientific Community* (New York: Basic Books, 1965), chap. 1.

6. N. W. Storer, *The Social System of Science* (New York: Holt, Rinehart and Winston, 1966), p. 106; also ibid., pp. 20–26, 103–106.

7. H. A. Zuckerman, *Scientific Elite.*

8. J. R. Cole and S. Cole, *Social Stratification in Science* (Chicago: University of Chicago Press, 1973).

tag is clear enough. The French Academy, it will be remembered, decided early that only a cohort of forty could qualify as members and so emerge as immortals. This limitation of numbers made inevitable, of course, the exclusion through the centuries of many talented individuals who have won their own immortality. The familiar list of occupants of this 41st chair includes Descartes, Pascal, Molière, Bayle, Rousseau, Saint-Simon, Diderot, Stendhal, Flaubert, Zola, and Proust.[9]

What holds for the French Academy holds in varying degree for every other institution designed to identify and reward talent. In all of them there are occupants of the forty-first chair, men outside the Academy having at least the same order of talent as those inside it. In part, this circumstance results from errors of judgment that lead to inclusion of the less talented at the expense of the more talented. History serves as an appellate court, ready to reverse the judgments of the lower courts, which are limited by the myopia of contemporaneity. But in greater part, the phenomenon of the forty-first chair is an artifact of having a fixed number of places available at the summit of recognition. Moreover, when a particular generation is rich in achievements of a high order, it follows from the rule of fixed numbers that some whose accomplishments rank as high as those actually given the award will be excluded from the honorific ranks. Indeed, their accomplishments sometimes far outrank those which, in a time of less creativity, proved enough to qualify talents for this high order of recognition.

The Nobel prize retains its luster because errors of the first kind—where scientific work of dubious or inferior worth has been mistakenly honored—are uncommonly few. Yet limitations of the second kind cannot be avoided. The small number of awards means that, particularly in times of great scientific advance, there will be many occupants of the forty-first chair (and, since the terms governing the award of the prize do not provide for posthumous recognition, permanent occupants of that chair). This gap in the award of the ultimate prize is only partly filled by other awards for scientific accomplishment, since these do not carry the same prestige either inside the scientific community or outside it. Furthermore, what has been noted about the artifact of fixed numbers producing occupants of the forty-first chair in the case of the Nobel prize holds in principle for other awards providing less prestige (though sometimes, nowadays, more cash).

Scientists reflecting on the stratification of honor and esteem in the world of science know all this; the Nobel laureates themselves know and empha-

9. I have adopted this term for the general phenomenon from the monograph on the French Academy by Arsene Houssaye, *Histoire du 41me Fauteuil de l'Académie Française* (Paris, 1886). See also chapter 19 in this volume.

size it, and the members of the Swedish Royal Academy of Science and the Royal Caroline Institute who face the unenviable task of making the final decisions know it. The latter testify to the phenomenon of the forty-first chair whenever they allude to work of "prize-winning calibre" which, under the conditions of the scarcity of prizes, could not be given the award. And so it is that, in the case of the Nobel prize, occupants of the forty-first chair comprise an illustrious company that includes such names as Willard Gibbs, Dmitri Mendeleev, W. B. Cannon, H. Quincke, J. Barcroft, F. d'Hérelle, H. De Vries, Jacques Loeb, W. M. Bayliss, E. H. Starling, G. N. Lewis, O. T. Avery, and Selig Hecht, to say nothing of the long list of still-living uncrowned Nobel laureates.[10]

In the stratification system of honor in science, there may be a "ratchet effect"[11] operating in the careers of scientists such that, once having achieved a particular degree of eminence, they do not later fall much below that level (although they may be outdistanced by newcomers and so suffer a *relative* decline in prestige). Once a Nobel laureate, always a Nobel laureate. Yet the reward system based on recognition for work accomplished tends to induce continued effort, which serves both to validate the judgment that the scientist has unusual capacities and to testify that these capacities have continuing potential. What appears from below to be the summit becomes, in the experience of those who have reached it, only another way station. The scientist's peers and other associates regard each of his scientific achievements as only the prelude to new and greater achievements. Such social pressures do not often permit those who have climbed the rugged mountains of scientific achievement to remain content. It is not necessarily the fact that their own Faustian aspirations are ever escalating that keeps eminent scientists at work. More and more is expected of them, and this creates its own measure of motivation and stress. Less often than might be imagined is there repose at the top in science.[12]

The recognition accorded scientific achievement by the scientist's peers is a reward in the strict sense identified by Parsons.[13] As we shall see, such recognition can be converted into an instrumental asset as enlarged facilities are made available to the honored scientist for further work.

10. This partial list of men who have done work of "prize-winning calibre" is derived from *Nobel: The Man and His Prizes* (London: Elsevier, 1962), an official publication of the Nobel prize-granting academy and institute, Nobelstiftelsen.

11. I am indebted to Marshall Childs for suggesting that this term, introduced into economics by James S. Duesenberry in quite another connection, could aptly refer to this pattern in the cumulation of prestige for successive accomplishments. For its use in economics, see Duesenberry, *Income, Savings, and the Theory of Consumer Behavior* (Cambridge, Mass.: Harvard Univ. Press, 1949), pp. 114–16.

12. This process of a *socially reinforced* rise in aspirations, as distinct from Durkheim's concept of the "insatiability of wants," is examined by R. K. Merton in *Anomie and Deviant Behavior*, ed. M. Clinard (New York: Free Press, 1964), pp. 213–42.

13. T. Parsons, *The Social System* (New York: Free Press, 1951), p. 127.

Without deliberate intent on the part of any group, the reward system thus influences the "class structure" of science by providing a stratified distribution of chances, among scientists, for enlarging their role as investigators. The process provides differential access to the means of scientific production. This becomes all the more important in the current historical shift from little science to big science, with its expensive and often centralized equipment needed for research. There is thus a continuing interplay between the status system, based on honor and esteem, and the class system, based on differential life-chances, which locates scientists in differing positions within the opportunity structure of science.[14]

The Matthew Effect in the Reward System

The social structure of science provides the context for this inquiry into a complex psychosocial process that affects both the reward system and the communication system of science. We start by noting a theme that runs through the interviews with the Nobel laureates. They repeatedly observe that eminent scientists get disproportionately great credit for their contributions to science while relatively unknown scientists tend to get disproportionately little credit for comparable contributions. As one laureate in physics put it[15]: "The world is peculiar in this matter of how it gives credit. It tends to give the credit to [already] famous people."

As we examine the experiences reported by eminent scientists we find that this pattern of recognition, skewed in favor of the established scientist, appears principally (1) in cases of collaboration and (2) in cases of independent multiple discoveries made by scientists of distinctly different rank.[16]

In papers coauthored by men of decidedly unequal reputation, another laureate in physics reports, "the man who's best known gets more credit, an inordinate amount of credit." In the words of a laureate in chemistry: "When people see my name on a paper, they are apt to remember *it* and not to remember the other names." And a laureate in physiology and medicine describes his own pattern of response to jointly authored papers:

14. Max Weber touches upon the convertibility of position in distinct systems of stratification in his classic essay "Class, Status, Party" in H. H. Gerth and C. Wright Mills, eds., *Max Weber: Essays in Sociology* (New York: Oxford Univ. Press, 1946).

15. Zuckerman, *Scientific Elite*, chapter 8. The laureates are not alone in noting that prominent scientists tend to get the lion's share of credit; similar observations were made by less eminent scientists in the sample studied by Hagstrom (*Scientific Community*, pp. 24, 25).

16. A third case can be inferred from the protocols of interviews, in which the view is stated that, had a paper written by a comparatively unknown scientist been presented instead by an eminent scientist, it would have had a better chance of being published and of receiving respectful attention. Systematic information about such cases is too sparse for detailed study. See Zuckerman, *Scientific Elite,* chapter 8, passim.

You usually notice the name that you're familiar with. Even if it's last, it will be the one that sticks. In some cases, all the names are unfamiliar to you, and they're virtually anonymous. But what you note is the acknowledgement at the end of the paper to the senior person for his "advice and encouragement." So you will say: "This came out of Greene's lab, or so-and-so's lab." You remember that, rather than the long list of authors.

Almost as though he had been listening to this account, another laureate in medicine explains why he will often not put his name on the published report of a collaborative piece of work: "People are more or less tempted to say: 'Oh yes, so-and-so is working on such-and-such in C's laboratory. It's C's idea.' I try to cut that down." Still another laureate in medicine alludes to this pattern and goes on to observe how it might prejudice the career of the junior investigator:

If someone is being considered for a job by people who have not had much experience with him, if he has published only together with some known names —well, it detracts. It naturally makes people ask: "How much is really his own contribution, how much [the senior author's]. How will he work out once he goes out of that laboratory?"

Under certain conditions this adverse effect on recognition of the junior author of papers written in collaboration with prominent scientists can apparently be countered and even converted into an asset. Should the younger scientist move ahead to do autonomous and significant work, this work *retroactively* affects the appraisals of his role in earlier collaboration. In the words of the laureate in medicine who referred to the virtual anonymity of junior authors of coauthored papers: "People who have been identified with such joint work and who then go on to do good work later on, [do] get the proper amount of recognition." Indeed, as another laureate implies, this retroactive judgment may actually heighten recognition for later accomplishments: "The junior person is sometimes lost sight of, but only temporarily *if* he continues. In many cases, he actually gains in acceptance of his work and in general acceptance, by having once had such association." Awareness of this pattern of retroactive recognition may account in part for the preference, described by another laureate, of some "young fellows [who] feel that to have a better-known name on the paper will be of help to them." But this is an expressive as well as a merely instrumental preference, as we see also in the pride with which laureates themselves speak of having worked, say, with Fermi, G. N. Lewis, Meyerhof, or Niels Bohr.

So much for the misallocation of credit in this reward system in the case of collaborative work. Such misallocation also occurs in the case of independent multiple discoveries. When approximately the same ideas or findings are independently communicated by a scientist of great repute and by one not yet widely known, it is the first, we are told, who ordinarily receives prime recognition. An approximation to this pattern is reported by a laureate who observes:

It does happen that two men have the same idea and one becomes better known for it. E—, who had the idea, went circling round to try to get an experiment for. . . . Nobody would do it and so it was forgotten, practically. Finally, A— and B— and C— did it, became famous, and got the Nobel Prize. . . . If things had gone just a little differently; if somebody had been willing to try the experiment when E— suggested it, they probably could have published it jointly and he would have been a famous man. As it is, he's a footnote.

The workings of this process at the expense of the young scientist and to the benefit of the famous one is remarkably summarized in the life history of a laureate in physics, who has experienced both phases at different times in his career. "When you're not recognized," he recalls,

it's a little bit irritating to have somebody come along and figure out the obvious which you've also figured out, and everybody gives him credit just because he's a famous physicist or a famous man in his field.

Here he is viewing the case he reports from the perspective of one who had this happen to him before he had become famous. The conversation takes a new turn as he notes that his own position has greatly changed. Shifting from the perspective of his earlier days, when he felt victimized by the pattern, to the perspective of his present high status, he goes on to say:

This often happens, and I'm probably getting credit now, if I don't watch myself, for things other people figured out. Because I'm notorious and when I say it, people say: "Well, he's the one that thought this out." Well, I may just be saying things that other people have thought out before.

In the end, then, a sort of rough-hewn justice has been done by the compounding of two compensating injustices. His earlier accomplishments have been underestimated; his later ones, overestimated.[17]

This complex pattern of the misallocation of credit for scientific work must quite evidently be described as "the Matthew effect," for, as will be remembered, the Gospel According to St. Matthew puts it this way:

For unto every one that hath shall be given, and he shall have abundance: but from him that hath not shall be taken away even that which he hath.

17. This compensatory pattern can only obtain, of course, among scientists who ultimately achieve recognition with its associated further rewards. But, as with all systems of social stratification involving differentials in life-chances, there remains the question of the extent to which talent among individuals in the deprived strata has gone unrecognized and undeveloped, and its fruits lost to society. More specifically, we have yet to discover whether or not the channels of mobility are equally open to talent in various institutional realms. Does contemporary science afford greater or less opportunity than art, politics, the practicing professions, or religion for the recognition of talent, whatever its social origins?

Put in less stately language, the Matthew effect consists of the accruing of greater increments of recognition for particular scientific contributions to scientists of considerable repute and the withholding of such recognition from scientists who have not yet made their mark. Nobel laureates provide presumptive evidence of the effect, since they testify to its occurrence, not as victims—which might make their testimony suspect—but as unwitting beneficiaries.

The laureates and other eminent men of science are sufficiently aware of this aspect of the Matthew effect to make special efforts to counteract it. At the extreme, they sometimes refuse to coauthor a paper reporting research on which they have collaborated in order not to diminish the recognition accorded their less-well-known associates. And, as Harriet Zuckerman has found,[18] they tend to give first place in jointly authored papers to one of their collaborators. She discovered, moreover, that the laureates who have attained eminence before receiving the Nobel prize begin to transfer first-authorship to associates earlier than less eminent laureates-to-be do, and that both sets of laureates—the previously eminent and not-so-eminent—greatly increase this practice *after* receiving the prize. Yet the latter effort is probably more expressive of the laureates' good intentions than it is effective in redressing the imbalance of credit attributable to the Matthew effect. As the laureate quoted by Zuckerman acknowledges: "If I publish my name first, then everyone thinks the others are just technicians. . . . If my name is last, people will credit me anyway for the whole thing, so I want the others to have a bit more glory."

The problem of achieving a public identity in science may be deepened by the great increase in the number of papers with several authors in which the role of novice collaborators becomes obscured by the brilliance that surrounds their illustrious coauthors. Even when there are only two collaborators, the same obscurant effect may occur for the junior who exhibits several "inferiorities" of status. The role ascribed to a doubly or trebly stigmatized coauthor may be diminished almost to the vanishing point so that, even in cases of later substantial achievements, there is little recognition of that role in the early work. Thus, to take a case close to home, W. I. Thomas has often been described as the sole author of the scholarly book *The Child in America*, although its title page unmistakably declares that it was written by both William I. Thomas and Dorothy Swaine Thomas. It may help interpret this recurrent misperception to consider the status of the collaborators at the time the book was published in 1928. W. I. Thomas, then 65, was president of the American Sociological Society in

18. H. Zuckerman, "Patterns of name-ordering among authors of scientific papers: a study of social symbolism and its ambiguity," *American Journal of Sociology* 74 (1968): 276–91. Dr. Zuckerman will not demean herself to give these practices their predestined tag, but I shall: plainly, these are instances of *nobelesse oblige*.

belated acknowledgement of his longstanding rank as dean of American sociologists, while Dorothy Swaine Thomas (not to become his wife until seven years later) was subject to the double jeopardy of being a woman of sociological science and still in her twenties. Although she went on to a distinguished scientific career (incidentally, being elected to the presidency of the American Sociological Society in 1952), the early book is still being ascribed solely to her illustrious collaborator even by ordinarily meticulous scholars.[19]

Once again, we can turn to the Scriptures to designate the status-enhancement and status-suppression components in the Matthew effect. These can evidently be described as "the Ecclesiasticus components," from the familiar injunction "Let us now praise famous men." in the noncanonical book of that name.

It will surely have been noted that the laureates perceive the Matthew effect primarily as a problem in the just allocation of credit for scientific accomplishment. They see it largely in terms of its action in enhancing rank or suppressing recognition. They see it as leading to an unintended double injustice, in which unknown scientists are unjustifiably victimized and famous ones, unjustifiably benefited. In short, they see the Matthew effect in terms of a basic inequity in the reward system that affects the careers of individual scientists. But it has other implications for the development of science, and we must shift our angle of theoretical vision in order to identify them.

The Matthew Effect in the Communication System

We now look at the same social phenomena from another perspective—not from the standpoint of individual careers and the workings of the reward system but from the standpoint of science conceived of as a system of communication. This perspective yields a further set of inferences. It leads us to propose the hypothesis that a scientific contribution will have greater visibility in the community of scientists when it is introduced by a scientist of high rank than when it is introduced by one who has not yet made his mark. In other words, considered in its implications for the reward system, the Matthew effect is dysfunctional for the careers of individual scientists who are penalized in the early stages of their development, but considered in its implications for the communication system, the Matthew effect, in cases of collaboration and multiple discoveries, may operate to heighten the visibility of new scientific communications. This is not the first instance

19. See the ascriptions of the book, for example, in Alfred Schutz, *Collected Papers*, 2 vols., edited and with an introduction by Maurice Natanson (The Hague: Martinus Nijhoff, 1962), 1:348, n71; Peter McHugh, *Defining the Situation* (Indianapolis: Bobbs-Merill Co., 1968), p. 7.

of a social pattern's being functional for certain aspects of a social system and dysfunctional for certain individuals within that system. That, indeed, is a principal theme of classical tragedy.[20]

Several laureates have sensed this social function of the Matthew effect. Speaking of the dilemma that confronts the famous man of science who directs the work of a junior associate, one of them observes:

> It raises the question of what you are to do. You have a student; should you put your name on that paper or not? You've contributed to it, but is it better that you shouldn't or should? There are two sides to it. If you don't [and here comes the decisive point on visibility], if you don't, there's the possibility that the paper may go quite unrecognized. Nobody reads it. If you do, it might be recognized, but then the student doesn't get enough credit.

Studies of the reading practices of scientists indicate that the suggested possibility—"Nobody reads it"—is something less than sheer hyperbole. It has been found, for example, that only about half of 1 percent of the articles published in journals of chemistry are read by any one chemist.[21] And much the same pattern has been found to hold in psychology:

> The data on current readership (i.e., within a couple [of] months after distribution of the journal) suggested that about one-half of the research reports in "core" journals will be read [or skimmed] by 1% or less of a random sample of psychologists. At the highest end of the current readership distribution, no research report is likely to be read by more than about 7% of such a sample.[22]

Several of the Coles's findings[23] bear tangentially on the hypothesis about the communication function of the Matthew effect. The evidence is tangential rather than central to the hypothesis since their data deal with the degree of visibility of the entire corpus of each physicist's work in the national community of physicists rather than with the visibility of particular papers within it. Still, in gross terms, their findings are at least consistent with the hypothesis. The higher the rank of physicists (as measured by the prestige of the awards they have received for scientific work), the

20. This pattern of social functions and individual dysfunctions is at variance with the vigorous and untutored optimism unforgettably expressed by Adam Smith, who speaks of "a harmonious order of nature, under divine guidance, which promotes the welfare of man through the operation of his individual propensities." If only it were that simple. One of the prime problems for sociological theory is that of identifying the special conditions under which men's propensities and the requirements of the social system are in sufficient accord to be functional for both individuals and the social system.

21. R. L. Ackoff and M. H. Halbert, *An Operations Research Study of the Scientific Activity of Chemists* (Cleveland: Case Institute of Technology Operations Research Group, 1958).

22. *Project on Scientific Information Exchange in Psychology* (Washington, D.C.: American Psychological Association, 1963), 1:9.

23. S. Cole and J. R. Cole, "Visibility and the Structural Bases of Observability in Science," a paper presented before the American Sociological Association, August 1967, and developed further in their *Social Stratification in Science*.

higher their visibility in the national community of physicists. Nobel laureates have a visibility score[24] of 85; other members of the National Academy of Sciences, a score of 72; recipients of awards having less prestige, a score of 38; and physicists who have received no awards, a visibility score of 17. The Coles also find that the visibility of physicists producing work of high quality is associated with their attaining honorific awards more prestigious than those they have previously received. Further investigation is needed to discover whether these same patterns hold for differences in the visibility (as measured by readership) of individual papers published by scientists of differing rank.

There is reason to assume that the communication function of the Matthew effect is increasing in frequency and intensity with the exponential increase[25] in the volume of scientific publications, which makes it increasingly difficult for scientists to keep up with work in their field. Bentley Glass[26] is only one among many to conclude that "perhaps no problem facing the individual scientist today is more defeating than the effort to cope with the flood of published scientific research, even within one's own narrow specialty." Studies of the communication behavior of scientists[27] have shown that, confronted with the growing task of identifying significant work published in their field, scientists search for cues to what they should attend to. One such cue is the professional reputation of the authors. The problem of locating pertinent research literature and the problem of authors' wanting their work to be noticed and used are symmetrical: the vastly increased bulk of publication stiffens the competition between papers for such notice. The American Psychological Association study[28] found that from 15 to 23 percent of the psychologist-readers' behaviors in selecting articles were based on the identity of the authors.

24. In the Coles' study (see fn. 23), the term *visibility scores* refers to percentages in a sample of more than 1,300 American physicists who indicated that they were familiar with the work of a designated list of 120 physicists. The study includes checks on the validity of these visibility scores.

25. See D. J. de Solla Price, *Little Science, Big Science*. Price has noted that "all crude measures, however arrived at, show to a first approximation that science increases exponentially, at a compound interest of about 7 per cent per annum, thus doubling in size every 10–15 years, growing by a factor of 10 every half-century, and by something like a factor of a million in the 300 years which separate us from the seventeenth-century invention of the scientific paper when the process began" (*Nature* 206 [1965]: 233–38).

26. B. Glass, *Science* 121 (1955): 583.

27. See, for example, H. Menzel, in *Communication: Concepts and Perspectives*, L. Thayer, ed. (Washington, D.C.: Spartan Books, 1966), pp. 279–95; and in *Amer. Psychologist* 21 (1966): 999. See also S. Herner in *Science* 128 (1958): 9, who notes that "one of the greatest stimulants to the use of information is familiarity with its source"; S. Herner, *Ind. Eng. Chem.* 46 (1954): 228.

28. *Project on Scientific Information Exchange in Psychology*, pp. 252, 254. Future investigations will require more detailed data on the actual processes of selecting scientific papers for varying kinds of "reading" and "skimming." But the data now available are at least suggestive.

The workings of the Matthew effect in the communication system require us to draw out and emphasize certain implications about the character of science. They remind us that science is not composed of a series of private experiences of discovery by many scientists, as sometimes seems to be assumed in inquiries centered exclusively on the psychological processes involved in discovery. Science is public, not private. True, the making of a discovery is a complex personal experience. And since the *making* of the discovery necessarily precedes its fate, the nature of the experience is the same whether the discovery temporarily fails to become part of the socially shared culture of science or quickly becomes a functionally significant part of that culture. But, for science to be advanced, it is not enough that fruitful ideas be originated or new experiments developed or new problems formulated or new methods instituted. The innovations must be effectively communicated to others. That, after all, is what we mean by a *contribution* to science—something given to the common fund of knowledge. In the end, then, science is a socially shared and socially validated body of knowledge. For the development of science, only work that is effectively perceived and utilized by other scientists, then and there, matters.

In investigating the processes that shape the development of science, it is therefore important to consider the social mechanisms that curb or facilitate the incorporation of would-be contributions into the domain of science. Looking at the Matthew effect from this perspective, we have noted the distinct possibility that contributions made by scientists of considerable standing are the most likely to enter promptly and widely into the communication networks of science, and so to accelerate its development.

The Matthew Effect and the Functions of Redundancy

Construed in this way, the Matthew effect links up with my previous studies of the functions of redundancy in science.[29] When similar discoveries are made by two or more scientists working independently ("multiple discoveries"), the probability that they will be promptly incorporated into the current body of scientific knowledge is increased. The more often a discovery has been made independently, the better are its prospects of being identified and used. If one published version of the discovery is obscured by "noise" in the communication system of science, then another version may become visible. This leaves us with an unresolved question: How can one estimate what amount of redundancy in independent efforts to solve a scientific problem will give maximum probability of solution

29. On the concept of functional redundancy as distinct from "wasteful duplication" in scientific research, see chapter 17 in this volume.

without entailing so much replication of effort that the last increment will not appreciably increase the probability?[30]

In examining the functions of the Matthew effect for communication in science, we can now refine this conception further. It is not only the number of times a discovery has been independently made and published that affects its visibility but also the standing, within the stratification system of science, of the scientists who have made it. To put the matter with undue simplicity, a single discovery introduced by a scientist of established reputation may have as good a chance of achieving high visibility as a multiple discovery variously introduced by several scientists no one of whom has as yet achieved a substantial reputation. Although the general idea is, at this writing, tentative, it does have the not inconsiderable virtue of lending itself to approximate test. One can examine citation indexes to find whether in multiple discoveries by scientists of markedly unequal rank it is indeed the case that work published by the scientists of higher rank is the more promptly and more widely cited.[31] To the extent that it is, the findings will shed some light on the unplanned consequences of the stratification system for the development of science. Interviews with working scientists about their reading practices can also supply data bearing on the hypothesis.

So much for the link between the Matthew effect and the functions of multiple discoveries in increasing both the probability and the speed of diffusion of significant new contributions to science. The Matthew effect also links up with the finding, reported in chapter 16 of this volume, that great talents in science are typically involved in many multiple discoveries. This statement holds for Galileo and Newton; for Faraday and Maxwell; for Hooke, Cavendish, and Stensen; for Gauss and Laplace; for Lavoisier, Priestley, and Scheele; *and* for most Nobel laureates. It holds,

30. One of the laureates questioned the ready assumption that redundancy of research effort necessarily means "wasteful duplication": "One often hears, especially when large amounts of money are involved, that duplication of effort should be avoided, that this is not an efficient way of doing things. I think that most of the time, in respect to research, duplication of effort is a good thing. I think that if there are different groups in different laboratories working on the same thing, their approach is sufficiently different [to increase the probability of a successful outcome]. On the whole, this is a good thing and not something that should be avoided for the sake of efficiency." Zuckerman, *Scientific Elite*, chapter 8.

31. So far as I know, no investigation has yet been carried out on precisely this question. At best suggestive is the peripheral evidence that papers of Nobel laureates-to-be were cited thirty times more often in the five years before their authors were awarded the prize than were the papers of the average author appearing in the Citation Index during the same period. See I. H. Sher and E. Garfield, "New Tools for Improving the Effectiveness of Research," paper presented at the 2nd Conference on Research Program Effectiveness, Washington, D.C., July 1965; H. Zuckerman, *Sci. Amer.* 217 (1967): 25.

in short, for all those whose place in the pantheon of science is largely assured, however much they may differ in the scale of their total accomplishment.

The greatness of these scientists rests in their having *individually* contributed a body of ideas, methods, and results which, in the case of multiple discoveries, has also been contributed by a sizable *aggregate* of less talented men. For example, we have found that Kelvin had a part in 32 or more multiple discoveries, and that it took 30 other men to contribute what Kelvin himself contributed.

By examining the interviews with the laureates, we can now detect some underlying psychosocial mechanisms that make for the greater visibility of contributions reported by scientists of established reputation. This greater visibility is not merely the result of a halo effect such that their personal prestige rubs off on their separate contributions. Rather, certain aspects of their socialization, their scheme of values, and their social character account in part for the visibility of their work.

Social and Psychological Bases of the Matthew Effect

Even when some of his contributions have been independently made by an aggregate of other scientists, the great scientist serves distinctive functions. It makes a difference, and often a decisive difference, for the advancement of science whether a composite of ideas and findings is heavily concentrated in the work of one scientist or one research group or is thinly dispersed among a great number of scientists and organizations. Such a composite tends to take on a structure sooner in the first instance than in the second. It required a Freud, for instance, to focus the attention of many psychologists upon a wide array of ideas which, as has been shown elsewhere (see footnote 29), had in large part also been hit upon by various other scientists. Such focalizing may turn out to be a distinctive function of eminent men and women of science.[32]

A Freud, a Fermi, and a Delbrück play a charismatic role in science. They excite intellectual enthusiasm among others who ascribe exceptional qualities to them. Not only do they themselves achieve excellence, they have the capacity for evoking excellence in others. In the compelling phrase of one laureate, they provide a "bright ambiance." It is not so much that these great men of science pass on their techniques, methods, information, and theory to novices working with them. More consequentially, they

32. Later in this discussion, I consider the dysfunctions associated with these functions of great men of science. Idols of the cave often continue to wield great influence even though the norms of science call for the systematic questioning of mere authority. Here, as in other institutional spheres, the problem is one of accounting for patterns of coincidence and discrepancy between social norms and actual behavior.

convey to their associates the norms and values that govern significant research. Often in their later years, or after their death, this personal influence becomes routinized, in the fashion described by Max Weber for other fields of human activity. Charisma becomes institutionalized, in the form of schools of thought and research establishments.

The role of outstanding scientists in influencing younger associates is repeatedly emphasized in the interviews with laureates. Almost invariably they lay great emphasis on the importance of problem-*finding*, not only problem-solving. They uniformly express the strong conviction that what matters most in their work is a developing sense of taste, of judgment, in seizing upon problems that are of fundamental importance. And, typically, they report that they acquired this sense for the significant problem during their years of training in evocative environments. Reflecting on his years as a novice in the laboratory of a chemist of the first rank, one laureate reports that he "led me to look for important things, whenever possible, rather than to work on endless detail or to work just to improve accuracy rather than making a basic new contribution." Another describes his socialization in a European laboratory as "my first real contact with first-rate creative minds at the high point of their power. I acquired a certain expansion of taste. It was a matter of taste and attitude and, to a certain extent, real self-confidence. I learned that it was just as difficult to do an unimportant experiment, often more difficult, than an important one."

There is one rough measure of the extent to which the laureates were trained and influenced in particularly creative research environments—the number of laureates each worked-under in earlier years. Of 84 American laureates, 44 worked in some capacity, as young scientists, under a total of 63 Nobel prize winners.[33] But apparently it is not only the experience of the laureates (and, presumably, other outstanding scientists) in these environments that accounts for their tendency to focus on significant problems and so to affect the communication function of the Matthew effect. Certain aspects of their character also play a part. With few exceptions, these are men of exceptional ego strength. Their self-assurance finds varied expression within the context of science as a social institution. That institution, as we know, includes a norm calling for autonomous and critical judgment about one's own work of others. With their own tendencies reinforced by such norms, the laureates exhibit a distinct self-confidence (which, at the extreme, can be loosely described as attractive arrogance). They exhibit a great capacity to tolerate frustration in their work, absorbing repeated failures without manifest psychological damage. One laureate alluded to this capacity while taking note of the value of psychological support by colleagues:

33. H. Zuckerman, *Scientific Elite*, chapter 5.

Research is a rough game. You may work for months, or even a few years, and seemingly you are getting nowhere. It gets pretty dark at times. Then, all of a sudden, you get a break. It's good to have somebody around to give a bit of encouragement when it's needed.

Though attentive to the cues provided by the work of others in their field, the Nobelists are self-directed investigators, moving confidently into new fields of inquiry once they are persuaded that a previous one has been substantially mined. In these activities they display a high degree of venturesome fortitude. They are prepared to tackle important though difficult problems rather than settle for easy and secure ones. Thus a laureate recalls having been given, early in his career, "a problem about which there was no risk. All I had to do was to analyze [the chemical composition of certain materials]. You could not fail because the method was well established. But I knew I was going to work on the t—— instead and the whole thing would have to be created because nothing was known about it." He then went on to make one of his prime contributions in the more risky field of investigation.[34]

This marked ego strength links up with these scientists' selection of important problems in at least two ways. Being convinced that they will recognize an important problem when they encounter it, they are willing to bide their time and not settle too soon for a prolonged commitment to a comparatively unimportant one. Their capacity for delayed gratification, coupled with self-assurance, leads to a conviction that an important problem will come along in due course and that when it does, their acquired sense of taste will enable them to recognize it and handle it. As we have seen, this attitude has been reinforced by their early experience in creative environments. There, association with eminent scientists has demonstrated to the talented novice, as didactic teaching never could, that he can set his sights high and still cope with the selected problem. Emulation is reinforced by observing successful, though often delayed, outcomes. Indeed, the idiom of the laureates reflects this orientation. They like to speak of the big problems and the fundamental ones, the important problems and the beautiful ones. These they distinguish from the pedestrian work in which they engage while waiting for the next big problem to come their way. As a result of all this, their papers are apt to have the kind of scientific significance that makes an impact, and other scientists tend to single out their papers for special attention.

The character structure of these leading scientists may contribute to the communication aspect of the Matthew effect in still another way, which

34. Germane results in experimental psychology shows that preferences for riskier work but more significant outcomes are related both to high motivation for achievement and to a capacity for accepting delay in gratification. See, for example, W. Mischel, *J. Abnormal Soc. Psychol.* 62 (1961): 543.

has to do with their mode of presenting their scientific work. Confident in their powers of discriminating judgment—a confidence that has been confirmed by the responses of others to their previous work—they tend, in their exposition, to emphasize and develop the central ideas and findings and to play down peripheral ones. This serves to highlight the significance of their contributions, raising them out of the stream of publications by scientists having less socially-validated self-esteem, who more often employ routine exposition.

Finally, this character structure and an acquired set of high standards often lead these outstanding scientists to discriminate between work that is worth publishing and that which, in their candid judgment, is best left unpublished though it could easily find its way into print. The laureates and other scientists of stature often report scrapping research papers that simply did not measure up to their own demanding standards or to those of their colleagues.[35] Seymour Benzer, for example, tells of how he was saved from going "down the biochemical drain": "Delbrück saved me, when he wrote to my wife to tell me to stop writing so many papers. And I did stop."[36] And a referee's incisive report on a manuscript sent to a journal of physics asserts a relevant consequence of a scientist's failure to exercise rigorous judgement in deciding whether to publish or not to publish: "If C—— would write fewer papers, more people would read them." Oustanding scientists tend to develop an immunity to *insanabile scribendi cacoēthes* (the itch to publish).[37] Since they prefer their published work to be significant and fruitful rather than merely extensive, their contributions are apt to matter. This in turn reinforces the expectations of their fellow scientists that what these eminent scientists publish (at least during their most productive period) will be worth close attention.[38] Once again this makes for operation of the Matthew effect, as

35. To this extent, they engage in the kind of behavior ascribed to physicists of the "perfectionist" type, who have been statistically identified by the Coles (see fn. 8), as those who publish less than they might but whose publications nevertheless have a considerable impact on the field, as indicated by citations. It is significant that this type of physicist was accorded more recognition in the form of awards for scientific work than any other types (including the "prolific" and the "mass producer" types).

36. S. Benzer, in *Phage and the Origins of Molecular Biology*, ed. J. Cairns, G. S. Stent, J. D. Watson (Cold Spring Harbor, N.Y.: Cold Spring Harbor Laboratory of Quantitative Biology, 1966), p. 165. This *Festschrift* clearly shows that Delbrück is one of those scientists who generally exercise this kind of demanding judgment on the publication of their own work and that of their associates.

37. For some observations on the prophylaxis for this disease, see R. K. Merton, *On the Shoulders of Giants* (New York: Harcourt, Brace and World, 1967), pp. 83–85.

38. It has been noted (G. Williams, *Virus Hunters* [New York: Knopf, 1959]) that the early confidence of scientists in the measles vaccine was a "paradoxical feedback of [Enders's] own scientific insistence, not on believing, but on doubting. His fellow scientists trust John Enders not to go overboard on anything."

scientists focus on the output of scientists whose outstanding positions in science have been socially validated by judgments of the average quality of their past work. And the more closely the other scientists attend to this work, the more they are likely to learn from it and the more discriminating their response is apt to be.[39]

For all these reasons, cognitive material presented by an outstanding scientist may have greater stimulus value than roughly the same kind of material presented by an obscure one—a principle which provides a socio-psychological basis for the communication function of the Matthew effect. This principle represents a special application of the self-fulfilling prophecy,[40] somewhat as follows: Fermi or Pauling or G. N. Lewis or Weisskopf sees fit to report this in print and so it is apt to be important (since, with some consistency, he has made important contributions in the past); since it is probably important, it should be read with special care; and the more attention one gives it, the more one is apt to get out of it. This becomes a self-confirming process, making for the greater evocative effect of publications by eminent scientists (until that time, of course, when their image among their fellows is one of leaders who have seen their best days—an image, incidentally, that corresponds with the self-image of certain laureates who find themselves outpaced by onrushing generations of new scientists).

Like other self-fulfilling prophecies, this one becomes dysfunctional under certain conditions. For although eminent scientists may be more *likely* to make significant contributions, they are obviously not alone in making them. After all, scientists do not begin by being eminent (though the careers of men such as Mössbauer and Watson may sometimes give us that mistaken impression). The history of science abounds in instances of basic papers having been written by comparatively unknown scientists, only to be neglected for years. Consider the case of Waterston, whose classic paper on molecular velocity was rejected by the Royal Society as "nothing but nonsense"; or of Mendel, who, deeply disappointed by the lack of response to his historic papers on heredity, refused to publish the results of his further research; or of Fourier, whose classic paper on the

39. This remains a moot conclusion. Hovland's experiments with laymen have shown that the *same* communications are considered less biased when attributed to sources of high rather than low credibility (C. I. Hovland, *Amer. Psychologist* 14 [1959]: 8). In an earlier study, Hovland and his associates found that, in the case of *factual* communications, there is "equally good learning of what was said regardless of the credibility of the communicator" (C. I. Hovland, I. L. Janis, H. H. Kelley, *Communication and Persuasion* [New Haven, Conn.: Yale Univ. Press, 1953], p. 270).

40. For an analysis of the self-fulfilling prophecy, see R. K. Merton, *Antioch Rev.*, (Summer 1948): 596; reprinted in R. K. Merton, *Social Theory and Social Structure* (New York: Free Press, 1957), pp. 421–36.

propagation of heat had to wait thirteen years before being finally published by the French Academy.[41]

Barber[42] has noted how the slight professional standing of certain scientists has on occasion led to some of their work, later acknowledged as significant, being refused publication altogether. And, correlatively, an experience of Lord Rayleigh's[43] provides an example in which an appraisal of a paper was reversed once its eminent authorship became known. Rayleigh's name "was either omitted or accidentally detached [from a manuscript], and the Committee [of the British Association for the Advancement of Science] 'turned it down' as the work of one of those curious persons called paradoxers. However, when the authorship was discovered, the paper was found to have merits after all."

When the Matthew effect is thus transformed into an idol of authority, it violates the norm of universalism embodied in the institution of science and curbs the advancement of knowledge. But next to nothing is known about the frequency with which these practices are adopted by the editors and referees of scientific journals and by other gatekeepers of science. This aspect of the workings of the institution of science remains largely a matter of anecdote and heavily motivated gossip.[44]

The Matthew Effect and Allocation of Scientific Resources

One institutional version of the Matthew effect, apart from its role in the reward and communication systems of science, requires at least short review. This is expressed in the principle of cumulative advantage that operates in many systems of social stratification to produce the same result: the rich get richer at a rate that makes the poor become relatively poorer.[45] Thus, centers of demonstrated scientific excellence are allocated far larger resources for investigation than centers which have yet to make their mark.[46] In turn, their prestige attracts a disproportionate share of the

41. See chapter 14 in this volume; also see R. H. Murray, *Science and Scientists in the Nineteenth Century* (London: Sheldon, 1925), pp. 346–48; D. L. Watson, *Scientists are Human* (London: Watts, 1938), pp. 58, 80; R. J. Strutt (Baron Rayleigh), *John William Strutt, Third Baron Rayleigh* (London: Arnold, 1924), pp. 169–71.

42. B. Barber, *Science* 134 (1961): 596, reprinted in B. Barber and W. Hirsch, eds., *The Sociology of Science* (New York: Free Press, 1962), pp. 539–56.

43. Quoted by B. Barber (see fn. 42) from R. J. Strutt, *John William Strutt*.

44. A subsequent investigation of the subject by Zuckerman and Merton will be found in chapter 21 of this volume.

45. Derek Price perceived this implication of the Matthew principle. See *Nature* 206 (1965): 233.

46. D. S. Greenberg, *Saturday Rev.*, 4 November 1967, p. 62; R. B. Barber, in *The Politics of Research* (Washington, D.C.: Public Affairs Press, 1966), p. 63, notes that "in 1962, 38 percent of all federal support went to just ten institutions and 59 per-

truly promising graduate students.[47] This disparity is found to be especially marked at the extremes: six universities (Harvard, Berkeley, Columbia, Princeton, the Johns Hopkins, and Chicago) which produced 24 percent of the doctorates in the physical and biological sciences produced fully 65 percent of the Ph.D.'s who later became Nobel laureates. Moreover, the 12 leading universities manage to identify early, and to retain on their faculties, these scientists of exceptional talent: they keep 60 percent of the future laureates in comparison with only 35 percent of the other Ph.D.'s they have trained.[48]

These social processes of social selection that deepen the concentration of top scientific talent create extreme difficulties for any efforts to counteract the institutional consequences of the Matthew principle in order to produce new centers of scientific excellence.

Summary

This account of the Matthew effect is another small exercise in the psychosociological analysis of the workings of science as a social institution. The initial problem is transformed by a shift in theoretical perspective. As originally identified, the Matthew effect was construed in terms of enhancement of the position of already eminent scientists who are given disproportionate credit in cases of collaboration or of independent multiple discoveries. Its significance was thus confined to its implications for the reward system of science. By shifting the angle of vision, we note other possible kinds of consequences, this time for the communication system of science. The Matthew effect may serve to heighten the visibility of contributions to science by scientists of acknowledged standing and to reduce the visibility of contributions by authors who are less well known. We examine the psychosocial conditions and mechanisms underlying this effect and find a correlation between the redundancy function of multiple discoveries and the focalizing function of eminent men of science—a function which is reinforced by the great value these scientists place upon finding basic problems and by their self-assurance. This self-assurance, which is partly inherent, partly the result of experiences and associations in creative

cent to just 25." See also H. Orlans, *The Effects of Federal Programs on Higher Education* (Washington, D.C.: Brookings Institution, 1962).

47. Thus, Allan M. Cartter reports that in 1960–63, 86 percent of (regular) National Science Foundation Fellows and 82 percent of Woodrow Wilson Fellows free to choose their place of study elected to study in one or another of the 25 leading universities (as rated in terms of the quality of their graduate faculties); See A. M. Cartter, *An Assessment of Quality in Graduate Education* (Washington, D.C.: American Council on Education, 1966), p. 108.

48. For this and other detailed information on the career patterns of laureates, see H. Zuckerman (in fns. 1 and 31).

scientific environments, and partly a result of later social validation of their position, encourages them to search out risky but important problems and to highlight the results of their inquiry. A macrosocial version of the Matthew principle is apparently involved in those processes of social selection that currently lead to the concentration of scientific resources and talent.[49]

49. Chancing to come upon the manuscript of this paper, Richard L. Russell, a molecular biologist of more than passing acquaintance, has informed me that a well-known textbook in organic chemistry (L. F. Fieser and M. Fieser, *Introduction to Organic Chemistry* [Boston: Heath, 1957]) refers to the "emperical rule due to Saytzeff (1875) that in dehydration of alcohols, hydrogen is eliminated preferentially from the adjacent carbon atom that is poorer in hydrogen." What makes the rule germane to this discussion is the accompanying footnote: "MATTHEW, XXV, 29, '. . . but from him that hath not shall be taken away even that which he hath.'" Evidently the Matthew effect transcends the world of human behavior and social process.

21 Institutionalized Patterns of Evaluation in Science

1971

[With Harriet Zuckerman]

The referee system in science involves the systematic use of judges to assess the acceptability of manuscripts submitted for publication. The referee is thus an example of status judges who are charged with evaluating the quality of role-performance in a social system. They are found in every institutional sphere. Other kinds of status judges include teachers assessing the quality of work by students (and, as a recent institutional change, students officially assessing the quality of performance by teachers), critics in the arts, supervisors in industry, and coaches and managers in sports. Status judges are integral to any system of social control through their evaluation of role-performance and their allocation of rewards for that performance. They influence the motivation to maintain or to raise standards of performance.

In the case of scientific and scholarly journals, the significant status judges are the editors and referees. Like the official readers of manuscripts of books submitted to publishers, or the presumed experts who appraise proposals for research grants, the referees ordinarily make their judgements confidentially, these being available only to the editor and usually to the author. Other judges in science and learning make their judgements public, as in the case of published book reviews and the often important review articles which assess the "credibility" of recent work in a special field of knowledge.

Originally published as "Patterns of Evaluation in Science: Institutionalisation, Structure and Functions of the Referee System," by Harriet Zuckerman and Robert K. Merton, in *Minerva* 9, no. 1 (January 1971): 66–100; reprinted with permission. The study was supported by a grant from the National Science Foundation to the Program in the Sociology of Science, Columbia University.

Although the referee system has its inefficiencies, practicing scientists see it even in its current form as crucial for the effective development of science. Professor J. M. Ziman puts the case emphatically:

The fact is that the publication of scientific papers is by no means unconstrained. An article in a reputable journal does not merely represent the opinions of its author; it bears the *imprimatur* of scientific authenticity, as given to it by the editor and the referees he may have consulted. The referee is the lynchpin about which the whole business of Science is pivoted.[1]

The chemist, Professor Leonard K. Nash, describes the "editors and referees of scientific journals" as "the main defenders of scientific 'good taste'."[2] Professor Michael Polanyi suggests that although there are of course many cases of disparate evaluative judgements about particular works in science, the structure of scientific authority has generally operated through the years so as to exhibit a remarkable degree of concurrence. He states, for example:

Two scientists acting unknown to each other as referees for the publication of one paper usually agree about its approximate value. Two referees reporting independently on an application for a higher degree rarely diverge greatly.[3]

Observations of this sort attest to the great significance scientists ascribe to the referee system. Yet until recently, the referee system itself has not

1. J. M. Ziman, *Public Knowledge: The Social Dimension of Science* (Cambridge: At the University Press, 1966), p. 148.

2. Leonard K. Nash, *The Nature of the Natural Sciences* (Boston: Little, Brown, and Co., 1963), p. 305.

3. Michael Polanyi, *Science, Faith and Society* (Oxford: Oxford University Press, 1946), p. 37. The evidence on the extent of agreement by referees has only begun to be assembled, but indications are that it varies appreciably among different fields of science and learning. We have found, for example, that in a sample of 172 papers evaluated by two referees for *The Physical Review* (in the period 1948–56), agreement was very high. In only five cases did the referees fully disagree, with one recommending acceptance and the other, rejection. For the rest, the recommended decision was the same, with two-thirds of these involving minor differences in the character of proposed revisions. In two biomedical journals, however, Orr and Kassab found that for 1,572 papers submitted over a five-year period and reviewed by at least two referees, "they agreed that a paper was either acceptable or unacceptable 75 per cent. of the time" (as compared with the 62 per cent. that could have occurred by chance) (Richard H. Orr and Jane Kassab, "Peer Group Judgments on Scientific Merit: Editorial Refereeing," presented to the Congress of the International Federation for Documentation, Washington, D.C., 15 October 1965). For one journal of sociology, agreement to accept or to reject occurred in 72.5 per cent of 193 pairs of independent editorial judgments (as compared with the 53.9 per cent that would have occurred by chance) (Erwin O. Smigel and H. Lawrence Ross, "Factors in the Editorial Decision," *The American Sociologist* 5 [February 1970]: 19–21). Systematic comparisons of variability in the extent of agreement in referee judgements would identify differences in the extent of institutionalization of different fields of science and learning.

been systematically examined and assessed. Professor Gordon Tullock, an economist, has remarked that "Given the importance of these editorial decisions for science, the absence of research into them is surprising."[4]

In this paper, we undertake an inquiry into four aspects of the referee system. We deal first with the faint beginnings in the latter seventeenth century of the institutionalization of evaluative judgements into a system of roles and procedures. We then examine and explore the implications of patterns of differences in the rates of rejecting manuscripts submitted to contemporary journals in fifteen fields of science and learning. In the greater part of the paper, we draw upon fairly recent archives of *The Physical Review* (which the editors kindly made available to us for the purpose) to identify and analyze patterns of decision by editors and referees. Finally, on the basis of these historical, comparative, and quantitative analyses, we consider the significance of the referee system for individual scientists, scientific communication, and the development of science.

Institutionalization of the Referee System

The referee system did not appear all at once as an integral part of the social institution of science. It evolved in response to the concrete problems encountered in working toward the developing goals of scientific inquiry and as a by-product of the emerging social organization of scientists.

The new scientific societies and academies of the seventeenth century were crucial for the social invention of the scientific journal,[5] which began to take an enlarged place in the system of written scientific interchange which had hitherto been limited to letters, tracts, and books. These organizations provided the structure of authority which transformed the mere *printing* of scientific work into its *publication*. From the earlier practice of merely putting manuscripts into print, without competent evaluation of their content by anyone except the author himself, there slowly developed the practice of having the substance of manuscripts legitimated, principally before publication although sometimes after, through evaluation by institutionally assigned and ostensibly competent reviewers. We see the slight beginnings of this in the first two scientific journals established just

4. Gordon Tullock, *The Organization of Inquiry* (Durham, N.C.: Duke University Press, 1966), p. 148.

5. First privately printed in 1913, the classic and still useful monograph by Martha Ornstein deals with the subject in chapter 7; see *The Role of Scientific Societies in the Seventeenth Century*, 3rd ed. (Chicago: University of Chicago Press, 1938); see also Harcourt Brown, *Scientific Organisations in Seventeenth Century France* (Baltimore: Williams and Wilkins, 1934).

300 years ago within two months of each other: the *Journal des Sçavans* in January 1665; the *Philosophical Transactions* of the Royal Society, in March of the same year. The *Journal* was a conglomerate periodical which catalogued books, published necrologies of famous persons, and cited major decisions of civil and religious courts as well as disseminating reports of experiments and observations in physics, chemistry, anatomy, and meteorology. The *Philosophical Transactions* was "a more truly scientific periodical . . . , excluding legal and theological matters, but including especially the accounts of experiments conducted before the [Royal] Society."[6]

Although not the official publication of the Royal Society until 1753, *Transactions* was first authorized by its council on 1 March 1664–65 in these sociologically instructive words:

> Ordered, that the *Philosophical Transactions*, to be composed by Mr. [Henry] Oldenburg [one of the two Secretaries of the Society], be printed the first Monday of every month, if he have sufficient matter for it; and that the tract be licensed under the charter by the Council of the Society, being first reviewed by some of the members of the same.[7]

Much relevant information is packed into this summary of an organizational decision. Prime responsibility for the new kind of periodical is assigned to one person, Oldenburg, for whom there does not yet exist the designation of editor, to say nothing of specifying his obligations in the editorial role. Before long, in trying to meet the problems of maintaining the journal, Oldenburg, together with concerned colleagues in the Society, introduced various adaptive expedients which ended up by defining the role of an editor. The council also recognized the immediate problem of having "sufficient matter" for this newly conceived periodical, and institutional devices were gradually evolved to induce scientists to contribute to the journal. What is perhaps most significant here is that the council, as sponsor of the *Transactions*, was involved with its fate and wanted to have a measure of control over its contents. These adaptive decisions provided a basis for the referee system.

6. J. R. Porter, "The Scientific Journal—300th Anniversary," *Bacteriological Reviews* 28 (September 1964): 211–30, at 221. In this short account of the institutionalization of the referee system, we have drawn upon S. B. Barnes, "The Scientific Journal, 1665–1730," *Scientific Monthly* 38 (1934): 257–60; F. H. Garrison, "The Medical and Scientific Periodicals of the 17th and 18th Centuries," *Bulletin*, Institute for the History of Medicine, 2 (1934): 285–343; D. McKie, "The Scientific Periodical from 1665 to 1798," *Philosophical Magazine* (1948), pp 122–32; D. A. Kronick, *A History of Scientific and Technical Periodicals* (New York: The Scarecrow Press, 1962).

7. Charles R. Weld, *A History of the Royal Society*, 2 vols. (London, 1848), 1:177; see also C. Webster, "Origins of the Royal Society," *History of Science* 6 (1967): 106–28.

As with the analysis of any case of institutionalization, we must consider how arrangements for achieving the prime goals—the improvement and diffusion of scientific knowledge—operated to induce or to reinforce motivations for contributing to the goals and to enlist those motivations for the performance of newly developing social roles. As we have noted, the first problem was to get enough work of merit for publication. In part this was a problem because of the comparatively small number of men seriously at work in science. But it also resulted from the circumstance that, intent upon safeguarding their intellectual property, many men of science still set a premium upon secrecy (as is evident in their correspondence with close associates). They maintained an attitude and continued a practice of (at least temporary) secrecy which, as Elizabeth Eisenstein has impressively suggested, was more appropriate to a scribal culture.[8] With the advent of printing, however, findings could be permanently secured, errors in the transmission of precise knowledge greatly reduced, and intellectual property rights registered in print. Printing thus provided a technological basis for the emergence of that component of the ethos of science which has been described as "communism": the norm which prescribes the open communication of findings to other scientists and correlatively proscribing secrecy.[9] But it appears that this norm did not fully develop in response to the new technology of printing; ancillary institutional inventions served to facilitate the shift from motivated secrecy to motivated public disclosure.

Before the *Transactions* were inaugurated, the Royal Society had adopted one such institutional device to encourage men of science to disclose their new work. The Society would officially establish priority of discovery by recording the date on which communications were first received. As Oldenburg put it, in reassuring terms to his friend and patron, Robert Boyle: "The Society alwayes intended, and, I think, hath practised hitherto, what you recommend concerning ye registring of ye time, when any

8. Elizabeth L. Eisenstein, "The Advent of Printing and the Problem of the Renaissance," *Past & Present*, no. 45 (November 1969), pp. 19–89, esp. pp. 55, 63, and 75–76. "Many forms of knowledge had to be esoteric during the age of scribes if they were to survive at all. . . . Advanced techniques could not be passed on without being guarded against contamination and hedged in by secrecy. To be preserved intact, techniques had to be entrusted to a select group of initiates who were instructed not only in special skills but also in the 'mysteries' associated with them."

9. For an analysis of "communism," universalism, organized skepticism and disinterestedness as basic institutional norms of science, see chapter 13 in this volume. For extended analyses of this normative structure, see Bernard Barber, *Science and the Social Order* (New York: Free Press, 1952), chapter 4; Norman Storer, *The Social System of Science* (New York: Holt, Rinehart and Winston, 1966), pp. 76–136; André Cournand and Harriet Zuckerman, "The Code of Science," *Studium Generale* 23 (1970): 941–62.

Observation or Expt is first mentioned." And, he adds, making the function of this practice altogether manifest, the Royal Society

> have declared it again, yt it should be punctually observed: in regard of wch Monsr. de Zulichem [Huygens] hath been written to, *to communicate freely to ye Society,* what new discoveries he maketh, or wt new Expts he tryeth, the Society being very carefull of registring as well the person and time of any new matter, imparted to ym, as the matter itselfe; *whereby the honor of ye invention will be inviolably preserved to all posterity.*[10]

Soon afterward, Oldenburg writes to Boyle again and even more emphatically reiterates the function of this institutional practice:

> This justice and generosity of our Society is exceedingly commendable, and doth rejoyce me, as often as I think on't, chiefly upon this account, yt I thence persuade myselfe, yt all Ingenious men will be thereby incouraged to impart their knowledge and discoveryes, as farre as they may, not doubting of ye Observance of ye Old Law, of Suum cuique tribuere [allowing to each man his own].[11]

Even before he became editor of the *Transactions*, then, Oldenburg had occasion to note that men of science might be induced to accept the new norm of free communication through a motivating exchange: open disclosure in exchange for institutionally guaranteed honorific property rights to the new knowledge given to others.

In the course of looking after Boyle's writings, the future editor of the *Transactions* came upon prompt publication as another device for preserving intellectual property rights. For, like other scientists of his time, Boyle was chronically and acutely anxious about the danger of what he described as "philosophicall robbery," what would be less picturesquely described today as plagiarism from circulated but unpublished manuscripts. Boyle felt that he had often been so victimized.[12] As his agent, Oldenburg arranged for quick publication of a batch of Boyle's papers, writing him reassuringly: "They are now very safe, and will be within this week in print, as [the printer] Mr. Crook assureth, who will also take care of

10. Henry Oldenburg, *Correspondence of Henry Oldenburg*, ed. and trans. A. Rupert Hall and Marie Boas Hall, 6 vols. (Madison: University of Wisconsin Press, 1966), 2:319, italics added. We have drawn extensively on the volumes of this correspondence, which provide an incomparable storehouse of information on the early days of the *Transactions*.

11. Ibid., p. 329.

12. It will be remembered that so disturbed by plagiarists of his work was Boyle that he prepared a document, later running to three folio pages of print, itemizing all the ingenious devices for thievery developed by the grand larcenists of seventeenth-century science. See *The Works of the Honourable Robert Boyle*, ed. J. Birch, 6 vols. (London, 1772), 1:cxxv–cxxviii, ccxxii–ccxxiv, and also Boyle's letter in Oldenburg, *Correspondence*, 4:94.

keeping ym unexposed to ye eye of a Philosophicall robber."[13] Later, as editor of the *Transactions*, Oldenburg could draw upon this motivation in having Boyle agree that he would "from time to time contribute some short Papers, to that Design you are monthly & happily prosecuting,"[14] Boyle all unknowing that in this way he was helping to institute a new form for the dissemination of knowledge which would eventually become identified as the "scientific paper."

Boyle did report, however, another motive for contributing to the newly invented journal. Almost in so many words, he saw this as a way for the scientist to have his work permanently secured in the archives of science, as he went on to say of Oldenburg's request to contribute to the *Transactions:*

> I mightly justly be thought too little sensible of my own Interest, if I should altogether decline so civil an Invitation, and neglect the opportunity of having some of my Memoirs preserv'd, by being incorporated into a Collection, that is like to be as lasting as usefull.

The fugitive nature of letters as the more familiar means of communicating short reports on scientific work may have emphasized by contrast the potentially enduring character of a journal, particularly one sponsored by a scientific society. In any case, we find in Boyle's remarks an early intimation of the scientific journal as a scientific archive.

Another motive could be harnessed to the developing innovation of a scientific periodical. Property rights in discovery were sought after by scientists primarily as individuals but occasionally also as nationals.[15] As A. R. Hall and M. B. Hall, the editors of the Oldenburg correspondence, observe, by 1667 Oldenburg was eager "to demonstrate English priority [on the filar micrometer] and careful to put it in print in the *Philosophical Transactions*. Similarly he took much care to insist in the *Philosophical Transactions* that it was the English, not the French or the Germans, who had invented the idea of injecting medicines into the veins and of practicing blood transfusion between animals."[16] Such interest in national priority could also be drawn upon to press reluctant scientists into contributing "sufficient matter" to the new publication. Thus the mathematician John Wallis, who played a large part in the early history of the *Transactions*, could argue the case for ensuring national priority in connection with the much-advertised claims of the French to having initiated blood transfusion:

13. Oldenburg, *Correspondence*, 2:291.
14. Ibid., 3:145.
15. On conflicting national claims to priority, see chapters 14 and 17 of this volume.
16. Hall and Hall, in the introduction to Oldenburg, *Correspondence*, 3:xxv.

Onely I could wish that those of our Nation; were a little more forward than I find them generally to bee (especially the most considerable) in timely publishing their own Discoveries, & not let strangers reape ye glory of what those amongst ourselves are ye Authors.[17]

Through these and kindred institutional devices, the new scientific society and the new scientific journal persuaded men of science to replace their attachment to secrecy and limited forms of communication with a willingness to disclose their newly found knowledge.[18] But institutionalization is more than a matter of changing values; it also involves their incorporation into authoritatively defined roles. As the organization sponsoring the *Transactions*, the Royal Society provided the power and authority which enabled it to institute new roles and associated rewards for acceptance of these roles. True, in its early days, the Royal Society included many members with little or no scientific competence. But, what was more consequential for the process of institutionalization, it included all English scientists (and many foreign ones) who were producing significant scientific work. As a result, it was widely identified, both in England and on the Continent,[19] as an authoritative body of scientists.

This authority based on demonstrated competence provided mutually reinforcing consequences for scientists in their triple roles as members of the Royal Society, as contributors to the *Transactions*, and as readers of it. These consequences shaped the early evolution of the scientific journal and the referee system in several ways. First, growing numbers of scientists seeking competent judgments of their work turned increasingly to the Royal Society. Thus the distinguished astronomer Hevelius wrote of his important work *Cometographia* that "as soon as it is published I will make it my first care to submit it to the high judgment and due consideration of the Royal Society."[20] The French astronomer and engineer, Pierre Petit, paid his respects to the "celebrated Society, to which judgment I submit all my ideas."[21] Nor were these merely polite phrases. As the Halls observe, it was not long before the "practice of writing for publication in the *Philosophical Transactions*" was greatly increasing among European men of science.[22] This new practice of writing *directly* for publication in a

17. Ibid., 3:373.

18. Through the process of socially induced displacement of goals, this value of open communication would eventually become transformed for appreciable numbers of scholars and scientists into an urge to publish in periodicals, all apart from the worth of what was being submitted for publication. This development would in turn reinforce a concern within the community of scholars for the sifting, sorting, and accrediting of manuscripts by some version of a referee system.

19. Hall and Hall, in the introduction to Oldenburg, *Correspondence*, 2:xxi.

20. Ibid., 2:138; see also ibid., 4:448.

21. Ibid., 2:595. For other cases in which the Royal Society was asked to sit as a court of scientific judges, see ibid., 3:6, 171, 219, 298.

22. Ibid., 4:xxiii.

journal constituted another appreciable change in the evolving role of the scientist. With the composite institution of learned society and learned journal at hand, scientists began to seize the new opportunity of having competent appraisals of their work by other authoritative scientists,[23] a pattern of attitude and behavior which is basic to the referee system.

The practice of having scientific communications assessed by delegated members of the Royal Society might have affected the quality of those communications. Communications intended for publication would ordinarily be more carefully prepared than private scientific papers, and all the more so, presumably, in the knowledge that they would be scrutinized by deputies of the Society.

The constituted representatives of the Royal Society, looking to its reputation, were in their turn motivated to institute and maintain arrangements for adequately assessing communications, before having them recorded or published in the *Transactions*. They repeatedly express an awareness that to retain the confidence of scientists they must arrange for the critical sifting of materials which in effect carry the *imprimatur* of the Society. Thus, the president of the Society, "before he will declare anything positively of ye figure of these Glasses, will by a gage measure ym; and if ye Invention bear his test, it will pass for currant, & be no discredit to ye Society, yt a member of theirs is ye Author thereof."[24] Or, as the editor-secretary Oldenburg later reported to Boyle, the matter could not be too carefully studied "before we give a publick testimony of it to ye world, as is desired of us."[25] The Society was also beginning to distinguish between evaluated and unevaluated work which came to its notice. On occasion this involved the policy of "sit penes authorem fides [let the author take responsibility for it]: We only set it downe, as it was related to us, without putting any great weight upon it."[26] In the course of establishing its legitimacy as an authoritative scientific body, the Royal Society was gradually developing both norms and social arrangements for authenticating the substance of scientific work.

23. In a series of papers Merton has developed the idea that this concern of scientists with having appropriate recognition of their work by peers is central to the workings of science as a social institution. See, for example, chapter 14 of this volume. This idea has been instructively advanced by Norman Storer as involving a concern with competent appraisal (see his *Social System of Science*, especially pp. 19–27 and 66–73) and by Warren Hagstrom as involving reciprocity and implicit exchange (see his *Scientific Community*, especially pp. 13–21).

24. Oldenburg, *Correspondence*, 4:223–24.

25. Ibid., 4:235. In the event, a short account of these optical instruments was soon published in the *Transactions*.

26. Ibid. It is of some interest that in response to the flood of manuscripts today, with its overloading of facilities for refereeing, some journals are adopting the same policy, allowing some papers to be published though unrefereed, providing that a note appended to the article testifies to its not having been refereed.

Ingredients of the referee system were thus emerging in response to distinctive concerns of scientists taken distributively and collectively. In their capacity as producers of science, individual scientists were concerned with having their work recognized through publication in forms valued by other members in the emerging scientific community who were significant to them. In their capacity as consumers of science, they were concerned with having the work produced by others competently assessed so that they could count on its authenticity. In providing the organizational machinery to meet these concerns, the Royal Society was concerned with having its authoritative status sustained by arranging for reliable and competent assessments.

There are intimations, even in this early period, that individual scientists in their role as informed consumers would begin to affect the process making for control of the quality of publications in journals. When the editor or the Royal Society slipped up by allowing dubious materials to be published in the *Transactions*, as they not infrequently did, readers would on occasion register their protest. The French astronomer Auzout, for example, censured the editor for printing unauthenticated and doubtful accounts:

> Some of our virtuosi are surprised at your speaking in your journal of parabolic lenses. Your students of dioptrics know that they are worthless and whatever fine promises are made, when these seem contrary to reason one ought not to speak of them until the results have been seen; for it is not very urgent to know what charlatans may promise.[27]

We have no evidence that such critical responses actually made for greater care in subsequent editorial decisions. The point is, however, that the newly instituted journal, unlike the printers of books at that time, provided an arrangement through which members of the scientific community could affect editorial practices. Through the emergence of the role of editor and the incipient arrangements for having manuscripts assessed by others in addition to the editor, the journal gave a more institutionalized form for the application of standards of scientific work.

Efforts to cope with immediate problems produced other adaptive changes in the learned journal. By the end of the seventeenth century, there were signs of role-differentiation, especially in journals dealing with diverse fields of knowledge, in the form of a staff or "board" of editors. The *Journal des Sçavans*, for one example, had by 1702 assigned responsibility for particular departments of learning to each member of a staff of editors who met weekly to review copy.[28] Other aspects of the journal

27. Ibid., 3:111 and, in the editors' translation quoted here, p. 114.

28. Sherman B. Barnes, "The Editing of Early Learned Journals," *Osiris* 1 (1936): 155–172, at pp. 157–159.

developed more slowly. It took a century for the format of the scientific papers to become more or less established and even longer for the scholarly apparatus of footnotes and citations to be generally adopted.[29]

Almost from their beginning, then, the scientific journals were developing modes of refereeing for the express purpose of controlling the quality of what they put in print.

Patterns of Evaluation in the Sciences and Humanities

Turning from those early days to the present, we find that some version of the referee system has been widely adopted. In the physical and biological sciences, for example, a recent survey of 156 journals in 13 countries found that 71 percent made some use of referees.[30]

What, then, are the gross outcomes of the evaluation process by editors and referees of journals in the principal fields of science and learning? Are there pronounced differences among the various disciplines? Are observed variations in outcome random or patterned? To explore these questions, we have compiled the rates of rejections in a sample of 83 journals in the humanities, the social and behavioral sciences, mathematics, and the biological, chemical, and physical sciences.[31] (The results are shown in table 1, with the disciplines ranked in order of decreasing rates of rejection.)

The figures exhibit marked and determinate variation. Journals in the humanities have the highest rates of rejection. They are followed by the social and behavioral sciences with mathematics and statistics next in line. The physical, chemical, and biological sciences have the lowest rates, running to no more than a third of the rates found in the humanities.

Confirming this empirical uniformity are subsidiary patterns of deviant rates within disciplines which virtually reproduce the major patterns. To begin with, consider the field of physics. The 12 journals had an average

29. Porter, "The Scientific Journal," p. 225; Derek J. de Solla Price, "Communication in Science: the Ends—Philosophy and Forecast," in Anthony de Reuck and Julie Knight, eds., *Ciba Foundation Symposium on Communication in Science* (London: J. and A. Churchill, 1967), pp. 199–200, at p. 200.

30. International Council of Scientific Unions, *A Tentative Study of the Publication of Original Scientific Literature* (Paris: Conseil International des Unions Scientifiques, 1962). There are marked variations by country: for example, only 2 of 49 journals published in the United States, in contrast to 9 of 30 French journals, made no use of referees.

31. A first list was drawn from Bernard Berelson's compilation of leading journals in his *Graduate Education in the United States* (New York: McGraw Hill, 1960). This list was supplemented by other research journals published under the auspices of the major associations of scholars and scientists. In all, the editors of 117 journals were queried by mail; responses were received from 97 of them and usable information from 83. The *Physical Review Letters* in physics and similar journals in other sciences are excluded from this list since they are especially designed for "rapid publication." On the special problems confronted by such publications, see S. A. Goudsmit, "Editorial," *Physical Review Letters* 21 (11 November 1968): 1425–26.

TABLE 1
Rates of Rejecting Manuscripts for Publication in Scientific and Humanistic Journals, 1967

	Mean rejection rate (%)	No. of journals
History	90	3
Language and literature	86	5
Philosophy	85	5
Political science	84	2
Sociology	78	14
Psychology (excluding experimental and physiological)	70	7
Economics	69	4
Experimental and physiological psychology	51	2
Mathematics and statistics	50	5
Anthropology	48	2
Chemistry	31	5
Geography	30	2
Biological sciences	29	12
Physics	24	12
Geology	22	2
Linguistics	20	1
Total		83

rejection rate of 24 percent, with the figures for 11 of them varying narrowly between 17 percent and 25 percent. But the twelfth journal, the *American Journal of Physics*, departs widely from this norm with a rejection rate of 40 percent. In the light of the general pattern of rejection rates, we suggest that this seemingly deviant case only confirms the rule. For this journal, alone among the twelve assigned to physics in table 1, is not so much a journal *in* physics as a journal *about* physics. It publishes articles dealing primarily with the humanistic, pedagogical, historical and social aspects of physics rather than articles presenting new research in physics. Accordingly, it diverges from the relatively low rate characteristic of the physical sciences in the direction of the substantially higher one characteristic of the humanities and social sciences.

We find similar patterns within other disciplines. The two journals in anthropology for example have an average rejection rate of 47.5 percent, considerably below that for the other social sciences. But this is a composite of drastically different rates for the two journals. The *American Anthropologist*, devoted largely to social and cultural anthropology, approximates the high rejection rates of the other social sciences with a figure of 65 percent, while the *American Journal of Physical Anthropology* with a figure of 30 percent approximates the low rates of the physical sciences. We find much the same difference in psychology. The journals devoted to social, abnormal, clinical, and educational psychology average a rejection rate of 70 percent while the journals in experimental, comparative, and physiological psychology diverge toward the physical sciences with an average of 51 percent. Consider only one more case of this confirming finer pattern within the gross pattern, this time for subjects ordinarily assigned to the humanities. The journals of language and literature in the humanistic tradition have an average rejection rate of 86 percent, whereas the journal, *Linguistics*, adopting mathematical and logical orientations in the study of language, has a rejection rate of 20 percent, much like that of the physical sciences.

The pattern of differences between fields and within fields can be described in the same rule of thumb: the more humanistically oriented the journal, the higher the rate of rejecting manuscripts for publication; the more experimentally and observationally oriented, with an emphasis on rigor of observation and analysis, the lower the rate of rejection.[32]

These variations in the institutional behavior of learned journals may in part reflect differences in the extent of agreement on standards of scholarship in the various disciplines. It appears to be the case that the journals with high rejection rates receive a larger proportion of manuscripts that in the judgement of the editor and his referees are not simply debatable border-line cases but fail by a wide margin to measure up even to minimum standards of scholarship. This suggests that these fields of learning are not greatly institutionalized in the reasonably precise sense that editors and referees on the one side and would-be contributors on the other almost always share norms of what constitutes adequate scholarship. In the case of one journal, for example, which rejects nine of every ten papers, about 40 percent are promptly turned down by the editor as hopelessly inept and unpublishable in any learned journal. The editor of a journal with a final rejection rate of about 80 percent perceives the standards employed by his referees as more demanding than those employed by other journals in the field. He himself rejected more than 40 percent of incoming manuscripts, explaining that they

> . . . were manuscripts which I judged to be extremely unlikely to survive our rigorous screening no matter who reviewed them, so I carefully reviewed them myself and typically sent the authors a one- to three-page, single-spaced letter explaining why we could not accept it here and how they might revise the

32. The empirical solidity of this rule of thumb is illustrated by an episode which occurred in the course of our survey of journals. The editor of a journal in chemical physics reported a rejection rate of 75 percent, far above figures for the other journals in physics and chemistry. Taking note of this anomalous figure, we asked the editor to account for it, only to have him report that it was simply a clerical error; he had reported the rate of acceptance, not the rule-like rejection rate of 25 percent.

manuscript for submission elsewhere or how they might improve on the present study so as to do some publishable research.

And the editor of another journal with a high rejection rate of 85 percent reports that about 20 percent of incoming papers "were so clearly unacceptable that I didn't want to waste a referee's time with them. . . . We still get a flow of articles of a thoroughly amateurish quality."

The influx of manuscripts judged to be beyond all hope of scholarly redemption testifies to the ambiguity and the wide range of dispersion of standards of scholarship in the discipline, all apart from the question whether the institutionally legitimated editors and referees or the would-be contributors are exercising better judgement. We do not know the comparative frequency of these reportedly unsalvageable manuscripts in different fields but the testimony of editors suggests that it is considerably higher in the humanities and the social sciences.

There are intimations in the data also that the editors and referees of journals with markedly different rates of rejection tend to adopt different decision-rules and so are subject, when errors of judgment occur, to different *kinds* of error. Editors and referees, of course want to avoid errors of judgement altogether. But recognizing that they cannot be infallible, they seem to exhibit different preferences. The editorial staff of high-rejection journals evidently prefer to run the risk of rejecting manuscripts which the wider community of scholars (or posterity) would consider publishable (or even, perhaps, important)—an error of the first kind—rather than run the risk of publishing papers that will be widely judged to be substandard. The editorial staff of low-rejection journals, where external evidence suggests that the decisions of scientists to submit papers are based on standards widely shared in the field, apparently prefer to risk errors, if errors there must be, of the second kind: occasionally to publish papers that do not measure up rather than to overlook work that may turn out to be original and significant. Thus the editor of a journal which rejects only one paper in five acts on the assumption that a manuscript is publishable until clearly proved otherwise. As he puts it, "If the first referee recommends publication, as received or with minor revision, that is usually sufficient. If the first referee's opinion is negative, or undecided, additional referee(s) will be consulted until a consensus is reached." Editors of another journal in this class note that they "have generally published 'borderline' papers—those on which referees' opinions differed." Put in terms reminiscent of another institutional sphere, the decision-rule in high-rejection journals seems to be: when in doubt, reject; in low-rejection journals, when in doubt, accept.

The actual distribution of these decision-rules and their consequences for the quality of scholarship in the various disciplines still remain to be determined. But even now it appears that the rules will have different consequences for scientists and scholars at different stages of their development. The Coles and Zuckerman have found that collegial recognition of the work of young scientists is important for their continued productivity.[33] This suggests that the discouragement of having papers rejected may be more significant for the novice than for the established scholar. The multiplicity of journals[34] need not entirely solve the problem for him. Since his research capabilities still require institutional certification, it can matter greatly to him whether his paper is published in a journal of higher or lower rank. Rejection of his paper by a high-ranking journal might be more acutely damaging, more often leading him to abandon his plans for publication altogether.

Whatever their consequences, the marked differences in the rejection rates of journals in the various disciplines can be tentatively ascribed only in part to differences in the extent of consensus with regard to standards of adequate science and scholarship. Beyond this are objective differences in the relative amount of space available for publication.[35] Editors of all

35. We are indebted to Dr. Jonathan R. Cole for suggesting this line of inquiry.

journals must allocate the scarce resources of pages available for print, but not all fields and journals are subject to the same degree of scarcity. Journals in the sciences can apparently publish a higher proportion of manuscripts submitted to them because the available space is greater than that found in the humanities. Take the case of physics. The articles in journals of physics are ordinarily short, typically running to only a few pages of print, so that the "cost" of deciding to publish a particular article is small, and the direct costs of publication are often paid by authors from research grants.[36] The increase in available journal space, moreover, has been outrunning the increase in the number of scientists. The number of pages published annually by *The Physical Review* (and *Physical Review Letters*),

33. Jonathan R. Cole and Stephen Cole, *Social Stratification in Science* (Chicago: University of Chicago Press, 1973); Harriet Zuckerman, *Scientific Elite: Nobel Laureates in the United States* (Chicago: University of Chicago Press, in press).

34. It has often been suggested that papers which are at all competent eventually find their way into print. See Tullock, *The Organisation of Inquiry*, p. 144; Storer, *Social System of Science*, pp. 132–33; Warren O. Hagstrom, *The Scientific Community* (New York: Basic Books, 1965), pp. 18 onwards. But only now are there the beginnings of evidence on the proportion of papers published by journals of differing rank which are first, second, or *n*th submissions for publication. See Nan Lin and Carnot E. Nelson, "Bibliographic Reference Patterns in Core Sociological Journals," *The American Sociologist* 4 (1969): 47–50. Beyond this, nothing is known about the use made of papers which have been published only after having circulated through the editorial offices of several journals.

36. The effects of "page charges" to authors on patterns of publication in scientific journals constitute a complex problem in its own right which is being studied by William Garvey, Belver Griffith, Frances Korten, and the Center for Research in Scientific Communication, at Johns Hopkins University.

for example, increased 4.6 times from 3,920 pages in 1950 to 17,060 in 1965; during the same interval, the number of members of the American Physical Society increased only 2.4 times. Preliminary counts for the humanities and social sciences do not show the same disproportionate increase in journal space beyond increase in the numbers of scholars. By way of comparison, the number of pages available in the official journal of the American Sociological Association remained about the same between 1950 and 1965, while the membership of the Association increased two and a half times.

Observations of this sort deal only with the final outcomes of the evaluative process as registered in comparative rates of rejecting manuscripts for publication. Of course this gross information tells next to nothing about the process of evaluation itself. This we can examine in some detail by turning to the scientific journal for which we have the needed archival evidence, *The Physical Review*.

Evaluative Behavior of Editors and Referees

First, a few words about *The Physical Review*. It publishes 72 issues a year (and two index volumes) in addition to weekly publication of short research reports in the *Physical Review Letters*. It makes up 6 percent of the world's journal literature in physics (together with the *Letters*, 9 percent). We can gauge the relative scale of this publication by noting that in 1965 *The Physical Review* itself—excluding the *PRL*—published more literature in physics than all 53 journals published in Germany, once the world center of physics.[37]

All this quantity need not, of course, make for high quality. But it turns out that in the 1950s, as now, *The Physical Review* ranked far ahead of all other journals of physics in the extent to which it was used in further research. Papers published in it were far more often cited than those published in any other journal of physics and cited more often than if it were simply holding its own—that is, getting the same share of citations as its share in the physics literature. In such leading journals as the Italian *Nuovo Cimento*, the Russian *Journal of Experimental and Theoretical Physics* (*JETP*), and the *Proceedings* of the Physical Society of London, the *Review* is cited far more often than these journals themselves:[38] 36 percent of the references in *Nuovo Cimento* are to *The Physical Review*

37. Stella Kennan and F. G. Brickwedde, *Journal Literature Covered by Physics Abstracts in 1965* (New York: American Institute of Physics, 1968), appendix 2.

38. M. M. Kessler, *Technical Information Flow Patterns* (Cambridge, Mass.: Lincoln Laboratories [Massachusetts Institute of Technology] 1958), pp. 247–57, reporting data for the year 1957; and M. M. Kessler, "The MIT Technical Information Project," *Physics Today* 10 (March 1965): pp. 28–36, at p. 30.

but only 17 percent to all Italian journals combined; 22 percent of all references in *JETP* are to the *Review*, compared with 15 percent going to the *JETP* itself; 34 percent of the references in the *Proceedings* are to the *Review*, compared with 9 percent to the *Proceedings* itself. This widespread use of work published in the *Review* is all the more notable since there is a general tendency for papers in each journal to cite other papers in the same journal. Kessler sums up his findings on patterns of use in the contemporary literature of physics by noting that "*The Physical Review* is truly a definitive journal for physicists. It commands overwhelming dominance over all other journals as a carrier of information between physicists of all lands."[39]

The behavior of physicists, both as consumers and producers of research, testifies to much the same judgement. As consumers, some 77 percent of the 1,300 American academic physicists queried by the Coles reported that the *Review* is among the journals they read most often (no other journal being mentioned by more than 25 percent of the sample).[40] As producers, the archives testify, physicists preferred to have their papers published in the *Review*, maintaining that this would give them greater visibility to their colleagues around the world. Plainly, we are dealing here with the outstanding scientific journal in its field. What, then, have been its patterns of editorial and referee evaluation?

between 1948 and 1956, containing correspondence between authors, editors, and referees, records of decisions made by the editors, the allocation of manuscripts to referees, their evaluations and the final disposition of the papers. This provides a rich body of materials, both quantitative and qualitative, for analyzing the infrastructure of scientific evaluation in a journal of the first class. More particularly, it enables us to find out how the workings of this structure are affected by the stratification system of science.

Consider first the population of physicists submitting manuscripts and the gross outcomes of the evaluative process. In this nine-year period, a

Sampling the Archives of *The Physical Review*

The basic data consist of the archives of the *Review*[41] for the nine years

39. Kessler, *Technical Information Flow Patterns*, p. 249.

40. Stephen Cole and Jonathan R. Cole, "Visibility and the Structural Bases of Awareness of Scientific Research," *American Sociological Review* 33 (June 1968): 397–413, at p. 412.

41. We are indebted to Professor Samuel A. Goudsmit, editor-in-chief of publications for the American Physical Society, for having made these archives available to us in 1966. He has recently described the editorial and refereeing procedures currently adopted by the *Physical Review* in "What Happened to My Paper?" *Physics Today* 22 (May 1969): 23–25.

total of 14,512 manuscripts were submitted (a little more than half of them had a single author). In this report we deal primarily with the papers with a single author, of which 80 percent were ultimately published. The sample we have drawn from these voluminous materials is based, on a conception of the stratification system of science as a distinctive compound of egalitarian values governing access to opportunity to publish and a hierarchic structure in which power and authority are largely vested in those who have acquired rank through cumulative scientific accomplishment. It is a status hierarchy, in Max Weber's sense, based on honor and esteem. Although rank and authority in science are *acquired* through past performance, once acquired they then tend to be *ascribed* (for an indeterminate duration). This combination of acquired and ascribed status introduces strains in the operation of the authority structure of science, as has been noted with great clarity by Michael Polanyi and Norman Storer.[42] These strains may be doubly involved in the processes of evaluating scientific work. In one direction, judgements *by* scientific authorities (whose status largely rests on their own past performance) may come to be assigned great or even decisive weight, and not simply because of their intellectual cogency. In the other direction, judgements *about* the work of ranking scientists may be systematically skewed by deference, by less careful appraisals involving less exacting criteria, by self-doubts of one's own sufficient competence to criticize a great man or by fear of affronting influential persons in the field. Although based on status acquired through assessed accomplishment, the hierarchy of excellence in science can militate in both ways against the unbiased, universalistic evaluation of scientific work.

With this stratification system of science in mind, we have drawn a sample of the contributors in the 1948–56 archives of *The Physical Review* which is stratified into three levels of institutionalized standing based on appraisals of past scientific work. In the first rank are *all* the physicists submitting manuscripts who, by the end of the period (1956), had received at least one of the ten most respected awards in physics (such as the Nobel prize, membership in the Royal Society and in the National Academy of Sciences).[43] These number 91 in all, with 55 of them having submitted papers of which they were the sole authors. The physicists of the second rank, although they had not been accorded any of the highest forms of recognition, had been judged important enough by the American Institute of Physics to be included in its archives of contemporary physicists. All 583 of the physicists in the American Institute of Physics list who had submitted manuscripts to the *Review* during this period make up this inter-

42. Michael Polanyi, *Personal Knowledge* (London: Routledge and Kegan Paul, 1958), especially chapters 6–7; Storer, *Social System of Science*, pp. 103–34.

43. See Cole and Cole, "Scientific Output and Recognition," p. 383, for the prestige-ranking of awards by a sample of 1,300 physicists.

mediate rank, with 343 of them having sent in manuscripts of which they were the sole authors. The remaining 8,864 contributors comprise the third rank in this hierarchy. They are not included in their entirety but are represented by two successive 10 percent random samples, yielding a total of 1,663 authors, with 659 of them having submitted manuscripts of which they were sole authors.[44]

For some special analyses, we also identified a mobile subgroup in the status hierarchy: the 49 contributors who were in the intermediate rank during the time covered by this study but who later moved into the most eminent stratum. In effect, these physicists were observed in the course of their ascent, after having achieved a measure of distinction but before receiving the highest recognition. It will be of interest to find out how the system for evaluating manuscripts dealt with physicists whose work was later to earn them great esteem.

The 354 referees who evaluated the manuscripts with a single author submitted by our sample of authors were stratified in the same way, with 12 percent of them turning up in the first rank, 35 percent in the second, and the remaining 53 percent in the third.

The sample of contributors and the derivative sample of referees were designed with an eye to the general problem of the interplay between the hierarchical structure of authority and the evaluation of scientific work. More specifically, we want to examine the extent to which universalistic and particularistic standards were utilized in evaluating the papers submitted to *The Physical Review* by physicists of differing rank. Since this is our purpose, we shall limit our analysis almost entirely to papers with one author, for reasons both substantive and procedural. Substantively, it turns out that papers with more than one author, largely reporting experimental results, have so high an acceptance rate (over 95 percent) that they can exhibit little variability in evaluations of the kind we want to investigate. Procedurally, it is the case that the rank of the single author can be unambiguously and realistically identified. But not so in the case of papers by several hands, with their varying numbers of authors, often of differing rank.

Drawing upon the samples of authors and referees, we want to examine four main sets of questions. First, do contributors variously located in the

44. A first 10 percent random sample was selected from the physicist-authors remaining in the files after all cases of top-ranking and intermediate authors were removed. This sample of third-rank authors numbered 866, with 355 of them having submitted papers which had a single author. Analysis of this first sample involving three or more variables led to results sometimes based on small numbers. To check these results, we drew a second 10 percent random sample of the remaining third rank authors, this yielding 797, of whom 304 had submitted papers with a single author. At it turns out, the results for the successive samples are so much the same—they vary by no more than three percentage points—that they are reported only in the aggregate.

stratification system differ in the rate at which they submit manuscripts for publication in the *Review*? Second, are there patterns of allocating manuscripts to referees variously situated in the status hierarchy and are these allocations related to the status of authors? This leads directly to the third question: are there differences in rates of acceptance depending upon the professional identity of the physicists submitting the manuscripts? And finally, are any such differences in rates of acceptance linked to the relative status of the referees and authors?

Status Differences in Submission of Manuscripts

It has long been known that eminent scientists tend to publish more papers and not only better ones than run-of-the-mill scientists. It comes as no surprise, then, that they also submit more manuscripts for publication. Among those physicists who submitted manuscripts, produced by themselves alone, to *The Physical Review* during the nine-year period, physicists of the highest rank averaged 4.09, the intermediaries, 3.46 and the physicists of the third rank averaged 2.02.[45] These differences between the strata are presumably all the greater in the population of physicists at large than in this self-selected population of would-be contributors.

The differences in rates of submission of papers are especially marked when it comes to the most prolific physicists in the sample. The physicists of the highest rank submitted 15 or more papers to this one journal at 12 times the rate of the rank-and-file, with 18 percent of the highest-rank physicists, 11 percent of the intermediates, and 1.5 percent of the third rank having sent that many (single- and multi-author) manuscripts.

This pattern of submission rates also contains a striking prognostic result. The 49 mobile physicists in the sample—those who had not attained eminence by the mid-1950s but did so afterwards—were the most prolific of all, with a whopping 47 percent of them having submitted as many as fifteen papers to the *Review*. Plainly these were physicists at a peak period of their productivity. Six of them have since received the Nobel prize, and from the look of things they will be joined by others from this group of the most prolific authors. We catch here, as with a camera, a phase in the process through which early productivity is converted into later recognition by the social system of science.

To this point, the data on submission of manuscripts merely confirm earlier findings on status differences in the number of published papers.

45. The differences in submission rates are greatly amplified for all manuscripts (of both single and multiple authorship), as might be expected in view of the greater facilities and opportunities for collaboration enjoyed by ranking physicists. When each author of a manuscript of multiple authorship is credited with a submission, the mean rates for the nine-year period run to 9.72 for the top-ranking physicists, 7.89 for the intermediates, and 1.97 for the third rank.

This, it might be said, is only to be expected. In general, the more manuscripts submitted, the more find their way into print. But this does not mean, of course, that the ratio of submitted papers to published papers is the same for the several strata of scientists. This would assume that scientists of every stripe adopt the same standards of what constitutes a paper worth submitting for publication and that the refereeing process results in uniform rates of acceptance for scientists at all levels of the stratification system.

A first intimation that these assumptions are unfounded is provided by the rates of submission and of acceptance for papers by physicists affiliated with the seventeen foremost university departments and with less distinguished ones.[46] Among the physicists submitting any single-author manuscripts at all, those in the leading departments submitted only slightly more, with an average of 2.62 compared with 2.49 for the others.[47] But when it comes to actual publication, not submission, the picture changes. Some 91 percent of the papers by physicists in the foremost departments were accepted as against 72 percent from other universities (producing average acceptances of 2.36 and 1.79 papers, respectively).

This result sets the general problem quite clearly. What patterns of evaluation intervene between the submission of papers and actual publication to produce this result? How does it happen that the physicists from the minor departments who are submitting almost as many single-author papers as their counterparts in the major departments end up by having significantly fewer of them published? The question is critical because the gross empirical finding lends itself to sharply different kinds of interpretation.

One interpretation would attribute the departmental differences in acceptance rates to the operation of the stratification system. It holds that the work of scientists in the upper strata is evaluated less severely, that these authors are given the benefit of the doubt by editors and referees, because of their standing in the field or affiliation with influential departments, and that all this is reinforced by particularistic ties between authors

46. See Heyward Keniston, *Graduate Study and Research in the Arts and Sciences at the University of Pennsylvania* (Philadelphia: University of Pennsylvania Press, 1959), for ranks of physics departments as judged by department chairmen in 1957. To Keniston's top fifteen departments, we added California Institute of Technology and Massachusetts Institute of Technology since technological institutions were not included in his survey. There are no comparable rankings of the quality of industrial laboratories or independent research organizations.

47. It should be emphasized here that these rates of submitting manuscripts do not, of course, register actual differences in the "per capita productivity" of departments of different rank. Since they are confined to physicists who contributed at least one manuscript to the *Physical Review* in this period, these figures take no account of the least productive physicists who are probably present in quite different proportions in departments of differing rank.

and referees. This hypothesis suggests that the status of both author and referee significantly affects the judgement of manuscripts, so that work of the same intrinsic worth will be differently evaluated according to these considerations of status.

Another interpretation would ascribe the different outcomes of the evaluation process principally to differences in the scientific quality of the manuscripts coming from different sources. This hypothesis maintains that universalistic standards tend to be rather uniformly applied in judging manuscripts but that, on the average, the quality of papers coming from the several strata actually differs. On this view, scientists in the departments of the highest rank tend to be positively selected in terms of demonstrated capacity, have greater resources for investigation, have more demanding internal standards before manuscripts are submitted, and are more apt to have their papers exactingly appraised by competent colleagues before sending them in for publication. On this hypothesis it is not a preferential bias toward the status of authors and their departments which makes for differing acceptance rates by referees, but intrinsic differences in the quality of manuscripts which in turn are the outcome of joint differences in the capabilities of scientists and in the quality of their immediate academic environments.

We should repeat that although the two interpretations differ in their conceptions of what goes on in the evaluative process, they are not contradictory in the sense that one necessarily excludes the other.[48] Both universalistic and particularistic standards might be *concretely* involved in the actual process of evaluation, but to varying extents and in different parts of the stratification system of science. We want to estimate the extent to which one or the other of these standards is adopted and the structural arrangements that make for use of one or the other.

It is no easy matter to disentangle these components of evaluation. The standing of physicists in their field, the Coles have found, is highly correlated with the quality of their previously published work, as this is assessed by fellow physicists on all levels of status.[49] This status, earned in part by

48. We note in passing that there seems to be a strong tendency to adopt one of these interpretations *to the exclusion* of the other. The first interpretation seems congenial to those who conceive of the social institution of science as dominated by influence and the exercise of power (decidedly not intellectual power), with the evaluative system having little to do with universally applied standards for judging validity and scientific significance. The second interpretation seems congenial to those who allow no place at all for social exchange in the institution of science, with the system of evaluation involving only the exercise of universal standards, subject to some margin of socially unpatterned errors in judgement. We hazard the guess that amongst those who seize exclusively upon one or the other interpretation, the first is more often adopted by scientists in the middle and lower reaches of the stratification system and the second, by those in the upper reaches.

49. Cole and Cole, "Scientific Output and Recognition," pp. 384–90.

past work, may be variously bound up with editorial judgements of the quality of their new work. If all papers submitted by Nobel laureates, for example, are accepted for publication, there remains the question whether some of these would have been rejected had they been submitted by scientists of distinctly lower standing. Correlatively, if scientists who enjoy the greatest prestige have been doing work of high quality in part because their critical associates and they themselves have demanding internal standards, then the manuscripts they decide to submit for publication are apt to be rigorously preselected, with consequently high rates of acceptance by referees applying similarly universalistic criteria.

These difficulties of analysis could be largely avoided if authors were altogether anonymous to referees. But arrangements designed for this purpose work imperfectly.[50] Various kinds of clues in manuscripts often provide unmistakable signatures of the authors, particularly, perhaps, the eminent ones. In any event, the *Physical Review* has not tried to provide for anonymity of manuscripts, for reasons emphatically set forth by Goudsmit:

> Removing the name and affiliation of the author does not make a manuscript anonymous. A competent reviewer can tell at a glance where the work was done and by whom or under whose guidance. One must also remove all references to previous work by the same author, all descriptions of special equipment and other significant parts of the paper. Nothing worth judging or publishing would be left.[51]

The archives of the *Review* nevertheless provide evidence enabling us to move a certain distance toward identifying basic patterns in the evaluation of scientific work.

Patterns of Allocation to Judges

A first phase in the evaluation process is crystal-clear. The higher the rank of authors in the prestige hierarchy, the greater the proportion of their papers which are judged by the two editors—either singly or in tandem—without going to outside referees. Of the manuscripts submitted by the physicists of the highest rank, 87 percent were judged exclusively by the editors, in contrast to 73 percent of those coming from the intermediate

50. In a study of social science journals, Professor Diana Crane concludes that the effort to maintain anonymity of authors does not affect differentials in rates of publication by authors from major and minor universities. As she implies, the findings are highly tentative since they are based entirely on actual patterns of publication, without taking into account patterned variations in the rates of submitting manuscripts. See Diana Crane, "The Gatekeepers of Science: Some Factors Affecting the Selection of Articles for Scientific Journals," *The American Sociologist* 2 (1967): 195–201.

51. S. A. Goudsmit, *Physics Today* 10 (January 1967): 12.

rank, and 58 percent of the rest. As we shall see, it is the more problematic papers which are sent to outside referees. All this has the immediate consequence that the higher the rank of the physicist, the more prompt the decision taken on his manuscript (table 2), a matter of concern to many scientists, especially those wanting to safeguard their priority.

TABLE 2
Duration of Editorial and Refereeing Process for Published Papers, by Rank of Author

	Rank of author		
Duration	Higher-rank physicists (%)	Intermediate physicists (%)	Third-rank physicists (%)
Less than 2 months	42	35	29
2–4 months	47	45	41
5 months +	11	20	30
Total	(202)	(1027)	(972)

SOURCE: This table and all subsequent ones are based on a sample of manuscripts with single authors submitted to the *Physical Review* from 1948 to 1956.

The referee system calls for evaluation of manuscripts by experts on their subject. It should come as no surprise, therefore, that the outside referees were drawn disproportionately from physicists of high rank. Compared with the 5 percent of the 1,056 authors (themselves in some measure a selected aggregate), almost 12 percent of the 354 outside referees assessing their papers were in the highest rank. Moreover, these 12 percent of the referees contributed one-third of all referee judgements. They refereed an average of 8.5 papers compared with 3.8 for the intermediates and 1.4 for the rank-and-file. And although some 45 percent of the referees were under the age of 40, thus giving major responsibility to the relatively young, it should also be noted that research physicists are a youthful aggregate, with fully 74 percent of the papers submitted to the *Physical Review* coming from men under 40. Much the same pattern of stratification is found when referees are classified according to the rank of the institution with which they are affiliated, rather than their individual rank. For example, about two-thirds of all referee judgements were made by physicists in the 17 major departments of physics in universities, the Bell Laboratories, and the Institute for Advanced Study.

The composite portrait of referees is clear enough. Whether gauged by their own prestige, institutional affiliations, or research accomplishments, they are largely drawn from the scientific elite, as would be expected from the principle of expertise and the principle of influence and authority.

We want now to consider patterns of allocating the referees to authors of varying rank. The possible patterns are describable in four models, which can be designated as the "oligarchical" model, the "populist" model, the "egalitarian" model, and the model of expertise. In the oligarchical model, the established elite of science alone has the power to judge the work of those beneath it in the status hierarchy. The second model corresponds to a populistic view which assigns power of judgement to "the people." The strictly egalitarian model, by contrast, calls for a policy in which papers are assessed only by juries of status peers. And last, the model of expertise calls for the allocation of manuscripts to referees who, regardless of rank, are especially competent to judge them.

It is easy enough to construct the distribution of cases in our data which would correspond to an oligarchical policy for allocating referees to authors. This would require all manuscripts to be evaluated by judges ranking higher than authors, with the exception of those submitted by the highest rank of scientists. Having reached the top of the status hierarchy, they would be exempt from oligarchy and judged by peers. Put in terms of our data, this model would have physicists of the highest rank evaluating all the manuscripts by the intermediate physicists, and these two ranks in turn would be charged with assessing the work of the third rank. As the lowest stratum, the third rank would do no refereeing at all. A glance at table 3 is enough to indicate that the actual pattern of allocation diverges greatly from this oligarchical model. "Status-inferiors" do much more refereeing and, by the same token, "status-superiors" far less than this model requires.

The second model, expressing a populist view, would have manuscripts judged exclusively by physicists ranking lower than authors. This model is of course at odds with the traditional ethos of science which holds that the quality of scientific accomplishment is the determinant of the status ascribed to scientists. It turns out that the data on actual allocations diverge very widely from the populistic model. Relative to their numbers, lower-ranking physicists do little refereeing altogether and also referee far fewer papers by the intermediate and highest-ranking physicists than would be the case under a populist allocation.

According to the strictly egalitarian model, papers would be assessed exclusively by status peers. The actual distribution, as table 3 shows, departs very widely from this model also. As can be seen by aggregating the cases in the left-to-right diagonal, only about a third of all judgements are made by status peers of authors, and there is a widening deviation from the model for the lesser ranks of physicists.

In short, the actual patterns of allocation of referees to authors approximate to none of these models.

The principle of expertise requires that referees should be assigned to manuscripts on the basis of their competence. The data presented in table

3 are at least consistent with the principle of expertise once it is assumed that demonstrated expertise is substantially (that is, imperfectly) correlated with rank in the hierarchy of prestige. On this assumption, the data would exhibit a preponderance but no monopoly of refereeing by physicists ranking higher than authors. Authors would occasionally out-rank referees in prestige (if not in competence), and judgement by peers would be relatively more frequent for the successively higher ranks of authors. These patterns turn up in the actual data recorded in table 3 as we see for the example of judgements by status peers accounting for 50 percent of the papers by top-ranking physicists, 41 percent for the intermediates, and 26 percent for the rank-and-file.

TABLE 3
Rank of Referees Assigned to Authors of Differing Rank

	Rank of referees			
Rank of authors	Higher-rank physicists (%)	Intermediate physicists (%)	Third-rank physicists (%)	Total judgments by referees
Higher-rank physicists	50	31	19	36
Intermediate physicists	38	41	21	394
Third-rank physicists	27	46	26	653
All authors	32	44	24	1083

This suggests, although it does not demonstrate, that expertise and competence were the principal criteria adopted in matching papers and referees. That papers by distinguished scientists were assigned for review to others of like stature need not mean, therefore, that an inner circle of physicists were being asked to pass judgements upon one another's work in a closed system of mutual support. The principle of expertise would lead to such allocations just as it would to the observed pattern of referees more often outranking authors than conversely.

In any case, we now know that the more highly placed physicists had power disproportionate to their number in deciding what was to enter into the pages of the *Physical Review*. How did they act in these positions of power?

Status Differences in Rates of Acceptance

Since the anonymity of authors cannot be uniformly assured, it would require a strict experimental design to find out decisively whether papers *of the same scientific quality* are assessed differently by referees according to the status of authors. Both ethics and practicality rule out the draconian experiment in which matched samples of referees, all unknowing, would independently judge the same manuscripts variously ascribed to physicists of different rank, in order to determine the extent of status-linked evaluations. Nor can we approximate the intent of that experimental design by adopting the number of citations to published papers as measures of quality to see whether papers rejected by the *Physical Review* but published elsewhere are of the same quality as those accepted by that journal.

At best, we can bring together data which provide cumulative intimations of the extent to which judgements by editors and referees relate to the status of authors. We begin by examining the successive disposition of manuscripts as this is summarized in the abbreviated flow chart of the refereeing process (chart 1). It turns out that 90 percent of the manuscripts submitted by top-ranking physicists have been accepted for publication, compared with 86 percent for the intermediates and 73 percent for the rank-and-file. These stratified rates are the outcome of a continuing process of evaluation (condensed into two phases in the chart). In each phase, the higher the rank of physicists, the better they fare. A larger proportion of their papers are accepted straightaway, a smaller proportion rejected outright, and a smaller proportion treated as problematic, requiring further assessments before final decision. Of the manuscripts judged to be problematic, moreover, a larger fraction by the high-ranking physicists ultimately get into print.

Once again, the observed patterns lend themselves to quite different interpretations. They are consistent with the opinion that physicists of the first rank submit better papers on the average and that they are also better able than the others to rehabilitate their problematic papers. But the data can also be interpreted in particularistic rather than universalistic terms. For the observed patterns would also obtain if the editors and referees were especially reluctant to reject papers submitted by the most distinguished physicists in their field and reluctant to judge them as needing further evaluation and revision.

Before turning to other evidence bearing on these alternative interpretations, we should consider the patterns of stratified differences within the context of other aspects of the refereeing process which can be reconstructed from the flow chart. We noted earlier that scientific journals with high rates of acceptance seem to prefer the decision-rule: when in doubt, accept. In the case of the *Physical Review*, this preference rule found

CHART 1
Evaluation of Manuscripts with Single Authors By Rank of Author

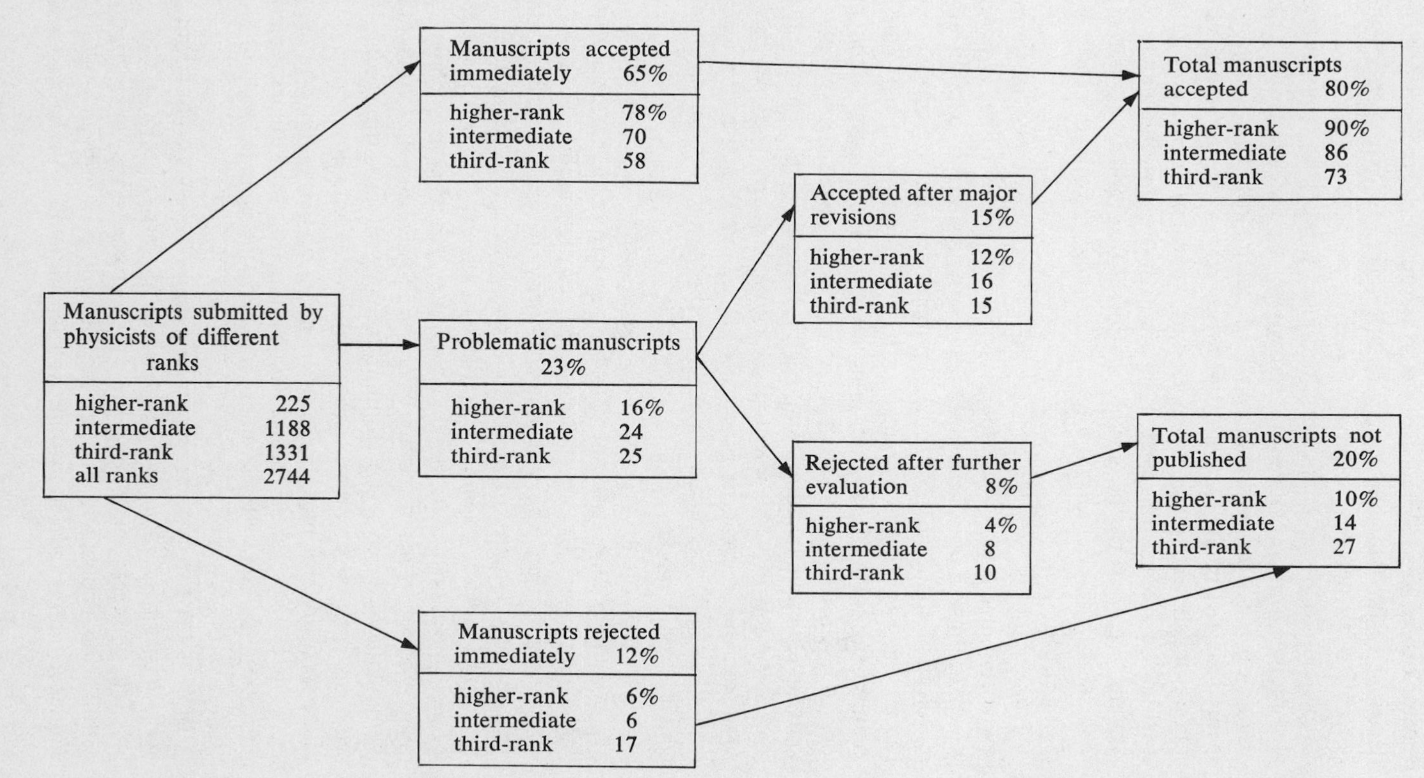

several expressions. When it came to acceptances, the ratio of immediate decisions to the later, more problematic ones was over 4 to 1 (that is, 65 percent to 15 percent) compared with a ratio of only 1.5 to 1 (that is, 12 percent to 8 percent) for rejections. Among the problematic papers undergoing further evaluation, moreover, acceptances still preponderate but at only 1.7 times the number of rejections. The decision-rule also seems reflected in the fact that the *Physical Review* mobilized more institutional machinery to reject papers than to accept them: more judges were used on the average for rejected papers than for those ultimately published. And in accord with the general pattern of stratification, the higher the rank of physicist-authors, the fewer the judges involved in accepting their manuscripts.

These patterns, we conjecture, are generally reversed in journals with low acceptance rates, where the decision-rule seems to be: when in doubt, reject. In those journals the early decisions presumably exhibit higher rates of rejection than the problematic papers sent on for further refereeing. For in the case of these journals, the presumption seems to be that the manuscripts they receive are not fit to print (at least in the particular journal), since they do in fact reject most manuscripts. Thus, for the journals in the humanities and social sciences, with their typically high rejection rates, it is the potentially acceptable paper which is problematic, while for the journals in physical science, such as the *Physical Review*, with their high acceptance rates, it is the potentially unacceptable paper which is problematic.

Another piece of evidence takes us a certain distance toward gauging the possibility that assessments of manuscripts in the *Physical Review* might have been affected by the standing of their authors. For this purpose, we note again that eminence and authority in science derive largely from the assessed quality of past and not necessarily continuing scientific accomplishments. We note also that in science, as in other institutional spheres, positions of power and authority tend to be occupied by older men. (Indeed, it has sometimes been said with mixed feelings that gerontocracy may even be a good thing in science; it leaves the young productive scientists free to get on with their work and helps to occupy the time of those who are no longer creative.) From these joint patterns, it would seem that if the sheer power and eminence of authors greatly affect refereeing decisions, then the older eminent scientists should have the highest rates of acceptance.

But, at least in physics, the young man's science, this is not what we find. It is not the older scientists whose papers were most often accepted but the younger ones. And these age-graded rates of acceptance hold within each applicable rank in the hierarchy of esteem (table 4). Both eminence and youth contribute to the probability of having manuscripts

TABLE 4
Rates of Acceptance of Manuscripts, by Age and Rank of Authors

	Rank of authors							
	Higher-rank physicists		Intermediate physicists		Third-rank physicists		All ranks	
Age of authors	%	No.	%	No.	%	No.	%	No.
20–29			91	287	83	385	87	672
30–39	96	80	89	519	77	440	85	1039
40–49	95	58	83	236	73	79	83	373
50+	80	87	71	126	50	14	73	227
No information on age							61	423
All ages							80	2734

accepted; youth to such a degree that the youngest stratum of physicists in the third rank had as high an acceptance rate as the oldest stratum of eminent ones whose work, we must suppose, was no longer of the same high quality it once was. Jonathan Cole's studies of citation and reference patterns of physicists lend support to this impression.[52] He finds that older physicists are less apt than younger ones to refer to currently influential work in their publications, this suggesting that their own work may no longer be as much in the mainstream. Evidently there comes a time in the life-cycle of physicists, even the most distinguished ones, when they can no longer count on having their papers almost invariably accepted in a major refereed journal such as the *Physical Review*. As Max Delbrück once observed, perhaps the chief function of unrefereed *Festschriften* is to provide a decent cemetery for oft-rejected manuscripts.

Relative Status and Differences in Acceptance Rates

Perhaps it is not the status of the author as such but his status relative to that of the referee which systematically influences appraisals of his manuscripts. Such biases in judgement might take various forms, depending on the pattern of relative status.

When referees and authors are status peers, an hypothesis of *status solidarity* would have it that referees typically give preferential treatment to manuscripts, just as a counter-hypothesis of *status competition* would have it that under the safeguard of anonymity, referees tend to undercut their rivals by unjustifiably severe judgements.

When authors outrank referees, an hypothesis of *status deference* would hold that the referees give preferential treatment to the work of physicists

52. Jonathan Cole, "The Social Structure of Science" (Ph.D. diss., Columbia University, 1969), chapter 6.

TABLE 5
Referees' Decisions to Accept, by Rank of Authors and Referees

	Rank of Referees							
	Higher-rank physicists		Intermediate physicists		Third-rank physicists		Total judgements by referees	
Rank of authors	%	No.	%	No.	%	No.	%	No.
Higher-rank physicists	*	18	*	11	*	7	50	36
Intermediate physicists	55	150	62	160	62	84	59	394
Third-rank physicists	54	179	61	302	59	172	59	653
All ranks							59	1083

* The number of manuscripts by higher-ranking physicists submitted to outside referees, as distinct from editorial judges, was too small for statistical analysis.

they respect or hold in awe just as a counter-hypothesis of *status envy* would have them be more exacting of the work of superiors.

And when referees outrank authors, an hypothesis of *status patronage* or sponsorship would maintain that referees are unduly kind and undemanding while a counter-hypothesis of *status subordination* would have them overly demanding.

Differing in other respects, these six hypotheses are alike in one: they all assume that the relative status of referee and author significantly biases judgements by referees, either in favor of the author or at his expense. More concretely, all assume that the rates of acceptance for each stratum of authors will differ according to the rank of the referees making the judgements.

The data assembled in table 5 run counter to all the hypotheses. Referees of each rank accept the same proportion of papers by authors from every stratum. As it happens, the highest-ranking referees accept somewhat smaller proportions of papers than their fellow referees, but, again, this they do uniformly for authors of every rank. There is, in short, no preferential pattern, as can be shown redundantly but emphatically by condensing the components of table 5 into three categories of relative status.

Relative status	Rate of acceptance (%)	Total judgements by referees (No.)
Referees outrank authors	58	631
Referees and authors: status peers	60	350
Authors outrank referees	59	102

All this suggests that referees were applying much the same standards to papers, whatever their source. This is confirmed further by patterns of even-handed evaluation in the case of other relative statuses of referees and authors. Referees affiliated with minor universities, for example, are no more apt to accept papers submitted by authors from universities of similar standing than were referees from the major universities. And whatever the academic rank of referees, it did not affect the rate at which they accepted papers by authors in the various academic ranks. For this journal, at least, the relative status of referee and author had no perceptible influence on patterns of evaluation.

We may conclude that the status-composition of the physicists engaged in refereeing manuscripts for the *Physical Review* during the period is one thing; what the referees did in exercising their authority is quite another.

Functions of the Referee System

As the prime journal in its field, the *Physical Review* can be assumed to apply exacting standards. All the same, the editorial and refereeing process results in as many as four of every five manuscripts being accepted for publication (a fair number of them, after greater or less revision). Does this mean that referees are largely superfluous? Like other observers of the referee system,[53] we think not. Referees, collectively engaged in sorting out good science from bad, serve diverse functions for the various members of their profession: for editors, authors, the referees themselves, and the relevant community of scientists.

For the editor(s), referees serve their prime function in the case of papers difficult to assess. At the extremes, as we have noted for the *Physical Review* and a variety of other journals, papers are comparatively easy to appraise and the editor(s) can sort them out. Manuscripts which, by the core standards of the field, provide sound, new, consequential ideas and information, clearly formulated and relevant to the particular journal, can be readily distinguished from their antitheses which are mistaken, redundant, trivial, obscure, and irrelevant. But not all manuscripts exhibit these neatly correlated arrays of intellectual virtues or vices. It is the often sizable number of more problematic manuscripts which particularly require examination by experts on their subjects. Apart from this manifest function of furnishing expert judgement, the corps of typically anonymous referees

53. The operation of the authority structure in science and the social structural basis of scientific objectivity have been most fully developed by Michael Polanyi, notably in his *Personal Knowledge* (London: Routledge and Kegan Paul, 1958); the discussion of the referee system is principally in chapter 6. See also, Ziman, *Public Knowledge*, pp. 111–17; Storer, *Social System of Science*, pp. 112–26; Hagstrom, *Scientific Community*, pp. 18–19.

sometimes serves the incidental and not altogether latent function of protecting the highly visible editor from the wrath of disappointed authors.[54] But what is helpful for the editor can of course be injurious to the author. The referee system is now under severe strain on the issue of enlarging the accountability of referees by removing their cloak of anonymity.[55] Since accountability is itself so much a component of the ethos of science, it may be that the practice of maintaining anonymity of referees will increasingly go by the board.

This will surely not be misunderstood to say that the interests of referees and authors are inherently at odds. Referees who conscientiously fulfil their role of course serve major functions for authors. They can and, as we have seen in the case of the *Physical Review*, often do suggest basic revisions for improving papers. They sometimes link up the paper with other work which the author happened not to know; they protect the author from unwittingly publishing duplications of earlier work; and, of course, as presumable experts in the subject, they in effect certify the paper as a contribution by recommending its publication. But like other people, referees are not uniformly conscientious in performing their roles. There are, it seems, differences in this respect among fields and among referees of differing kind so that the functions of refereeing for authors and consequently for the discipline are imperfectly realized. This is scarcely the first time that an institution devoted to evaluation confronts the problem of who judges the judges. A sorting and sifting of referees would seem as much a functional requirement of the referee system as the sorting and sifting of papers for publication.

The role of referee also serves functions and creates difficulties for the referees themselves. As experts in the subject, many referees are already informed of developments at its frontier. But especially in fields without efficient networks of informal communication or in rapidly developing fields, referees occasionally get a head start in learning about significant new work. Moreover, as some referees report, the role-induced close scrutiny of manuscripts, in contrast to the often perfunctory scanning of possibly comparable articles already in print, sometimes leads them to perceive potentialities for new lines of inquiry which were neither stated by the author nor previously considered by the referee. This unplanned evocative function of the paper often puts both referee and author under

54. Based on our sample of the archives of the *Physical Review*, a qualitative analysis of the tacit rules involved in rejecting a manuscript has been set out by Stanley Raffel, "The Acceptance of Rejection," a paper presented at the meetings of the American Sociological Association in 1968.

55. The pros and cons of referee anonymity are being strenuously debated in various fields; for examples, see the letters by Rustum Roy and H. K. Henisch, in *Physics Today* 23 (August 1970): 11; Werner J. Cahnman, in *The American Sociologist* 2 (May 1967): 97–98; A. G. Steinberg, in *Science* 148 (23 April 1965): 444.

stress. What the referee defines as an instance of his having legitimately and appreciatively borrowed or learned from the manuscript, the author, not surprisingly, may define as an instance of pilfering or downright plundering, as he observes the referee going on to pursue and so, perhaps, to preempt the new line of investigation.

The basic and, it would seem, thoroughly rational practice of selecting experts as referees makes for its own stresses in the system. Some scientists have argued that it is particularly the experts who can exploit their fiduciary role to advance their own interests and so are most subject to possible conflict of interest. Here is one among many recent expressions of this view:

> The referee, or more often a member of his group or one of his graduate students, may be working on the very problem he is asked to judge. Of course we must rely upon his personal integrity not to "sit on" the submitted paper, take unfair advantage of the pre-publication information or be unduly critical of the work, thus "buying time" for his own people. He could, in fact, return the paper to the editor citing conflict of interest as his reason for no recommendation, but he cannot avoid the fact of being informed. The point becomes crucial in rapidly developing competitive fields and for publications such as *Physical Review Letters* or *Applied Physics Letters* where priority claims are important.[56]

Plainly, the institutionalized concern with intellectual property[57] in science provides the context for these stresses on the referee system. Neither the context nor the stresses are anything new. The concern with intellectual property, which we found to play its distinctive part in the beginnings of the scientific journal, has created difficulties for the developing referee system right along. Here, for example, is the young T. H. Huxley emphatically expressing his conviction that should "the great authority" on his subject serve as referee, he would never allow Huxley's paper to see print:

> You have no idea of the intrigues that go on in this blessed world of science. Science is, I fear, no purer than any other region of human activity; though it should be. Merit alone is very little good; it must be backed by tact and knowledge of the world to do very much.
>
> For instance, I know that the paper I have just sent in [to the Royal Society] is very original and of some importance, and I am equally sure that if it is referred to the judgement of my "particular friend" ——— that it will not be published. He won't be able to say a word against it, but he will pooh-pooh it to a dead certainty.
>
> You will ask with some wonderment, Why? Because for the last twenty years ——— has been regarded as the great authority on these matters, and has had

56. A. G. Prinz, in *Physics Today* 23 (August 1970): 11–12.

57. On intellectual property as a significant context for the behavior of scientists, see the essays in parts 3 and 4 of this volume.

no one to tread on his heels, until at last, I think, he has come to look upon the Natural World as his special preserve, and "no poachers allowed." So I must manoeuvre a little to get my poor memoir kept out of his hands.[58]

With all its imperfections, old and new, the developing institution of the referee system provides for a warranted faith that what appears in the archives of science can generally be relied upon. As Professor Michael Polanyi in particular has observed,[59] the functional significance of the referee system increases with the growing differentiation of science into arrays and extensive networks of specialities. The more specialized the paper, the fewer there are who can responsibly appraise its worth. But while only a few may be fully competent to assess, many more on the periphery of the subject and in other related fields may find the paper relevant to their work. It is for them that the role of the referee as deputy takes on special importance. When a scientist is working on a problem treated in a published article, he can serve as his own referee. He may, in fact, be better qualified to assess its worth than the official referee who helped usher it into print. It is not so much the fellow specialist as the others making use of published results in fields tangential to their own who particularly depend upon the referee system.

Scientists also benefit from the refereeing of papers in their own special fields but for somewhat different reasons. They may often be equipped to test for themselves the substance of the papers on which they draw, but to do so repeatedly would only subvert their motivation. The fun and excitement in doing science comes largely from working on problems not yet solved. The continuing rather than occasional need to recheck the observations, experimental results, and theories advanced by others would seem an excellent means for depleting creative energies. By providing for generally warranted confidence in the research reported in accredited publications, the system of expert referees helps scientists get on with their own imaginative inquiries.

Editors of journals in many fields of learning remark, sometimes with an air of puzzlement, upon the willingness of scientists and scholars to serve in the anonymous and often exacting role of referee. In some fields such participation is widely diffused. Almost 30 percent of a sample of high energy theorists in physics, for example, had engaged in refereeing and editorial work for journals.[60] A sense of reciprocation for benefits received from the referee system probably supports the motivation for

58. Leonard Huxley, *Life and Letters of Thomas Henry Huxley* (London: Macmillan and Co., 1900), 1:97.

59. Polanyi, *Personal Knowledge*, p. 163.

60. Miles A. Libbey and Gerald Zaltman, *The Role and Distribution of Written Informal Communications in Theoretical High Energy Physics* (New York: American Institute of Physics, 1967), p. 49.

serving in the role of referee as it becomes recognized that the maintenance of standards is a collective responsibility. For young scientists and scholars there may also be the further symbolic reward of having been identified as enough of an expert to serve as a referee.

The very existence of the referee system, Dr. Simon Pasternack has suggested,[61] makes for quality control of scientific communications. In part this control works by anticipation. Knowing that their papers will be reviewed, authors take care in preparing them before submission, all the more so, perhaps, for papers sent to high-ranking journals with a reputation for thorough refereeing. This would also make for the scientists' internalization of high standards. Furthermore, Pasternack points out, even the "scientific journals that have little or no refereeing or editing . . . exist within a framework of the edited journals, which set the pattern and the standard." The referee system may thus be raising standards adopted by journals ostensibly outside that system.

These observations on the functions of the referee system do not at all imply the contrary-to-fact assumption that it works with unfailing effectiveness. Errors of judgment, of course, occur. But the system of monitoring scientific work before it enters into the archives of science means that much of the time scientists can build upon the work of others with a degree of warranted confidence. It is in this sense that the structure of authority in science, in which the referee system occupies a central place, provides an institutional basis for the comparative reliability and cumulation of knowledge.[62]

61. Simon Pasternack, "Is Journal Publication Obsolescent?," *Physics Today* 19 (May 1966): 38–43, at p. 40 and p. 42. Dr. Pasternack has been editor of the *Physical Review* since 1956 (which will be remembered as the end of the nine-year-period examined in this paper) and on its staff since 1951.

62. Several articles bearing on the subject of this paper have appeared since it was completed. Most directly relevant is the work of Richard Whitley on the operation of science journals. His study of an interdisciplinary journal and one in social science found that in both cases editorial decisions on manuscripts were unrelated to the rank and institutional affiliation of contributors. See Richard D. Whitley, "The Operation of Science Journals: Two Case Studies in British Social Science," *Sociological Review* n.s. 18 (July 1970): 241–58. In his study of 32 journals in social science, Whitley found that the older journals and those devoted to fundamental rather than applied science had tended, more than the others, to develop criteria for judging manuscripts. This is consistent with the hypothesis advanced in the present paper that differences among the disciplines in rates of rejection are associated with the extent of consenus on the criteria of adequate scholarship in the various disciplines. See Richard D. Whitley, "The Formal Communication System of Science: A Study of the Organisation of British Social Science Journals," *The Sociological Review: Monograph No. 16* (September 1970), pp. 163–79. Whitley also found that the extent of control by professional associations over the communication system in social science was significantly related to the use of formal procedures for evaluating manuscripts. (Ibid., p. 175).

Two studies based on surveys of journals in clinical, personality and educational psychology report substantial agreement among the editors of these journals on the criteria for judging the acceptability of manuscripts. Since these studies are not based on investigation of the archives, however, they cannot determine the possibility of socially patterned differences in the application of these criteria. See Wirt M. Wolff, "A Study of Criteria for Journal Manuscripts," *American Psychologist* 25 (July 1970): 636–39; T. T. Frantz, "Criteria for Publishable Manuscripts," *Personnel and Guidance Journal* 47 (1968): 384–86.

Bearing directly upon the findings on differences in rejection rates by journals in the humanities and sciences reported in this paper is a survey of the importance assigned to various criteria for good scientific writing by members of 16 departments of social and natural science at a major university. The results indicate that "the harder natural sciences stress precise mathematical and technical criteria, whereas the softer social sciences emphasise less defined logico-theoretical standards." See Janet M. Chase, "Normative Criteria for Scientific Publication," *American Sociologist* 5 (August 1970): 262–65. We owe the information in this footnote to Mr. Aron Halberstam.

22 Age, Aging, and Age Structure in Science

1972

[With Harriet Zuckerman]

Critical overviews of a particular subject are ordinarily occasioned by the need to consolidate rapid advance in knowledge about it and to identify new directions of research. That is surely not so here. The best case that can be made for a chapter dealing with age stratification in science is that so little is known about it. In point of fact, systematic research over the years has been devoted to only one problem in this field: the patterns and sources of changes in the productivity of scientists during their life course. Beyond that, just about any methodical research on age, age cohorts, and age structure in science would qualify, through prior default, as a "new" direction.

The character of this chapter is therefore practically dictated by the twin circumstance that few theoretical questions have been addressed to the phenomena of age stratification in science and that only scattered investigations have dealt with these questions. Since the field is so short on facts, we must here be long on conjectures. Still the path of acknowledged conjecture, leading from mere guess and surmise to the neighborhood of grounded hypothesis, has its uses too. It can take us to a problematics[1] of our subject: the formulation of principal questions that should be

Harriet Zuckerman and Robert K. Merton, "Age, Aging, and Age Structure in Science," reprinted, with permission, from Matilda White Riley, Marilyn Johnson, and Anne Foner, eds., *A Sociology of Age Stratification*, vol. 3 of *Aging and Society* (New York: Russell Sage Foundation, 1972). We are indebted to Ian Maitland for his assistance. The writing of this paper was supported in part by a grant from the National Science Foundation to the Program in the Sociology of Science, Columbia University.

1. On the concept of problematics and of problem-finding, see R. K. Merton, L. Broom, and L. S. Cottrell, eds., *Sociology Today: Problems and Prospects* (New York: Basic Books, 1959), pp. v, ix–xxxiv.

investigated together with the rationale for considering these as questions worth investigating. For this purpose, we draw upon both the scattered evidence in hand and the model of age stratification developed by Matilda White Riley, Marilyn Johnson, and Anne Foner.[2]

Put in utterly plain language, the aim of this chapter is to set out the little that we know and the much that we do not know but could know about the sociology of age differentiation in science and to explain why we should want to know it.

The paucity of materials on age differentiation in science appears to be the product of two distinct kinds of neglect in sociology. The first is the strangely slight attention given to age stratification generally even though age is recognized as one of the universal components of social structure. The second is the perhaps greater and now equally strange neglect of the entire sociology of science which, until the last decade or so, had enlisted the sustained interest of a remarkably small number of investigators. Neglect of this subject is odd if only because both advocates and critics of science are agreed that science is one of the major dynamic forces in the modern world. The convergence of the two kinds of neglect has meant that it was most improbable that sociologists would methodically investigate age differentiation in science. Nevertheless, enough work has been done to furnish a basis for formulating the problematics of the subject.

In this essay we deal with only a limited range of those problems. The first section is concerned with the connections between the growth and age structure of science; the second, with the relations between one aspect of the cognitive structure of science, here called "codification," and age-patterned behavior in science; the third, with age-related allocations of roles and with role-sequences in science; the fourth, with the operation of age-stratified structures of authority, principally in the form of gerontocracy; the fifth, with age- and prestige-stratified patterns of collaboration in science; and the sixth, final section dealing with age-patterned foci of problems selected for scientific investigation.

1. Growth and Age Structure of Science

By all the rough-and-ready measures now available, science, with its various interacting components, has been growing more rapidly than most other fields of human activity. For example, the population of scientists, with a doubling time of about fifteen years, is far outrunning the accelerating rate of increase in the general population. The science population of the United States, comprised of scientists, engineers, and technicians, grew from just under half a million in 1940 to 1.8 million in 1967. (About

2. We refer in the main to chapter 1 of *A Sociology of Age Stratification.*

300,000 of these in the latter period are estimated to be scientists.)[3] The number of doctorates annually awarded in science has increased from about 60 in 1885 to some 16,000 in 1969, a growth rate of some 7 percent per annum through this long period.[4] And as Price[5] has observed, the number of scientists and of scientific journals, articles, and abstracts, and, more recently, the funds available for research have all been increasing exponentially at varying rates.

It is of immediate interest to us that this growth results from the recruitment into science of youthful cohorts, principally in their twenties, and practically not at all from the migration into it of older persons drawn from other occupations. And once in science, they tend to stay there. One now hears much about the so-called flight from science, but the historical pattern still remains. Scientists continue to be a tenacious lot, seldom leaving science for an entirely different occupation,[6] although there is an appreciable amount of transfer between fields of science at various phases in the life course. This raises a significant sociological problem in its own right, one that has been little explored in the sociology of occupations.

Query: How are occupations arrayed in terms of rates of growth in personnel, and how do these rates and their components of entries and exits compare at different phases in the life course? What structural properties of occupations and characteristics of people recruited by them make for observed patterns of turnover?

Such detailed and systematic comparisons between age-stratified rates of turnover in various occupations are in short supply.[7] All the same, we suppose that classes of occupation have their exits and their entrances, distinctively patterned by age. In the case of science, transfers of personnel

3. United States Bureau of Census, Department of Commerce, *Statistical Abstract of the United States for 1970* (Washington: Government Printing Office, 1970), p. 525.

4. National Academy of Sciences–National Research Council, *Doctorate Recipients from the United States Universities 1958–66* and *Summary Reports for 1967, 1968, 1969 and 1970* (Washington, D.C.: Government Printing Office, 1970).

5. Derek de Solla Price, *Little Science, Big Science* (New York: Columbia University Press, 1963), chap. 1, and *Science Since Babylon* (New Haven: Yale University Press, 1961), chap. 5.

6. This statement refers to those who have actually begun to work in science, not to those who have taken an undergraduate degree in science. In the United States, about half of all college graduates majoring in science and engineering go on into other occupational fields. See Wallace R. Brode, "Manpower in Science and Engineering Based on a Saturation Model," *Science* 173 (1971): 206–13.

7. In the absence of suitable comparative investigations, Donald E. Super (*The Psychology of Careers* [New York: Harper & Row, 1957], chap. 4) touches the periphery of this problem by impressionistically examining "occupational life spans."

to other fields have been so conspicuously small that we take it to be in this respect near one extreme in the family of occupations. Once in the profession, scientists generally exit only through death or reluctant retirement.[8]

The patterns of growth and turnover in science are probably affected in at least three ways by the long period of socialization required to qualify for scientific work. Even in recent years, with the substantial funding of scientific education through fellowships and training grants in the United States, it takes an average period of eleven years after entry into college to obtain the doctorate in science.[9] First, the prolonged period devoted to acquiring the knowledge, skills, attitudes, values, and behavior patterns of the scientist means, as we have noted, that few enter the profession in later years after they have been at work in some other occupation. Second, this entails a considerable investment in the prospects of a career in science. The investment is not merely economic, with its financial costs of education and its foregone income during the period of preparation, but is affective as well. It is investment in a preferred way of life and commitment to it. Third, the high rate of attrition before entering the profession probably reflects a stringent process of social selection. Some 40 percent of candidates for the doctorate fail to complete their program of graduate study,[10] possibly leaving a more deeply committed and intellectually apt residue to go on into science.

Such modal patterns of entry and exit, coupled with the exponential increase in numbers of scientists, leave their stamp upon the age structure of the profession; and, as we shall indicate, this structure in turn affects both the normative system of science and its intellectual development. Figure 1, which compares the age distribution of scientific personnel with that of all employed personnel in the United States, suggests the ways in which the life-course processes of entry and exit combine with growth through cohort succession to yield the current age structure of science. On the one hand, the impact of delayed entry of scientific personnel into the labor force, necessitated by long years of prior training, is apparent from comparison of the age categories under 24 and 25–34. The relative size of the older strata, on the other hand, reflects the *juvenescent effect* of rapid growth on the age structure, as the size of entry cohorts expands. Despite the tendency of scientists to delay their retirement, the proportion of the scientific population in the older age strata is relatively small. Thus

8. There is, however, a fair amount of mobility through the statuses and roles *within* the social system of science, as we shall see in the later section of this essay.

9. National Science Foundation, *Science and Engineering Doctorate Supply and Utilization* (Washington, D.C.: National Science Foundation, 1971), p. iii.

10. Bernard Berelson, *Graduate Education in the United States* (New York: McGraw-Hill, 1960), p. 154.

the age structure reflects the counteracting patterns of life-course and cohort flow.

The rate of increase in the population of scientists far outruns that of the American population at large, and this results in an increasingly youthful population of scientists, both absolutely and comparatively. Their higher rate of increase means that scientists in the aggregate are also more youthful than the other professions. For example, more than half of those listed in the 1968 National Register of Scientific and Technical Personnel are under the age of 40 and 81 percent of them under 50, compared with 63 percent for lawyers and physicians.[11]

Education, age structure, and allocation

As with all social statuses and group memberships,[12] the status of "scientist" is not merely a matter of self-definition and ascription but also, and more significantly, of social definition and ascription. It is not enough to declare oneself to be a chemist or psychologist or space scientist; in order for the given status to have social reality it must be validated by "status judges," those institutions and agents charged with authenticating claims. Licensing boards for many professions and specialty boards in medicine are only the more visible and firmly institutionalized specimens of status judges. The criteria for the status of scientists are typically educational attainment ("earned academic degrees") and role-performance ("experience").

Since formal education is typically confined to the early years in life, the changing age structure of science is closely related to a concurrent educational upgrading of the scientific population. As Harvey Brooks has highlighted this trend:

> A striking fact is that in 1968 the percentage of the age cohort receiving Ph.D.'s in science and engineering and medical or dental degrees was higher than the percentage of the corresponding age cohort that received bachelor's degrees in science and engineering in 1920.[13]

This rapid change in the educational composition of the science population is a special and illuminating case of the general rise in the level of education. As Riley and Foner have noted,[14] the long-range rise in formal

11. United States Bureau of the Census, Department of Commerce, *Statistical Abstract*, pp. 65, 155, 525.

12. For a theoretical analysis of group membership and nonmembership, see Robert K. Merton, *Social Theory and Social Structure* (New York: The Free Press, 1968), pp. 338–51; the concepts of status-judges and status-imputation are treated in Merton, *Status-Sets and Role-Sets: Structural Analysis in Sociology* (unpublished ms.) and in chapter 19 of this volume.

13. Harvey Brooks, "Thoughts on Graduate Education," *Graduate Journal* 8, no. 2 (1971): 319.

14. Matilda White Riley and Anne Foner, eds., *Aging and Society*, vol. 1 (New York: Russell Sage Foundation, 1968), chap. 5.

FIGURE 1
Comparative Age Structure: Scientific Personnel and Total Employed Personnel, United States

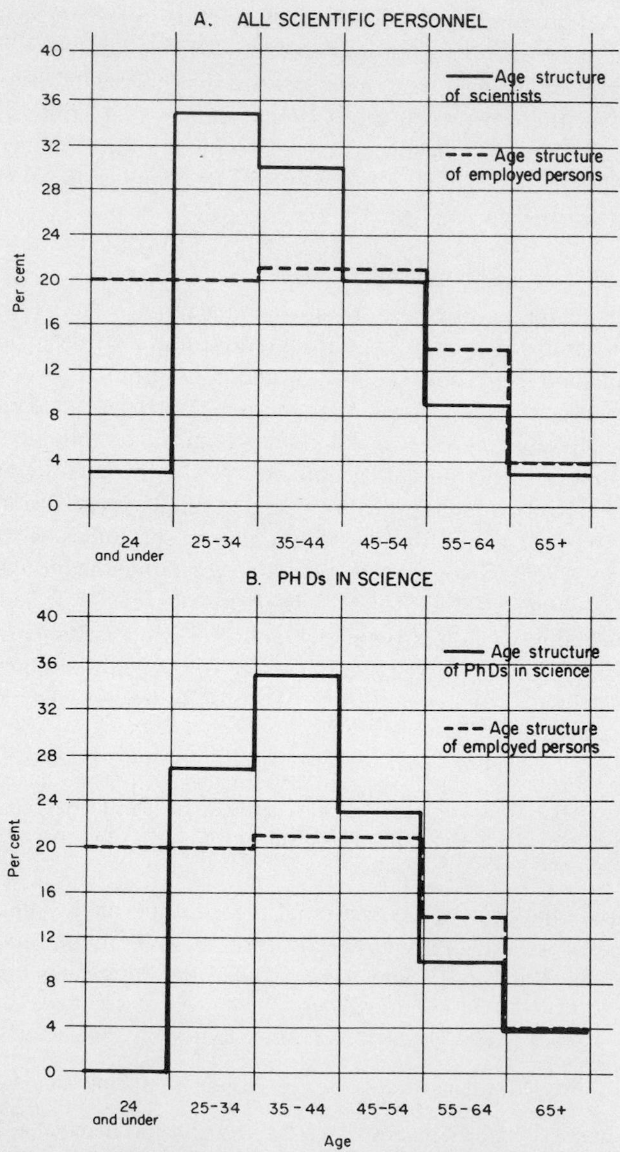

SOURCE: National Science Foundation, *American Science Manpower 1968: A Report of the National Register of Scientific and Technical Personnel* (Washington, D.C.: National Science Foundation, 1969), p. 50: and United States Department of Labor, *Employment and Earnings for June 1968* (Washington, D.C.: Government Printing Office, 1968), p. 30.

education for the population at large has meant that, at any given time, older people are educationally disadvantaged in comparison with younger ones. In the special case of science, we deal of course with the uppermost reaches of the educational distribution, and this introduces distinctive complexities. It may not be so much the social change in the level of attained education but the cultural change in the extent and character of scientific knowledge that presents the age cohorts of scientists with distinctive ranges of difficulties and opportunities. All this may lead to problems of career obsolescence and kindred difficulties for scientists in their life course. It also introduces various problems for the social organization of science; for example:

Query: How do the social and cultural changes in the education of scientists and in the growth of scientific knowledge affect the working relations between cohorts of scientists? Do they make for social cleavage and discontinuity between them? What social mechanisms, if any, bind the cohorts together and make for continuity in scientific development?

Thus the changing age and educational structure of science is congruent with the self-image, prevalent among scientists, that pictures "science as a young man's game." But this expression, repeated on every side by scientists young and old, does not refer primarily to the age (or sex) distribution of scientists. Rather, it only announces a widespread belief that the best work in science is done at a comparatively early age. This posited linkage between age and significant productivity is still the focus of the little research that has been done on age stratification in science, and we shall be examining that research in detail. But here, the imagery of science as the prerogative of the young generates another kind of sociological question, this one dealing with the linkages between age structure, prevalent values, and social organization.

Query: How do the age structures of occupations and their component specialties relate to age-connected values (for example, youth as asset or liability)? How do they relate also to other forms of stratification within the occupation (for example, allocation of authority-statuses and roles, of rewards, and so forth)?

Age and education in the component sciences

Turning from science in the large to the constituent sciences, we find that the median age of their personnel varies considerably, as appears in table 1. These divergences reflect sociologically interesting differences in the social definitions of boundaries of disciplines and of criteria for admission into them. In part, the age differences in the various science

TABLE 1
Median Age of Ph.D's in Selected Science Fields

Field	Median age[a]
All scientific fields	41.2
Mathematics	38.2
Physics	38.3
Chemistry	41.0
Biological sciences	41.8
Psychology	42.2
Sociology	44.1

SOURCE: National Science Foundation, *American Science Manpower 1968*, pp. 50 ff. (adapted).
NOTE: When medians are computed for all personnel, with or without the Ph.D., the rank order is roughly similar except that sociology rises to third place.
[a] Computed from grouped data by straight line interpolation.

populations merely reflect a differing mix of scientists, who often hold doctorates; engineers, who rarely do; and technicians who almost never do.[15] As can be seen from table 2, some 37 percent of American scientists held doctorates in 1968, but the figures ranged widely from 95 percent (anthropology) to 7 percent (computer sciences) among the disciplines. Thus the social science populations, with comparatively fewer technicians, have far higher proportions holding doctorates. Still, as we have seen in table 1, when comparisons are confined to those holding doctorates, substantial differences remain in the median ages of workers in the various sciences.

Very likely, the differing age distributions of component sciences result largely from differing rates of recruitment of new scientists and interchanges of personnel between the sciences at various stages in their educational and occupational careers. Although we assume that age-specific death and retirement rates are much the same for the various sciences, rates of transfer, which vary among age-strata, affect their age structures.[16] Substantial transfers from one science to others have been found in the course of undergraduate education, and longitudinal studies have identified streams of interchange among practicing scientists.[17] Yet, despite some

15. The Committee on Utilization of Scientific and Engineering Manpower of the National Academy of Sciences reported in 1964 that fewer than 2 percent of engineers had a Ph.D. No comparable counts have been published for technicians, no doubt because it would be a waste of funds, public or private, to search for what is known not to be there.

16. National Science Foundation, *Science and Engineering Doctorate Supply*, pp. 16, 30.

17. Jeffrey G. Reitz, "The Flight from Science" (Ph.D. diss., Columbia University, 1972); and Lindsey R. Harmon, *Profiles of Ph.D.'s in the Sciences: Summary Report on Follow-Up of Doctorate Cohorts, 1935–60* (Washington, D.C.: National Academy of Sciences–National Research Council, 1965), chap. 5.

TABLE 2
Percentages of Scientists Holding Doctorates, According to Fields of Science, United States, 1968

Field	Percentage holding doctorate
Anthropology	95
Psychology	64
Linguistics	62
Political science	59
Economics	53
Sociology	51
Biological sciences	48
Physics	44
Statistics	35
Chemistry	31
Mathematics	28
Earth and Marine Sciences	21
Agricultural sciences	18
Atmospheric and space sciences	9
Computer sciences	7
All fields	37

SOURCE: National Science Foundation, *American Science Manpower 1968*, adapted from table, p. 23.

strands of evidence, the sources of age differences among the sciences remain almost as much unexplored territory as the range of their consequences for the character and rate of development in the various sciences.

Slowing growth rates of science

It has long been recognized in principle that the exponential growth of science at a rate greater than that of society could not continue indefinitely. Manifestly, the rate of expansion in the numbers of scientists and the amount of resources allocated to science and in the volume of scientific publication would have to slow down. Otherwise, as Derek Price is fond of writing, and as many others are fond of quoting, "We should have two scientists for every man, woman, child and dog in the population, and we should spend on them twice as much money as we had."[18]

There are signs that, at least in societies with a substantial science base, the rate of expansion has tapered off. In the United States, for example, the resources expended for research and development have remained at about 3 percent of GNP since 1964.[19] There are indications, also, that the proportion of college graduates going into science has reached a ceiling.[20]

18. Price, *Little Science, Big Science*, p. 19.

19. Joseph P. Martino, "Science and Society in Equilibrium," *Science* 165 (1969): 769.

20. Brode, "Manpower in Science and Engineering," pp. 206–8.

Such equilibrium growth rates have direct consequences for the age and opportunity structures of science, as Martino has indicated.[21] A direct consequence, of course, is an increase in the median age of scientists. Slackened increase in resources will mean fewer new research installations and new departments in new universities. In due course, the age structure of research groups will probably change with possible consequences for their productivity.[22] The rate of increase in the volume of scientific publication will tend to diminish. But if Weiss and Ziman are correct in suggesting that the proliferation of scientists and the big expansion in funds for research have resulted in an even more rapid proliferation of trivial research and trivial publications,[23] this decline in volume of publication need not mean a corresponding decline in the growth of knowledge. Finally, since the various changes do not occur uniformly in the individual sciences, the relative rates of development of long established and newly emerging disciplines will probably continue to differ.[24]

2. Age Stratification and the Codification of Scientific Knowledge

In principle, the sociology of knowledge and, more narrowly, the sociology of science are concerned with the *reciprocal* relations between social structure and cognitive structure. In practice, however, sociologists of knowledge have dealt almost exclusively with the influences of the social structure upon the formation and development of ideas. And when sociologists of science have investigated the "impact of science upon society," this has been principally in the form of examining the social consequences, chiefly unanticipated, of science-based technology. In neither case is the effort made to trace the consequences of the cognitive structure of the various sciences for their distinctive social structures.

21. Martino, "Science and Society in Equilibrium," p. 772.

22. Jan Vlachý ("Remarks on the Productive Age," *Teorie a Metoda* [Czechoslovakia] 11 [1970]: 121–50) and Hugo Thiemann ("Changing Dynamics in Research and Development," *Science* 168 [1970]: 1427–31) maintain that the age structure of research groups affects their productivity, but little is yet known about this question. Available data reflect differences in the productivity of age strata rather than in various distributions of these strata.

23. See Paul A. Weiss, *Within the Gates of Sciences and Beyond: Science in Its Cultural Commitments* (New York: Hafner, 1971), p. 135. John Ziman puts the thesis vigorously: "the consequences of flabbiness [in science policy] are all too sadly evident in all quarters of the globe—the proliferation of third-rate research which is just as expensive of money and materials as the best, but does not really satisfy those who carry it out, and adds nothing at all to the world's stock of useful or useless knowledge" ("Some Problems of the Growth and Spread of Science into Developing Countries," Rutherford Memorial Lecture, in *Proceedings of the Royal Society*, 1969, A.311, p. 361).

24. See Hilary Rose and Steven Rose, *Science and Society* (London: Allen Lane, Penguin Press, 1969), pp. 250–51.

Yet the problem of the significance of cognitive structure for social structure is with us still. We propose to touch upon a limited case in point by examining one aspect of the cognitive structures of the sciences—what we call "codification"—in its relations to their distinctive age structures. Since so little preparatory work has been done on this problem, our observations are altogether tentative, designed to raise questions rather than to answer them.

Codification refers to the consolidation of empirical knowledge into succinct and interdependent theoretical formulations. The various sciences and specialties within them differ in the extent to which they are codified. It has often been remarked, for example, that the intellectual organization of much of physics and chemistry differs from that of botany and zoology in the extent to which particulars are knit together by general ideas. The extent of codification of a science should affect the modes of gaining competence in it. Experience should count more heavily in the less codified fields. In these, scientists must get command of a mass of descriptive facts and of low-level theories whose implications are not well understood. The comprehensive and more precise theoretical structures of the more codified field not only allow empirical particulars to be derived from them but also provide more clearly defined criteria for assessing the importance of new problems, new data, and newly proposed solutions. All this should make for greater consensus among investigators at work in highly codified fields on the significance of new knowledge and the continuing relevance of old.

The notion that the sciences and specialties differ in the extent of their codification is not, of course, new. Some twenty years ago, James Conant introduced the counterpart idea that the various branches of science differ in "degree of empiricism." As he put it, "Where wide generalizations and theory enable one to calculate in advance the results of an experiment or to design a machine (a microscope or a telescope, for example), we may say that the degree of empiricism is low."[25] In our terms, the more codified a science or scientific specialty, the lower its degree of empiricism.[26]

25. James B. Conant, foreword to *Harvard Case Studies in Experimental Science* (Cambridge, Mass.: Harvard University Press, 1950), p. 9.

26. The recent distinction between "hard" and "soft" science also bears a family resemblance to the concept of codification. Norman W. Storer uses that pair of terms to characterize sciences given more or less to quantification and to rigor. Derek Price has suggested various features which are distinctive of the literatures of hard science, soft science, technology, and other fields of learning. See Storer, "The Hard Sciences and the Soft: Some Sociological Observations," *Bulletin of the Medical Library Association* 55 (1967): 75–84; and Price, "Citation Measures of Hard Science and Soft Science, Technology and Non-Science," in Carnot E. Nelson and Donald K. Pollack, eds., *Communication Among Scientists and Engineers* (Lexington, Mass.: Heath Lexington Books, 1970). So far as we know, the "somewhat vague distinction between what we may call 'hard' data and 'soft' data" was introduced, with these words, by Bertrand Russell in his exemplary Lowell Lectures, *Our Knowledge of the External World* (New York: W. W. Norton, 1929), p. 75.

By consolidating both data and ideas in theoretical formulations, the more highly codified fields tend to obliterate the original versions of past contributions, incorporating their essentials in the newer formulations. As Paul Weiss noted in his influential paper, "Knowledge: A Growth Process," "Each field of knowledge must be accorded its own merit ratio between generalization and particularization, taking for granted that assimilation will be driven to the utmost limits compatible with the nature of the field."[27] In this sense, the more highly codified the field, the higher the rate of "obsolescence" of publications in it. Weiss measured differentials in obsolescence by the age distributions of references appearing in publications in various sciences, a procedure which has since been described as "citation analysis."[28] Weiss found that references to recent work were much more frequent in "analytical physiology than in its more descriptive biological sister sciences" of zoology and entomology.

The general pattern has since been confirmed for a wider variety of disciplines.[29] The journals in fields we intuitively identify as more highly codified—physics, biophysics, and chemistry—show a larger share of reference to recent work; they exhibit a greater "immediacy," as Derek Price calls it.[30] By way of illustration,

27. Paul A. Weiss, "Knowledge: A Growth Process," in American Philosophical Society, *Proceedings* 104 (1960): 247.

28. In his monumental *Introduction to the History of Science* (3 vols. in 5 parts [Baltimore: Williams and Wilkins, 1927–48]), George Sarton often quantified the citations of a major scientific work to previous works as one way of establishing its intellectual heritage. Eugene Garfield suggested the use of systematic citation indexing for historical research in 1955 and developed the computerized Science Citation Index in 1964. With its more than 20 million bibliographic citations, the SCI data base has greatly advanced citation analysis in historical and sociological investigations of science. See Eugene Garfield, "Citation Indexes for Science," *Science* 122 (1955): 108–11; and Eugene Garfield, L. H. Sher, and R. J. Torpie, *The Use of Citation Data in Writing the History of Science* (Philadelphia: Institute of Scientific Information, 1964).

29. Derek Price has done most to extend the use of "citation-and-reference-analysis" to distinguish modes of development in the various fields of learning. For early work, see Price, *Little Science, Big Science*, pp. 78ff, and "Networks of Scientific Papers." *Science* 149 (1965): 510–15; for recent work, see his "Citation Measures of Hard Science and Soft Science, Technology and Non-Science." The abundance of citation studies includes R. N. Broadus, "An Analysis of Literature Cited in the American Sociological Review," *American Sociological Review* 17 (1952): 355–57, and "A Citation Study for Sociology," *American Sociologist* 2 (1967): 19–20; P. E. Burton and R. W. Keebler, " 'Half-life' of Some Scientific and Technical Literatures," *American Documentation* 11 (1960): 18–22; Duncan MacRae Jr., "Growth and Decay Curves in Scientific Citations," *American Sociological Review* 34 (1969): 631–35; J. Cole and S. Cole, "Measuring the Quality of Sociological Research: Problems in the Use of the *Science Citation Index*," *The American Sociologist* 6 (1971): 23–29.

30. "Immediacy is much increased use of the last few years of papers over and above the natural growth of the literature and its normal slow aging." "A literature growing at the rate of 5 per cent per annum," Price goes on to calculate, "doubles

72% of the references in *The Physical Review** are to papers published within the preceding five years, as are
63% in the *Cold Spring Harbor Symposium* on quantitative biology;
58% in *Analytical Chemistry*;
50% in the *Anatomical Record*; and
47% in the *American Zoologist*.*[31]

Similar citation data suggest that the social sciences are intermediate to the physical sciences and the humanities in degree of codification. The findings are notably consistent. As we have just seen, in the more analytical physical sciences, about 60 percent or more of the citations refer to publications appearing within the preceding five years. In the humanities—represented by such journals as the *American Historical Review, Art Bulletin*, and the *Journal of Aesthetics and Art Criticism*—the corresponding figures range from 10 to 20 percent. In between are the social sciences—represented by such journals as the *American Sociological Review* and the *Journal of Abnormal and Social Psychology*—where from 30 to 50 percent of the citations refer to equally recent publications.

It appears, moreover, that the citation pattern of a science transcends national and cultural boundaries. At any rate, American, European, and Soviet journals of physics have been found to exhibit almost identical age distributions of citations.[32]

One limited aspect of the cognitive structure of the various sciences, then, is the extent of their codification. We want now to explore possible relations between codification and age-patterned behavior and processes in the sciences, along the following lines:

1. The extent to which codification affects opportunities for scientific discovery by different age strata;
2. age differentials in responsiveness to new scientific ideas in fields that are variously codified;
3. the effects of codification upon the visibility of scientific contributions; and
4. the linkages between codification, changing foci of research and opportunities for discovery.

---

in size every 13.9 years and contains about 22 per cent of all that has been published in its last five years of publication." This means, of course, that the extent to which a field focuses on new work and so makes for obsolescence of earlier work is measured by the excess of recent citations beyond what would be expected on the basis of growth in the literature; see Price, "Citation Measures of Hard Science and Soft Science, Technology and Non-Science," pp. 9–10.

31. The data identified by an * are from Price; we have compiled the rest.

32. See Stevan Dedijer, "International Comparisons of Science," *New Scientist* 379 (1964): 461.

Codification and age differentials in discovery

We begin with the premise that up to a given age, older and more experienced scientists have an edge on their much younger colleagues in the opportunities for discovery. After all, they know the field as the novice does not. What needs to be explained, in our view, is not so much discovery by experienced and knowledgeable elders as discovery by newly trained youth. In this connection, we need to ask whether discovery by young scientists is more frequent in some sciences and, if so, how this comes to be.

Codification facilitates mastery of a field by linking basic ideas in a theoretical framework and by reducing the volume of factual information that is required in order to do significant research. This should lead scientists in the more codified fields to qualify earlier[33] for work at the research front—at least, as collaborators of more mature investigators. And early achievement in science may give an enduring advantage by providing both increasingly abundant facilities for research and early access to the social networks of scientists at the research front where information and criticism are exchanged and motivation for getting on with one's work is maintained.[34] The best-known because heavily publicized specimen of this process is that of the twenty-five-year-old James D. Watson[35] soon finding his way into the center of work on the structure of DNA once he was sponsored by his intellectually influential teachers, Salvador Luria and Max Delbrück. In this process, intellectual mobility and social mobility (of a jointly sponsored and contest variety) are mutually reinforcing.

The organization of scientific inquiry and of training in science also promotes early entrance into the research role. In a sense, young scientists are more apt than their expert teachers to be abreast of the range of knowledge in their field. Since advanced research in science demands concentration on a narrow range of problems at hand, the established specialist experts, intent on moving ahead with their own research, tend to fall be-

33. This early start may also reflect early recruitment to the more codified sciences. We know of no data on the matter but share the widespread impression that decisions to go into mathematics and physics are made much earlier than decisions to enter the soft sciences. The significance of age at time of the decision to enter the field of medicine is examined by Natalie Rogoff, "The Decision to Study Medicine," in Robert K. Merton, George G. Reader, and Patricia Kendall, eds., *The Student-Physician* (Cambridge, Mass.: Harvard University Press, 1957), and by Wagner Thielens, Jr., "Some Comparisons of Entrants to Medical and Law School," in ibid., pp. 109–22 and 131–52.

34. See the informed and astute account of how young scientists achieve entry into the "Invisible College" of their specialty by the physicist and sociologist of science, John Ziman, in *Public Knowledge: The Social Dimension of Science* (Cambridge: At the University Press, 1968), pp. 130–34.

35. Watson's detailed account of how all this worked out for him is one of the many features of *The Double Helix* that make it an unexampled personal document in the sociology of science. See *The Double Helix: Being a Personal Account of the Discovery of the Structure of DNA* (New York: Atheneum Press, 1968).

hind on what others are doing outside their own special fields. But, at least in the best departments, students are trained by an *aggregate* of specialists at work on the research front of their specialties. This brings them up to date, if only for a time, in a wider variety of fields than their older and temporarily more specialized teachers.

This pattern we believe to hold in all the sciences, but we suspect that it is more marked and efficacious in the more highly codified ones, those that provide more powerful means for acquiring competence in current knowledge. The opportunity structure confers two advantages on the young in the more codified fields: the chance to begin research early as qualified junior colleagues and the chance to have training that is both up-to-date *and* relatively diversified. Both should advance their opportunities for making significant research contributions early in their careers.

It is from this standpoint that we come upon the one problem that has almost monopolized discussions of the, in fact, multiform connections between age and scientific activity: the time of life at which scientists do their most important work or, as it is sometimes put, the relations between age and scientific productivity. The data bearing on this subject are faulty or severely limited but, on first inspection, they seem to confirm our expectations. Various investigators report lower median ages for discoverers in physics and chemistry than in the more descriptive biological sciences, with these being lower in turn than in the behavioral sciences.[36] Nobel laureates in physics, for example, were on the average 36 at the time of doing their prize-winning work; laureates in chemistry, 38; and those in medicine and physiology, 41.[37] This does not mean, of course, that a higher rate of discovery in youth is the norm in the more codified sciences, much less for them all. Apparently, we sometimes need to be reminded that median ages at time of discovery tell us that half of the discoveries were made after the median age as well as before. In contrast to the usual emphasis, Lehman's findings could be reported, for example, as indicating that "fully half" of the discoveries listed in Magie's *Source Book of Physics* were made by scientists *over* the age of 38 or that "fully half" of the discoveries listed in genetics were by scientists *over* 40.

Beyond this, we need only mention other caveats in the use of these data on age and scientific achievements. They are faulty in two basic respects. First, they do not take into account the age structure of the scientific

36. See H. C. Lehman, "The Creative Years in Science and Literature," *Scientific Monthly* 43 (1936): 162, and *Age and Achievement* (Princeton: Princeton University Press, 1953), p. 20; C. W. Adams, "The Age at which Scientists Do Their Best Work," *Isis* 36 (1946): 116–69.

37. Data for 1901–50 are drawn from E. Manniche and G. Falk, "Age and the Nobel Prize," *Behavioral Science* 2 (1957): 301–7, and for the period 1951–69 from Harriet Zuckerman, *Scientific Elite: Studies of Nobel Laureates in the United States* (Chicago: University of Chicago Press, in press).

population. As we know from the exponential growth in the numbers of scientists, the young ones make up a hefty percentage at any given time and so will produce a large aggregate of contributions. But of course this does not provide the evidence on comparative rates of scientific contributions at various ages. What is required are data not on the proportion of contributors in each age stratum but on the proportion of each age stratum making contributions. This requirement holds even in comparing age-linked rates of productivity among the various sciences, since, as we have seen, the age composition of the sciences does differ appreciably.

Second, Lehman's studies do not take into account, as Wayne Dennis has emphatically demonstrated, the biasing effects of differing life spans on the distribution of achievements at various ages.[38] The fact that short-lived scientists are cut off from making any contributions in later years factitiously enlarges the proportions assigned to young scientists in data on age and achievement that do not take longevity into account. The essential issue is caught up in the lament of Newton over the premature death at age 34 of his protégé, the mathematician, Roger Cotes: "If he had lived, we might have known something." Or the similar observation by John Maynard Keynes about the brilliant young logician and mathematician, Frank Ramsey, robbed of his future at the age of 27.

We suggest, moreover, that the statistical bias in apparent age-specific productivity differs among the variously codified sciences and humanities. In doing so, we depart a bit from the ancient adage that the good die young. We propose a less crisp but more germane version: comparatively more good mathematicians and physicists than good historians and sociologists die young. In saying this, we do not propose the improbable hypothesis that rates of premature mortality are in fact higher in the more codified fields than in the less codified ones. We assume the same rates of age-specific mortality but an earlier age of prime achievement. Mathematicians of genius who died prematurely, such as a Galois dead at 21 or a Niels Abel dead at 26, have been enduringly identified by their early work as being of the first rank. But by hypothesis, sociologists or botanists of genius who die young will have fulfilled less of their potential in their early years and so do not even appear in the standard histories of their fields. This reconstruction is consistent with the Dennis data, which suggest that a far larger share of the total life output of long-lived mathematicians and chemists than of equally long-lived historians, botanists, and geologists

38. See Wayne Dennis, "Age and Productivity among Scientists," *Science* 123 (1956): 724–25; "Age and Achievement: A Critique," *Journal of Gerontology* 11 (1956): 331–37; "The Age Decrement in Outstanding Scientific Contributions," *American Psychologist* 13 (1956): 457–60; and "Creative Productivity between the Ages of 20 and 80," *Journal of Gerontology* 21 (1966): 1–8. For a discussion of the fallacies in the interpretations drawn from the Lehman data and the implications of the Dennis data, see Riley and Foner, *Aging and Society*, vol. 1, chap. 18.

is completed during their first decade of work. This sort of thing can thus foster the illusion that good mathematicians die young but that, say, good sociologists linger on forever.[39]

To this point, we have centered on differentials in rates of scientific contributions by age-strata in sciences codified in varying degree. We have now to touch upon the further question whether the truly transforming ideas in science, the fundamental reconceptualizations, are more apt to be the work of youthful minds rather than older ones. T. S. Kuhn, in his vastly influential book on scientific revolutions, suggests that creators of fundamental new paradigms are almost always young or very new to the field.[40] A long and familiar roster of cases can be provided to illustrate his suggestion. Newton wrote of himself that at 24, when he had begun his work on universal gravitation, and the calculus and the theory of colors: "I was in the prime of my age for invention, and minded Mathematics and Philosophy more than at any time since." Darwin was 22 at the time of the Beagle voyage and 29 when he formulated the essentials of natural selection. Einstein was 26 in the year of three of his great contributions, among them the special theory of relativity; and finally, eight of the ten physicists generally regarded as having produced quantum physics were under the age of 30 when they made their contributions to that scientific revolution.[41]

Arresting illustrations of this kind are of course not enough to show that young scientists are especially apt to revolutionize scientific thought. But in the absence of systematic data on the age composition of scientists in various historical periods, they remain the basis for the generalization being widely accepted as commonplace. Yet, as Kuhn himself goes on to say, it is a generalization that "badly needs systematic investigation."

39. It will be noted that this gives special meaning to the old adage that "science is a young man's game," which, plainly, should now be coupled with the ostensibly older adage that "history is an old man's game" (as Robert Graves, in his *I, Claudius*, has the Roman historian Caius Asinius Pollio saying). However, the crisp literary contrast drawn in these coupled aphorisms should not be allowed to obscure the quantitative contrast drawn in our argument. Still, we can illustrate our argument in part by recalling that "the Nestor of historians" in the nineteenth century, Leopold von Ranke, did much of his *most important* work while in his fifties, sixties, and seventies, setting out in his eighty-sixth year "on his life-long ambition to write a world history," just as the Nestor of historians in the twentieth century, Friedrich Meinicke wrote two of *his* most important treatises—*Die Idee der Staatsräson* and *Die Entstehung des Historismus*—while in his sixties and seventies, reserving the, for him, significant but comparatively minor work, *The German Catastrophe* (1946) for his eighty-fifth year. On Ranke and Meinicke, see Fritz Stern, *The Varieties of History*, rev. ed. (New York: World Publishing Co., 1972), pp. 55, 267–68; on a Nestor among Nobel laureates in science whose work during his seventies and eighties was of secondary importance, see page 530 in this chapter.

40. See Thomas S. Kuhn, *The Structure of Scientific Revolutions* (Chicago: University of Chicago Press, 1962), pp. 89–90.

41. George Gamow, *Thirty Years that Shook Physics* (Garden City, N.Y.: Doubleday, 1966).

Codification and age differentials in receptivity to new ideas

As we have suggested, new developments in the more codified sciences are closely linked to work done just before. These sciences grow, as Price puts it, "from the skin." As a result, awareness of new ideas and critical acceptance of them is especially important in the codified sciences. If certain age strata are more responsive to them, this increases their chances for making further discoveries.

It has often been said that aging leads to growing resistance to novelty and, specifically, that older scientists tend to resist new ideas.[42] As Planck, who did not develop the idea of the quantum until he was 42, remarked: "a new scientific truth does not triumph by convincing its opponents and making them see the light, but rather because its opponents eventually die, and a new generation grows up that is familiar with it."[43]

Observations of this sort, based on lore rather than systematic evidence, raise the perennial questions: is it really so? and if so, how does it come to be?

If there are these age-stratified differences in receptivity to new (and sound) ideas, this should be reflected in various behaviors. For one example, younger scientists should rely more on recently published research and cite it more frequently than older scientists in the same fields. For another, the age strata of scientists should differ on what they take to be the most significant contributions to their field. Some limited data in hand lend credence to these inferences.

We find for a small sample of scientists that those under 30 are more given to citing recent work (papers published in the five years preceding their own) than men in the same fields who were on the average thirty years older: 71 percent of the references in the work of the younger investigators and 58 percent by the older ones being to the recent literature.[44] But the pattern does not hold for all older scientists. Nobel laureates do

42. But see changes over the life course in political attitudes reported by Foner in chapter 4.2 of *Aging and Society*, vol. 3.

43. Max Planck, *Scientific Autobiography and Other Papers* (New York: Philosophical Library, 1949), pp. 33–34. This must surely be one of the most frequently quoted observations of its kind in recent years. It was put to good use by Bernard Barber in his paper "Resistance by Scientists to Scientific Discovery," *Science* 134 (1961): 592–602; and, among others, by Kuhn in *Structure of Scientific Revolutions*, p. 150; by Warren Hagstrom in *The Scientific Community*, p. 283; and by Daniel S. Greenberg in *The Politics of Pure Science* (New York: New American Library, 1967), p. 45.

44. Harriet Zuckerman, unpublished data. Stephen Cole ("Age and Scientific Behavior," read before the American Sociological Association, 30 August 1972) found the same pattern for a variety of scientific fields. Since the process of refereeing and editing scientific papers before publication probably tends to homogenize initial age differences in citations, even small differences between age strata in the published papers represent an a fortiori case.

not exhibit the same citation behavior as their less distinguished age peers. They appear to be just as responsive to new research as men thirty years their junior, with 73 percent of their references being to the most recent literature. This finding bears only tangentially on the Planck doctrine but it does suggest that attentiveness to new developments in science is stratified by both age *and* scientific achievement.

Substantive differences in judgments of what constitutes important work may crystallize, as Hagstrom has noted, into "generational disputes." But, as he goes on to say, "even if some disputes are generational, they need not be simply 'innovative youth' versus 'conservative age.' Rather, the outlook of a generation is strongly influenced by events occurring when its members embark upon their careers. Age may be more radical than youth."[45]

The evidence on age differentials in receptivity to new ideas in science remains thin and uncertain. But should further investigation find, as the widespread belief has it, that older scientists are indeed more resistant to new ideas, that would only raise a series of questions of how that comes to be. It would not follow, for example, that it results from physiological aging or senescence.[46] As Barber has noted in this context, *aging* "is an omnibus term which actually covers a variety of social and cultural sources of resistance." He goes on to suggest several possible social and cultural components in such resistance:

> As a scientist gets older he is more likely to be restricted in his response to innovation by his substantive and methodological preconceptions and by his other cultural accumulations; he is more likely to have high professional standing, to have specialized interests, to be a member or official of an established organization, and to be associated with a "school." The likelihood of all these things increases with the passage of time, and so the older scientist, just by living longer, is more likely to acquire a cultural and social incubus. But this is not always so, and the older workers in science are often the most ardent champions of innovation.[47]

This provides in effect, a formidable agenda for investigation of age-associated differences in receptivity and resistance to new conceptions in science.

Codification and visibility of scientific contributions

We turn now from considering possible age-patterned responses to new ideas, irrespective of their source, to consider possible differences in the

45. Hagstrom, *Scientific Community*, pp. 284–85.

46. For systematic review of the evidence on the multiple interpretations of such types of data, see Riley and Foner, *Aging and Society*.

47. Barber, "Resistance by Scientists to Scientific Discovery," p. 602. On much the same point, see also Hagstrom, *Scientific Community*, pp. 283–84.

responses to new ideas advanced by scientists of differing age. The visibility of new ideas in a discipline may be affected by the sheer volume of its literature, which may, in turn, be related to its degree of codification. That is not of interest here. However that may be, our interest lies instead in the direct implications of codification itself for the visibility of ideas introduced by scientists of differing age.

It would seem that new ideas are more difficult to identify as important in disciplines that are largely descriptive and only spottily and loosely organized by theory. In these less codified disciplines, the personal and social attributes of scientists are more likely to influence the visibility of their ideas and the reception accorded them. As a result, work by younger scientists who, on the average, are less widely known in the field, will have less chance of being noticed in the less codified sciences. Put another way, the "Matthew effect" (see chapter 20 of this volume)—the tendency for greater recognition to be accorded contributions by scientists of great repute—is apt to operate with special force in the less codified fields.

Correlatively, in the more codified sciences, new ideas, whatever their source, can better carry their own credentials. Important contributions by young scientists, or older ones, for that matter, are not only more visible in the codified fields; they are taken more seriously since their theoretical importance can be more readily assessed. This tends to put the young on a par with eminent seniors in communicating ideas and in having them noticed.

Although Stephen Cole's studies of the Matthew effect have been confined to physics, his findings are suggestive in this regard. In this highly codified science, the work of eminent investigators is incorporated into ongoing research only a little more quickly than contributions of comparable quality by less distinguished investigators. The age of physicists also has had little effect on the speed with which their ideas diffuse.[48] In physics, then, the merits of the investigation seem to govern its reception with the attributes of the investigator playing only a small part. Comparative study is now needed to find out whether the strength of the Matthew effect does in fact vary inversely with the extent to which sciences are codified.

Codification, inter-science transfers, and discovery

Although, as we have seen, scientists seldom leave the occupation of science altogether, a considerable number transfer from one field to another. About a quarter of American scientists have made shifts of this kind, the

48. Stephen Cole, "Professional Standing and the Reception of Scientific Discoveries," *American Journal of Sociology* 76 (1970):297, 299.

rates of transfer differing among the various sciences of origin and, of course, among age cohorts.[49]

Scientists who have left the field in which they were trained, to work in another where they were not, constitute a special class of newcomers. In certain respects, these older neophytes are functionally like the younger novices in that field. Both are being rapidly introduced to the research front of the field although the older newcomers differ from the younger indigenous recruits in having been less comprehensively trained in the new field. Both include people bringing new perspectives on old problems to the degree that they have not acquired commitments to the definitions of problems or to the form of their probable solutions that are conventionally adopted in that field. The transfers, unlike their young colleagues just beginning to work in science, also bring with them styles of research new to their adopted field, as was dramatically the case when physicists turned their attention to biology and created the field of molecular biology.[50] In part, the contributions of newcomers derive from their transferring to the new field standards and modes of investigation customary in their field of origin.

Some of the attributes of the more codified sciences that facilitate the fairly rapid learning of essentials by the incoming student should thus also facilitate effective transfer into the field by older, experienced scientists. Historical and statistical data[51] suggest that transfers tend to occur among sciences codified to about the same extent with a subsidiary pattern of movement toward less codified fields. There is little interchange between the extremes of codification: between physics and, say, botany or zoology.

The general pattern of such transfers has been connected by Joseph Ben-David to what he describes as "role-hybridization": applying the means usual to role A in trying to achieve the goals of role B. As he sums it up:

49. Harmon, *Profiles of Ph.D.'s in the Sciences*, pp. 50–52; National Research Council, Office of Scientific Personnel, *Careers of PhD's: Academic versus Nonacademic. A Second Report on Follow-up of Doctorate Cohorts 1935–60* (Washington, D.C.: National Science Foundation, 1968), pp. 59–62.

50. The *Festschrift* for Max Delbrück, one of the founders of molecular biology, consists largely of a remarkable series of lively and informative personal accounts of the beginnings of the field. See John Cairns, Gunther S. Stent, and James D. Watson, eds., *Phage and the Origins of Molecular Biology* (Cold Spring Harbor: Laboratory of Quantitative Biology, 1966). *What is Life?*, a short book by the physicist Erwin Schrödinger, proved decisive in transforming physicists into biologists. For a further account of the emergence of molecular biology, see Donald Fleming, "Emigré Physicists and the Biological Revolution," in Donald Fleming and Bernard Bailyn, eds., *The Intellectual Migration* (Cambridge, Mass.: Harvard University Press, 1969), pp. 152–89.

51. Harmon, *Profiles of Ph.D.'s in the Sciences*, p. 51.

Scientific disciplines differ in the degree of their theoretical closure and methodological precision. The phenomena most similar to role hybridization would be shifts from a theoretically and methodologically more advanced discipline to one less advanced. These must be distinguished from shifts between two disciplines of the same level and from less to more advanced disciplines.[52]

The processes and consequences of patterns of transfer among variously codified sciences have only begun to be investigated. But something of the process can be pieced together for the eminent men who have changed fields. They often exhibit an almost playful arrogance about the time required for retooling. Symbolic stories abound. Leo Szilard is said to have taken all of three weeks at Cold Spring Harbor in order to effect his transformation from physicist to biologist, this at the age of 47. Francis Crick's leap from physics to biology has been twice chronicled.[53] The same theme of the rapid acquisition of fundamentals appears in Waddington's account of the European origins of molecular biology. He writes of the journey to the first conference of geneticists and crystallographers:

> Most of us tried to sleep on the benches in the general saloon, but Darlington and [the crystallographer] Bernal kept sea-sickness at bay by the former teaching the latter "all the genetics and cytology anyone needs to know" throughout the course of the night. Before dawn, Bernal had already decided that the mitotic spindle must be a positive tactoid.[54]

This suggests not only that much can be learned in short order but also that knowledgeable newcomers accustomed to being in command of their field, even if they are not quite of Bernal's caliber, can achieve enough understanding of fundamentals to introduce new ideas at the outset.

These topflight migrants from one science to another seem unworried by their ignorance of the problematics prevailing in the new field. This keeps them from some stale preconceptions. Maria Goeppert Mayer, the Nobel laureate, provides an apt example. Her work on "the magic numbers problem" (a problem of such profound interest that it had been given a name) and her subsequent development of the shell model of the nucleus, she reports, depended on a specific kind of ignorance. Trained as a physicist but working mainly in physical chemistry, she was brought up on the

52. Joseph Ben-David, "Roles and Innovations in Medicine," *American Journal of Sociology* 65 (1960):557–68; see also, Ben-David and Randall Collins, "Social Factors in the Origins of a New Science: The Case for Psychology," *American Sociological Review* 31 (1966): 557–68.

53. Some fourteen years earlier, Szilard had been told by the physiologist A. V. Hill that he could pick up the essentials of physiology by simply setting himself the task of teaching it; see Leo Szilard, "Reminiscences," in Fleming and Bailyn, *Intellectual Migration*, p. 98. For accounts of Crick's transition, see Watson, *Double Helix*, and Robert Olby, "Francis Crick, DNA and the Central Dogma," *Daedalus* 99 (Fall 1970): 938–87.

54. C. H. Waddington, "Some European Contributions to the Prehistory of Molecular Biology," *Nature* 221 (1969):318.

"Bethe Bible"[55] and so was unhampered by knowing "what everyone knew" about spin-coupling.[56] Focused naïveté and focused ignorance evidently have their functions in science—especially for anything but naïve and otherwise immensely informed scientists.[57]

Like geographic migration, intellectual migration in the form of transfers from one science to another should lend itself to cohort analysis. Are there certain times in the careers of scientists at which they tend to make the change? Do these patterns persist among successive cohorts or are they fairly constant? Have there been historical changes in the frequency and patterns of transfers? Are the migrants in science representative of the field, more able on the whole, less so, or bimodal in their distribution of capacity and achievement?

Whatever the patterns of transfer, they need to be examined within the context of the intellectual organization of the sciences of origin and sciences of destination.

3. Scientific Roles

Inventory of roles in science

Like other domains of social life, the social structure of science has its distinctive array of statuses and roles, allocated to members through complex processes of social selection. We focus here on the status of scientist. But we should note in passing that the social structure of science, especially as we know it today after centuries of institutionalization and social differentiation, contains a variety of other statuses and roles. Often indispensable to the effective advancing of scientific inquiry, these parascientific roles include technicians of every stripe, the builders of experimental apparatus and instruments, and the broad spectrum of assistants engaged in facilitating scientific work (for example, by preparing and taking care of experimental materials).[58]

55. In a series of papers published in the late 1930s in *Reviews of Modern Physics*, Hans Bethe attempted to consolidate what was known then about the atomic nucleus. That these papers had an immense impact on physics is registered in the fondly respectful title by which they are known.

56. J. H. D. Jensen, who independently solved the same problems, was, interestingly enough, also unaware of prevailing ideas of spin-coupling.

57. On the general idea of the uses of ignorance under certain conditions, see Wilbert E. Moore and Melvin M. Tumin, "Some Social Functions of Ignorance," *American Sociological Review* 14 (1949):787–95. On the "outsider" in science and technology, see S. Colum Gilfillan, *The Sociology of Invention* (Chicago: Follett, 1935), pp. 88–91; Ben-David, "Roles and Innovations in Medicine," pp. 557–59; and chapter 5 of this volume.

58. For a short inventory of roles in science, see Weiss, *Within the Gates of Science and Beyond*, pp. 29–30. It will be remembered from chapter 16 of this volume that Francis Bacon described a variety of scientific roles in his *Solomon's House*.

Like other statuses, the status of scientist involves not a single role but, in varying mixture, a complement of roles. These are of four principal kinds: research, teaching, administrative, and gatekeeper roles.[59] Each of these is differentiated into subroles, which we only note here but do not consider in detail.

The research role, which provides for the growth of scientific knowledge, is central, with the others being functionally ancillary to it. For plainly, if there were no scientific investigation, there would be no new knowledge to be transmitted through the teaching role, no need to allocate resources for investigation, no research organization to administer, and no new flow of knowledge for gatekeepers to regulate. Possibly because of its functional centrality, scientists apparently place greater value on the research role than any of the others. As is generally the case in maintaining a complex of mutually sustaining roles, ideology does not fully reflect this differential evaluation of roles in the role-set: scientists will often insist on the "indispensability" and consequently equal importance of the ancillary roles. Yet, almost in a pattern of revealed preference, the working of the reward system in science testifies that the research role is the most highly valued. The heroes of science are acclaimed in their capacity as scientific investigators, seldom as teachers, administrators or referees and editors.

The research role divides into subroles, distinguished to varying degree in the different sciences. In research, scientists define themselves and are defined by others as experimentalists (or, more generally, empirical investigators) or as theorists, with occasional high-yield hybrids such as Enrico Fermi or Linus Pauling embodying both subroles effectively. The differentiation seems more marked in the more codified sciences. Little is known about the processes leading scientists to adopt one or another of these subroles. In the lore of science, this is not even problematic. Scientists are assumed to become either experimentalists or theorists as their highly specific capacities dictate. But it seems that the process is more complex than the simple matching of roles to self-evaluated capacities. It presumably involves, at the least, interaction between developing self-images of aspirants to scientific investigation, socialization by peers and mentors, and continuing evaluation of their role performance by peers, superiors, and themselves.

To the extent that the research role in science involves interaction between scientists, it also makes for some, often reciprocal, teaching and learning. The teaching role, particularly in the sciences, calls not only for explicit didactics but, probably much more in the sciences than in the

59. For the general conception that each status has its distinctive complement of roles, or its role-set, see Merton, *Social Theory and Social Structure*, enlarged ed. (New York: Free Press, 1968), pp. 422–38.

humanities, for tacit instruction through observed example. The master-apprentice relation is central to socialization in the sciences, particularly in laboratories which provide for mutual observability by master and apprentice. This structural difference between the sciences and the humanities is reflected in the fact that the status of postdoctoral student is widespread in the sciences and rare in the humanities.

There is, in the normative system of science, an ambivalence toward the preferred relations between the research and teaching roles. For some, the norm requires the scientist to recognize his prime obligation to train up new generations of scientists, *but* he must not allow teaching to preempt his energies at the expense of advancing knowledge. For others, the norm reads just as persuasively in reverse. We have only to remember the complaints about Faraday that he had never trained a successor as Davy had trained him, yet consider the frequent criticism of scientists who give up research for teaching. There are indications, as we shall see, that the time scientists allocate to the roles of teaching and research changes during the life course.

A third major role of scientists is ordinarily (and not very instructively) caught up in the term "administration." The term often covers a wide gamut of quite distinct structural conditions, ranging from occasional service on advisory or policy-making committees, through direction of a small-scale research inquiry to full specialization in the one role as with full-time "science administrators" or "R & D administrators." What is described as the increasing bureaucratization of science often refers to the growing number of full-time administrative roles and their growing power to affect the course of scientific development. And such bureaucratization, precisely because it involves allocation of resources to the various sciences and to groups and individuals within them, also tends to engage more of the "nonadministrator" scientists in administrative activities: the preparation of prospectuses on work planned and of reports on work done, this in addition to the dissemination of the actual results of scientific investigation.

Although it is often (and loosely included under "administration," a fourth role of the scientist needs to be distinguished from the others since it is basic to the systems of evaluation and the allocation of roles and resources in science. This is the gatekeeping role.[60] Variously distributed

60. As is well known, the notion of the gatekeeper role was introduced into social science by Kurt Lewin in "Forces behind Food Habits and Methods of Change," *Bulletin of the National Research Council* 108 (1943): 65. Alfred de Grazia ("The Scientific Reception System and Dr. Velikovsky," *American Behavioral Scientist* 7 [1963]:38–56) and Diana Crane ("The Gatekeepers of Science: Some Factors Affecting the Selection of Articles for Scientific Journals," *American Sociologist* 2 [1967]: 195–201) refer to editors of journals as "the" gatekeepers of science. This usage is too restrictive; gatekeepers also regulate scientific manpower and the allocation of resources for research.

within the organizations and institutions of science, it involves continuing or intermittent assessment of the performance of scientists at every stage of their career, from the phase of youthful novice to that of ancient veteran and providing or denying access to opportunities.

The operation of the gatekeeper role affects contemporary science in its every aspect. First, with regard to the input and distribution of personnel, these scientists are asked to evaluate the promise and limitations of aspirants to new positions, thus affecting both the mobility of individual scientists and, in the aggregate, the distribution of personnel throughout the system. In American science, at least, and probably in other national communities of science, this gatekeeping function seems to involve a mixture of Turner's types of mobility: contest mobility based on role performance and reinforced by the norm of universalism and sponsored mobility in which elites or their agents help recruit their successors fairly early.[61]

Second, with regard to the allocation of facilities and rewards, the gatekeeper role, at least in the American social structure of science, operates largely through broad- or narrow-spectrum "panels of peers." These panels recommend and, in the usual event, determine the distribution of fellowships, research grants, and honorific awards. The term "panel of peers" refers to the fellow-scientists of assumed competence in the fields in question and not, of course, to age peers, as we shall see in examining the age structures of groups of gatekeepers.

Third, with regard to the outputs of the variously allocated resources, the gatekeeper role is organized principally in the subroles of referees, charged with gauging the validity and worth of manuscripts submitted for publication, and of editors and editorial staff who make the final determination of what shall enter this or that archive of science.[62] Here again we shall want to identify phases of their careers in which scientists tend to be most involved in these roles that help shape the permanent record of scientific work, and to find out whether there are distinctive age-related patterns in their performance of these roles.

61. For the general concepts, see Ralph Turner, "Sponsored and Contest Mobility and the School System," *American Sociological Review* 25 (1960):855–67; for their pertinence to the case of science, see Lowell Hargens and Warren Hagstrom, "Sponsored and Contest Mobility of American Academic Scientists," *Sociology of Education* 40 (1967):24–38; also see Harriet Zuckerman, "Stratification in American Science," *Sociological Inquiry* 40 (1970):243–47.

62. Research on the operation of this role has lately burgeoned; for example, see Crane, "Gatekeepers of Science"; Richard D. Whitley, "The Operation of Science Journals: Two Case Studies in British Social Science," *Sociological Review* n.s. 18 (1970): 241–58; and Zuckerman and Merton, "Patterns of Evaluation in Science: Institutionalization, Structure and Functions of the Referee System," reprinted as chapter 21 of this volume.

Role-sequence and role-allocation in science

As we have noted, individual scientists have their own mixtures of these four roles, according different amounts of time and energy to each of them. At the extremes of specialization, scientists are engaged in one of these roles to the full exclusion of the others; more commonly, they will perform all of them in varying mix. For individual scientists, the question arises whether there are patterned sequences in preponderating roles during the life course. And for successive cohorts of scientists, the correlative question arises whether there are historical changes in the distribution of scientific roles.

In considering these questions, we should note that what is *role-sequence*[63] from the standpoint of the individual moving along the phases of his life course is *role-allocation* from the standpoint of the social system of science. Role-sequences—that is, the succession of roles or role-configurations through which appreciable proportions of people move in the course of their lives—are presumably affected by role-allocation—that is, patterned access to the structure of opportunity to engage in the various roles. The first deals with patterned career-lines, the other with (historically changing) processes and structures of role distribution. Concretely, individual preferences and social system pressures interact to produce the observed historical patterns of role-sequences.

Systematic data on role-sequences are in short supply generally and all the more so in the sociology of science. An approximation to investigation of role-sequences is provided by studies of "occupational careers,"[64] which, however, tend to deal with patterns of mobility from one occupation to another rather than with patterned sequences of role-configurations for individuals remaining within the same occupation. For the field of science, there is a unique set of data assembled by Lindsey Harmon,[65] which traces the succession of role-complements for six cohorts of American scientists receiving their doctorates in ten major fields of science at five-year intervals from 1935 to 1960. Substantively, the Harmon Report provides incomparable clues to the patterned sequences of roles for American scientists in the last generation or so; procedurally, the Harmon data exemplify the great difficulties of disentangling from such cohort analysis the

63. For the general conception of status- and role-sequences, see Merton, *Social Theory and Social Structure*, pp. 434–38.

64. For instructive examples of the difficult art of "career analysis," see Harold L. Wilensky, "Work, Careers and Social Integration," *International Social Science Journal* 12 (1960):543–60; idem, "Orderly Careers and Social Participation," *American Sociological Review* 26 (1961): 521–39; Delbert C. Miller and William H. Form, *Industrial Sociology*, rev. ed. (New York: Harper & Row, 1964).

65. See *Profile of Ph.D.'s in the Sciences.*

TABLE 3
Distribution of Time Assigned to Their Various Roles by Selected Cohorts of American Scientists

Date of Ph.D.	Work period					
	1935	1940	1945	1950	1955	1960
a. Percentage of time devoted to teaching						
1960						(0) 33
1955					(0) 34	(5) 33
1950				(0) 40	(5) 34	(10) 31
1945			(0) 41	(5) 42	(10) 36	(15) 34
1940		(0) 42	(5) 30	(10) 33	(15) 28	(20) 28
1935	(0) 47	(5) 44	(10) 35	(15) 36	(20) 33	(25) 32
b. Percentage of time devoted to research						
1960						(0) 48
1955					(0) 48	(5) 43
1950				(0) 45	(5) 41	(10) 37
1945			(0) 42	(5) 36	(10) 34	(15) 32
1940		(0) 42	(5) 40	(10) 36	(15) 33	(20) 28
1935	(0) 36	(5) 33	(10) 32	(15) 29	(20) 28	(25) 26

components of role-sequences in life course and of historical shifts in role-allocations, difficulties to which we have been alerted by the Riley-Johnson-Foner model of age stratification.[66]

Table 3, drawn from the Harmon report, summarizes the average proportions of time which six cohorts estimate[67] they have assigned to various

66. The cohorts of scientists in this study are thus of one kind identified in chapter 1.1.D of *Aging and Society*, vol. 3: aggregates of people who share not a common date of birth but a common date of entry into the field. Also see ibid., chapter 2 and appendix.

67. Grateful for these incomparable data, we do not here discuss the question of possible response error deriving from the fact that these retrospective estimates of the allocation of time cover periods ranging up to twenty-five years. Harmon is thoroughly aware of the problem, and internal evidence suggests that errors in reporting are random rather than systematic.

Date of Ph.D.	Work period 1935	1940	1945	1950	1955	1960
c. Percentage of time devoted to administration						
1960						(0) 10
1955					(0) 8	(5) 15
1950				(0) 7	(5) 16	(10) 24
1945			(0) 11	(5) 16	(10) 22	(15) 26
1940		(0) 8	(5) 18	(10) 22	(15) 30	(20) 34
1935	(0) 8	(5) 14	(10) 23	(15) 28	(20) 30	(25) 32
d. Percentage of time devoted to other functions						
1960						(0) 8
1955					(0) 9	(5) 8
1950				(0) 7	(5) 8	(10) 9
1945			(0) 6	(5) 7	(10) 8	(15) 8
1940		(0) 8	(5) 12	(10) 9	(15) 9	(20) 10
1935	(0) 9	(5) 8	(10) 10	(15) 8	(20) 9	(25) 10

SOURCE: Harmon, *Profiles of Ph.D.'s in the Sciences*, p. 65 (adapted).
NOTE: Figures in parenthesis indicate number of years since Ph.D.

role-activities at successive periods in their careers. The evidence bears upon both role-sequences in individual life courses and upon social changes in the role structure of science.

First of all, for each cohort, the figures across the rows show a steady decline during the *life course* in the proportion of time assigned to research and a steady increase in the time assigned to administration.[68] Teaching, like research, also tends to decline over the life course, with the interesting exception of 1950, when each cohort increased the relative time devoted to teaching. This deviation from the general life course pattern apparently

68. We report the full table from Harmon but attach no significance to the category of "other functions." At best, this is a catchall with unidentified ingredients. Moreover, it does not appear to be patterned by age in any systematic fashion.

reflects the impact of unique historical events upon the age strata. For the observed historical bump in role-sequence probably represents the additional teaching that came with the rapid expansion in "GI" programs of education just after World War II.

Second, table 3 also reflects *historically changing* patterns in the distribution of time American scientists assign to their several roles at each stage of their careers. This can be seen by inspecting the left-to-right diagonals in the first two parts (a and b) of table 3. Consider, for example, the top diagonal of those having just received the doctorate. Each more recent cohort tends to devote less of its aggregate time to teaching and more to research. The historical trend at each of the other stages of the scientists' careers also approximates this decrease in aggregate cohort time assigned to teaching and the increase in that assigned to research, except for the dramatic departure of the 1945 cohort (note that the 1945 row tends to fall out of line with each of the diagonals in parts a and b of the table). The experience of the great influx of World War II veterans into colleges and universities seems to have left an enduring imprint upon the 1945 cohort which, just entering upon their careers at the time, continue at each succeeding time period to interrupt the general cohort trend of less teaching and more research.

Interestingly enough, the cohort trends in teaching and research are not accompanied by complementary trends in administrative activity (shown in the diagonals of table 3, part c). The absence of any consistent trend here raises some question about the nature of the historically increasing bureaucratization of science. To be sure, in each cohort scientists devote relatively more of their aggregate time to administration as they age. But, age for age, contemporary scientists devote no larger proportion of their aggregate time to administration than did scientists in past years.[69]

A third type of comparison of the figures in table 3, comparison down the columns, reveals for each time period the combined effects of the life-course patterns, cohort trends, and unique historical events we have described. Thus the column for a particular year represents the structuring of scientific roles among the *age strata*. Here we find that although the age strata do not differ substantially in time devoted to teaching, they do show striking differences in research and administration.

At any given time in the past quarter-century, the younger the stratum of scientists, the more of their aggregate time they devote to research and

69. Aggregate data of this kind do not allow us to distinguish between role-specialization in the form of full-time administrators and other changes in the distribution of time among the several role-activities by individual scientists. As noted earlier in this section, both types of change are often and indiscriminately caught up in the phrase "the bureaucratization of science." It should be noted also that changes in the proportions of scientists employed by universities, government, and industry affect the observed historical patterns.

the less to administration. In 1960, for example, this ranges in linear progression from 26 percent of all work time being assigned to research by the 1935 cohort (then twenty-five years past the Ph.D. and presumably in their early 50s) to 48 percent by the most recent cohort of 1960 (then having just received the doctorate and presumably in their late 20s). Put more generally, these results suggest that the social system of science provides more time for the research role to younger than to older scientists. Like the youthful age structure of science generally, this distribution of roles accords with the widespread ideology[70] that holds that "science is a young man's game."

These data, representing aggregate averages for cohorts, of course provide only a rough approximation of individually patterned role sequences among scientists. They do not indicate the *composite* patterns of time that scientists allocate to their various roles at each phase in their careers. Nor, unlike panel data, do they indicate changes in these patterns for individual scientists during the course of their careers.

Partial but suggestive evidence on the individual patterning of roles can be found in the Harmon data.[71] Of particular interest are the age patterns of specialization that can be identified. While there are no notable age differences in the proportion spending full time on some one activity in their current jobs[72]—on teaching or research or administration—there are striking differences among the age strata in the *type* of specialization that does tend to occur. These differences parallel those found in the cohort analysis of table 3 above. Thus the young are far more likely than the old to give the major portion of their time to research; conversely, the older strata are more likely to specialize in administrative roles (while age differences in teaching are not pronounced). For example, Harmon finds that

1. The proportion of scientists devoting no time at all to research on their present job is twice as large in the oldest age category as in the youngest; the proportion devoting full time to research is half as large.
2. Among the oldest, the percentage spending full time in administration is four times as large as in the youngest stratum.

The same general patterns emerge in examining the *composites* of roles performed by scientists in the several age strata. For example:

70. We describe this as ideology since it includes both idea and norm, both what is assumed to be and what should be. It is, of course, only one component in the ideology about the roles of young and old in science.

71. *Profile of Ph.D.'s in the Sciences,* tables 9–11, pp. 19–21, and appendix 6, tables A, B, C, and D.

72. As can be observed by adding the relevant percentages in Harmon's table 9 (ibid., p. 19), 31 percent of the youngest age category, 27 percent of the middle stratum, and 30 percent of the oldest category devote full time to either teaching or research or administration.

1. In the oldest age category, about half who do no teaching also do no research, most of these specializing in administration.

2. In the youngest category, of those who do no teaching, 70 percent spend their time predominantly in research.

The Harmon data contain additional clues that the greater emphasis on research among younger scientists reflects some attrition over the life course of the research role within the composite of roles performed by individuals. Comparing allocations of time on the first job held by scientists with allocations on the current job, Harmon emphasizes the modal tendency toward *persistence* of role-patterning by individuals over the life course.[73] His data also show, however, that among those who do shift, those decreasing the proportion of time devoted to research far outnumber those enhancing their research roles.[74]

Data such as these identify major patterns of role-specialization. But they tell us nothing, of course, about the kinetics of role-sequences and role-retention, the social and psychological mechanisms through which and the structural contexts within which the observed patterns come about. These largely remain matters for speculation.

Mechanisms of role-attrition and role-retention

The ideological accent on youth in science provides part of the context for shifts in roles. In its extreme form, the doctrine holds that the best scientific work is done early in the career with nothing of consequence to be expected after that. P. A. M. Dirac, one of the more powerful minds in theoretical physics, found occasion to express this gloomy version, partly in parody, partly in sadness:

> Age is, of course, a fever chill
> that every physicist must fear.
> He's better dead than living still
> when once he's past his thirtieth year.[75]

On this view, the scientists who have made significant contributions early in their careers burn out soon afterwards. And the many more who have done little in their early years can count on doing even less later on. For

73. Harmon, ibid., pp. 19–20. Note that all cohorts are combined in this portion of the analysis.

74. Derived by comparison of the summated frequencies in the upper-right diagonal (increasers) with the lower-left diagonal (decreasers) of Harmon's table 11 (ibid., p. 21). A similar, though less pronounced, tendency to attrition is apparent through parallel analysis of individual shifts in the teaching role (table 10, p. 20), clearly pointing to administration as the activity compensating for declines in research.

75. It is appropriate that Dirac should have formulated his mathematical theory describing the relativistic electron when he was 26 and that he became a Fellow of the Royal Society at 28 and a Nobel laureate just over the watershed age, at 31.

both, continuing to do research with advancing age is at best an act of self-deception. This extreme form of the ideology typically includes the premise that each scientist has within him a fixed quantum of contributions to make and that this is soon exhausted.[76]

A less severe version of the ideology, reinforced perhaps by Lehman's widely publicized and somewhat misleading data, holds that creative work peaks early in the scientist's life and diminishes more or less rapidly both in extent and consequence. The edge is sometimes taken off this version by noting that what age loses in creative powers, it can gain in mature experience. This provides a rationale for continuing in the research role with a degree of restrained optimism. John von Neumann, known for contributions of the first order to several branches of mathematics, took this to be the case for his own field:

> When a young man, he [von Neumann] mentioned to me several times that the primary mathematical powers decline after the age of about twenty-six, but that a certain more prosaic shrewdness developed by experience manages to compensate for this gradual loss, at least for a time. Later the limiting age was slowly raised.[77]

Without assuming that ideology determines behavior, we should note that the extreme version of the ideology of youth in science thoroughly undermines the case for continuing in the research role, while the moderate version provides it with no great support. All apart from other considerations, this is the sort of ideological climate that we would expect to make for attrition of the research role, just as the Harmon data indicate.

There is reason to believe that the general pattern of shifts from research to other roles holds more for journeymen scientists than for the more accomplished scientists. Sociological theory leads us to expect and scattered evidence leads us to believe that the more productive scientists, recognized as such by the reward system of science, tend to persist in their research roles, forcing death rather than retirement to spell the end to their research careers. One piece of evidence deals with the scientific ultra-elite, the Nobel laureates. Compared with less distinguished scientists matched with them in age, specialty, and type of organizational affiliation, the laureates begin publishing research earlier in their career and continue to publish longer.[78] On the average, the laureates were not quite 25 years old at the

76. The idiom puts it that scientists have no more than a paper or book "in them." Without at all subscribing to the total ideology of youth, Derek J. de Solla Price and Donald DeB. Beaver ("Collaboration in an Invisible College," *American Psychologist*, 1966, pp. 1011–18) press the idiom further by referring to coauthors who manage to "squeeze out" of themselves the fraction of a paper that is in them.

77. Stanislaw Ulam, "John von Neumann, 1903–57," in Fleming and Bailyn, *Intellectual Migration*, p. 239.

78. Harriet Zuckerman, "Nobel Laureates in Science: Patterns of Productivity, Collaboration and Authorship," *American Sociological Review* 32 (1967):392–93.

time of their first papers, while scientists in the matched sample were past 28. What is more in point in the matter of role-retention is the publishing record toward the other end of the career. Of the nine laureates and their matches who had passed the age of 70, all the laureates but only three of the paired scientists continued to publish, indicating that they have more staying power in the research role. In part, this may result from their being subject to consistently greater expectations, both from others in the immediate and extended environment, to remain productive in research and in part, from their having established routines of work, also supported by the environment. One laureate, then past 80, reports that he feels no obligation to continue doing research—as he puts it, "After all, enough is enough"; nevertheless, his papers continue to appear in the scientific journals.[79] The oldest laureate, F. P. Rous, was described as "still hard at work" at the michelangelical age of 87.

Retention of the research role, or its attrition, among scientists ranked high in accomplishment seems to be affected also by their selection of reference individuals and reference groups for self-appraisal. Some take their own prior achievements as a benchmark and conclude that the prospects are slight for their maintaining that standard. They become more receptive to the opportunities for taking up other roles: administering research organizations, serving as elder statesmen to provide liaison between science and other institutional spheres, or, occasionally, leaving the field of science altogether for ranking positions in university administration or international diplomacy. Other eminent scientists take the run of scientists as their reference group. They conclude that even if youthful peaking has occurred for them, they will continue to be far more productive, even on the assumed down slope of their careers, than most other scientists at the peak of their careers.

Query: The generic problem of the determinants of selection of reference groups remains unsolved. Taking the matter of role-retention by scientists as a strategic case in point, we ask: What leads some scientists, highly productive in their youth, to take this as a reference mark and to anticipate *relative unproductivity* in the future, while other scientists, equally productive in their youth, anticipate *relative productivity* as they compare their work with that of most scientists even in their most productive years.

79. In her restudy of fifty-four eminent scientists, seventeen of them over the age of 65 at the time of her revisit, Anne Roe ("Changes in Scientific Activities with Age," *Science* 150 [1965]:313–18) also found that they tended to persist in their research, even when taking up administrative roles.

The value system of science can make for a retention of the research role in spite of the ideology of research as essentially a young man's game. Of the various roles in the institution of science, greatest value is attached to research, theoretical and experiential. As a result, the self-esteem of scientists once effectively engaged in research depends greatly upon their continuing to do research, even though they may be plagued by doubts stemming from the ideology of youth. Beyond that, many scientists, precisely because they are minds trained in scientific inference, realize that even if scientific productivity or creativity does decline with aging for most scientists, it of course remains unsound to assume that this must hold for any particular scientist.

Conducing to retention of the research role is the comparative ambiguity about the kind and number of contributions to knowledge that would justify one's continuing in research. Since few make pathbreaking contributions, even an occasional craftsmanlike piece of work may be enough to maintain the self-conception of being engaged in research. This is particularly the case for academic scientists who, in the aggregate, appear to devote much the same proportion of their time—about one-fifth—to research during the greater part of their active career. It is the scientists in nonacademic employment, whose research productivity is presumably gauged in more utilitarian terms, that successively devote less of their time to research and more to administration.[80] This pattern suggests that the criteria of what constitutes "satisfactory research" differ within the social subsystems of science in academia, industry and government with consequent differences in rates of role-retention.[81]

Patterns of role-retention and attrition probably differ also among the various levels in the social stratification of science. For there are socially stratified differences in opportunity-structure and in socially patterned pressures in science as in other departments of social life. Eminent research scientists are often subject to cross-pressures. On the one hand, in accord with the principle of cumulative advantage, their earlier achievements in research ordinarily provide them with enlarged facilities for research. On the other, the prestige they have gained in the research role often leads them to be sought out for alternative roles as advisors, sages, and statesmen, both within the domain of science and in the larger society.

In the main, however, the socially reinforced commitment to research seems to prevail in the upper reaches of the stratification system. This

80. The data on these patterns are set out in the second career patterns report following up the Harmon report; see the National Research Council, *Careers of PhD's*, p. 53.

81. For apposite observations, see Simon Marcson, *The Scientist in American Industry* (New York: Harper & Row, 1960), and Barney G. Glaser, *Organizational Scientists: Their Professional Careers* (Indianapolis: Bobbs-Merrill, 1964).

occurs even though there appears to be a ratchet effect operating in the careers of scientists such that, once having achieved substantial eminence, they do not later fall much below that level (although they may be outdistanced by newcomers and so suffer a *relative* decline in prestige). Once a Nobel laureate, always a Nobel laureate. But the reward system of science makes it difficult for laureates, if we may put it so, to rest on their laurels. What appears from below to be the summit of accomplishment becomes, in the experience of those who have reached it, only another way station. Each contribution is defined only as a prelude to other contributions. Emphatic recognition for work accomplished, in this context, tends to induce continued effort, serving both to validate the judgment that the eminent scientist has unusual capacities and to testify that these capacities have continuing potential. Such patterned expectations make it difficult for those who have climbed the rugged mountains of scientific achievement to call a halt. It is not necessarily their own escalating Faustian aspirations that keep the more accomplished scientists at work. They are subject to the enlarged expectations of their peers and reference groups. More is expected of them, at least for the time, and this environment of expectation creates its own measure of motivation and stress. Less often than might be imagined is there repose at the top.[82]

Although socially reinforced motivation for continuing in research may be greater for high-ranking scientists, they are far from absent for the rest of us who know ourselves to be, at best, journeymen of science. For one thing, our own more modest contributions can be compared with those of even less distinction, as we select reference groups and individuals that sustain our self-esteem. For another, the prevailing imagery of science as a vast collectivity in which each contributes his bit to build the cathedral of knowledge also helps to maintain the ordinary scientist in his research.[83] Nevertheless, this would not seem to provide the same degree of social reinforcement that accrues to outstanding scientists.

All these observations suggest the hypothesis that attrition of the research role and enlargement of teaching, administrative, and other roles will tend to occur earlier and relatively more frequently among scientists lower in the stratification system of science. So far as we know, there has been no systematic investigation of this conjecture, although Harmon's cohort data for 10,000 scientists could be adapted to the purpose by incorporating

82. This account of the process of socially reinforced aspirations and of consequent role-retention draws verbatim upon "The Matthew Effect in Science," chapter 20 in this volume.

83. On this imagery of science and for evidence bearing on its validity, see Jonathan Cole, "Patterns of Intellectual Influence in Scientific Research," *Sociology of Education* 43 (1970):377–403. And for evidence putting this imagery in question, see Jonathan R. Cole and Stephen Cole, "The Ortega Hypothesis," *Science* 178 (27 October 1972): 368–75.

indicators of standing in the field in the body of data already in hand. However, some evidence does bear tangentially upon the conjecture. Zuckerman[84] provides qualitative evidence for the reinforcing character of recognition in the early years of research for Nobel laureates, and Stephen Cole and Jonathan Cole[85] have found for a sample of American university physicists that the more recognition for their early work received in the form of citations by variously productive physicists, the more often they continued to be productive in research. Since degrees of recognition by the community of scientists make for location in the stratification system, this evidence is at least consistent with the hypothesis.

The patterns of shifting from research to other roles are not all of a kind. They differ phenomenologically and in their social and psychological mechanisms. In one pattern, the shift expresses a change in the values of the scientist or an enlarged access to alternative roles, which, in some sense, are more highly rewarding to him than research. In either case, the change represents a pull of the new role rather than a push from the old. The scientist searches out the shift rather than having it imposed upon him. He does not doubt his continuing competence to do research. He simply prefers another role that seems more significant to him. He may be responding to rapidly changing values in the larger society or modifying his values in more idiosyncratic fashion or simply finding an administrative post, with its better pay and greater power, more attractive to him. He perceives the change as one of extending his scope, perhaps by helping to shape the changing place of science in the society or by helping strategic publics to understand the risks, costs, and benefits of science and science-based technology.

In other cases, the scientist finds that his research no longer measures up to his standards and so takes little satisfaction in continuing with it.[86] He turns to an alternative role. This class of changing roles is sociologically unproblematical, requiring little interpretation.

A superficially similar but actually quite different pattern of shift in roles is that of the private self-fulfilling prophecy. In these cases, the scientist would prefer to go on with research. But he has become per-

84. *Scientific Elite*, chap. 6.

85. Stephen Cole and Johnthan R. Cole, "Scientific Output and Recognition: A Study in the Operation of the Reward System in Science," *American Sociological Review* 32 (1967): 388–89; see also Jonathan Cole and Stephen Cole, *Social Stratification in Science* (Chicago: University of Chicago Press, 1973), chap. 5.

86. We should perhaps remind ourselves that to note the faults in the Lehman kind of data on age and scientific productivity does not mean that there is no relation between the two. The quantity and quality of scientific output may in fact decline for the aggregate of scientists in the later years and, in any case, such declines are known to occur for individual scientists (just as for occasional others, research continues unabated or occasionally expands). We refer here to those scientists who experience a declining research output and so are motivated to take up other roles in science.

suaded that he is approaching an age at which his creative potential, great or small, is bound to decay. Rather than continue in a role in which he believes himself destined to fall increasingly short, he makes a preemptive shift. He assumes new administrative responsibilities, turns more of his attention to teaching, becomes active in the public business of science. Once the premise of his prospective diminishing capacity for research is accepted, the preemptive adaptation becomes entirely sensible. But as with every kind of self-fulfilling prophecy, the question is, of course, whether the original premise leading to the behavior which seemingly validates that premise was sound in the first place.

Another pattern of shift from the research role also involves a self-fulfilling prophecy but in its more consequential public form. Here the shifts in role are system-induced, not personally induced. The process is set in motion not by the individual scientist's own definition of his capacity to continue doing research but by the institutionalized belief that the amount and quality of scientific output generally deteriorate badly after a certain age. To the extent that this belief is incorporated in policy, some older research scientists reluctantly find themselves elevated into administrative posts and others find their facilities for research limited. Subsequent declines in research output with age seem only to confirm the soundness of the policy and are taken as fresh evidence for the general validity of the belief in declining productivity with age.[87]

Both kinds of self-fulfilling prophecy, the self-generated and the socially generated, interact and reinforce each other. Social assessments of role-performance come to be reflected in self-images, and behavior in accord with those self-images tends to make for the patterned social assessments. What interests us here is the possibility that appraisals of the research may turn out to be stratified by age, with younger scientists being especially critical of older scientists. Consider as a case in point the psychological and sociological bases for the ambivalence of apprentices to their masters.[88] In the psychological analysis of the pattern, the apprentice esteems the master and takes him as a role-model while also aiming to replace the master who, after a time, stands in his way. Without assuming that such ambivalence is typical, we can readily identify many instances in the

87. For a statement of the social costs involved in policies of "premature retirement" from research in socialist countries, see Władysław Szafer, "Creativity in a Scientist's Life: An Attempt of Analysis from the Standpoint of the Science of Science," *Organon* 5 (1968):33–34.

88. The following passage on ambivalence draws almost verbatim on Robert K. Merton and Elinor Barber, "Sociological Ambivalence," in Edward A. Tiryakian, ed., *Sociological Theory, Values and Sociological Change: Essays in Honor of Pitirim Sorokin* (New York: Free Press, 1963), pp. 92–93; see also chapter 18 in this volume. A detailed analysis of master-apprentice relations involving Nobel laureates is provided in Zuckerman, *Scientific Elite*, chapter 5.

history of science: Kepler's strong ambivalence toward Tycho Brahe; Sir Ronald Ross's toward his master Manson in the quest for the malarial parasite, his devotion to his teacher pushing him to extravagant praise, his need for autonomy pushing him to excessive criticism. Or consider, appropriately enough, the checkered history of psycho-analysis itself with the secessionists Jung and Adler displaying their ambivalence toward Freud; in sociology (to come no closer home to our own day), the mixed feelings of the young Comte toward Saint-Simon; in psychiatry, the mixed feelings of Bouchard toward Charcot; and in medicine, of Sir Everard Home toward John Hunter; and so on through an indefinitely long list of apprentice-master ambivalence in science.

The probabilities of such ambivalence of apprentices toward masters—or, to put the matter more generally, of younger toward older scientists—presumably differ according to the context provided by the social structure of science. For example, ambivalence may be more apt to develop when, owing to the paucity of major chairs in a field, the talented apprentice finds that he "has no (appropriate) place to go" after he has made his mark other than the place occupied by the master (or others like him). But if the social system of his science provides an abundance of other places, some as highly esteemed as that currently occupied by the stratum of masters, there is less structurally induced motivation for ambivalence. And by the same token, the masters, in the reciprocity of relations, may be less motivated to develop ambivalence toward the apprentices who, in more restricted circumstances, might be competing with them as "premature" successors.

Query: Do age-stratified cohorts of scientists tend to adopt criteria differing in stringency if not kind in assessing the research of others or does a commonality of criteria transcend age differences? To what extent do cohorts agree in judging the research accomplishments and continuing potential of leading scientists in their field? Do these patterns differ according to the "market situation" in the various sciences and within the same science in various social systems that differ in the opportunity-structure for young scientists?

The various patterns of role-change in the life course involve an interplay between the individual's own expectations and those prevailing in the relevant social environment. This means, of course, that role-changes are affected both by developments distinctive to individuals and by trends in their environment. Individuals experience the social correlates of their own aging in particular social contexts. The contexts affect the meaning they attach to those changes and their adaptations to them. Early retirement from the research role should thus have different consequences for successive cohorts of scientists who come upon this experience at differing points in the historically evolving social structure of science. For scientific investigators to turn to the role of science administrator or science educator at a time in which such changes are relatively infrequent is quite another kind of experience than doing so when it has become common. In complementary fashion, both the probability and consequences of such shifts from research to other roles differ according to the changing degree of support —economic, technical, and social—available for research. The rapid growth in such resources has meant, for example, that the advanced graduate student or newfledged Ph.D. can now obtain technical help and services not available to seasoned investigators a generation ago.[89] This change may directly affect the age of entry into consequential research and indirectly affect the competitive positions of the various age cohorts of research scientists.

The Riley-Johnson-Foner model and Pinder's striking formulation of "the noncontemporaneity of the contemporaneous"[90] both suggest to us that the various age cohorts of scientists will tend to perceive the allocation of resources and the role structure of science from differing perspectives. For the newest cohorts, coming into science at a time of abundance, the availability of resources is largely a matter of ordinary expectation. After all, this is all they know from their own direct experience. The older cohorts tend to see this as drastic change, and not necessarily all for the better, as they nostalgically and sometimes invidiously contrast the current affluence to their own difficult days as novice investigators when outside resources were meager and inner resources all-important.

Other age-connected differences in perspective may derive from the allocation of roles within the changing status-structure of science. Younger scientists often come to see the positions of power practically monopolized

89. Polykarp Kusch, "Style and Styles in Research," *Robert A. Welsh Foundation Research Bulletin* 20 (1966): 12.

90. "Die 'Ungleichzeitigkeit' des Gleichzeitigen" is the seemingly paradoxical phrasing adopted by the art historian, Wilhelm Pinder, to introduce his distinction between Gleichzeitigkeit (contemporaneity or temporal coexistence) and Gleichaltrigkeit (coevality, coetaneity or the condition of age cohorts). Consider this germane passage: "Jeder lebt mit Gleichaltrigen und Verschiedenaltrigen in einer Fülle gleichzeitiger Möglichkeiten. Für jeden ist die gleiche Zeit eine andere Zeit, nämlich *ein anderes Zeitalter seiner selbst*, das er nur mit Gleichaltrigen teilt. Jeder Zeitpunkt hat für Jeden nicht nur dadurch einen anderen Sinn, dass er selbstverständlich von Jedem in individueller Färbung erlebt wird, sondern—als wirklicher 'Zeitpunkt,' unterhalb alles individuellen—schon dadurch, dass das gleiche Jahr für einen Fünfzigjährigen ein anderer Zeitpunkt seines Lebens ist, als für einen Zwanzigjährigen—und so fort in zahllosen Varianten" (Wilhelm Pinder, *Das Problem der Generation in der Kunstgeschichte Europas*, 2d ed. [Berlin: Frankfurter Verlags-Anstalt, 1928], chap. 1 at p. 11). This sort of observation on contemporaneous age-cohorts and their perspectives is fully caught up in the Riley-Johnson-Foner model.

by older scientists. For although the professionalization and institutionalization of science and the great growth in the resources of science have multiplied the number of policy-making roles, it may be that the exponential increase in numbers of scientists, all apart from other processes, has tended actually to decrease the proportions of the newer cohorts in these positions and to raise the age at which they enter them.

These few observations on the differing perspectives of younger and older scientists might seem to imply that the relations between these cohorts are dominated by stress, strain, and conflict. But to focus on the structure and processes making for tension and conflict is not, of course, to say that these are all. We have noted the integrative aspects of complementary age-connected roles in the process of socialization in science where, perhaps more often than in other disciplines, the roles of teacher and student soon become transformed into those of research colleagues. The differentiated age structure of research groups provide bases for cooperation as well as conflict. It is probably in the politics of science that conflict between the age strata of scientists runs deep.[91]

4. Gerontocracy in Science

The claim that the organization of science is controlled by gerontocracy is anything but new. Complaints to this effect appeared as early as the seventeenth century and perhaps before. But the vast historical changes in the scale and power of science greatly intensify and complicate the problems of social control.

Dysfunctions of gerontocracy

Neither empirical evidence nor theoretical reason leads us to suppose that rule by elders is more characteristic of science than of other institutional spheres. Gerontocracy may turn out, in fact, to be less marked in science. But it can be argued that rule by elders is apt to be more dysfunctional for science than for other institutions. For although the Ogburnian notion that science and technology develop and change more rapidly than other parts of civilization and culture has yet to be empirically demonstrated,[92] we do know that the values of science call for maximizing the rate of developing knowledge and the procedures and equipments required to advance that knowledge. And with the great expansion of the personnel and resources

91. Greenberg, *Politics of Pure Science*, bk. 1.

92. Ogburn advanced this idea in his classic *Social Change* and developed it further in several monographs. Its validity has been questioned, principally by Sorokin; see William F. Ogburn, *Social Change* (New York: Viking Press, 1922), and Pitirim A. Sorokin, *Social and Cultural Dynamics*, 4 vols. (New York: American Book Co., 1937), vol. 4.

of science, scientific knowledge has for some time been growing at an accelerating rate. Now it is an old and plausible sociological maxim, though one more often announced than confirmed by actual investigation, that the higher the rate of social and cultural change, the less the advantage of age, with its obsolescent experience.[93]

It has been further argued, most emphatically by the distinguished physicist J. D. Bernal, whose own major contributions to crystallography have continued in his seventh decade, that "the advances in basic conceptions have become so rapid that the majority of older scientists are incapable of understanding, much less of advancing, their own subjects. But nearly the whole of what organization of science exists, and the vital administration of funds, is in the hands of old men."[94] Bernal suggests, moreover, that as science expands in numbers, complexity, and influence, and as it becomes more closely linked with government, industry, and finance, its control is increasingly exercised by older scientists.

Plausible as these observations are, they have yet to be systematically investigated. The fact is that we do not know the comparative extent of gerontocracy in science and in other principal institutional spheres. Nor do we know whether the vast expansion of science has brought with it enlarged control by gerontocrats. Nor, finally, do we know whether gerontocracy is more dysfunctional[95] for the development of science and for the society than alternative forms of age-patterned control, such as proportional age representation or, at the other extreme, juvenocracy.[96] Since comprehensive evidence on these complicated questions is absent, here are a few straws in the wind.

93. The observation has been made in one form or another through the centuries. Here is how Roberto Michels put it in 1911: "The ancient Greeks said that white hairs were the first crown which must decorate the leaders' foreheads. Today, however, we live in an epoch in which there is less need for accumulated personal experience of life, for science puts at every one's disposal [such] efficient means of instruction that even the youngest may speedily become well instructed. Today everything is quickly acquired, even that experience in which formerly consisted the sole and genuine superiority of the old over the young. Thus, not in consequence of democracy, but simply owing to the technical type of modern civilization, age has lost much of its value, and therefore has lost, in addition, the respect which it inspired and the influence which it exercised" (Michels, *Political Parties* [1st German ed., 1911; New York: Free Press, 1949], p. 76).

94. J. D. Bernal, *The Social Function of Science* (New York: Macmillan, 1939), p. 116 and also pp. 290–91.

95. Recall the remark, advanced in wry or acerb mood, that gerontocracy may even be a good thing in science; it leaves the young productive scientists free to get on with their research and helps to occupy the time of those who are no longer creative.

96. To the extent that existing social structures tend to be reflected in language, there is perhaps a certain interest in noting that while the word "gerontocracy" has been around for at least two centuries, the word "juvenocracy" appears here, so far as we know, for the first time. It is, unfortunately, a hybrid. But a language which has absorbed such inelegant hybrids as "electrocution" and even "sociology" can surely make room for a much-needed another, such as "juvenocracy."

Evidence of gerontocracy: the National Academy of Sciences

Consider the age composition of the National Academy of Sciences, the influential organization of scientists established during the Civil War and designated by Congressional charter to advise the federal government on matters of science. Apart from its own membership of some 900, the Academy draws upon thousands of other scientists through its operating adjunct, the National Research Council. Designed as an honorary society as well as an advisory body, the Academy is not likely to have a membership numerically representative of the entire national population of scientists: in regional distribution, university affiliation, age, or anything else.

The average age of Academy members is 62, with about a quarter of them being 70 or older. In 1969 three-quarters of the members of advisory committees and panels of the National Research Council were over 45; a third over 55. This contrasts with the median age of 41 for all scientists (holding doctorates) in the United States in 1968, with a quarter of these being under the age of 35. Contrasting age distributions such as these give high visibility to the pattern of gerontocracy.[97]

The elite character of the National Academy plainly affects its age composition. Scientists are seldom elected to membership on the basis of a single contribution to science, however outstanding; a continued record of contributions is ordinarily required. Young talented scientists are left to ripen on the vine before they are picked for membership. Moreover, scientists drawn from the various sectors of employment are evidently judged to meet the Academy's criteria for membership at different ages, as can be seen from figures in table 4.

The scanty data we have assembled on the National Academy indicate no continuing historical trend toward recruiting older scientists. As early as the turn of the century, when the astronomer George Ellery Hale was elected to the Academy at the very early age of 35, he described it, in the words of a friend, "as more interested in keeping young men out of its membership than in acting as a vital force in the scientific development of the United States."[98] The mean age at time of election continued to rise until 1940 but has since remained fairly constant at a somewhat lower level as we see in table 5.

Contrary to first impression, historical patterns of this sort may help account for the belief that positions of prestige and power in science are

97. Concerned with this and kindred problems, the National Research Council has established a panel to examine the composition of advisory committees. These figures are drawn from the preliminary report of that panel. [The full report has since been published: Committee on the Utilization of Young Scientists and Engineers in Advisory Services to Government, *The Science Committee* (Washington, D.C.: National Academy of Sciences, 1972.]

98. Frederick W. True, *A History of the First Half-Century of the National Academy of Sciences* (Washington, D.C.: National Academy of Sciences, 1913), p. 73.

TABLE 4
Mean Age at Election to National Academy of Sciences According to Organizational Affiliation of Scientists, 1863–1967

Affiliation	Mean age	Number
Major universities	48.9	843
Government	51.5	141
Other universities and colleges	51.8	285
Industry	53.3	70
No affiliation	53.7	54
Retired	66.8	12
		1405
No information		8
Total		1413

SOURCE: Zuckerman, *Scientific Elite*, chapter 6.

TABLE 5
Mean Age at Election to National Academy of Sciences, 1863–1967

Time of election	Mean age	Number
Before 1900	47.0	195
1900–19	49.2	158
1920–39	51.7	252
1940–59	50.5	522
1960–67	50.7	286
		1413

SOURCE: Zuckerman (unpublished analysis).

increasingly held by older people. For even in the past half-century when the average age at which they acquire these positions has stabilized or declined somewhat, this has been occurring in a period when exponential growth has been producing an increasingly youthful population of scientists. This results in widened discrepancies of age between the governing and the governed and might be enough to produce a sense of increasing gerontocracy. Moreover, when, as in the case of the National Academy, the status, once acquired, is retained for life,[99] the aging of the group is encouraged by the increasing longevity of its members.[100]

The age-distribution of those occupying the positions of power in science does not tell us, of course, how that power is exercised. Systematic inquiry, rather than swift assumption, is needed to find out whether there are age-

99. Members of the Academy recently rejected the proposal that they should become emeriti after the age of 75.

100. The longevity of college graduates has been increasing in the United States for at least the past century.

patterned differences in policies and the exercise of power. Such studies have yet largely to be made.

Exercise of power: the referee system

One recent study of the referee system in science[101] touches upon the question. Drawing upon the archives for the nine years 1948–56 of the *Physical Review*, the outstanding journal in physics, the study examines the behavior of scientists of differing rank and age in the gatekeeper's role. The referee system calls for evaluation of manuscripts by experts on their subject. It comes as no surprise, therefore, that referees for the *Physical Review* were drawn disproportionately from physicists of high rank.[102] Compared with the 5 percent of the 1,056 authors (themselves in some measure a selected aggregate),[103] almost 12 percent of the 354 referees assessing their papers were in the highest rank. Moreover, these 12 percent contributed one-third of all referee judgments. They refereed an average of 8.5 papers compared with 3.8 for the referees of intermediate rank and 1.4 for the rank-and-file. And although 45 percent of the referees were under the age of 40, thus giving major responsibility to the relatively young, we know that physicists are altogether a youthful aggregate and research physicists particularly so.[104] Fully 74 percent of the papers submitted to the *Physical Review* came from physicists under the age of 40.

The referees, then, are older and higher in prestige and rank than the authors or the general population of physicists. But, as we have noted, such a skewed age distribution among those holding power is simply a

101. Harriet Zuckerman and Robert K. Merton, "Patterns of Evaluation in Science," *Minerva* 9 (1971): 92–94; reprinted as the preceding chapter in this volume, it does not examine age-related patterns of referee behavior.

102. In the first rank are those physicists submitting manuscripts who, by the end of the period (1956), had received at least one of the ten most respected awards in physics (such as the Nobel prize, or membership in the Royal Society or in the National Academy of Sciences). Physicists of the second rank, although they had not been accorded any of the highest forms of recognition, had been judged important enough by the American Institute of Physics to be included in its archives of contemporary physicists. The remaining contributors comprise the third rank in this hierarchy. Referees are ranked by the same criteria. (For details see "Institutionalized Patterns of Evaluation in Science," chapter 21 in this volume.)

103. The special nature of this sample of authors must be understood as (a) resulting from considerable preselection through decisions to produce and submit manuscripts; (b) consisting only of sole authors of manuscripts, with joint authors excluded; (c) consisting of a 20 percent sample of third-ranked contributors, but of *all* first- and second-ranked physicists submitting singly authored manuscripts during the study period. Had the sample included all single authors of every rank, fewer than 2 percent would be included among the first rank.

104. It will be remembered from the first part of this paper that physics has the lowest median age among the several major fields of science and from our discussion of role-sequence that the physicists engaged substantially in research are the youngest of the lot.

static indicator of structure; it provides no information about the functioning and consequences of that structure. Age-distribution does not in itself represent gerontocracy. For even when used descriptively rather than invidiously, the word "gerontocracy" ordinarily carries with it the notion that power disproportionately placed in the hands of the elders comes to be used to their own advantage or, in more moderate version, that it results in policies and decisions that differ drastically from those that are or would be adopted by younger power-holders. In the case of the gatekeeper role, we want to know, then, whether the behavior of referees is systematically affected by their own age and rank as well as by the age and rank of authors.

One piece of evidence takes us a certain distance toward gauging the extent to which the rejection and acceptance of manuscripts for publication was affected by the standing of referees and authors. In examining this evidence, we should note again that eminence in science derives largely from the assessed quality of past and not necessarily continuing scientific accomplishments. And we have found that, in science, as in other institutional spheres, positions of power and authority tend to be occupied by older men. From these joint patterns, it would seem that if sheer power and eminence greatly affect the decisions of referees, then manuscripts submitted by older eminent scientists should have the highest rate of acceptance.[105]

But at least in physics, the distinctively young man's science, this is not what we find. As we have seen in chapter 21 of this volume, "it is not the older scientists whose papers were most often accepted but the younger ones. And the age-graded rates of acceptance hold within each rank in the hierarchy of esteem. . . . Both eminence and youth contribute to the probability of having manuscripts accepted; youth to such a degree that the youngest stratum of physicists in the third rank had as high an acceptance rate as the oldest stratum of eminent ones whose work, we must suppose, was no longer of the same high quality it once was."[106]

105. On the general hypothesis, see Storer, *The Social System of Science* (New York: Holt, Rinehart & Winston, 1966), pp. 132–34. This hypothesis assumes that the identity of the authors of manuscripts is known to the referees; this is the case for the *Physical Review*, which does not try to provide for anonymity of authors, since, it is maintained, this cannot be achieved in most cases. Referees, however, are generally anonymous.

106. See above, pp. 488–89; also see table 4 of chapter 21 in this volume. Some caution must be exercised in this comparison, however, since most of the scientists for whom no information on age was available are in the third rank. These scientists have a relatively low acceptance rate and, should they also be disproportionately young, could depress the acceptance figure for the younger third-ranking scientists below that for the oldest first-ranking.

This is a first indication that the "gerontocratic" body of gatekeepers does not exercise its power by denying or restricting access of younger physicists to publication in the most widely read and most influential journal in the field. This still leaves open the possibility that it is not the age of the author as such but his age *relative* to that of his referees which systematically influences appraisals of his manuscripts. Such biases in judgment might take various forms, depending upon the pattern of relative age,[107] just as we have seen to be the case, in chapter 21, with the pattern of relative rank. (To emphasize the identity of patterns in relative prestige and relative age, we adopt the same language in this formulation of alternative hypotheses in both papers.)

When referees and authors are age peers, an hypothesis of *age-stratum-solidarity* would have it that referees typically give preferential treatment to manuscripts just as a counterhypothesis of *age-stratum-competition* would have it that, under the safeguard of anonymity, referees tend to undercut their rivals by unjustifiably severe judgments.

When authors are older than referees, an hypothesis of *age-status-deference* would hold that the referees give preferential treatment to the work of the older established scientists, just as a counterhypothesis of *age-status-envy* would have them downgrade the work of older scientists.

And when referees are older than authors, an hypothesis of *sponsorship or patronage* would maintain that referees are unduly kind and undemanding, while a counterhypothesis of *age-oppression* would have them overly demanding of the young.

Differing in other respects, these six hypotheses are alike in one: they all assume that the relative age of referee and author significantly biases the role-performance of the gatekeepers, either in favor of the author or at his expense. More concretely, all the hypotheses assume that the rates of acceptance of manuscripts submitted by each age stratum of authors will differ according to the age of the referees passing judgment on them.

The data assembled in table 6 seem to run counter to most of these hypotheses.[108] For the most part, the relative age of author and referee has no perceptible influence on patterns of evaluation. With one exception, both younger and older referees are more likely to accept the work of younger authors. And each age stratum of authors, again with one excep-

107. On the concept of relative age, see the Riley-Johnson-Foner model (chapters 1.1.D and 10.1; also 9.2 by Hess, in vol. 3 of Riley and Foner, *Aging and Society*) and the theoretical analysis by S. N. Eisenstadt in *From Generation to Generation: Age Groups and Social Structure* (New York: The Free Press, 1956), chap. 1 passim.

108. The separate effects of each of these hypothetical processes cannot, of course, be weighed by examination of a single set of cross-sectional data, since several tendencies may operate in tandem or in opposition to produce observed proportions.

TABLE 6
Referees' Decisions to Accept Manuscripts, by Age of Authors and Referees

	Age of referees					
	Under 40		40 and over		Total judgments by referees	
Age of authors	% Acceptances	No. judgments	% Acceptances	No. judgments	% Acceptances	No. judgments
20–29	59	106	76	136	68	242
30–39	63	193	63	189	63	382
40–49	63	65	58	71	60	136
50+	43	42	43	61	43	103
No information	53	106	52	96	52	202
All ages	58	512	61	553	60	1065

SOURCE: *Physical Review*, 1948–56.

NOTE: The data refer to the number of judgments, not papers, made by 344 external referees and do not include judgments by the two editors. The table omits 18 cases in which there is no information on the age of the 10 referees judging them. Since papers judged exclusively by the editors are omitted, the analysis is based on judgments of fewer than half the total papers reported in table 4 in chapter 21 of this volume.

tion, has the same proportion of papers accepted by referees of differing age. Interestingly enough, the pattern of evenhanded treatment by older and younger referees holds even for that class of physicists who were not well enough known to have their ages listed in any of the standard registries of scientists.

The one exception to the general pattern appears in the youngest stratum of authors who have more of their papers accepted by older referees than younger ones. This tentatively identified exception is consistent with the hypothesis of acute competition within the age-cohort in the earliest phase of their careers and with the hypothesis that older scientists are less demanding in assessing the work done by progressively younger scientists. The data allow no unequivocal choice between the hypotheses. Either or both may obtain. In any case, neither the general pattern nor the limited departure from it exhibits distinctively gerontocratic patterns of evaluation among the gatekeepers of this major scientific journal.

This particular case is enough to suggest the general point: a skewed age distribution of scientists assigned authoritative roles is one thing; what they do in exercising their authority can be quite another. But, of course, we cannot conclude from this one inquiry that there are no patterns of age-graded evaluations or policy decisions in science. Much more research will be needed to examine and develop this major question. For example, just as the several sciences differ in age structure, so they may differ both in the age distribution of authoritative roles and in age-patterned performance of those roles. We hazard the conjecture that the more theoretically codified the science the greater the consensus among the age strata in their patterns of evaluation. The more codified disciplines such as physics should exhibit less disparity between age cohorts than less codified disciplines such as sociology on all manner of evaluations: the comparative significance of problems requiring investigation and of contributions to the field as well as such questions of science policy as the allocation of resources to various kinds of research.

It should be noted also that the question of gerontocracy in the formation of broad science policy remains as much a matter of conjecture as the question of gerontocracy within scientific disciplines. We know, for example, that the mean age of members of PSAC (President's Science Advisory Committee) has been about 50, with the Eisenhower advisors being somewhat older and the Kennedy advisors somewhat younger. But we do not know how far the age composition of this and of other advisory and policy-making groups affects the substance of science policy.[109] In exploring this

Year		Mean ages of members of PSAC	No. of advisors
1958	Eisenhower	53.8	(18)
1962	Kennedy	49.0	(17)
1965	Johnson	50.3	(15)
1969	Nixon	50.5	(11)

question, we need to distinguish the age composition and the rates of turnover of these influential groups, recognizing that each may have its independent effect. It would come as no surprise to find that optimum science policy is apt to be developed neither by gerontocracy nor by juvenocracy but, like the community of scientists itself, by age-diversified meritocracy.

5. Age, Social Stratification, and Collaboration in Science

The extent to which significant interaction takes place within age strata and between them, together with the consequences of such patterns, are no better known for the domain of science than for most other institutional spheres. Although there has been no investigation to determine which activities in science tend to be age-segregated or age-integrated, it is evident that some basic functions of science are served through institutional arrangements involving interaction between age strata rather than separation of them. First among these is, of course, education and training, both in the narrow sense of transmission of knowledge and skills and in the

109. On the general issue, see Rose and Rose, *Science and Society*, pp. 266–68.

broader sense of socialization involving the transmission of values, attitudes, interests, and role-defined behavior. Since he has himself been variously engaged in the process of professional socialization, just about every scientist has his own opinion about how that process actually works. Yet the plain fact is that there has been little methodical investigation of that process in science.[110]

Queries: Which components in the culture of science are principally transmitted by older to younger scientists? And which are largely acquired from age peers? Do these age-channeled streams of socialization merge or diverge? Which of the values, interests, and patterns of behavior derived from differing age strata are mutually supporting, complementary, or at odds? Do the observed patterns of socialization tend to persist so as to be much the same for successive age cohorts or are they subject to change, among other things, in response to the changing boundaries, technology, problematics, and substance of the sciences? How does the age-patterned process of socialization differ among the various sciences and these, in turn, from socialization in other fields of learning (such as the humanities and technology)?

Growth of research collaboration in science

Just as the early years of education in science generally provide in varying degree for interaction between age strata, so in particular with that advanced form of socialization that takes place through collaboration in research.[111] This form of socialization takes on growing importance as the social organization of scientific inquiry has greatly changed, with collaboration and research teams becoming more and more the order of the day. One pale reflection of this change is the sustained growth in the proportion of scientific articles published by two or more authors. Table 7 shows each successive decade of this century registering a higher percentage of multi-authored papers in the physical and biological sciences.[112] The social sciences begin this practice later and sparingly but then rapidly increase the

110. But see Howard S. Becker and James Carper, "The Elements of Identification with an Occupation," *American Journal of Sociology* 21 (1956): 341–48, for the case of physiologists; Ralph Underhill, "Values and Post-College Career Change," *American Journal of Sociology* 72 (1966): 163–72, for physical, biological, and social scientists; and Lydia Aran and Joseph Ben-David, "Socialization and Career Patterns as Determinants of Productivity of Medical Researchers," *Journal of Health and Social Behavior* 9 (1968): 3–15.

111. This discussion of age and collaboration in research draws extensively on Harriet Zuckerman, "Nobel Laureates in the United States: A Sociological Study of Scientific Collaboration" (Ph.D. diss., Columbia University, 1965), pp. 394–96.

112. For data based on other samples showing similar results, see E. L. Clarke, "Multiple Authorship: Trends in Scientific Papers," *Science* 143 (1964): 822–24. We do not examine here the relation between actual patterns of research (individual,

TABLE 7
Percentage of Multiauthored Papers in the Physical and Biological Sciences, Social Sciences, and Humanities, 1900–59

Date of publication	Physical and biological sciences	Social sciences	Humanities
1900–09	25 (928)	—	—
1910–19	31 (1,685)	—	—
1920–29	49 (2,148)	6 (2,643)	1 (1,822)
1930–39	56 (3,964)	11 (3,905)	2 (2,088)
1940–49	66 (4,918)	16 (4,328)	2 (1,972)
1950–59	83 (9,995)	32 (6,605)	1 (2,304)
Total	66 (23,639)	20 (17,481)	1 (8,186)

SOURCE: Zuckerman, "Nobel Laureates in the United States," pp. 76–77 (adapted).
NOTE: The figures were compiled by counting the number of authors of articles appearing in a sample of journals for two of every ten years. The physical and biological sciences include the fields of physics, chemistry, biology; the social sciences include anthropology, economics, political science, psychology, and sociology; the humanities, history, language and literature, and philosophy.

rate of collaboration. Both contrast with the humanities which have practically no place for collaborative research reported in scholarly articles.

Rank-stratified rates of collaboration

Imperfect indicators of the actual organization of research as they are, the data on multiple authorship nevertheless raise some questions and suggest some conjectures. Does the practice of collaborative research obtain to the same extent on all levels in the social stratification system of science? More concretely, how does this stand with an institutionally identified elite, such as Nobel laureates, compared with the collaborative practices found in a sample of scientists matched with them in terms of age, field of specialization, and type of organizational affiliation? And how do the rates of collaboration compare at various ages and phases in the scientific career?

As a crude approximation to an answer—crude since the data do not allow us to compare age cohorts throughout their life course—table 8 presents age-specific rates of collaboration for the laureates and their less elevated counterparts. Laureates are somewhat more given to collabora-

with and without assistants; collaboration of varying numbers of peers; small-team research and large-scale research) and the number of authors of papers published by these types of research formations. There is evidence that research patterns and size of author-sets are correlated, though not as closely as is sometimes assumed by the many investigators who have used authorship data as indicators of research practices. The question is considered in some detail in Zuckerman, "Nobel Laureates in the United States," chap. 5.

TABLE 8
Percentage of Multiauthored Papers by Age of Publication, for Nobel Laureates and a Matched Sample of Scientists

Age at publication	Laureates	Matched sample
20–29	58 (523)	40 (288)
30–39	65 (1,382)	55 (756)
40–49	66 (1,641)	53 (590)
50–59	60 (1,198)	51 (622)
60 or over	55 (768)	46 (264)
Total	62 (5,512)	51 (2,520)

SOURCE: Zuckerman, "Nobel Laureates in the United States," p. 395.

tion, with 62 percent of their papers being multiauthored compared with 51 percent of those by the age-matched sample. The difference holds, moreover, at every age. Table 8 also exhibits a slight curvilinear relationship between age and publication and multiauthorship for both laureates and the matched sample. We cannot explain these patterns, but other evidence enables us to speculate on their sources.

Consider first the seeming curvilinear pattern in which collaboration appears to be more frequent in the middle years. We say "seeming" pattern not because we really doubt this set of data but only because we know that more extensive data on age cohorts are required to *establish* the authenticity and to discover the generality of the pattern. It must be admitted also that we give credence to the numerical data in hand for the worst of reasons: they tally with our conjectures about age-connected processes making for collaboration.

These processes can be reconstructed in terms of age-patterned opportunities and age-patterned motivations for collaboration. As novices being inducted into the mysteries of the craft, young scientists who collaborate at all with mentors will typically do so with only one of them at any given time. Beyond that, the young scientists, sometimes at the urging of their sponsors, are motivated to do their own work and to establish a public identity within the field by publishing papers of their own. Not quite paradoxically, the motivation of beginning scientists to publish single-authored papers may only be strengthened by the great historical increase in papers with several, sometimes many, "authors." For the distinctive contributions of the individual get lost in the crowd of scientists putting their names to the paper and this, as they know, is especially damaging for young scientists who have not published independent work that testifies to their abilities. On this view, the smaller proportion of collaborative papers published in the early years of the career results in part from the often stressful operation of the reward system that has developed in science.

Toward the other end of the career, the dropoff in published research, which we have noted in the data put together by Wayne Dennis[113] may mean that collaborators are no longer as available as before. It is in this phase also that scientists often turn to broader "philosophical" or "sociological" subjects of a kind that have little place for collaboration.

It is the middle years, then, that presumably provide both the greatest opportunity and deepest role-induced motivation for collaborative work. Should the curvilinear pattern of collaboration turn out to be fairly general, it need not be counteracted by the historical trend toward more and more collaboration in the sciences; for the reasons we have indicated, it may even become more marked in successive age cohorts.

Consider now the consistently higher rates of collaboration among laureates at each phase of their life course. Throughout our interpretation, we make the rather undemanding assumption that, *on the average*, laureates exhibited evidence of greater talent for research than a random assortment of other scientists of their age in the same field. This perceptible difference would have set certain consequential processes in motion—processes such as self-selection and selective recruitment. In their twenties, as their capacities became identified, laureates-to-be were more often selected as apprentices by scientists of assured standing. (Although it would be too much to say that laureates are bred by laureates, it is the case that forty-four of the eighty-four American prizewinners worked, as younger men, under a total of sixty-three older laureates.) There is reason to assume that these masters were more willing to grant coauthorship to their apprentices than were those of less elevated and secure standing.[114] This would make for the higher rate of published collaboration by laureates-to-be in their youth.

The life chances of scientists are greatly improved by having their substantial abilities identified early.[115] By the time they were in their thirties, every laureate had a position in a major university or research laboratory

113. See Dennis: "Age and Productivity Among Scientists"; "Age and Achievement"; "Age Decrement in Outstanding Scientific Contributions"; and "Creative Productivity between the Ages of 20 and 80."

114. Only one laureate complained that he had been deprived of authorship by his senior collaborator when, in his judgment, it was deserved. Far more often, the laureates reported what they perceived as generous treatment in the matter of coauthorship with their typically eminent sponsors.

115. On the bias in favor of precocity built into current institutions for detecting and rewarding talent, see Alan Gregg *For Future Doctors* (Chicago: University of Chicago Press, 1957). The crucial point, which holds in the domain of science as well as in the field of medical practice of which Gregg writes, is this: *"once you have most of your students of the same age*, the academic rewards—from scholarships to internships and residencies—go to those who are uncommonly bright for their age. In other words, you have rewarded precocity which may or may not be the precursor of later ability. So, in effect you have unwittingly belittled man's cardinal educational capital—time to mature" (italics added). For further sociological implications of this institutionalized bias, see chapter 19 of this volume.

providing a micro-environment of other scientists in his specialty. By that time many of the future laureates were making the status transition from junior to senior collaborator. They had acquired resources for research enabling them to surround themselves with younger scientists wanting to work with them on the problems in hand.

Re-enacted roles in age cohorts

As they move into the role of senior collaborators, the laureates seem to reproduce the same patterns of collaborative work with youngsters that they themselves experienced when they were young. This may turn out to be one of several kinds of *reenactment of role-defined patterns of behavior* at successive stages in the careers of scientists, especially those who occupy statuses comparable to those of their masters in the past. They are in a position to attract promising young scientists whose contributions are sufficient to merit coauthorship, much as the laureates, when they were young, were also included among the authors of papers. They are also in a position, even before receiving the Nobel prize (since most of them were eminent before being accorded that ultimate symbol of accomplishment), to exercise cost-free noblesse oblige, the generosity expected of those occupying undisputed rank, by granting authorship even to junior collaborators who, in the given case, may not have contributed much.

We cannot demonstrate that the laureates are more apt than less distinguished scientists to acknowledge the contributions of junior associates, since we do not know how much the younger men had actually contributed. We can, however, compare the *degree* of recognition given to collaborators on jointly authored papers. We can approximate a check on this model of the laureates' re-enactment of collaborative roles as they move through their career by comparing the variously *visible name-orders* of authors of joint papers published by the laureates and by scientists in the matched sample (who, we now report, were matched not only for age, specialty, and organizational affiliation but also for the initial letter of their last name).[116] A prevalent type of name-ordering gives prime visibility to the first author.[117]

The evidence is consistent with our model of complementary roles being re-enacted in the course of the life-work-cycle.[118] The laureates-to-be, when

116. This is designed, of course, to control for variations that would otherwise occur in cases of alphabetical ordering of authors.

117. On the social symbolism of name-ordering among authors of scientific papers, see Zuckerman, "Patterns of Name-Ordering among Authors of Scientific Papers: A Study of Social Symbolism and Its Ambiguity," *American Journal of Sociology* 74 (1968): 276–91.

118. It should be emphasized that the samples consist of scientists working within a particular historical and institutional context: one sample comprises forty-one of the fifty-five laureates at work in the United States (in 1963); the matched sample was drawn from *American Men of Science*. Obviously the attribution of authorship

they were in their twenties, were first authors on nearly half of all their collaborative papers at the same time that scientists in the matched sample were first only a third of the time. In papers coauthored with their laureate masters, the pattern is even more marked, with the young scientists being first authors in 60 percent of the papers and the laureates in only 16 percent.[119] Moving into the role of senior collaborator, the laureates, by the time they were in their forties, reduplicate the pattern taking first authorship on only 26 percent of their collaborative papers at the same time that the scientists in the matched sample do so in 56 percent of their collaborative papers. The negligible cost of this kind of noblesse oblige for scientists who have made their mark is put in so many words by a laureate in biochemistry: "It helps a young man to be senior author, first author, and doesn't detract from the credit I get if my name is farther down on the list."

Substantial qualitative evidence obtained in interviews with laureates confirms that this kind of re-enactment of complementary roles often occurs when they have attained a status like that of their own masters. But this fact does not rule out other, not necessarily incompatible, interpretations of the numerical evidence on first-authorship. The dual pattern of authorship might also reflect age-associated changes and rank-stratified differences in the extent of contributions to collaborative papers. In their youth, the laureates-to-be might in fact have contributed more to jointly published papers than their age peers in the matched sample and so appear as first author more of the time. And, in their maturity, the laureates, having attracted talented youngsters, might simply be re-experiencing the same phenomenon, this time from the perspective of the senior role, with their young collaborators making prime contributions and so being accorded first place. Correlatively, the age-matched scientists of less distinction, in their youth, will have contributed less and received first authorship less often in their collaborative papers with their less distinguished mentors, just as, in their maturity, they re-enact the pattern by attracting, on the average, less talented youthful collaborators than those coming to the laureates and so would themselves turn up more often as first authors. It is the processes of self-selection and selective recruitment operating within the context of the reward system of science, rather than any autarchic scientist-playwright, which recreate this drama in many times and places

---

would be quite different in institutional frameworks where all or much of the research done in a laboratory or department is regularly ascribed to their chiefs. This is just another instance of institutional contexts serving to pattern interpersonal relations.

119. Neither laureates-to-be nor their laureate masters were first authors in the remaining 24 percent of papers. This situation is in marked contrast to papers coauthored by scientist peers, both of whom later became laureates. For these papers, one future laureate is just as apt as the other to be first author in papers having at least three authors.

with the plot and roles remaining intact, and the only change being that the inevitably aging members of the cast now play, in the style of their mentors, roles complementary to the ones they played in their youth.

Age, recognition, and the structure of authority in science

The observed patterns of authorship might therefore involve the re-enactment of complementary roles at different phases in the career in an apparently different way than we had at first supposed. They might reflect the objective situation of differing extent of contribution rather than the exercise of noblesse oblige that came with established standing. In broader theoretical perspective, however, these hypotheses turn out to be much the same. They bring us back to the general idea, much emphasized in our discussion of gerontocracy, that the age distribution of power and authority in science, as elsewhere, is only a static structural fact and does not, in itself, tell us much about how that power and authority are actually exercised.

In the matter of deciding on authorship and name-order as symbolic of contribution, it is generally the senior investigator who has the authority. The exercise of that authority is hedged in by norms and by constraints of maintaining a degree of cooperation in the research group. As the laureates and the matched sample of scientists take over control of these decisions, they apparently do not exercise raw power by putting themselves uniformly in the forefront. At the least, the data suggest, they tend to accede to the norms governing authority; at most, and especially when secure in high rank, they engage in cost-free supererogation.

This much can be said then about our specimen of interplay between social stratification and age stratification in science. Having moved early into the higher reaches of the opportunity-structure, the laureates are more apt to collaborate at every age than other investigators of less eminence. Their tendency toward collaboration we take to be reinforced by their ability to contribute enough to merit association and coauthorship with masters in the field when they are young and by status-supported dispositions to share authorship with the young when they are mature or old. And, to repeat, these patterns obtain within an institutional framework that calls for ascribing credit for research on the basis of contribution rather than having it uniformly assigned, as in an authoritarian framework, to the head of a department or laboratory.

But, as we have suggested earlier in this section, the change toward Big Science, partly reflected in the growth of multiauthorship, makes for a change in the structure of power and authority in science where, goodwill, noblesse oblige, and normative constraints notwithstanding, it becomes increasingly difficult and sometimes impossible to gauge the contributions of individual scientists to the collective product of ever larger groups of investigators. The possible consequences of this structural change on the information- and reward-system of science have been strongly formulated by Ziman:

> It is obvious, in the first place, that there is a grave threat to the convention of awarding promotion, or other forms of recognition, on the strength of published work. The mere fact that a candidate for a lectureship in elementary particle physics has his name amongst the dozens of 'authors' of some significant discovery says little about his scientific skill. In the long run, the leader of such a team gets the credit for its contributions to knowledge, but he must be already the selected and tested boss of a big group. Evidence of ability at a more junior level can only be assessed within the framework of the project itself, just as it would be in an army, a civil service or other bureaucracy. This . . . puts direct power into the hands of the seniors, and opens the way to careerism, personal autocracy, and other evils, as well as giving the advantage to the 'other-directed' personality, at the expense of those protestant virtues of being 'inner-directed' which have contributed so much, in the past, to the scientific attitude. . . . [O]ne of the primary functions of the conventional communication system is losing weight. The necessity of maintaining an open market for the creations of the individual scholar, as objective evidence of achievement and promise, is no longer evident.[120]

Without going into the matter further, it becomes evident that the apparently bland subject of age patterns of collaboration opens up into a large array of basic questions about the operation of contemporary science. Here, as elsewhere in the field, we are still long on demonstrable questions and short on demonstrated answers. But we have seen enough, in this section of our chapter and in the section touching upon gerontocracy, to identify a variety of related problems.

Queries: How much and in which respects do the structures of authority in Big Science and in Smaller Science actually differ? To the extent that they do, how does this affect the operation of the communication system and reward system? What are the consequences of various authority-structures for scientists of differing age and at different phases in their careers? To what degree are power and authority correlated with age in different sciences and in various national scientific establishments? Which decision points, at every level in the social organization of science, are most consequential for the advance of scientific knowledge? How can more headway be made in investigating the process of decision-making in science at the several levels of its organization?

120. John Ziman, "The Light of Knowledge: New Lamps for Old," Fourth Aslib Annual Lecture. *Aslib Proceedings*, May 1970, pp. 191–92.

6. Age Strata and Foci of Scientific Interest

Historical changes in the foci of scientific work are a matter of experience familiar to sufficiently long-lived scientists and a commonplace among historians and sociologists of science. But how these changes come about and how they are distributed through the community of scientists remains a long-standing and knotty problem,[121] which has lately attracted a renewed interest. As much else in the history, philosophy, and sociology of science, this recent development is a *self-exemplifying pattern* in which workers in these fields are registering a sort of shift in research interests much like that of scientists whose comparable behavior they are trying to interpret or explain.

Both reflecting and deepening the renewed interest in this problem is Thomas S. Kuhn's book, *The Structure of Scientific Revolutions*, which in less than a decade has given rise to a library of criticism and appreciative applications.[122] To judge from the assorted use of this book in just about every branch of learning, it has become something of a complex projective test, meaning all things to all men and women. We do not propose to add still another interpretation of the book in referring to its at least symptomatic relevance here. For our purposes, it is enough to note that Kuhn puts forward three relevant points in his book and supplementary papers. One, he joins Popper in a major concern with "the dynamic process by which scientific knowledge is acquired rather than . . . the logical structure of the products of scientific research." Two, central to this kind of inquiry is an understanding of "what problems [scientists] will undertake." And three, it "should be clear that the explanation must, in the final analysis, be psychological or sociological. It must, that is, be a description of a value system, an ideology, together with an analysis of the institutions through which that system is transmitted and enforced."[123]

Kuhn thus reinstitutes as a concern central to the history and sociology of science an understanding of the changing foci of attention among scientists; more specifically, the question of how it is that scientists seize upon

121. Among philosophers of science, Karl Popper has been concerned with the problem in a long series of books and papers at least since his *Logik der Forschung* (1935). See the translation in its second edition, *The Logic of Scientific Discovery* (New York: Basic Books; London: Hutchinson, 1960). For an early sociological effort to investigate what are described as the "foci and shifts of interest in the sciences and technology," see chapter 8 of this volume.

122. For a recent and, in some of its essays, penetrating examination of Kuhn's ideas, see Imre Lakatos and Alan Musgrave, eds., *Criticism and the Growth of Knowledge* (Cambridge: At the University Press, 1970). For an energetic attack on both Kuhn and Lakatos, see Joseph Agassi, "Tristram Shandy, Pierre Menard and All That: Comment on *Criticism and the Growth of Knowledge*," *Inquiry* 14 (1971): 152–64.

123. Thomas S. Kuhn, "Logic of Discovery or Psychology of Research?" in Lakatos and Musgrave, *Criticism and the Growth of Knowledge*, pp. 1, 21.

some problems as important enough to engage their sustained attention while others are regarded as uninteresting. But Kuhn seems to us too restrictive in saying that the sociological form of the answer to questions of this kind must ultimately be in terms of a value system and the institutions that transmit and enforce it. Sociological interpretations of extratheoretical influences upon the selection of problems for investigation in a science include more than its norms and institutional structure. They also include exogenous influences upon the foci of research adopted by scientists that come from the environing society, culture, economy, and polity, influences of a kind put so much in evidence these days in the heavily publicized form of changing priorities in the allocation of resources to the various sciences and to the problem areas within them as to become apparent even to the most cloistered of scientists. All apart from such exogenous influences, there is the question, of immediate concern to us here, of (the largely unintended) influences upon the foci of research that derive from the social structure as distinct from the normative structure of science, that is, that derive from the social composition and relations of scientists at work in the various disciplines.

In this paper, we happen to deal with the problematics of the age structure of the sciences and specialties within them as one part of their respective social structures just as others might deal with the problematics of their religious, ethnic, or political composition. We have noted in section 1 of this essay that the age structures of the several sciences vary. This at least suggests that the age cohorts entering science at different times may have tended to find different sciences of prime interest to them. The further question whether the age strata *within* a science tend to focus on different problems and to approach the same problems in different ways remains, of course, moot. It is an exemplary question for the sociology of science directing us to one form of interaction between the social structure and the cognitive structure of science and inviting the thought that, in some of its aspects, the cognitive structure of a field may appreciably differ for subgroups of scientists within it.

Scraps of evidence as well as speculation suggest that there are age-patterned foci of research interest and theoretical orientations in the sciences. That this is the case is so fully implied as almost to be expressed in Kuhn's own observation, which we have encountered earlier, that "Almost always the men who achieve [the] fundamental inventions of a new paradigm have been either young or very new to the field whose paradigm they change."[124] But Kuhn turns out to be instructively ambivalent about this statement. At one moment he considers it common enough a generalization to qualify as a cliché and a point so obvious that he should hardly

124. Kuhn, *Structure of Scientific Revolutions*, pp. 89–90.

have made it explicit, while at the next moment he thinks it a generalization much in need of systematic inquiry. In the conspicuous absence of methodical evidence as distinct from much anecdotage bearing on the matter, we are inclined to dissolve the ambivalence by plumping for further investigation.

The question of there being age-patterned foci of attention and theoretical perspectives in science need not be limited to the rare cases of fundamental changes in the structure of prevailing theory. There is reason to suppose that such age-stratified differences obtain more generally. Although the modal pattern is probably one in which the several age cohorts of investigators in a field center on much the same problems, we should not be surprised to find a subsidiary pattern in which young and older scientists tend to focus their work on different problems and so to attend to somewhat different segments of the work going on in the field. It should not be difficult to find out whether such subsidiary age-stratified patterns do occur. A systematic content analysis[125] of papers published by scientists of differing age would yield the information needed on the foci of research just as systematic citation analyses[126] would yield the correlative information needed on the range of work to which they are paying attention.

Consider briefly how patterns of citation might reflect age-stratified differences in the foci of scientific attention. Our conjectural model of the sources and consequences of such differences involves a re-enactment of complementary role-behavior by successive age cohorts much like that which we have provisionally identified for patterns of scientific collabora-

125. For rather primitive instances of such content-analyses of scientific work, which did not, however, go on to examine possible differences among scientists of differing age or rank, see the classification of papers in the *Philosophical Transactions* 1665–1702 and the research recorded in the minutes of the Royal Society in the seventeenth century in chapter 8 in this volume.

126. Ever since the invention of the Science Citation Index, citation studies have been increasing at such a rapid rate that they threaten to get out of hand. Many methodological problems are being neglected in their frequently uncritical use. Moreover, the very existence of the SCI and the growing abundance of citation analyses (even for such matters as aids in deciding upon the appointment and promotion of scientists) may lead to changes in citation practices that will in due course contaminate or altogether invalidate them as measures of the quality of research. This would not be the first case where the introduction of statistical records of role performance has led to a *displacement of goals* in which the once-reliable statistical indicator rather than the actual performance becomes the center of manipulative concern. On the early use of citation analyses, see Garfield, Sher, and Torpie, *The Use of Citation Data in Writing the History of Science*; for a critical overview of methodological problems in citation-analysis, see J. Cole and S. Cole, *Social Stratification in Science*, chap. 2; S. Cole and J. Cole, "Measuring the Quality of Sociological Research: Problems in the Use of the Science Citation Index," *American Sociologist* 6 (1971): 23–29; and Richard D. Whitley and Penelope A. Frost, "The Measurement of Performance in Research," *Human Relations* 24 (1971): 161–78. On displacement of goals in statistical measures of performance, see Peter Blau, *The Dynamics of Bureaucracy* (Chicago: University of Chicago Press, 1955), chap. 3.

tion. We begin with one well-worn assumption and one familiar fact. The assumption (which is also adopted by Kuhn) holds that the time in their career at which scientists encounter ideas will significantly affect their responses to them. The familiar fact is the strong, and perhaps increasing, emphasis in science on keeping up with work on the frontiers of the field, that is, with new work.

Whether the intensity of concern with keeping abreast of new work is age-stratified or not—we know of no evidence bearing on this—it should have somewhat different consequences among the age cohorts. Plainly, the work that scientists come to know as new when they enter the field ages along with them. As incoming cohorts move toward what in science is a swiftly approaching middle age, the work they had focused on in their youth has grown "old," as age of publications is judged in much of contemporary science.[127] The problems, new or old, which members of the older cohorts are investigating will often reactivate memories of pertinent work in the literature which they had encountered as new in years gone by. Meanwhile, the younger cohort working at the same time turn their attention primarily to new work (just as their middle-aged colleagues did in their youth). But not having the same immediate knowledge about work which had been done in the, for them, remote past of fifteen or twenty years before, they are less apt to be put in mind of earlier germane investigations.

In this model, scientists in each successive cohort re-enact much the same citation behavior at the same phases of their career. By doing so, younger and older scientists *to a degree* contribute differently to the development of science: the older scientists providing somewhat more for intellectual continuity by linking current work with work done some time before; the younger scientists pushing ahead somewhat more on their own, less "encumbered" by past formulations. We suggest that although the norms governing the communication of scientific knowledge are much the same for all, leeway in their observance is enough to allow for the occurrence of such unplanned, and often unnoticed, variation in the age-patterned reporting of scientific work.

If these conjectured differences do in fact obtain, they should be reflected in various ways. For one thing, younger scientists should be given more than older ones to making rediscoveries: findings and ideas, independently arrived at, that are substantively identical with earlier ones or functionally equivalent to them.[128] The Santayana dictum that those who fail to remember history are destined to repeat it should hold with special

127. It will be remembered from section 2 of this chapter that the half-life of references in many of the sciences is five years or less.

128. On patterns of rediscovery, see Merton, *Social Theory and Social Structure*, chap. 1.

force in the domain of science—subject more than other spheres of culture to the objective constraints of finding authentic solutions to designated problems.[129] And if the conjectured differences obtain, we should also find age-connected patterns of references and citations along the lines suggested by the preliminary and tentative investigations by Stephen Cole and by Zuckerman,[130] which, it will be remembered from section 2 of this chapter, found that older scientists are more likely than younger ones to cite older publications.

When scientists themselves do not understand this reiterative pattern of age-related foci of attention, they are ready to pass invidious judgments upon the behavior of those in "the other" age stratum. Older scientists then describe younger ones as parochial if not downright barbarian in outlook, little-concerned to read and ponder the classical work of some years back and even less concerned to learn about the historical evolution of their field (the judges forgetting all the while that the new youth in science are only reproducing the attitudes and behavior they had exhibited in their own youth). In turn, younger scientists deride the orientation of older ones to the past as mere antiquarianism, as a sign that they are unable to "keep up" and so are condemned to repeat the obsolete if not downright archaic stuff they learned long ago (the judges being all the while unable to anticipate their own future behavior that they will then likely perceive as providing needed lines of continuity in scientific development).

Historical changes in patterns of citation may provide context for these sequential patterns of citation during the life course. The exponential growth in the numbers of scientists and of scientific publication may result in changed norms and practices with respect to linking up work on the research front with that which has been done some time before. More specifically, this raises the question whether it is the case that, as science becomes bigger in every aspect, a not altogether functional adaptation develops in which successive cohorts of scientists give less and less attention to pertinent work of the past while scientific publications provide less and less space for it. This may be still another instance in which historical changes in the social and cognitive parameters of science interact with sequential patterns in the life course to produce both similarities and differences in the behavior of successive cohorts of scientists.

129. In suggesting this, we scarcely subscribe to positivistic or Whig doctrine.

130. Stephen Cole, "Age and Scientific Behavior"; Zuckerman, unpublished data. We should also note again that age-patterned differences in references and citations represent a finding a fortiori (that is, under conditions tilted against the hypotheses). For any such differences observed in print are there in spite of countervailing suggestions by referees and colleagues (often differing in age from the author). Further investigation would compare the age distributions of references in collaborative papers written by age peers (old or young) and by authors of substantially differing age.

Concluding Remarks

An exploratory paper like this one has no place for "conclusions," but it does call for a few afterthoughts.

Plainly we have only touched upon the problematics of our subject, not having discussed a variety of questions which even now could be examined to good purpose. Here is a scattering of examples:

How is the *age of research groups* related to their scientific productivity?[131]

What is the age distribution of the "founders" of new formations in the various sciences: for example, new specialties, new forms of investigation (laboratories), new journals, scientific societies, and so forth?

What is the relation between the quantity and quality of scientific output at various phases in the scientific career?

How is age variously associated with intellectual authority[132] and with bureaucratic authority in science, and what are the consequences of such differences for the development of scientific disciplines?

What historical changes have occurred in the span of research careers, and how do these relate to the durability of the intellectual influence of scientists?

To what extent do age cohorts in science develop into age-sets with their continued interaction and solidarity to produce old-boy networks (not unlike new-boy networks)?

Perhaps enough has been said to indicate what we take to be the principal aim of investigating questions of this sort. That aim is to find out how age and age structure variously interact with the cognitive structure and development of science.

131. For a critical examination of the question since the first studies by H. A. Shepard, W. P. Wells, D. C. Pelz, and F. M. Andrews, see Clagett G. Smith, "Age of R and D Groups: A Reconsideration," *Human Relations* 23 (1970): 81–96; also Vlachý, "Remarks on the Productive Age."

132. Most emphatically, peers in science need not be age peers. A century ago, William Perkins was, at 23, the world authority on dyes just as a Joshua Lederberg or a Murray Gell-Mann were authorities in their subjects at a comparable age today. This sort of thing should lead us to abandon the practice, common in the jargon of sociology and psychology, of having the word "peers" refer elliptically only to "age peers." As everyone else seems to know, "peer" refers to one who is of equal standing with another, in whatever terms that standing is gauged: political rank, esteem, authority, *and* age.

Bibliography

Writings of Robert K. Merton in the Sociology of Science

Books (listed in order of publication)

Science, Technology and Society in Seventeenth-Century England. In *Osiris: Studies on the History and Philosophy of Science.* Bruges, Belgium: Saint Catherine Press, Ltd., 1938. With new preface. New York: Howard Fertig, Inc., 1970; paperback edition, New York: Harper & Row, 1970.

Social Theory and Social Structure. New York: The Free Press, 1949. Revised edition, 1957. Enlarged edition, 1968.

The Student-Physician: Introductory Studies in the Sociology of Medical Education. [Coedited with G. G. Reader and P. L. Kendall] Cambridge: Harvard University Press, 1957.

On the Shoulders of Giants: A Shandean Postscript. New York: The Free Press, 1965; paperback edition, New York: Harcourt Brace Jovanovich, Inc., 1967.

On Theoretical Sociology. New York: The Free Press, 1967.

Papers (listed in order of publication)

*Papers marked with an * are included in this volume.*

"The Course of Arabian Intellectual Development, 700–1300 A.D." [With P. A. Sorokin] *Isis* 22 (February 1935): 516–24.

"Fluctuations in the Rate of Industrial Invention." *The Quarterly Journal of Economics* 49 (May 1935): 454–70.

*"Science and Military Technique." *Scientific Monthly* 41 (December 1935): 542–45.

"The Unanticipated Consequences of Purposive Social Action." *American Sociological Review* 1 (1936): 894–904.

"Puritanism, Pietism, and Science." *Sociological Review* 28 (January 1936): 1–30.

"Civilization and Culture." *Sociology and Social Research* 21 (November-December 1936): 103–13.

"Some Economic Factors in Seventeenth Century English Science." *Scientia: Revista di Scienza* 62 (1937): 142–52.
"Science, Population and Society." *The Scientific Monthly* 44 (February 1937): 165–71.
"Social Time: A Methodological and Functional Analysis." [With Pitirim A. Sorokin] *The American Journal of Sociology* 42 (March 1937): 615–29.
"Sociological Aspects of Invention, Discovery and Scientific Theories." [With P. A. Sorokin] In P. A. Sorokin, *Social and Cultural Dynamics*. New York: American Book Company, 1937.
"The Sociology of Knowledge." *Isis*, 27 (November 1937): 493–503.
"Social Structure and Anomie." *American Sociological Review* 3 (1938): 672–82.
*"Science and the Social Order." *Philosophy of Science* 5 (1938): 321–37.
"Science and the Economy of Seventeenth Century England." *Science and Society* 3 (Winter 1939): 3–27.
"Karl Mannheim and the Sociology of Knowledge." *The Journal of Liberal Religion* 2 (Winter 1941): 125–47.
*"Znaniecki's *The Social Role of the Man of Knowledge*." *American Sociological Review* 6 (February 1941): 111–15.
*"Science and Technology in a Democratic Order." *Journal of Legal and Political Science* 1 (October 1942): 115–26.
"Role of the Intellectual in Public Bureaucracy." *Social Forces* 23 (May 1945): 405–15.
*"Sociology of Knowledge." In *Twentieth Century Sociology*, edited by Georges Gurvitch and Wilbert E. Moore. New York: Philosophical Library, 1945.
"Mass Persuasion: A Technical Problem and a Moral Dilemma." Chapter 7 in Robert K. Merton, Marjorie Fiske, and Alberta Curtis, *Mass Persuasion*. New York: Harper & Brothers, 1946; Westport: Greenwood Press, 1971.
"The Machine, the Worker, and the Engineer." *Science* 105 (24 January 1947): 79–84.
"Selected Problems of Field Work in the Planned Community." *American Sociological Review* 12 (June 1947): 304–12.
"The Self-Fulfilling Prophecy." *The Antioch Review*, Summer 1948, 193–210.
*"The Role of Applied Social Science in the Formation of Policy." *Philosophy of Science* 16 (July 1949): 161–81.
"Election of Polling Forecasts and Public Images of Social Science." [with Paul K. Hatt] *The Public Opinion Quarterly* 13 (Summer 1949): 185–222.
"Comparison of *Wissenssoziologie* and Mass Communications Research." In Robert K. Merton, *Social Theory and Social Structure*. 1949; enlarged edition, New York: The Free Press, 1968.
"Introduction: Sociology of Science." In Robert K. Merton, *Social Theory and Social Structure*. 1949; New York: The Free Press, 1968.
"Social Scientists and Research Policy." [With Daniel Lerner] In *The Policy Sciences*, edited by Daniel Lerner and Harold D. Lasswell. Stanford: Stanford University Press, 1951.
"The Research Budget." In *Research Methods in Social Relations*, edited by Marie Jahoda, Morton Deutsch and Stuart W. Cook. New York: Dryden Press, 1951. Pp. 342–51.
*"Foreword" to *Science and the Social Order*, by Bernard Barber. New York: The Free Press, 1952.

"Brief Bibliography for the Sociology of Science." [With Bernard Barber] In American Academy of Arts and Sciences. *Proceedings* 80 (May 1952): 140–54.

"Foreword" to *Character and Social Structure*, by Hans Gerth and C. W. Mills. New York: Harcourt Brace & World, 1953.

"The Knowledge of Man." In *The Unity of Knowledge*, edited by Lewis Leary. New York: Doubleday, 1955.

"Studies in the Sociology of Medical Education" [With Samuel Bloom and Natalie Rogoff] *Journal of Medical Education* 31 (August 1956): 552–65.

*"Priorities in Scientific Discovery: A Chapter in the Sociology of Science," *American Sociological Review* 22 (December 1957): 635–59.

"Some Preliminaries to a Sociology of Medical Education." In *The Student-Physician*, edited by R. K. Merton, G. G. Reader and P. Kendall. Cambridge: Harvard University Press, 1957.

"Bibliographical Postscript to 'Puritanism, Pietism and Science'." In Robert K. Merton, *Social Theory and Social Structure*. 1957; New York: The Free Press, 1968.

"Procedures for the Sociological Study of the Value Climates of Medical Schools." [with Richard Christie] In *The Ecology of the Medical Student*, edited by H. H. Gee and R. J. Glaser. Evanston, Ill.: Association of American Medical Colleges, 1958.

"Medical Education as a Social Process." [With P. L. Kendall] In *Patients, Physicians and Illness*, edited by E. G. Jaco. Glencoe: Free Press, 1958.

"The Scholar and the Craftsman." In *Critical Problems in the History of Science*, edited by Marshall Clagett. Madison: University of Wisconsin Press, 1959.

"Notes on Problem-Finding in Sociology." In *Sociology Today*, edited by R. K. Merton, L. Broom and L. S. Cottrell. New York: Basic Books, 1959.

*" 'Recognition' and 'Excellence': Instructive Ambiguities." In *Recognition of Excellence*, edited by A. Yarmolinsky. New York: Free Press, 1960.

"The Mosaic of the Behavioral Sciences." In *The Behavioral Sciences Today*, edited by Bernard Berelson. New York: Basic Books, 1960.

*"Singletons and Multiples in Scientific Discovery: A Chapter in the Sociology of Science." In American Philosophical Society. *Proceedings* 105 (13 October 1961): 470–86.

*"Social Conflict over Styles of Sociological Work." In Fourth World Congress of Sociology. *Transactions*. Louvain, Belgium: International Sociological Association, 1961. 3:21–36.

"Notes on Sociology in the U.S.S.R." [With Henry Riecken] *Current Problems in Social-Behavioral Research*. Symposia Studies Series No. 10. Washington, D.C.: The National Institute of Social and Behavioral Science, 1962. Pp. 7–14.

*"Resistance to the Systematic Study of Multiple Discoveries in Science." *European Journal of Sociology* 4 (1963): 237–82.

*"The Ambivalence of Scientists." *Bulletin of the Johns Hopkins Hospital*. 112 (1963): 77–97.

*"Sorokin's Formulations in the Sociology of Science." [With Bernard Barber] In *P. A. Sorokin in Review*, edited by P. J. Allen. Durham, N.C.: Duke University Press, 1963.

"Basic Research and Potentials of Relevance." *American Behavioral Scientist* 6 (May 1963): 86–90.

"Practical Problems and the Uses of Social Science." [With Edward C. Devereux, Jr.] *Trans-action* 1 (1964): 18–21.

"Foreword" to *The Technological Society*, by Jacques Ellul. New York: A. Knopf, 1964.

"The Environment of the Innovating Organization." In *The Creative Organization*, edited by Gary Steiner. Chicago: University of Chicago Press, 1965.

"On the History and Systematics of Sociological Theory" and "On Sociological Theories of the Middle Range." In Robert K. Merton, *On Theoretical Sociology*. New York: The Free Press, 1967.

*"The Matthew Effect in Science: the Reward and Communication Systems of Science." *Science* 199 (5 January 1968): 55–63.

"Observations on the Sociology of Science." *Japan-American Forum* 14 (April 1968): 18–28.

"Seminars Without Constraint." *The Columbia University Forum* 11 (Winter 1968): 38–39.

*"Behavior Patterns of Scientists." Copublished in *American Scientist* 57 (Spring 1969): 1–23, and *American Scholar* 38 (Spring 1969): 197–225.

"Insiders and Outsiders: An Essay in the Sociology of Knowledge." 1st ed. in *Conspectus of Indian Society*, edited by R. N. Saxena. Agra, India: Satish Book Enterprise, 1971. 2d ed. in *Essays on Modernization of Underdeveloped Societies*. Bombay, India: Thacker & Co., Ltd., 1971.

*"Insiders and Outsiders: A Chapter in the Sociology of Knowledge." 3d ed. *American Journal of Sociology* 77 (July 1972): 9–47.

"The Precarious Foundations of Detachment in Sociology." In *The Phenomenon of Sociology*, edited by Edward A. Tiryakian. New York: Appleton-Century-Crofts, 1971.

*"The Competitive Pressures (1): The Race for Priority." [With Richard Lewis] *Impact of Science on Society* 21 (1971): 151–61.

*"Patterns of Evaluation in Science: Institutionalisation, Structure and Functions of the Referee System." [With Harriet Zuckerman] *Minerva* 9 (January 1971): 66–100.

*"Age, Aging, and Age Structure in Science." [With Harriet Zuckerman] In *Aging and Society*. Vol. 3, *A Sociology of Age Stratification*, edited by Matilda W. Riley, Marilyn Johnson, and Anne Foner. New York: Russell Sage Foundation, 1972.

"On Discipline-Building: The Paradoxes of George Sarton." [With Arnold Thackray] In *Isis* 63 (1972): 473–95, and in George Sarton, *Introduction to the History of Science*. Cambridge, Mass.: M.I.T. Press, in press.

Commentaries, Continuities, and Complementaries

Works marked with a † are publications of the Columbia University Program in the Sociology of Science.

Agassi, Joseph. "The Origins of the Royal Society." *Organon* 7 (1970): 117–35.

———. "Revolutions in Science, Occasional or Permanent?" *Organon* 3 (1966): 47–61.

———. "Towards an Historiography of Science." *History and Theory*, suppl. 2. The Hague: Mouton, 1963.

Baldamus, W. "The Role of Discoveries in Social Science." In *The Rules of the*

Game, edited by Teodor Shanin. London: Tavistock Publications, 1972.

Barbano, Filippi. "H. Marcuse, R. K. Merton ed il Pensiero Critico: 'Sociologia Negativa' e Sociologia Positiva." *Sociologia: Revista di studi sociali dell' Instituto Luigi Sturzo* 1 (September 1967): 31–58.

———. "Social Structures and Social Functions: The Emancipation of Structural Analysis in Sociology." *Inquiry* 11 (1968): 40–84.

———. "Teoria e metodo nell' indagine delle scienze sociali: introduzione a saggi di R. K. Merton e Talcott Parsons." In *Antologia di Scienze Sociali*, edited by Angelo Pagani. Bologna: Società Editrice Il Mulino. Vol. 1, *Teoria e Ricerca nelle Science Sociali.* Chapter 1.

Barber, Bernard. *Science and the Social Order.* New York: The Free Press, 1952.

———. "The Structure of Scientific Competition and Reward and Its Consequences for Ethical Practices in Bio-Medical Research." Paper read at meeting of American Sociological Association, August 1972.

Barber, Bernard, and Hirsch, Walter, eds. *The Sociology of Science.* New York: The Free Press, 1962.

Barber, Bernard; Lally, John J.; Makarushka, Julia; and Sullivan, Daniel. *Research on Human Subjects: Problems of Social Control in Medical Experimentation.* New York: Russell Sage Foundation, 1973.

Barnes, Barry, ed. *Sociology of Science: Selected Readings.* Hammondsworth, Eng.: Penguin Books, 1972.

Barnes, S. B., and Dolby, R. G. A., "The Scientific Ethos: A Deviant Viewpoint." *European Journal of Sociology* 11 (1970): 3–25.

Ben-David, Joseph. "The Profession of Science and Its Powers." *Minerva* 10 (1972): 362–83.

———. "Roles and Innovations in Medicine." *American Journal of Sociology* 65 (1960): 557–68.

———. "Scientific Growth: A Sociological View." *Minerva* 3 (1964): 455–76.

———. "Scientific Productivity and Academic Organization in Nineteenth-Century Medicine." *American Sociological Review* 25 (1960): 828–43.

———. "The Scientific Role." *Minerva* 4 (1965): 15–54.

———. *The Scientist's Role in Society: A Comparative Study.* Englewood Cliffs, N.J.: Prentice-Hall, 1971.

Ben-David, Joseph, and Collins, Randall. "Social Factors in the Origins of a New Science: The Case of Psychology." *American Sociological Review* 31 (1966): 451–65.

Bernard, M. Russell. "Scientists and Policy Makers: A Case Study in the Ethnography of Communication." La Jolla: Scripps Institution of Oceanography, 1973. Manuscript.

Blisset, Marlan. *Politics in Science.* Boston: Little, Brown & Co., 1972.

Brown, Paula. "Bureaucracy in a Government Laboratory." *Social Forces* 32 (1954): 259–68.

Brown, Theodore M. "Institutional Challenge and the Advance of Science in Seventeenth Century England." Manuscript.

———. "The Rise of Baconianism in the Interregnum: An Attempt at a New Perspective on Science and Society in Seventeenth Century England." Manuscript.

Burstyn, Harold L., and Hand, Robert S. "Puritanism and Science Reinterpreted." *Actes du XIe Congrès International d'Histoire des Sciences*, pp. 139–43.

Carroll, James W. "Merton's Thesis on English Science." *American Journal of Economics and Sociology* 13 (July 1954): 427–32.

Cohen, I. Bernard. "Essay Review: *Science, Technology and Society in Seventeenth Century England.*" *Scientific American* 228 (February 1973): 117–20.

†Cole, Jonathan. "Patterns of Intellectual Influence in Scientific Research." *Sociology of Education* 43 (Fall 1970): 377–403.

†Cole, Jonathan, and Cole, Stephen. "Measuring the Quality of Sociological Research." *American Sociologist* 6 (February 1971): 23–29.

†———. "The Ortega Hypothesis." *Science* 178 (27 Oct. 1972): 368–75.

†———. "The Reward System of the Social Sciences." Presented at American Academy of Arts and Sciences, 16 February 1973. Manuscript.

†———. *Social Stratification in Science.* Chicago, Ill.: University of Chicago Press, 1973.

†Cole, Stephen. "Continuity and Institutionalization in Science: A Case Study of Failure." In *The Establishment of Empirical Sociology*, edited by Anthony Oberschall. New York: Harper and Row, 1972.

———. "In Defense of the Sociology of Science." *The G. S. S. Journal*, Columbia University Graduate Sociological Society, 1965, pp. 30–38.

†———. "Professional Standing and the Reception of Scientific Discoveries." *American Journal of Sociology* 76 (September 1970): 286–306.

†Cole, Stephen, and Cole, Jonathan. "Scientific Output and Recognition: A Study in the Operation of the Reward System in Science." *American Sociological Review* 32 (June 1967): 391–403.

†———. "Visibility and the Structural Bases of Awareness of Scientific Research." *American Sociological Review* 33 (June 1968): 397–413.

Collins, Randall. "Competition and Social Control in Science." *Sociology of Education* 41 (1968): 123–40.

Conant, James B. "The Advancement of Learning during the Puritan Commonwealth." *Proceedings of the Massachusetts Historical Society* 66 (1942): 3–21.

Connor, Patrick E. "Scientific Research Competence: Two Forms of Collegial Judgment." *Pacific Sociological Review* (July 1972): 355–66.

Coser, Lewis A. "Georg Simmel's Style of Work: A Contribution to the Sociology of the Sociologist." *American Journal of Sociology* 63 (May 1958): 635–41.

———. *Masters of Sociological Thought: Ideas in Historical and Social Context.* New York: Harcourt, Brace, Jovanovich, 1971.

———. *Men of Ideas.* New York: Free Press, 1965.

———. "Social Involvement or Scientific Detachment: The Sociologist's Dilemma." *Antioch Review* 28 (1968): 108–13.

Cotgrove, Stephen, and Box, Steven. *Science, Industry and Society: Studies in the Sociology of Science.* London: Allen & Unwin, 1970.

†Cournand, André, and Zuckerman, Harriet. "The Code of Science: Analysis and Some Reflections on Its Future." *Studium Generale* 23 (1970): 941–62.

Crane, Diana. "Fashion in Science." *Social Problems* 16 (1969): 433–40.

———. "The Gatekeepers of Science: Some Factors Affecting the Selection of Articles for Scientific Journals." *The American Sociologist* 2 (1967): 195–201.

———. *Invisible Colleges: Diffusion of Knowledge in Scientific Communities.* Chicago: University of Chicago Press, 1972.

———. "Scientists at Major and Minor Universities." *American Sociological*

Review 30 (October 1965): 699–714.
Crespi, Pietro. "Robert K. Merton e la sociologia americana." *Il Pensiero Critico* 2 (1960): 1–11.
Crowther, J. G. *Science in Modern Society*. London: Cresset Press, 1967. Pp. 290–99.
Daniels, George. "The Process of Professionalization in American Science." *Isis* 58 (1967): 151–66.
Dedijer, Stevan. *An Attempt at a Bibliography of Bibliographies in the Science of Science*. Lund, Sweden: Science Policy Center, 1969.
———. "International Comparisons of Science." *New Scientist* (February 1964): 461–64.
De Gré, Gerard. *Science as a Social Institution*. New York: Doubleday, 1955.
†Dietrich, Lorraine. "Communication in Five Scientific Fields: A Comparative Analysis." Presented at American Sociological Association meeting, September 1972.
Dillenberger, John. "Religious Stimulants and Constraints in the Development of Science." *Continuum* 5 (1967): 6–11.
Dolby, R. G. A., "Sociology of Knowledge in Natural Science." *Science Studies* 1 (1971): 3–21.
Downey, Kenneth J. "The Scientific Community: Organic or Mechanical?" *Sociological Quarterly* 10 (1969): 438–58.
———. "Sociology and the Modern Scientific Revolution." *Sociological Quarterly* 8 (1967): 239–54.
Elkana, Yehuda. "The Conservation of Energy: A Case of Simultaneous Discovery." *Archives Internationales d'Histoire des Sciences* 90–91 (1970): 31–60.
———. "Hemholtz' *Kraft*: An Illustration of Concepts in Flux." *Historical Studies in the Physical Sciences* 2 (1970): 263–98.
———. "Science, Philosophy of Science and Science Teaching." *Educational Philosophy and Theory* 2 (1970): 15–35.
———. *The Theory and Practice of Cross-Cultural Contacts in Science: Queries and Presuppositions*. Manuscript, 1973.
Feuer, Lewis S. *The Scientific Intellectual*. New York: Basic Books, 1963.
Garvey, W. D., and Griffith, B. C. "Scientific Communication as a Social System." *Science* 157 (1967): 1011–16.
———. "Scientific Communication: Its Role in the Conduct of Research and Creation of Knowledge." *American Psychologist* 26 (1971): 349–62.
Garvey, W. D., Lin, Nan, and Nelson, C. E. "Communication in the Physical and Social Sciences." *Science* 170 (1970): 1166–73.
Garvey, W. D.; Lin, Nan; Nelson, C. E.; and Tomita, Kazuo. "Research Studies in Patterns of Scientific Communication." *Information Storage and Retrieval* 8 (1972): 111–22; 159–69; 207–21.
Gaston, Jerry. "Big Science in Britain: A Sociological Study of the High Energy Physics Community." Ph.D. diss., Yale University, 1969.
———. "The Reward System in British Science." *American Sociological Review* 35 (August 1970): 718–32.
———. "Secretiveness and Competition for Priority of Discovery in Physics." *Minerva* 9 (October 1971): 472–92.
Gillispie, Charles C. "Physick and Philosophy: A Study of the Influence of the College of Physicians of London upon the Foundations of the Royal Society." *Journal of Modern History* 19 (September 1947): 210–25.

Glaser, Barney G. "Differential Association and the Institutional Motivation of Scientists." *Administrative Science Quarterly* (1965): 82–97.

———. *Organizational Scientists*. Indianapolis: Bobbs-Merrill, 1964.

———. "Variation in the Importance of Recognition in Scientists' Careers." *Social Problems* 10 (1963): 268–76.

Gordon, Gerald and Marquis, Sue. "Freedom, Visibility of Consequences and Scientific Innovation." *American Journal of Sociology* 72 (Sept. 1966): 195–202.

Gouldner, Alvin W. *The Coming Crisis of Western Sociology*. New York: Basic Books, 1970.

———. "Cosmopolitans and Locals: Towards an Analysis of Latent Social Roles." *Administrative Science Quarterly* 2 (1957): 281–306; 2 (1958): 444–80.

———. *Enter Plato: Classical Greece and the Origins of Social Theory*. New York: Basic Books, 1965.

Greaves, Richard L. "Puritanism and Science: The Anatomy of a Controversy." *Journal of the History of Ideas* 30 (1969): 345–68.

Greenberg, Daniel A. "The Ethos of Science." Presented to the Seventeenth Conference on Science, Philosophy, and Religion, 1966. Manuscript.

Griffith, Belver C. and Miller, A. J. "Networks of Informal Communication among Scientifically Productive Scientists." In *Communication Among Scientists and Engineers*, edited by C. E. Nelson and D. K. Pollock. Lexington, Mass.: Lexington Heath Books, 1970.

Griffith, Belver C., and Mullins, Nicholas C. "Coherent Social Groups in Scientific Change." *Science* 177 (15 September 1972): 959–64.

Haberer, Joseph. *Politics and the Community of Science*. New York: Van Nostrand Reinhold, 1969.

Habermas, Jürgen. *Knowledge and Human Interests*. Boston: Beacon Press, 1971.

Hagstrom, Warren O. "Inputs, Outputs and the Prestige of American University Science Departments." *Sociology of Education* 44 (Fall 1971): 375–97.

———. *The Scientific Community*. New York: Basic Books, 1965.

Hall, A. Rupert. "Merton Revisited, or, Science and Society in the Seventeenth Century." *History of Science: An Annual Review* 2 (1963): 1–16.

———. "Science, Technology and Utopia in the Seventeenth Century." In *Science and Society, 1600–1900*, edited by Peter Mathias. Cambridge: Cambridge University Press, 1972.

Hall, Marie Boas. "Sources for the History of the Royal Society in the Seventeenth Century." *History of Science* 5 (1966): 62–76.

Hall, Michael G., "Renaissance Science in Puritan New England." In *Aspects of the Renaissance*, edited by Archibald R. Lewis. Austin: University of Texas Press, 1967.

Halmos, Paul, ed. *The Sociology of Science*. The Sociological Review: Monograph No. 18, 1972.

———, ed. *The Sociology of Sociology*. The Sociological Review: Monograph No. 16, 1970.

Hargens, Lowell, and Hagstrom, W. O. "Sponsored and Contest Mobility of American Academic Scientists." *Sociology of Education* 40 (1967): 24–38.

Herpin, Nicholas. *Théorie et expérience chez Robert K. Merton*. Ph.D. dissertation, Sorbonne, 1967.

Hetherington, Robert W. "Local and Cosmopolitan Physicians." *Canadian Review of Sociology and Anthropology*, February 1971, pp: 32–46.

Hill, Christopher. *Intellectual Origins of the English Revolution.* Oxford: Clarendon Press, 1965.

———. "The Intellectual Origins of the Royal Society—London or Oxford." *Notes and Records of the Royal Society of London* 23 (December 1968): 144–56.

———. "Puritanism, Capitalism and the Scientific Revolution." *Past & Present* 29 (1964): 88–97.

Hirsch, Walter. *Scientists in American Society.* New York: Random House, 1968.

Hooykaas, R. "Answer to Dr. Bainton's Comment on 'Science and Reformation'." *Journal of World History* 3 (1956): 781–84.

———. *Humanisme, Science et Reforme.* Leiden, 1958.

———. *Religion and the Rise of Modern Science.* Edinburgh: Scottish Academic Press, 1972.

———. "Science and Reformation." *Journal of World History* 3 (1956): 109–39.

———. "Science and Religion in the Seventeenth Century." *Free University Quarterly* 1:169–83.

Kaplan, Norman, ed. *Science and Society.* Chicago: Rand McNally, 1965.

———. "Sociology of Science." In *Handbook of Modern Sociology*, edited by R. E. L. Faris. Chicago: Rand McNally, 1964.

Kearney, Hugh F., ed. *Origins of the Scientific Revolution.* London: Longmans, 1964.

———. "Puritanism and Science: Problems of Definition." *Past & Present* 31 (July 1965): 104–10.

———. "Puritanism, Capitalism and the Scientific Revolution." *Past & Present* 28 (July 1964): 81–101.

Kemsley, Douglas S. "Religious Influences in the Rise of Modern Science: A Review and Criticism, Particularly of the 'Protestant-Puritan Ethic' Theory." *Annals of Science* 24 (1968): 199–226.

King, M. D. "Reason, Tradition and the Progressiveness of Science." *History and Theory: Studies in the Philosophy of History* 10 (1971): 3–32.

Klima, Rolf. "Einige Widersprüche im Rollen-Set des Soziologen." In *Thesen zur Kritik der Soziologie*, edited by B. Schäfers. Frankfurt: Suhrkamp, 1969, pp. 80–95.

———. "Scientific Knowledge and Social Control in Science." In *Social Processes of Scientific Development*, edited by Richard D. Whitley. London: Routledge & Kegan Paul, in press.

———. "Theoretical Pluralism, Methodological Dissension and the Role of the Sociologist: The West German Case." *Social Science Information* 11 (1972): 69–108.

Kocher, Paul Harold. *Science and Religion in Elizabethan England.* San Marino, Calif.: The Huntington Library, 1953.

Kröber, Günter, and Lorf, Marianne, eds. *Wissenschaft: Studien zu ihrer Geschichte, Theorie und Organisation.* Translated from the Russian. Berlin: Akademie-Verlag, 1972.

———, eds. *Wissenschaftliches Schöpfertum.* Translated from the Russian. Berlin: Akademie-Verlag, 1972.

Krohn, Roger G. *The Social Shaping of Science.* Westport, Conn.: Greenwood Publishing Co., 1971.

Kuhn, Thomas S. "Energy Conservation as an Example of Simultaneous Discovery." In *Critical Problems in the History of Science*, edited by Marshall Clagett. Madison: University of Wisconsin Press, 1959.

———. "The History of Science." In the *International Encyclopedia of the Social Sciences* (New York: Macmillan, 1968), 14:74–83. See especially "The Merton Thesis," 79 ff.

———. *The Structure of Scientific Revolutions.* Chicago: University of Chicago Press, 1962. Second edition, enlarged, 1970.

Ladd Jr., Everett C. and Lipset, Seymour Martin. "Politics of Academic Natural Scientists and Engineers." *Science* 176 (9 June 1972): 1091–1100.

Lakatos, Imre. "Criticism and the Methodology of Scientific Research Programmes." Aristotelian Society. *Proceedings* 69 (1968): 149–86.

Lakatos, Imre, and Musgrave, A., eds. *Criticism and the Growth of Knowledge.* Cambridge: University Press, 1970.

Lécuyer, Bernard. "Histoire et sociologie de la recherche sociale empirique: problèmes de théorie et de la méthode." *Epistémologie Sociologique* Cahiers Semestriels 6 (Deuxieme Semestre 1968): 119–31.

Lemaine, Gérard, and Matalon, Benjamin, with the collaboration of Provansal, Bernard. "La lutte pour la vie dans la cité scientifique." *Revue Française de Sociologie* 10 (1969): 139–65.

Lepenies, Wolf. "Melancholie als Unordnung bei R. K. Merton." In *Melancholie und Gesellschaft.* Frankfurt am Main: Suhrkamp Verlag, 1969.

Lipset, Seymour Martin and Ladd Jr., Everett C. "The Politics of American Sociologists." *American Journal of Sociology* 78 (July 1972): 67–104.

Lipset, Seymour Martin. "Academia and Politics in America." In *Imagination and Precision in the Social Sciences*, edited by T. J. Nossiter. London: Faber, 1972.

Lodahl, Janice B. and Gordon, Gerald. "The Structure of Scientific Fields and the Functioning of University Graduate Departments." *American Sociological Review* 37 (February 1972): 57–72.

MacCrae, Duncan. "Growth and Decay Curves in Scientific Citations." *American Sociological Review* 34 (1969): 631–35.

Marcson, Simon. *The Scientist in American Industry.* New York: Harper, 1960.

Mathias, Peter. "Who Unbound Prometheus? Science and Technical Change, 1600–1800." In *Science and Society, 1600–1900*, edited by Peter Mathias. Cambridge: University Press, 1972.

Matthijssen, M. Katholiek middelbaar onderwijs en intellectuele emancipatie. Een sociografische facetstudie van het emancipatie—vraagstuk der Katholieken in Nederland. Nijmegen Thesis, 1958 [cited by R. Hooykaas, *Religion and the Rise of Modern Science*].

Menard, Henry W. *Science: Growth and Change.* Cambridge: Harvard University Press, 1971.

Mendelsohn, Ernest. "The Biological Sciences in the Nineteenth Century." *History of Science* 3 (1964): 39–59.

———. "The Emergence of Science as a Profession in Nineteenth-Century Europe." In *Management of Scientists*, edited by K. Hill. Boston: Beacon Press, 1963.

Menzel, Herbert. "Planned and Unplanned Scientific Communication." Reprinted from *Proceedings*, International Conference on Scientific Information, 1959, in *The Sociology of Science*, edited by Bernard Barber and Walter Hirsch. New York: Free Press, 1962.

———."Planning the Consequences of Unplanned Action in Scientific Communication." In *Communication in Science*, edited by A. de Reuck and J. Knight. London: J. & A. Churchill, 1967.

Mirsky, E. M. "Science Studies in The USSR: History, Problems, Prospects." *Science Studies* 2 (1972): 281–94.

Moscovici, Serge, "L'histoire des sciences et le science des historiens." *Archives Européennes de Sociologie* 7 (1966): 116–26.

Mulkay, Michael. *The Social Process of Innovation: A Study in the Sociology of Science*. London: Macmillan, 1972.

———. "Some Aspects of Cultural Growth in the Natural Sciences." *Social Research* 36 (1969): 22–53.

Mullins, Nicholas C., "The Distribution of Social and Cultural Properties in Informal Communication Networks Among Biological Scientists," *American Sociological Review* 33 (October 1968): 786–97.

———. "Social Networks among Biological Scientists." Ph.D. dissertation, Harvard University, 1966.

———. "The Structure of an Elite: the Advisory Structure of the Public Health Service." *Science Studies* 2 (1972): 3–29.

Musson, A. E. *Science, Technology and Economic Growth in the Eighteenth Century*. London: Methuen & Co., 1972.

Musson, A. E., and Robinson, Eric. *Science and Technology in the Industrial Revolution*. Manchester: University Press, 1969.

Nagi, Saad Z., and Corwin, Ronald G. *The Social Contexts of Research*. New York: Wiley, 1972.

Needham, Joseph. *The Grand Titration: Science and Society in East and West*. London: Allen & Unwin, 1969.

———. "Science and Society in East and West." In *The Science of Science*, edited by M. Goldsmith and A. Mackay. New York: Simon and Schuster, 1965. Esp. pp. 146–49.

———. *Time: The Refreshing River*. New York: Macmillan, 1945.

Nelson, Benjamin. "The Early Modern Revolution in Science and Philosophy." In *Boston Studies in the Philosophy of Science*, edited by R. S. Cohen and M. Wartofsky. Dordrecht, Holland: D. Reidel, 1967. Vol. 3.

———. "Review Essay: *Science, Technology and Society in Seventeenth-Century England*." *American Journal of Sociology* 78 (July 1972): 223–31. Reprinted in *Varieties of Political Expression in Sociology*. Chicago: University of Chicago Press, 1972, pp. 202–10.

Opp, Karl-Dieter. *Methodologie der Sozialwissenschaften*. Hamburg: Rowohlt, 1970.

———. "Theories of the Middle Range as a Strategy for the Construction of a General Sociological Theory: A Critique of a Sociological Dogma." *Quality and Quantity* 4 (1970): 243–54.

———. "Zur Anwendung Sozialwissenschaftlicher Theorien für Praktisches Handeln." *Zeitschrift für die Gesamte Staatswissenschaft* 123 (July 1967): 393–418.

Oromaner, Mark Jay. "Comparison of Influentials in Contemporary American and British Sociology: a Study in the Internationalization of Sociology." *British Journal of Sociology* 13 (1970): 324–42.

Parsons, Talcott. "The Institutionalization of Scientific Investigation." In Parsons, *The Social System*. New York: The Free Press, 1951. Pp. 335–48.

———. "On Building Social System Theory: A Personal Memoir." *Daedalus* (Fall 1970): 826–81.

Pelseneer, J. "Les influences dans l'histoire des sciences." *Archives Internationale d'Histoire des Sciences* 1 (1948): 348–53.

———. "L'origine protestante de la science moderne." *Lychnos*, 1947, pp. 246–48.

Polanyi, Michael. *Personal Knowledge*. Chicago: University of Chicago Press, 1968; London: Routledge & Kegan Paul, 1958.

———. *Science, Faith and Society*. Chicago: University of Chicago Press, 1964.

———. *The Study of Man*. London: Routledge & Kegan Paul, 1959.

———. *The Tacit Dimension*. London: Routledge & Kegan Paul, 1967.

Price, Derek J. de Solla. "Is Technology Historically Independent of Science? A Study in Statistical Historiography." *Technology and Culture* 6 (1965); 553–68.

———. *Little Science, Big Science*. New York: Columbia University Press, 1963.

———. "Nations Can Publish or Perish." *Science and Technology*, no. 70 (October 1967): 84–90.

———. "Networks of Scientific Papers." *Science* 149 (1965): 510–15.

———. *Science Since Babylon*. New Haven: Yale University Press, 1961.

Price, Derek J. de Solla, and Beaver, Donald DeB. "Collaboration in an Invisible College." *American Psychologist* 21 (November 1966): 1011–18.

Rabb, Theodore K. "Puritanism and the Rise of Experimental Science in England." *Journal of World History* 7 (1962): 46–67.

———. "Religion and the Rise of Modern Science." *Past & Present* 31 (July 1965): 111–26.

Rattansi, P. M. "Paracelsus and the Puritan Revolution." *Ambix* 11 (1963): 24–30.

———. "The Social Interpretation of Science in the Seventeenth Century." In *Science and Society, 1600–1900*, edited by Peter Mathias. Cambridge: At the University Press, 1972.

Ravetz, J. R. "Francis Bacon and the Reform of Philosophy." In *Science, Medicine and Society in the Renaissance* (Walter Pagel Festschrift), edited by A. Debus. New York: Neale Watson Academic Publications Inc., 1972.

———. *Scientific Knowledge and Its Social Problems*. Oxford: Clarendon Press, 1971.

Reif, F. "The Competitive World of the Pure Scientist." *Science* 134 (15 December 1961): 1957–62.

†Reitz, Jeffrey G. *Choice of Science Careers Among College Men*. Ph.D. dissertation, Columbia University, 1972.

Richter, Maurice N., *Science as a Cultural Process*. Cambridge: Schenkman Publishing Co., 1972. Esp. chapter 6.

Rose, Hilary, and Rose, Steven. *Science and Society*. London: Allen Lane, Penguin Press, 1969. Chapter 13.

Rosen, George. "Left-Wing Puritanism and Science." *Bulletin of the Institute of the History of Medicine* 15 (1944): 375–80.

Rossi, Paolo. *Francis Bacon: From Magic to Science*. Chicago: University of Chicago Press, 1968.

Rothman, Robert A. "A Dissenting View on the Scientific Ethos." *The British Journal of Sociology* 23 (March 1972): 102–8.

Rule, John C., Dowd, David, and Snell, John L., eds. *Critical Issues in History*. Boston: D. C. Heath, 1966. Vol. 2, pt. IV 2 ("The Scientific Revolution of the Seventeenth Century: A Question of Causes").

Salomon, Jean-Jacques. "La Politique de la Science et Ses Mythes." *Diogène* 70 (1970): 3–30.

Shapiro, Barbara J. "Latitudinarianism and Science in 17th-Century England." *Past & Present* 40 (1968): 16–41.

Scheffler, Israel. *Science and Subjectivity*. Indianapolis: Bobbs-Merrill, 1967.

———. "Vision and Revolution: A Postscript on Kuhn." *Philosophy of Science* 39 (September 1972): 366–74.

Shils, Edward. *The Intellectuals and the Powers*. Chicago: University of Chicago Press, 1972.

———. "The Profession of Science." *The Advancement of Science* 24 (June 1968): 469–80.

———. "Tradition, Ecology and Institution in the History of Sociology." *Daedalus* (Fall 1970): 760–825.

Solomon, Susan Gross. "Controversy in Social Science: Soviet Rural Studies in the 1920s." Ph.D. dissertation, Columbia University, 1973.

Solow, Robert M. "Merton's *Science, Technology and Society in 17th-Century England*." Harvard University, 1942. Manuscript.

Solt, Leo F. "Puritanism, Capitalism, Democracy and the New Science." *American Historical Review* 73 (1967): 18–29.

Sorokin, P. A. *Sociological Theories of Today*. New York: Harper & Row, 1966. Pp. 445–56.

Statera, Gianni, "La sociologia della scienza di Robert K. Merton." *La Critica Sociologica* 3 (Autumn 1964): 19–33.

Stearns, Raymond P. "The Relations Between Science and Society in the Later Seventeenth Century." In *The Restoration of the Stuarts, Blessing or Disaster?* Washington, D.C.: The Folger Shakespeare Library, 1960. Pp. 67–75.

———. "The Scientific Spirit in England in Early Modern Times." *Isis* 34 (1943): 293–300.

Stent, Gunther S. "Prematurity and Uniqueness in Scientific Discovery." *Scientific American* 227 (December 1972): 84–93.

———. "What They Are Saying about Honest Jim." *The Quarterly Review of Biology* 43 (June 1968): 179–84.

Stone, Lawrence. "Prosopography." *Daedalus* 100 (1971): 46–79.

Storer, Norman. "Basic versus Applied Research: The Conflict between Means and Ends." *Indian Sociological Bulletin* 2 (October 1964): 34–42.

———. "Comparative Study of Scientific Disciplines: Opportunities for Research." Paper read at meetings of American Sociological Association, August 1972.

———. "The Hard Sciences and the Soft: Some Sociological Observations." *Bulletin*, The Medical Library Association, 55 (January 1967): 75–84.

———. "Relations among Scientific Disciplines." In *The Social Contexts of Research*, edited by Saad Z. Nagi and Ronald G. Corwin. New York: Wiley, 1972.

———. *The Social System of Science*. New York: Holt, Rinehart and Winston, 1966.

Storer, Norman, and Parsons, Talcott. "The Disciplines as a Differentiating Force." In *The Foundations of Access to Knowledge*, edited by E. B. Montgomery. Syracuse, N.Y.: Syracuse University Press, 1968.

Strauss, Anselm L., and Rainwater, Lee. *The Professional Scientist: A Study of American Chemists*. Chicago: Aldine Press, 1962.

Sullivan, Daniel. "Competition in Bio-Medical Science: Its Extent and Some of Its Consequences." Manuscript.

Swatez, G. M. "The Social Organization of a University Laboratory." *Minerva* 8 (1970): 36–58.

Thackray, Arnold. *John Dalton: Critical Assessments of His Life and Science*. Cambridge: Harvard University Press, 1972.

———. "Reflections on the Decline of Science in America and on Some of Its Causes." *Science* 173 (2 July 1971): 27–31.

———. "Science: Has Its Present Past a Future?" In *Historical and Philosophical Perspectives of Science*, edited by R. Steuwer. Minneapolis: University of Minnesota Press, 1970.

———. "Science, Technology and the Industrial Revolution." *History of Science* 8 (1970): 76–89.

Thackray, Arnold, and Merton, Robert K. "On Discipline-Building: The Paradoxes of George Sarton." *Isis* 63 (1972): 473–95, and in George Sarton, *Introduction to the History of Science*. Cambridge, Mass.: M.I.T. Press, 1974.

Thorner, Isidor. "Ascetic Protestantism and the Development of Science and Technology." *American Journal of Sociology* 58 (1952): 25–33.

Wanderer, Jules J. "An Empirical Study in the Sociology of Knowledge." *Sociological Inquiry* 39 (Winter 1969): 19–26.

Webster, C. "The Origins of the Royal Society: Essay Review." *History of Science* 6 (1967): 106–28.

Weingart, Peter, ed. *Wissenschaftssoziologie I: Wissenschaftliche Entwicklung als sozialer Prozess*. Frankfurt am Main: Athenäum Fischer Verlag, 1972.

Whitley, Richard D. "Black Boxism and the Sociology of Science: A Discussion of the Major Developments in the Field." *The Sociological Review* Monograph no. 18 (1972): 61–92.

———. "The Formal Communication System of Science." *The Sociological Review* Monograph no. 16 (September 1970): 163–79.

———. "The Operation of Science Journals: Two Case Studies in British Social Science." *Sociological Review*, n.s. 18 (July, 1970): 241–58.

———, ed. *Social Processes of Scientific Development*. London: Routledge & Kegan Paul, in press.

Yellin, Joel. "A Model for Research Problem Allocation among Members of a Scientific Community." *Journal of Mathematical Sociology* 2 (1972): 1–36.

Zaltman, Gerald. *Scientific Recognition and Communication Behavior in High Energy Physics*. New York: American Institute of Physics, 1968.

Ziman, John. "The Light of Knowledge: New Lamps for Old." The Fourth Aslib Annual Lecture. *Aslib Proceedings* (May 1970): 186–200.

———. *Public Knowledge: The Social Dimension of Science*. Cambridge: At the University Press, 1968.

———. "Some Problems of the Growth and Spread of Science into Developing Countries." The Rutherford Memorial Lecture. *Proceedings* of the Royal Society, A. 311 (1969): 349–69.

†Zuckerman, Harriet A. "Interviewing an Ultra-Elite." *Public Opinion Quarterly* 36 (Summer 1972): 159–75.

†———. "Knowledge and Social Structure." In *Society Today*. Del Mar, Calif.: CRM Books, 1970. Chapter 28.

†———. "Nobel Laureates in Science: Patterns of Productivity, Collaboration and Authorship." *American Sociological Review* 32 (1967): 391–403.

†———. "Patterns of Name-ordering among Authors of Scientific Papers: A Study of Social Symbolism and its Ambiguity." *American Journal of Sociology* 74 (November 1968): 276–91.

†———. *Scientific Elite: Studies of Nobel Laureates in the United States.* Chicago: University of Chicago Press, in press.

†———. "The Sociology of the Nobel Prize." *Scientific American* 217, no. 5 (1967): 25–33.

†———. "Stratification in American Science." *Sociological Inquiry*, Spring 1970, pp. 235–57.

†———. "Women and Blacks in American Science and Engineering." In *Women and Minorities in Science and Engineering*, edited by Daniel Kevles. In press.

†Zuckerman, Harriet A., and Merton, Robert K. "Age, Aging, and Age Structure in Science." In *Aging and Society*. Vol. 3, *A Theory of Age Stratification*, edited by Matilda W. Riley, Marilyn Johnson, and Anne Foner. New York: Russell Sage Foundation, 1972.

†———. "Patterns of Evaluation in Science: Institutionalization, Structure and Functions of the Referee System." *Minerva* 9 (January 1971): 66–100.

Index of Names

Index of Subjects